The Interstellar Medium

The Interstellar Medium

S. A. Kaplan and S. B. Pikelner

Harvard University Press Cambridge, Massachusetts, 1970

Library of Congress Catalog Card Number 70-85076
Standard Book Number 674-46075-8
Printed in the United States of America

Translated from *Mezhzvezdnaya Sreda* (Moscow, 1963),
with corrections and additions by the authors

Preface

The study of the interstellar medium occupies an important place in astrophysics. This can be attributed on the one hand to the peculiarity of the physical conditions, which often permit a simple interpretation; on the other hand, especially recently, there is the important role of the interstellar medium in the over-all picture of the formation and evolution of the Galaxy. The interstellar medium in the widest sense of the word is not limited to the "classical" components—gas and dust. Radiation fields, cosmic rays, and magnetic fields must all be equally considered. All these components contain energy comparable to the thermal energy of the gas and all interact closely with one another. At the same time, methods for investigating each component vary, and so does the extent of our knowledge of each.

This book presents in an up-to-date manner, as far as possible, all aspects of the physics of the interstellar medium. Differences in apparatus employed and in the degree of our knowledge are indicated by the heterogeneous nature of the relevant chapters. The physical conditions in interstellar gas, its emission, degree of ionization, and temperature, can be studied by sufficiently reliable methods. Observations yield sound, quantitative data, while the theory linking these processes is even classical. A study of the properties of cosmic dust offers less certain results. For example, it is still impossible to say with confidence what interstellar dust particles consist of. Here, theory uses different hypotheses and the application of refined mathematical methods is not always justified. This is even more true in the case of gaseous motion, cosmic rays, and magnetic fields, where, as a rule, only qualitative methods can be employed. In the chapter on cosmic rays and radio emission, only aspects of these problems connected with the interstellar medium are considered. There are more detailed studies of these problems in special monographs, such as V. L. Ginzburg and S. I. Syrovatskii, *Origin of cosmic rays* (Izd. AN SSSR, 1963; Pergamon, New York, 1965), and I. S. Shklovsky, *Cosmic radio waves* (Fizmatgiz, 1956; trans. by R. B. Rodman and C. M. Varsavsky, Harvard University Press, Cambridge, Massachusetts, 1960). The latter book is directed to the expert reader. As an introduction,

S. B. Pikelner's book *The physics of interstellar space* (Izd. AN SSSR, 1959; Foreign Language Publishing House, USSR, 1961; Philosophical Library, New York, 1963, "Soviet science of interstellar space") may be used.

Considerable assistance was given to the authors by their colleagues in this work who read through particular chapters and made many useful comments: A. A. Boyarchuk, V. L. Ginzburg, N. S. Kardashov, Yu. N. Pariiskii, I. I. Pronik, S. I. Syrovatskii, and I. S. Shklovskii. Special help in every chapter of the book was given by R. E. Hershberg, V. V. Ivanov, and V. I. Pronik. The authors extend to all of them their hearty thanks.

Contents

The Interstellar Medium

1. Interstellar Hydrogen

1. The Ionization of Interstellar Hydrogen

The physical properties of the interstellar medium are marked above all by its extreme diluteness—1 cm^3 contains only a few atoms. A further characteristic feature is that interstellar gas is found in a very attenuated interstellar radiation field, the density of which is of the order of a few photons per cubic centimeter. The concept of thermodynamic equilibrium is generally inapplicable to interstellar gas, although it may be employed in isolated cases, so that kinetic methods form the basis for the physical study of the interstellar medium.

Fortunately, because of the rarefaction of the interstellar gas and the radiation mentioned above, the kinetics of elementary processes are here relatively simple. For example, atoms absorb quanta only from ground states and cascades of permissible downward transitions are virtually never interrupted by collisions or absorption of quanta, except at very high levels. The simplicity of the kinetic processes has made it possible to form a more or less complete theory of emission from interstellar gas, in both the optical and the radio bands, and to obtain sufficiently reliable determinations of its temperature and degree of ionization.

The theory of kinetic processes in interstellar gas depends essentially on the accuracy of determining the probability of elementary transitions in atoms. These values are known best for hydrogen and for hydrogenlike atoms. Moreover, hydrogen is a basic component of the interstellar gas—the hydrogen content exceeds 80 percent of the total number of atoms. Therefore, the state of the hydrogen—primarily, its degree of ionization—frequently determines the physical condition of the interstellar medium in general. For these reasons we think it best first to investigate the kinetic processes of and the emission from hydrogen atoms in interstellar space, and then to take into account the presence of other atoms and molecules. In the first chapter, we shall study only interstellar hydrogen. The essential data on the potentials of elementary processes in hydrogen atoms are given in Appendix I. (References to the equations in the appendix will be found in the text; in numbering these equations, a Roman numeral is used, indicating the appendix number; for example: (I.20) is Eq. (20), Appendix I.)

Ionization Zones of Interstellar Hydrogen

The determination of the degree of ionization of interstellar hydrogen is one of the main problems in the physics of the interstellar medium. It was first made by Strömgren in 1939 [22]. His main conclusion was that interstellar space is divided into rather sharply defined regions of ionized (H II) and nonionized (H I) hydrogen; he was in many respects responsible for the further development of the physics of the interstellar medium.

Interstellar hydrogen is ionized by radiation in the far-ultraviolet region (Lyman continuum, $\lambda < 912$ Å) of the spectrum of hot stars with high luminosities about which little is known at present. Therefore, it seems expedient to consider first a simple estimate of the degree of ionization of interstellar hydrogen, and then to develop more accurate equations.

Suppose a hot star is situated in a region of interstellar hydrogen of constant density. It is clear that radiation in the Lyman continuum travels a distance s_0 that corresponds to an optical thickness $\tau = a_1(\nu)\overline{N}_1 r$ equal approximately to unity. Here $a_1(\nu)$ is the absorption coefficient and $\overline{N}_1$ is the average number of hydrogen atoms per unit volume in the first level. Using Eq. (I.20) with $g_1^{(0)} = 0.8$ and neglecting temporarily the dependence of τ on frequency, we obtain $\tau = 6.3 \times 10^{-18} N_1 s_0$, or, expressing s_0 in parsecs,

$$\overline{N}_1 s_0 \approx \tfrac{1}{20}. \tag{1.1}$$

We note that the value of N_1 may change appreciably over the distance s_0 from the star to the zone boundary; however, the use of the mean value $\overline{N}_1$ in Eq. (1.1) is entirely justified.

It is easy to see that L_c radiation is rarely present beyond the limits of the zone of radius s_0, and the hydrogen remains nonionized. Actually, most of the L_c quanta are changed after absorption into quanta of the Balmer and other series and therefore disappear. We shall show later that transition from almost complete ionization to the neutral state occurs within a thin layer. Therefore, the idea of an H II zone of radius s_0 and an H I zone outside this sphere suggests itself. The numerical value of s_0 is obtained as follows. In equilibrium the number of L_c quanta emitted per second by a star and absorbed in a sphere of volume $\frac{4}{3}\pi s_0^3$ must be equal to the number of recombinations in the same volume at the second and higher levels. Recombinations at the first level must be excluded, since an L_c quantum is again formed and again ionizes a hydrogen atom. Thus, calling N_L the number of L_c quanta emitted per square centimeter of stellar surface per second, we obtain

$$4\pi R_*^2 N_L = 4\pi R_*^2 \int_{\chi_1/h}^{\infty} \pi F_\nu \frac{d\nu}{h\nu} = N_e N_p \frac{4\pi s_0^3}{3} \sum_{n=2}^{\infty} a_n(T)$$

$$= N_e N_p \frac{4\pi s_0^3}{3} (\alpha_t - \alpha_1). \qquad (1.2)$$

Here πF_ν is the energy flux at frequency ν per square centimeter of stellar surface, R_* is the stellar radius, α_n is the recombination probability at the nth level, α_t is the recombination probability at all levels (see Appendix I), and N_e and N_p are the numbers of electrons and of protons per cubic centimeter. The temperature in ionized hydrogen regions is about 10,000°. Then, in agreement with Eq. (I.25), we have $\alpha_t - \alpha_1 = 2.45 \times 10^{-3}$ cm^3/sec, and hence

$$s_0 N_e^{2/3} = \left[\frac{3R_*^2 \mathrm{N_L}}{\alpha_t - \alpha_1}\right]^{1/3} = U\,(\text{sp. cl.}) = 1.23 \times 10^{-7} \left(\frac{R_*}{R_\odot}\right)^{2/3} \mathrm{N_L}^{1/3} \text{ pc/cm}^2, \qquad (1.3)$$

where, when hydrogen is almost completely ionized, $N_e \approx N_p$. The expression $s_0 = U(\text{sp. cl.}) N_e^{-2/3}$ means that the radius of the ionized region depends upon the star's spectral class (R_* and $\mathrm{N_L}$). The magnitude of s_0 is called the *radius of the Strömgren sphere*.

So far, there are no direct observations of the far-ultraviolet spectrum of stars; the value of $\mathrm{N_L}$ can be determined only theoretically. Furthermore, because of absorption by interstellar hydrogen, we shall probably not be able to observe stellar radiation beyond the Lyman limit. The study of ionization zones can provide a check on the theoretical calculations of spectra of hot stars.

A first approximate estimate may be obtained by assuming a Planck distribution at $\lambda < 912$ Å with a stellar surface temperature T_*. Actually, the absorption coefficient in the Lyman continuum is so great that the radiation comes from a relatively thin surface layer. Using Planck's formula, we obtain

$$\mathrm{N_L} = \int_{\nu_c}^{\infty} \pi F_\nu \frac{d\nu}{h\nu} = \frac{2\pi}{c^2} \int_{\nu_c}^{\infty} \frac{\nu^2 d\nu}{e^{h\nu/kT_*} - 1} \approx \frac{2\pi k T_*}{hc^2} \nu_c^2 e^{-h\nu_c/kT_*}. \qquad (1.4)$$

Here ν_c is the frequency of the limit of the Lyman continuum ($\nu_c = 3.29 \times 10^{15}$ sec^{-1}). It is assumed in the integration that $h\nu_c \gg kT_*$. The value of $\mathrm{N_L}$ was originally estimated from this equation. However, calculations of model atmospheres of hot stars [23–31] show that the distribution of energy in the ultraviolet part of the spectrum of these stars differs markedly from the Planck distribution. Examples are shown in Figs. 1(*a*) and 1(*b*), where the calculated and the Planck distributions are plotted for two model atmospheres. It is clear that radiation beyond the Lyman limit, directly responsible for the ionization zone H II, is considerably less than that given by the Planck distribution. Observational verification of these calculations is not possible, of course, since all Lyman radiation is observed in the interstellar medium.

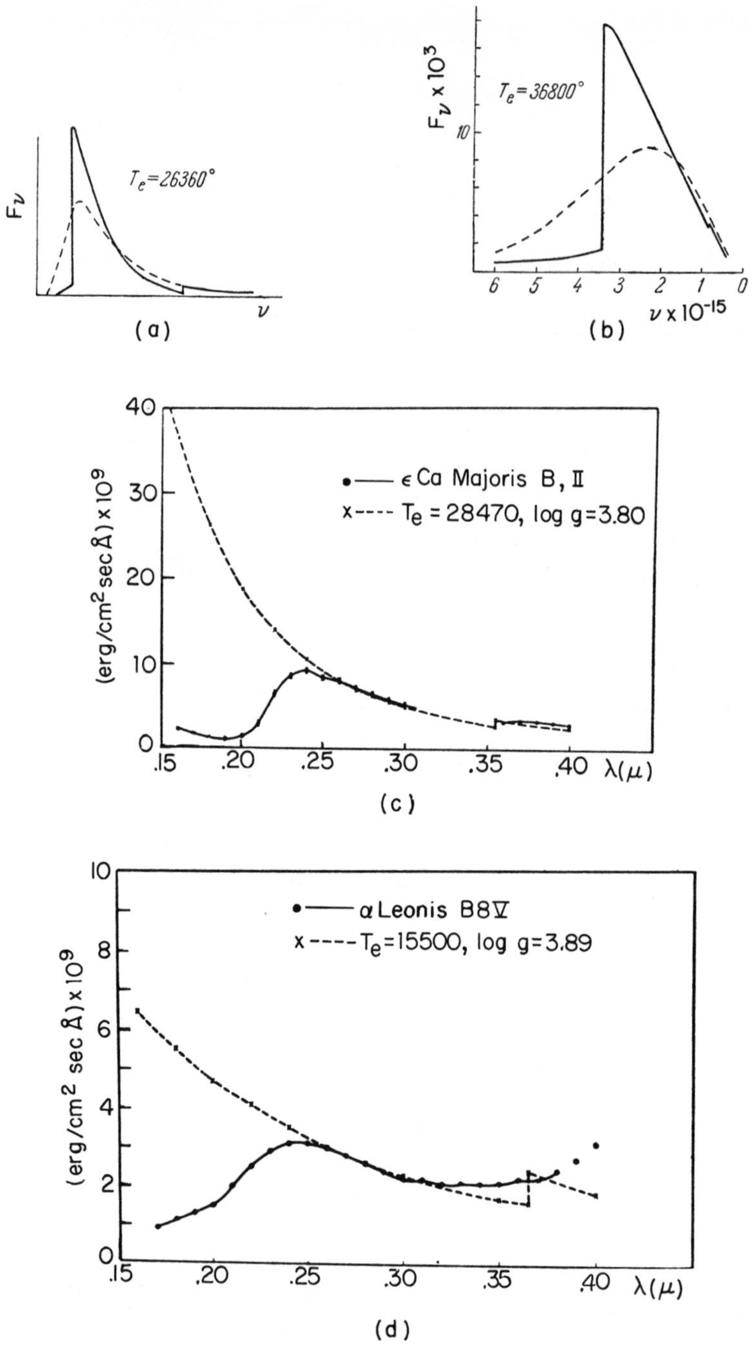

FIG. 1. Energy distribution in the spectra of hot stars and black-body radiation at the effective star temperature : (*a*) calculated [26]; (*b*) observed [24]; (*c*) *solid curve*, ε Canis Majoris B1 II; *broken curve*, $T_e = 28{,}470°$; log $g = 3.80$ [64]; (*d*) *solid curve*, α Leonis B8 V; *broken curve*, $T_e = 15{,}500°$; log $g = 3.89$ [64].

Spectra of hot stars up to the Lyman limit can be measured from rockets. The first results of this kind [64] showed that for most stars, in particular for B stars, a depression occurs at wavelengths $\lambda < 2400$ Å. The observed intensity is less, sometimes by several orders of magnitude, than the calculated value. Examples of observed and calculated stellar energy distributions are shown in Figs. 1(*c*) and 1(*d*). This discrepancy is explained, obviously, by neglect in the calculation of absorbers, particularly quasi molecules. When two atoms or ions approach each other, a system is established that is similar to a molecule, even though it is unstable. From the moment of approach, such a system can absorb a quantum, which corresponds to a molecular electronic transition. Quasi

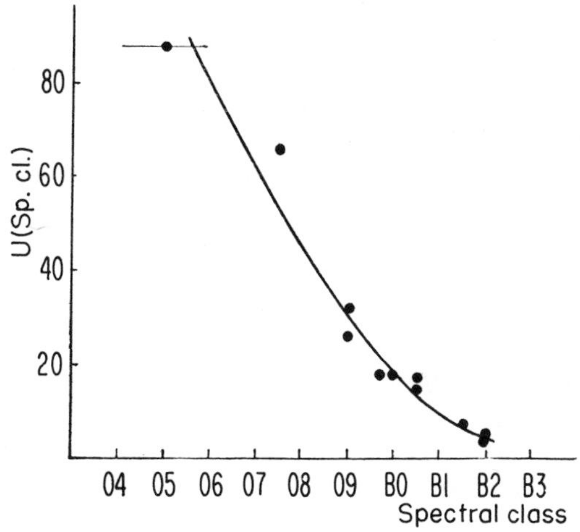

FIG. 2. The function U(sp. cl.) [32].

molecules may arise from atoms and ions of H, He, and other abundant elements. Other analogous absorption mechanisms have been suggested for continuous spectra, such as collisions between hydrogen atoms in the ground and excited state [64a]. On the other hand, Pecker [64b] considered absorption by circumstellar dust. A decrease in flux in the near ultraviolet region must lead to an increase in flux beyond the Lyman limit. Although O stars were not observed, additional absorption is in principle possible for them as well. Thus F_ν and hence U(sp. cl.) may be increased. It is true that since $s_0 \sim N_L{}^{1/3}$ a change in N_L has little effect on the determination of s_0.

In reference [32], the total number of L_c quanta radiated per second was calculated for given models [23–31], which yielded by means of Eq. (1.3) a value of the function U(sp. cl.) for stars of various spectral classes. Results of the calculation are shown in Fig. 2 and tabulated in Table 1.1. The points correspond to the different models and the average values are given in the table (the solid curve in Fig. 2). The last two lines of Table 1.1 give the observed values of the quantity

U(sp. cl.) from the flux of thermal radio emission [33], and from optical measurements from diffuse nebulae (a comparison is made in references [34] and [33a]). We see that the agreement between the theoretical calculations and the observed data obtained by independent methods is quite satisfactory. This means that the estimates of N_L are more or less correct.

Table 1.1

Spectral class	O5	O6	O7	O8	O9	O9.5	B0	B1	B2
$N_L \times 10^{-23}$	8.7	—	5.8	—	2.3	0.92	0.36	—	0.015
U (sp. cl.) (pc/cm^2)									
Calculated	88	80	62	46	31	25	20	11	4.6
Observed, radio	90	73	54	39	27	22	17.5	13	—
Observed, optical	80	67	44	48	—	—	15	16	—

If the gas density in H II zones is of the order of 10–100 particles per cubic centimeter, they are observed as emission nebulae, and if the density is somewhat lower, as extended weak hydrogen fields. In the following section we shall show that the intensity of the internal emission from a nebula is proportional to the so-called emission measure ME [35]. For an H II zone

$$\mathrm{ME} = \int_0^{2s_0} N_e^2\, dr \approx \overline{N}_e^2\, 2s_0 \approx 2U(\text{sp. cl.})\overline{N}_e^{4/3}, \qquad (1.5)$$

where the integral is over the line of sight through the whole H II region and r is expressed in parsecs. Regions with ME < 50–100 are difficult to observe against the background of the night sky. Therefore, it follows from Eq. (1.5) that only the denser H II regions can be observed. We note that, since Eq. (1.3) or (1.5) connects one stellar parameter (the spectrum) with two parameters of the interstellar gas (s_0 and N_e), if we can measure two of them a third can be found. For example, by measuring s_0 and N_e, the spectral class of the ionizing star can be determined if we can be assured that the measured s_0 does in fact represent the boundary of the H II zone. The average degree of hydrogen ionization is found by comparing Eqs. (1.1) and (1.3);

$$\left(\frac{\overline{N}_p}{N_1}\right) \approx \left(\frac{\overline{N}_e}{N_1}\right) \approx 20N_e s_0 \approx 20U(\text{sp. cl.})N_e^{1/3}. \qquad (1.6)$$

In typical H II zones, where $N_e \approx 1$–$10\ \mathrm{cm}^{-3}$, the ratio of the number of protons to neutral atoms is of the order of 10^3.

In the transition layer between the ionized (H II) and nonionized (H I) hydrogen regions, the degree of ionization falls very rapidly. Since τ increases with distance, N_1 increases correspondingly, which in turn leads to a still further rapid increase in τ. In the transition zone of thickness Δs_0, the number of neutral atoms increases from $N_1 \approx 10^{-3}N_H$ to $N_1 \approx N_H$; here N_H is the total number of neutral and ionized atoms per unit volume. Let us determine the thickness of the transient layer in which τ changes, for example, by two units, and where the number of neutral atoms is on the average of the order of $\frac{1}{2}N_H$. Obviously, we have

$$\Delta s_0 \approx \frac{2}{6.3 \times 10^{-18} N_1} \approx \frac{0.2}{N_H} \text{ pc.} \tag{1.7}$$

Hence, the transition layer is indeed relatively thin ($\Delta s_0 \ll s_0$, when $N_H > 1$) for stars of spectral classes O and B. The thickness of the transition layer, in contrast to s_0, depends very little on the radius of the ionization zone or on the general characteristics of the ionizing star. The structure of the transition layer is discussed again in Chapter 5.

The mass of hydrogen M that a single star can ionize is determined in the following way. Assuming constant density, we have obviously

$$M = \frac{4\pi}{3} s_0^3 N_p m_H = \frac{4\pi m_H}{3N_e}[U(\text{sp. cl.})]^3 = \frac{0.1 M_\odot}{N_e}[U(\text{sp. cl.})]^3, \tag{1.8}$$

where m_H is the mass of a hydrogen atom. For stars of spectral class O7, for example, $U(\text{sp. cl.}) = 62$ pc/cm^2 and Eq. (1.8) yields

$$\frac{M}{M_\odot} = \frac{24{,}000}{N_e}.$$

Masses of H II regions are comparatively large. The inverse relation of M to density is explained by the fact that a reduction in N_e in a given mass reduces the number of recombinations. Therefore, a flux of L_c quanta from a star can maintain a larger number of hydrogen atoms in the ionized state.

Direct observations show that the distribution of gas in interstellar space, and in H II regions in particular, is extremely heterogeneous. The foregoing relations were obtained under the assumption of a homogeneous hydrogen distribution. The effect of heterogeneity (or "porosity" as it is sometimes called) in the medium can be allowed for by the following method [34]. Assume that a certain volume δ is occupied by ionized hydrogen of density N_p (in this case, for purposes of neutrality, $N_e = N_p$), while the remaining volume $1 - \delta$ is free from hydrogen. Then the average electron density $\bar{N}_e = N_e\delta$ and the average number of recombinations per unit volume is equal to $\alpha_t N_e^2 \delta = \alpha_t \bar{N}_e^2/\delta$. Hence, for the radius of the H II zone, we have

$$s_0 = \frac{U(\text{sp. cl.})}{\bar{N}_e^{2/3}} \delta^{1/3} = \bar{s}_0 \delta^{1/3}. \tag{1.9}$$

Since the emission measure is proportional to $N_e^2 s_0 \delta^{1/3} \approx \bar{N}_e^2 \bar{s}_0 \delta^{-4/3}$, its dependence on the "porosity" of the medium is determined by the factor $\delta^{-4/3}$.

The Ionization Equation of Interstellar Hydrogen

We turn now to the derivation of quantitative relations used in calculating the distribution of hydrogen ionization within the limits of an H II zone [22, 35]. The development of the ionization equation is virtually the same as for Eq. (1.2), but with the difference that we must now equate the number of L_c quanta absorbed per second at a given point in space (and not in the total H II region) with the number of recombinations in the same time and in the same volume. In this way we obtain

$$4\pi N_1 \int_{\nu_c}^{\infty} a_1(\nu)\bar{I}_\nu \frac{d\nu}{h\nu} = N_p N_e \sum_{n=2}^{\infty} \alpha_n(T) = N_p N_e(\alpha_t - \alpha_1). \tag{1.10}$$

Diffuse L_c radiation is taken into account in Eq. (1.10) by assuming that the quanta of diffuse radiation originating by recombination at the first level are always absorbed by other hydrogen atoms. Generally speaking, this assumption is correct only on the average, since a re-emitted L_c quantum is also absorbed in other places. Therefore Eq. (1.10) is applicable if the optical thickness of the H II region for L_c quanta is not small and the density distribution is homogeneous. If these conditions are not fulfilled, we must begin to sum on the right-hand side of Eq. (1.10) from the first level (that is, to retain only α_t) and add to $\bar{I}_\nu$ on the left-hand side the average intensity of diffuse L_c emission. The method of calculating this value is explained in Sec. 3. In Eq. (1.10) $\bar{I}_\nu$ is the average (over direction) intensity of stellar radiation in a given volume of the H II zone.

If πF_ν is the radiation flux at frequency ν near the surface of a star (F_ν is the average intensity on the surface), then at a distance r the average intensity $\bar{I}_\nu$ at this frequency will be determined by the expression

$$\bar{I}_\nu = F_\nu W e^{-\tau_\nu} = \frac{R_*^2 F_\nu}{4r^2} \exp\left[-\int_{R_*}^{r} a_1(\nu) N_1 dr\right]. \tag{1.11}$$

Here W is the dilution coefficient, that is, the ratio of the area of the visible stellar surface πR_*^2 to the surface of a hypothetical sphere at a distance r from the observation point for $R_* \ll r$. We emphasize that $\bar{I}_\nu$ defines the intensity of radiation averaged over the whole sphere. Therefore, the density of this radiation is $\rho_\nu = 4\pi\bar{I}_\nu/c$. The factor $e^{-\tau_\nu}$ takes into account absorption of radiation along the path between the star and the element of volume under consideration.

Substituting Eq. (1.11) in Eq. (1.10), we obtain

$$N_1 \frac{\pi R_*^2}{r^2} \int_{\nu_c}^{\infty} a_1(\nu) F_\nu e^{-\tau_\nu} \frac{d\nu}{h\nu} = N_p N_e(\alpha_t - \alpha_1). \tag{1.12}$$

Replacing $a_1 N_1$ by $d\tau_\nu/dr$ in Eq. (1.12), multiplying both sides of the equation by $r^2 dr$, and integrating with respect to τ_ν and r from zero to infinity, we obtain

$$R_*^2 \int_{\nu_c}^{\infty} \pi F_\nu \frac{d\nu}{h\nu} = \int_0^{\infty} (\alpha_t - \alpha_1) N_e N_p r^2 dr. \qquad (1.13)$$

If we assume now that outside the H II zone $N_p \approx 0$ and inside the zone $N_p \approx N_e$ and that density and temperature are constant inside the zone, Eq. (1.13) leads directly to Eq. (1.2). Equation (1.13) may be used to determine the radius of an H II region when the temperature or density distribution inside such a region is not homogeneous (bearing in mind, as we discussed before, the role of diffuse L_c radiation). Returning to Eq. (1.12), we obtain the equation of ionization equilibrium,

$$\frac{N_p N_e}{N_1} = \frac{4\pi W}{\alpha_t - \alpha_1} \int_{\nu_c}^{\infty} a_1(\nu) F_\nu e^{-\tau_\nu} \frac{d\nu}{h\nu}. \qquad (1.14)$$

From Eq. (1.14), it is easy to obtain an ionization equilibrium equation similar to Saha's equation, but valid only for conditions in interstellar space. To do this, we use the following simplifications and assumptions:

(1) Let us consider the energy distribution in the ultraviolet region of the spectrum to be a Planck distribution (more precisely, to satisfy Wien's law):

$$F_\nu = \frac{2h\nu^3}{c^2} e^{-h\nu/kT_*}. \qquad (1.15)$$

(2) We take the factor $e^{-\tau_\nu}$ outside the integral, replacing τ_ν by the optical thickness at the limit of the Lyman series.

(3) For $\alpha_t - \alpha_1$ we substitute the value of the recombination probability at the first level. The basis for such a formal substitution rests in the fact that for $T \approx 10{,}000°$K (absolute temperature will be used henceforth in this book), the recombination probability at the first level is approximately equal to the sum of the recombination probabilities at all remaining levels.

Considering the explicit expression for α_1, Eq. (I.22), we can rewrite Eq. (1.14) under these conditions in the form

$$\frac{N_p N_e}{N_1} = W \times 2 \frac{(2\pi mkT)^{3/2}}{h^3} \frac{\mathrm{Ei}(h\nu_c/kT_*)}{\mathrm{Ei}(h\nu_c/kT)} e^{-(h\nu_c/kT)-\tau}$$
$$= 2\left(\frac{2\pi mkT_*}{h^2}\right)^{3/2} e^{-h\nu_c/kT} \left[\left(\frac{T}{T_*}\right)^{1/2} We^{-\tau}\right]. \qquad (1.16)$$

In the second equation we take into account that $h\nu_c \gg kT_* \gtrsim kT$. Equation (1.16) is the usual equation for hydrogen-ionization equilibrium and is accurate to the factor in braces. Specifically in interstellar space the very small factor $We^{-\tau}$ ($\sim 10^{-15}$) accounts for the weakness of ionizing radiation (dilution and absorption). Equation (1.16), together with the equation $N_H = N_p + N_1$, where N_H is the total number of hydrogen atoms per unit volume, and with the definition of optical depth τ,

$$\tau = \int_0^r a_1^{(0)} N_1 dr, \tag{1.17}$$

forms a system whose solution gives the degree of ionization as a function of distance. When $\tau \ll 1$, we see from Eq. (1.16) that $N_p/N_1 \sim W/N_e \sim 1/r^2$. The distribution of N_1 near the transition layer was first calculated by Strömgren [22] by numerical integration, but it may also be found by an analytic procedure if the curvature of layers is neglected [7]. In an implicit form this solution is

$$2 \ln \frac{x_0(1-x)}{x(1-x_0)} + \frac{1}{1-x_0} - \frac{1}{1-x} = N_H a_1^{(0)} r, \tag{1.18}$$

where $x = N_p/N_H$ is the degree of ionization and x_0 is the degree of ionization near a star when $\tau = 0$. It immediately follows from Eq. (1.18) that the thickness of the transition layer agrees in order of magnitude with Eq. (1.7), when x is changed by an appreciable amount. Since s_0 diminishes with a decrease in T_*, the relative thickness of the transition zone increases as we go to later spectral classes. We should note that in these calculations scattered L_c emission, which plays such an important role in the transition layer, was not taken into account. As was indicated earlier, Eq. (1.16) was obtained with essential restrictions, of which the most serious is the assumption concerning the Planck energy distribution in the ultraviolet region of the stellar spectrum. Therefore, we require for more accurate calculations the degree of ionization calculated from Eq. (1.14). As we shall see below, the temperature in H II zones is almost constant and is equal to 10,000°. Taking $\alpha_t - \alpha_1 = 2.45 \times 10^{-13}$ cm^3/sec and using theoretical models of the ultraviolet spectra of stars to determine the functions U(sp. cl.) for stars of each spectral class, the magnitude of the quantity

$$\frac{N_p N_e}{N_1 W} e^{\tau} = \frac{4\pi}{\alpha_t - \alpha_1} \int_{\nu_1}^{\infty} a_1(\nu) F_\nu \frac{d\nu}{h\nu}. \tag{1.14'}$$

can be calculated. Here, as before, we neglect the dependence of optical depth on frequency. The results of this calculation are given in Table 1.2. Using this table, we can easily determine the degree of ionization at any point in an H II region.

Table 1.2

Spectral class	B1.5	B0.5	O9	O7.5	O5	Wolf-Rayet
Effective temperature (10^3 °K)	26.4	31	37	41	50	62
$(N_p N_e/N_1 W)e^{\tau}$	0.08	0.7	1.5	8.5	29	60

Although interstellar hydrogen is mainly ionized by ultraviolet radiation from stars, examples occur (although they are observationally rare) where ionization of hydrogen is caused by electron collisions. If ionization equilibrium also takes place, we may use the obvious equation

$$N_1 N_e q_{1i} = N_p N_e (\alpha_t - \alpha_1) \tag{1.19}$$

to determine the degree of ionization; here q_{1i} is the probability of ionization by electron collision at the first level. From Eqs. (1.19) and (I.43) we obtain

$$\frac{N_p}{N_1} = \frac{q_{1i}}{\alpha_t - \alpha_1} = \frac{5.25\pi e^4}{(2\pi m k T)^{1/2}} \frac{e^{-\chi_1/kT}}{(\alpha_t - \alpha_1)\chi_1}. \tag{1.20}$$

This equation is applicable only at high electron temperatures where appreciable ionization by electron collisions is expected.

A similar problem arises, for example, in studying ionization in the sun's corona (see [36], where Eq. (1.20) is studied in detail). When electron temperatures change from 156,000° to 1,560,000°, the value of N_p/N_1 changes from 10^6 to 10^7. Proportionality to the first power of the temperature is explained by the fact that at high temperatures the recombination coefficients are proportional to $T^{-2/3}$.

It is convenient for practical calculations to use the approximate formula

$$\frac{N_p}{N_1} \approx 4 \times 10^5 \frac{kT}{\chi_1} e^{-\chi_1/kT} = 2.5 \times 10^{-69{,}000/T}, \tag{1.21}$$

which follows from Eq. (1.20). This formula, with different numerical coefficients, was obtained by Elwert [37]. The coefficient was corrected to correspond with the latest data on the probabilities of ionization by electron collision.

From rocket investigations it is possible to find ultraviolet and x-ray spectra of the sun and of the interplanetary medium. Calculations [37a] and observations of L_α [37b] emission and radioemission at 4–5 MHz [37c, d] give for the dimensions of the solar H II zone some hundreds of astronomical units and for its emission measure about 5.

2. Emission from Interstellar Hydrogen

The theory of the emission from hydrogen in the optical-wavelength band, in both discrete and continuous spectra, was originally developed (starting with the work of Zanstra [38, 39]) for application to the study of planetary nebulae. At the present time this theory has attained a sufficiently high degree of perfection and can probably be considered one of the most developed areas in the physics of nebulae and the interstellar medium.

Population of Levels

Hydrogen emission in both line and continuous spectra is usually of a recombination character. An electron is captured by a proton at any excited level (whereupon a quantum of the continuous spectrum beyond the limit of the series is emitted) and then a cascade (chain) of spontaneous transitions occurs down to different states as far as the 1*s* ground state. During these transitions quanta are emitted from the various series of the discrete spectrum. Applying this process to H II regions, we shall consider them to be opaque for lines of the Lyman series. Thus, quanta of the Lyman series, emitted during transitions to the ground state, are quickly absorbed by other atoms. Absorption and re-emission of Lyman quanta continue until they are all changed into a single L_α quantum ($2p \rightarrow 1s$ transition), a single quantum of the Balmer series, and possibly a few quanta of other series. If during the process of cascade transitions an electron falls into the 2*s* state, from there it falls to the 1*s* state, either through two-quantum emission or through the 2*p* state during the collision of the excited atom with an electron.

Thus under all conditions each recombination finally results in the emission of a single L_α quantum (or two quanta with the combined frequency $\nu_1 + \nu_2 = \nu_\alpha$) and a single Balmer quantum, which may be a quantum of the Balmer continuum. In addition, it is possible (but not necessary) that quanta from other series may be emitted as well.

If the H II region under discussion is transparent to the Lyman lines, quanta of this series that are formed can escape freely, whereupon the radiation intensity in the lines of the Balmer and other series will be reduced. However, in the actual conditions of interstellar space this case is very unlikely.

To determine the intensity of interstellar hydrogen radiation in a particular line of the recombination spectrum, we must first find the population of the levels (or terms) N_n and the sublevels N_{nl}, that is, the number of hydrogen atoms per unit volume found in these levels or sublevels. We shall first describe the method of determining the population of levels, without paying attention to the azimuthal quantum numbers. The corresponding general equations for finding N_{nl} are not difficult.

It is evident that in an equilibrium state the number of atoms in the *n*th level must be determined by the balance between the number of "upward" transitions (direct recombinations to the *n*th level and spontaneous transitions to levels $n'' > n$) and the number of atomic transitions downward from this level (spontaneous transitions to levels $n' < n$). From this condition we obtain the so-called steady-state equation,

$$N_e N_p \alpha_n(T) + \sum_{n''=n+1}^{\infty} A_{n''n} N_{n''} = N_n \sum_{n'=2}^{n-1} A_{nn'}, \qquad (2.1)$$

$$n = 3, 4, 5, \ldots, \infty.$$

Summation in the right-hand side of Eq. (2.1) is started from the second level, since a transition to the first level gives a Lyman quantum, which is immediately absorbed by another atom and therefore, on the average, returns the system to its original state.

An infinite system of equations (2.1) with an infinite number of unknowns N_n/N_eN_p may be solved approximately if we terminate the system at some level, $n = m$, and assume that values of N_n for $n > m$ are equal to zero. Then in the equation $n = m$, only N_m occurs, for which we obtain

$$N_m = N_e N_p \alpha_m(T) / \sum_{n'=2}^{m-1} A_{mn'} .$$

Substituting the value obtained for N_m in the equation for $n = m - 1$, we find N_{m-1}, and so on. The first solution of the system (2.1) was obtained in this way by Cillié in 1932 [40, 41]. However, if we neglect high levels, we obtain values for N_n that are too low.

The general solution of (2.1) can be written down directly by using Seaton's cascade matrices (I.8) [42],

$$N_n = N_e N_p \sum_{n''=n}^{\infty} \alpha_{n''}(T) C_{n''n} / \sum_{n'=2}^{n-1} A_{nn'} . \tag{2.2}$$

Indeed, by definition of the cascade matrix, $C_{n''n}$ defines the probability of an atom's falling into the nth level from the n''th level by any path. Therefore, $\alpha_{n''}C_{n''n}$ is the probability of recombination to the n''th level and the subsequent transition of the atom to the nth level. Summation over all values of n'' (when $C_{n''n''} = 1$) yields the total number of atoms falling to the nth level from above. The reciprocal of the value of the sum $\sum A_{nn'}$ is the lifetime of an atom in the nth level. Having analytic expressions for $\alpha_n(T)$ and knowing the elements of the cascade matrices, it is easy to evaluate Eq. (2.2), taking into account an infinite number of levels (for large values of n, summation is replaced by integration). Menzel and Baker [9] solved this problem by a similar method, but without introducing cascade matrices. However, their solution was numerically in error. The accurate solution of Eq. (2.1) was obtained by Seaton [42]. The results of his calculations will be given below.

The cascade-matrix method is extremely effective. It can also be employed in those cases in which levels are excited, not by recombinations, but by electron collisions or by absorption of quanta. For example, in the case of hydrogen-atom excitation by electron collision, the populations of the levels are determined by the expression

$$N_n = N_e N_1 \sum_{n''=n}^{\infty} q_{1n''}(T) C_{n''n} / \sum_{n'=2}^{n-1} A_{nn'} . \tag{2.3}$$

A similar problem (correct for a restricted number of levels and without the use of cascade matrices) was examined by Miyamoto [43, 44], Chamberlain [45], and S. A. Kaplan and S. I. Gopasyuk [46].

The parameter b_n offers a more graphic idea of the population of levels, and indicates the deviation of the populations from the thermodynamic-equilibrium population

$$N_n = b_n N_e N_p \frac{n^2 h^3}{(2\pi m k T)^{3/2}} e^{\chi_n/kT}. \tag{2.4}$$

Here T is the temperature of the electron gas. In a system in thermodynamic equilibrium at this temperature, $b_n = 1$ for any value of n. In the actual conditions of interstellar space, thermodynamic equilibrium clearly does not exist, and therefore $b_n \neq 1$. However, we can expect the population of very high levels ($n \to \infty$, $\chi_n \to 0$) to approximate the values that occur at thermodynamic equilibrium in an electron gas, since continuous transitions from discrete states to the continuum must take place. In this way $b_n \to 1$ as $n \to \infty$. The smaller n becomes, the more the population level deviates from equilibrium. The greatest deviation is for the first level ($n = 1$), where $b_1 \approx (1/W)$ (1.16). We stress that in speaking of thermodynamic equilibrium within an electron gas we are concerned only with the electron velocity distribution. There is no true thermodynamic equilibrium in an electron gas, since the radiation density in this case does not correspond to its temperature.

The approximation $b_n = 1$ at very high n levels is justified by the fact that transitions between adjacent levels under the action of "weak" electron collisions (distant encounters) become more probable than radiative transitions. Indeed, comparing Eqs. (I.46) and (1.7), we find

$$N_e q_{n,\,n-1} = \frac{6.5 \times 10^{-6} n^4}{T^{1/2}} N_e > \sum_{k=1}^{n-1} A_{n,\,k}$$
$$\approx \frac{n-1}{2}(A_{n1} + A_{n(n-1)}) \approx \frac{9 \times 10^9}{n^4}. \tag{2.5}$$

In this way, when $n > n_k = 78T^{1/16}N_e^{-1/8} \approx 140N_e^{-1/8}$, the populations of levels must be determined by collisions with free electrons possessing a thermodynamically stable (Maxwellian) velocity distribution. Therefore, the population of these levels also appears as a characteristic condition of thermodynamic equilibrium. In other words, b_n approaches unity with increase in n faster than predicted by the solution of the system (2.1). Without serious loss in accuracy, we can set $b_n = 1$ for $n > 2n_k$. We note that in dense nebulae, where, for example, $N_e = 10^4$ cm^{-3}, we have $n_k \approx 46$. A more detailed study of the system (2.1) with additional terms allowing for transitions under the influence of electron collisions was given by Seaton [49d]. It was shown that for $n < n_k$ the role of collisions is not great, whereas for $n > n_k$ the parameter is $0.9 \leqslant b_n \leqslant 1$. However, this increased precision has practically no influence on the intensity of the lines in the optical region of the spectrum. Transitions between neighboring levels when $n > 60$ fall into the radio band, and here we know that $b_n = 1$ approximately without even solving the equilibrium equations.

Before proceeding to the numerical results of the solution of (2.1), let us briefly consider azimuthal degeneracy. We can readily see (remembering that $l' = l \pm 1$) that in this case the system of equations analogous to (2.1) has the form

$$N_e N_p \alpha_{nl}(T) + \sum_{n''=n}^{\infty} (N_{n''(l+1)} A_{n''(l+1)nl} + N_{n''(l-1)} A_{n''(l-1)nl}) = N_{nl} \left(\sum_{n'=l+2}^{n} A_{nln'(l+1)} + \sum_{n'=l}^{n} A_{nln'(l-1)} \right), \tag{2.6}$$

$$n = 3, 4, 5, 6, \ldots, \infty, \qquad l = 0, 1, 2, \ldots, n - 1.$$

Here also parameters b_{nl}, which characterize the degree of deviation from thermodynamic equilibrium, may be introduced:

$$N_{nl} = b_{nl} N_e N_p \frac{(2l + 1)h^3}{(2\pi mkT)^{3/2}} e^{\chi_n/kT}. \tag{2.7}$$

Equation (2.7) follows directly from Eqs. (1.3) and (2.4). At thermodynamic equilibrium, $b_{nl} = 1$.

System (2.6), and similarly system (2.1), were approximately solved initially [47, 48, 49] as a system of algebraic equations, where, because of their clumsiness, only a small number of equations were usually taken. A solution was obtained by Pengelly [49b] with the aid of cascade matrices.

The role of electron collisions described above remains valid here. What is more, since these collisions are more effective in the case of transition with no change in principal quantum number, the populations of sublevels conform to thermodynamic equilibrium at lower values of n [49c]. Using Eq. (I.47) and proceeding as in the derivation of Eq. (2.5), we find that for $n > n'_k \approx 45 N_e^{-1/8}$ the distribution over sublevels must correspond to that for thermodynamic equilibrium. That is, for $n'_k < n < n_k$ we have $b_{nl} = b_n \neq 1$. The solution of system (2.1) in this region of principal quantum numbers gives a more accurate result than the solution of system (2.6). Values for the parameters $b_n e^{\chi_n/kT}$ at different temperatures, obtained by the numerical solution of Eqs. (2.1) [42], and parameters b_{nl} when $T = 10{,}000^\circ$, but for different values of l [49b], are given in Tables 2.1 and 2.2.

From these tables we can see that the population of levels (more accurately, the degree of deviation from thermodynamic equilibrium) depends relatively little on temperature. The dependence on azimuthal quantum number is much stronger; s states are overpopulated, while p and d states, on the contrary, are underpopulated. This is explained by the high probability of $p \to s$ and $p \to d$

transitions and the low probability of $s \to p$ transitions. However, since the mean populations

$$b_n = \sum_{l=0}^{n-1} \frac{2l+1}{n^2} b_{nl},$$

calculated from the data of Table 2.2, differ by not more than 10–20 percent from b_n obtained by solving (2.1), the error caused by neglecting azimuthal degeneracy is not great and has no effect on the interpretation of the observational results.

Table 2.1

n	T(°K)			
	2500	5000	10,000	20,000
3	0.257	0.422	0.668	1.013
4	.218	.350	.540	0.792
5	.220	.346	.519	.739
6	.231	.355	.520	.725
7	.245	.369	.529	.722
8	.259	.384	.540	.725
9	.274	.398	.552	.730
10	.287	.412	.563	.735
15	.344	.466	.605	.756
20	.386	.503	.635	.772
25	.417	.531	.656	.785
30	.442	.552	.673	.795

To determine the populations of levels excited by electron collisions we can rewrite Eq. (2.3) in the form

$$\begin{aligned} \frac{N_3}{N_e N_1} &= \frac{1}{A_{32}} \sum_{n''=3}^{\infty} q_{1n''} C_{n''3} \\ &= 2.28 \times 10^{-8}(q_{13} + 0.516 q_{14} + 0.484 q_{15} + \cdots), \\ \frac{N_4}{N_e N_1} &= \frac{1}{A_{43} + A_{42}} \sum_{n''=4}^{\infty} q_{1n''} C_{n''4} \\ &= 5.78 \times 10^{-8}(q_{14} + 0.363 q_{15} + 0.306 q_{16} + \cdots), \end{aligned} \tag{2.8}$$

and so on. The introduction of the parameters b_n does not lead to any simplification here, since values of $q_{1n''}$ are given in tabular form. In the case of high temperatures, by taking into account the condition (I.44), that is, the proportionality of q to the corresponding oscillator strengths, these expressions may be written in the form

$$\frac{N_3}{N_e N_1} \approx 2.28 \times 10^{-8}(1.45 q_{13}) = 3.3 \times 10^{-8} q_{13},$$
$$\frac{N_4}{N_e N_1} \approx 5.78 \times 10^{-8}(1.41 q_{14}) = 8.2 \times 10^{-8} q_{14}. \tag{2.9}$$

These relations are approximately correct at temperatures of the order of a few tens of thousands of degrees. In particular, at $T = 50{,}000^\circ$, third- and fourth-level populations are equal respectively to

$$N_3 \approx 9.9 \times 10^{-18} N_e N_1 \text{ cm}^{-3}, \qquad N_4 \approx 8.2 \times 10^{-18} N_e N_1 \text{ cm}^{-3}. \tag{2.10}$$

For lower temperatures, the population falls off rapidly with an increase in T. For example, when $T \approx 10{,}000^\circ$

$$N_3 \approx 7 \times 10^{-23} N_e N_1 \text{ cm}^{-3}, \qquad N_4 \approx 2.5 \times 10^{-23} N_e N_1 \text{ cm}^{-3}. \tag{2.11}$$

Here "experimental" values of the excitation coefficients are used; "theoretical" coefficients would yield populations reduced by a factor 2.

Table 2.2

n	l					
	0	1	2	3	4	5
3	1.01	0.265	0.0855			
4	1.12	.320	.146	0.269		
5	1.19	.384	.189	.332	0.460	
6	1.24	.429	.221	.372	.524	0.599

Let us compare the values of Eqs. (2.10) and (2.11) with the populations of levels during recombination excitations. Substituting in Eq. (2.4) the numerical values of the atomic constants, we find

$$N_n = 4.2 \times 10^{-16} \frac{n^2 N_e N_p}{T^{3/2}} b_n e^{\chi_n/kT}, \tag{2.12}$$

or for $T = 10{,}000^\circ$,

$$N_3 \approx 2.5 \times 10^{-21} N_e N_p \text{ cm}^{-3}, \qquad N_4 \approx 3.6 \times 10^{-21} N_e N_p \text{ cm}^{-3}. \tag{2.13}$$

Considering that in H II regions the degree of ionization is high ($N_p \approx 10^3 N_1$), we find that, when $T \approx 10{,}000^\circ$, excitation by electron collision plays no role, but when $T \gtrsim 50{,}000^\circ$ the role of electron collisions increases substantially. However, in the steady state of interstellar gas, such temperatures are not reached.

Equations for determining the populations of levels are linear in N_n. Therefore, if both excitation mechanisms are present and $N_n \ll N_1$, we can find separately for each of them the corresponding population and the resulting values for $N_{n'}$ (recombinations) and $N_{n''}$ (electron collisions), which are then combined. First of all, of course, we must determine the degree of ionization.

So far we have been discussing the third and higher levels. Populations of the $2s$ and $2p$ sublevels are estimated in another way, since the optical thickness in the L_α line is large. The number of atoms in the $2p$ sublevel is found by studying the diffusion of L_α quanta. This will be treated in the next section. We must remember that the results obtained above are correct only in the case of small optical thickness in lines of the Balmer series.

The population of the $2s$ sublevel is determined by the condition of equilibrium between, on one hand, the number of atoms arriving at this level by means of "upward" spontaneous transitions, by direct recombination into the $2s$ state, and by excitation by electron collisions, and, on the other hand, the number leaving the level by means of "downward" transitions to the $1s$ state during two-quantum emission and by $2s \to 1s$ excitation due to electron collisions. In this way we obtain

$$(N_e q_{2s1s} + N_e q_{2s2p} + A_{2s1s})N_{2s} = N_e N_p \alpha_{2s}(T) + N_e N_1 q_{1s2s} + \sum_{n=3}^{\infty} N_{np} A_{np2s}. \quad (2.14)$$

Here we also take into account the probable $2s \to 2p$ transition due to electron collision. It is convenient for solution to introduce a parameter X, which describes the fraction of all the recombinations that bring atoms to the $2s$ state [50]. Neglecting excitation and de-excitation due to electron collisions, we can write

$$N_{2s} = XN_e N_p \sum_{n=2}^{\infty} \alpha_n(T)/A_{2s1s} = XN_e N_p \frac{\alpha_t - \alpha_1}{A_{2s1s}}. \quad (2.15)$$

The parameter X may be calculated from the data of Table 2.2. It turns out that $X \approx 0.34$, while its value is almost independent of temperature [49].

Hence we find at once the population of the $2s$ sublevel ($T = 10{,}000^\circ$):

$$N_{2s} \approx \frac{2.45 \times 10^{-13}}{0.227} 0.34 N_e N_p \approx 1.0 \times 10^{-14} N_e N_p \text{ cm}^{-3}. \quad (2.16)$$

Thus, the metastable $2s$ sublevel is overpopulated compared to all the remaining sublevels of the hydrogen atom (except $2p$) by five or six orders of magnitude. In dense nebulae the population of the $2p$ sublevel may be large. In this case $2p \to 2s$ transitions must also be taken into account [51].

Line Emission

Calculation of radiation intensity for known populations offers no difficulty. For example, the energy emitted in the H_β line (transition $4 \to 2$) per unit volume per second per steradian (the so-called volume emission coefficient), is determined by the expression

$$\varepsilon(\mathrm{H}_\beta) = \frac{h\nu_{42}}{4\pi} A_{42} N_4 = \frac{h\nu_{42}}{4\pi} \frac{A_{42} 16h^3}{(2\pi mkT)^{3/2}} b_4 e^{\chi_4/kT} N_e N_p$$

$$= \frac{h\nu_{42}}{4\pi} \frac{h^3}{(2\pi mkT)^{3/2}} (A_{4s2p} b_{4s} + 3A_{4p2s} b_{4p} + 5A_{4d2p} b_{4d}) e^{\chi_4/kT} N_e N_p. \quad (2.17)$$

The remaining transitions ($4d \to 2s$, $4f \to 2p$) are forbidden by selection rules. Calculations with Eq. (2.17), both with and without taking into account azimuthal degeneracy (Tables 2.1 and 2.2, respectively), give the same result; for $T_e = 10{,}000°$,

$$\varepsilon(\mathrm{H}_\beta) = 0.97 \times 10^{-26} N_e N_p \text{ erg/cm}^3 \text{ sec ster.} \quad (2.18)$$

When the temperature is doubled ($T = 20{,}000°$), the emission coefficient $\varepsilon(\mathrm{H}_\beta)$ is reduced approximately to half, while at a temperature of 5,000°, $\varepsilon(\mathrm{H}_\beta)$, on the contrary, is $1\frac{1}{2}$ times as large [$\varepsilon(\mathrm{H}_\beta) \sim T^{-3/2} b_4 e^{\chi_4/kT}$.

In the majority of cases, absorption in the H_β line is negligible since the optical thickness in this line is much less than unity. The intensity of the H_β line is

$$I(\mathrm{H}_\beta) = \int \varepsilon(\mathrm{H}_\beta) dr = 0.97 \times 10^{-26} \int N_e N_p dr$$

$$= 3 \times 10^{-8} \mathrm{ME} \text{ erg/cm}^3 \text{ sec ster.} \quad (2.19)$$

Here the integral is taken along the line of sight. The radiation intensity at a given temperature is determined by a single parameter—the emission measure [Eq. (1.5)]. Therefore, the concept of the emission measure, introduced for the first time by Strömgren in 1948 [35], proved to be very useful in the physics of interstellar gases.

At the present time, with the aid of wide-band filters, nebulae can be observed that have an emission measure of about 400, and with narrow-band filters nebulae with emission measures of 50–100 are seen. The possibility of observing very weak emitting regions is limited, not by apparatus, but by the radiation from circumterrestrial hydrogen (geocorona). Its emission measure, as P. V. Shcheglov showed, varies over the range from 5 to 50.

The intensity of other hydrogen lines and also the intensity of the continuous spectrum are usually given relative to H_β. It is obvious that when the ratio $I_{nn'}/I(\mathrm{H}_\beta)$ is calculated common factors cancel, the emission measure among them, and this ratio turns out to be dependent only on temperature.

In particular, the ratio of intensities of lines of the Balmer series, the so-called Balmer decrement, is determined by the equation

$$\frac{I_{n2}}{I(\mathrm{H}_\beta)} = \frac{\nu_{n2}}{\nu_{42}} \frac{A_{n2}}{A_{42}} \frac{N_n}{N_4} = \frac{64}{n^3} \frac{g_{n2}}{g_{42}} \frac{b_n}{b_4} e^{(\chi_n - \chi_4)/kT}. \quad (2.20)$$

In this case, Eqs. (I.2) and (I.5) are used. Calculation of the ratio (2.20) from the data of Table 2.1 is not difficult. Results of the calculation of relative line intensities for two temperatures are given in Table 2.3 ([42, 49a]; the mean

Table 2.3

Determination	Temperature (°K)	H_α	H_β	H_γ	H_δ	H_ε
Neglecting azimuthal	10,000	2.71	1.00	0.506	0.298	0.192
quantum number	20,000	2.79	1.00	.421	.262	.178
Considering azimuthal	11,000	2.87	1.00	.466	.256	.158
quantum number	20,000	2.76	1.00	.474	.262	.162
Observed		2.55	1.00	.50	.29	—

observational results are also taken from these papers). The data show clearly that the Balmer decrement is sensitive neither to temperature change nor to azimuthal quantum number. Therefore it is impossible to use the observed Balmer decrement to determine the temperature of an electron gas.

The mean observed Balmer decrement agrees well with theoretical calculations. However, in a number of cases, especially in emission nebulae, considerable deviations are detected—the observed ratio $I(H_\alpha)/I(H_\beta)$ is greater than the theoretical value. In principle, such deviations can be explained by the fact that the optical thickness of these nebulae for the H_α and H_β lines is greater than unity. This situation was analyzed by Pottasch [53, 54], who showed that the ratio $I(H_\alpha)/I(H_\beta)$ increased with increase in optical thickness, since in this case some of the H_β quanta are divided into H_α and P_α quanta. Generally speaking, the above-calculated population of the $2s$ sublevel is inadequate to make a nebula optically thick in the H_α line under usual conditions. For example, in a nebula of about 3 pc in extent, the optical thickness for the H_α line is

$$\tau(H_\alpha) \approx 10^{-14} N_e N_p \times 10^{-13} \times 10^{19} \approx 10^{-8} N_e N_p \ll 1.$$

Here the absorption coefficient for H_α is assumed equal to 10^{-13} cm^2. In denser nebulae, $N_e > 10^4$, but as a rule such nebulae are not large. Moreover, in dense nebulae, the population of the $2p$ sublevel increases rapidly because of the repeated scattering of L_α quanta, which can also increase the optical thickness for H_α and H_β lines. We return to this question in Sec. 3.

We must stress that observations are still not reliable enough to allow us to propose a hypothesis concerning the large optical thickness of nebulae for lines of the Balmer series. In particular, it is difficult to take into account correctly selective interstellar absorption, which also makes the observed Balmer decrement steeper.

If hydrogen emission is excited by electron collisions, then the Balmer decrement is also extremely steep. Using Eqs. (2.20) and (2.11), we obtain for the ratio $I(H_\alpha)/I(H_\beta)$ 10.8 at 10,000°, 4.7 at 50,000°, and 4.2 in the limit temperatures ($q_{13}/q_{14} = f_{13}/f_{14} = 0.079/0.029$).

However, it is necessary to recall that in the case of high temperatures emission will occur from recombination (ionization by electron collision and subsequent recombination) as well, and the actual Balmer decrement will be flatter. Moreover, at high temperatures, the absolute brightness of the radiation is in general low, since the probability of recombination, and consequently the number of neutral atoms, diminish.

It is also possible to calculate the relative intensities of lines from other series, or the relative intensities of lines belonging to different series. For example, the relative intensity of P_α and H_α (first Paschen and Balmer lines) is 0.260 at $T = 10{,}000°$ and 0.224 at $T = 20{,}000°$. The values of the intensity ratios P_β/H_β are 0.311 and 0.273, respectively [49]. Here the dependence on temperature is relatively weak. Comparing the observed ratio of these lines with the numbers presented above, we can evaluate selective absorption along the line of sight to a nebula.

The Continuous Spectrum

Besides its line spectrum, interstellar hydrogen has a continuous spectrum also. It is true that for a large number of diffuse nebulae and H II emission regions its intensity is usually feeble, but in the hottest nebulae, for example the Orion nebula, radiation in the continuum is extremely strong.

Three mechanisms are known for the emission of quanta of the continuous spectrum by hydrogen atoms and ions in interstellar space: free-free transitions, recombinations, and two-quantum transitions.

(1) *Radiation due to free-free transitions* (collisions between a proton and an electron). This is best considered by using Kirchhoffs law: the ratio of the volume emission coefficient $\varepsilon(\nu)$ to the absorption coefficient $\kappa(\nu)$ per unit volume is the Planck function

$$\frac{\varepsilon(\nu)}{\chi(\nu)} = B_\nu(T) = \frac{2h\nu^3}{c^2}\frac{1}{e^{h\nu/kT} - 1}. \tag{2.21}$$

Here T is the temperature of the electron gas. The applicability of Eq. (2.21) in a given case is determined by whether the distribution function of electron velocities is Maxwellian, that is, whether the distribution of electrons in the levels of the continuous spectrum is the same as at thermodynamic equilibrium.

The hydrogen absorption coefficient per unit volume for free-free transitions (taking into account stimulated emission) is calculated from Eq. (I.34) for $Z = 1$:

$$\chi(\nu) = k'(\nu)N_e N_p = \frac{8\pi}{3}\frac{e^6 g(\nu)}{mhc(6\pi mkT)^{1/2}}\frac{1 - e^{-h\nu/kT}}{\nu^3}N_e N_p. \tag{2.22}$$

Substituting Eq. (2.22) into Eq. (2.21) and denoting the volume emission coefficient for free-free transitions by $\varepsilon_{ff}(\nu)$, we obtain

$$\varepsilon_{ff}(\nu) = \frac{16\pi}{3} \frac{e^6 g N_e N_p}{mc^3(6\pi mkT)^{1/2}} e^{-h\nu/kT}$$

$$= 5.44 \times 10^{-39} g T^{-1/2} e^{-h\nu/kT} N_e N_p \text{ erg/cm}^3 \text{ sec ster Hz.} \quad (2.23)$$

If the nebula is transparent at frequency ν, the ratio of the emission intensity in the continuous spectrum to the intensity of H_β (we set the temperature T of the electron gas in the factor preceding the exponential in Eq. (2.23) equal to 10,000°) is

$$\frac{I_\nu(d\nu)}{I(H_\beta)} = \frac{\varepsilon_{ff}(\nu)d\nu}{\varepsilon(H_\beta)} = 5.6 \times 10^{-15} g e^{-h\nu/kT} d\nu. \quad (2.24)$$

Here the exponential dependence on temperature is preserved. The fact is that the factor preceding the exponential depends very little on temperature and therefore Eq. (2.24) can be used at $T \neq 10{,}000°$, but not too far from this value. Remember also that when $h\nu/kT = 1$–7 the Gaunt factor $g = 1.1$–1.4. Integrating Eq. (2.23) over all frequencies, we obtain for the total emission in all directions

$$4\pi\varepsilon_{ff} = 4\pi \int_0^\infty \varepsilon_{ff}(\nu)d\nu = 1.43 \times 10^{-27} T^{1/2} \bar{g} N_e N_p \text{ erg/cm}^3 \text{ sec.} \quad (2.25)$$

In this way, the total emission from free-free transitions over the whole spectrum has the same order of magnitude as the emission from a single H_β line. For $\bar{g}$ the value 1.2 may be used.

(2) *Continuous emission from recombination.* In the visible and near-ultraviolet regions we find the Balmer and Paschen continua ($n = 2$ and 3). To calculate the corresponding emission coefficients it is simplest to use the definition of the effective recombination cross section [see Appendix I, Eqs. (I.16)–(I.20)]:

$$\varepsilon_n(\nu) = \frac{h\nu}{4\pi} \sigma_n^i(v) v f(v) \frac{dv}{d\nu} N_e N_p$$

$$= h\nu \left(\frac{m}{2\pi kT}\right)^{3/2} \sigma_n^i(v) v^3 e^{-mv^2/2kT} \frac{dv}{d\nu} N_e N_p, \quad (2.26)$$

where $\frac{1}{2}mv^2 + \chi_n = h\nu$. The factor $dv/d\nu = h/mv$ embodies the fact that $\varepsilon_n(\nu)$ is calculated per frequency interval and the distribution function of electron velocities $f(v)$ per velocity interval. Substituting Eq. (I.21) in Eq. (2.26), we obtain for hydrogen

$$\varepsilon_n(\nu) = \frac{2^7 \pi^4 e^{10} m g_n}{c^3 h^2 (6\pi mkT)^{3/2}} \frac{1}{n^3} e^{-(h\nu - \chi_n)/kT} N_e N_p$$

$$= 1.7 \times 10^{-33} g_n T^{-3/2} n^{-3} e^{-(h\nu - \chi_n)/kT} N_e N_p \text{ erg/cm}^3 \text{ sec ster Hz.} \quad (2.27)$$

The ratio of the emission coefficient in the Balmer continuum to the emission coefficient in the H_β hydrogen line (in the factor preceding the exponential $T = 10{,}000°$) is

$$\frac{\varepsilon_2(\nu)d\nu}{\varepsilon(H_\beta)} = 2.2 \times 10^{-14} e^{-h(\nu-\nu_2)/kT} d\nu, \tag{2.28}$$

where ν_2 [$= 8.2 \times 10^{14}$ sec^{-1}] is the frequency of the Balmer limit. Equation (2.28) determines the radiation intensity only at $\nu > \nu_2$ or at $\lambda < 3647$ Å. At longer wavelengths, $\varepsilon_2(\nu) = 0$. Because of the presence of the Balmer discontinuity, this emission mechanism may be separated from other emission mechanisms in continuous spectra. The Paschen continuum is observed at frequencies $\nu_2 > \nu > \nu_3 = \chi_3/h$. The radiation intensity in this case is determined by an equation like Eq. (2.28), in which it is necessary, in the first place, to replace ν_2 by ν_3, and in the second place to multiply the numerical coefficient by $(\frac{2}{3})^3 = 0.296$.

The total energy emitted in the continuous spectrum from recombinations to the nth level is

$$4\pi\varepsilon_n = 4\pi \int_{\nu_n}^{\infty} \varepsilon_n(\nu)d\nu = \frac{2.15 \times 10^{-32}\bar{g}_n}{T^{3/2}n^3} \int_{\nu_n}^{\infty} e^{-h(\nu-\nu_n)/kT} d\nu \times N_e N_p$$

$$= \frac{4.5 \times 10^{-22}}{T^{1/2}n^3} N_e N_p \text{ erg/cm}^3 \text{ sec}. \tag{2.29}$$

When $T = 10{,}000°$, total emission in the Balmer continuum is approximately four times the total emission from free-free transitions. Total emission from all recombinations to the second and higher levels is

$$0.202 \times 4.5 \times 10^{-22} T^{-1/2} N_e N_p = 0.91 \times 10^{-22} T^{-1/2} N_e N_p.$$

Lyman continuous radiation converts itself to radiation of other series, to L_α quanta, and to the two-quantum emission from the transition $2s \rightarrow 1s$.

(3) *Emission from two-quantum transitions* $2s \rightarrow 1s$ [55, 50]. Employing the definition of the probability of spontaneous two-quantum transitions (I.15), we may write

$$\varepsilon_{2q}(\nu)d\nu = \frac{h\nu}{4\pi} dA_{2s1s} N_{2s} = 1.0 \times 10^{-42} \nu\Psi(\nu/\nu_\alpha) N_{2s} d\nu, \tag{2.30}$$

or, considering Eq. (2.16),

$$\varepsilon_{2q}(\nu)d\nu = 1.0 \times 10^{-56} \nu\Psi(\nu/\nu_\alpha) N_e N_p d\nu \text{ erg/cm}^3 \text{ sec ster}. \tag{2.31}$$

The ratio of the intensity for two-quantum transitions to the H_β intensity is determined by the equation

$$\frac{\varepsilon_{2q}(\nu)d\nu}{\varepsilon(H_\beta)} = 1.03 \times 10^{-30} \nu\Psi(\nu/\nu_\alpha)d\nu. \tag{2.32}$$

The total radiation from two-quantum transitions is most readily calculated from the formula

$$4\pi\varepsilon_{2q} = 4\pi \int_0^{\nu_\alpha} \varepsilon_{2q}(\nu)d\nu = h\nu_\alpha A_{2s1s}N_{2s} = 1.4 \times 10^{-24}N_e N_p. \tag{2.33}$$

It is comparable in order of magnitude with those given by Eqs. (2.25) and (2.29).

Thus, all three continuous-emission mechanisms play approximately equal roles in the total emission. However, their spectral distributions are different [56]. If in the visible region of the spectrum free-free and two-quantum transitions produce radiation whose intensity changes little with frequency, the spectral distribution of the radiation from the Balmer and Paschen recombinations will have a discontinuity.

The total hydrogen radiation in the continuous spectrum is determined by the sum

$$\varepsilon(\nu) = \varepsilon_{ff}(\nu) + \varepsilon_{2q}(\nu) + \sum_{n=2}^{\infty} \varepsilon_n(\nu),$$

while the relative value, for example, of the Balmer discontinuity is

$$D_B = \log \left.\frac{\varepsilon(\nu)}{\varepsilon(\nu) - \varepsilon_2(\nu)}\right|_{\nu=\nu_2}. \tag{2.34}$$

Two-quantum transitions are responsible for approximately 50 percent of the continuous hydrogen radiation at $\lambda \approx 5000$ Å, 70 percent immediately before the Balmer discontinuity, but only 17 percent beyond this discontinuity when $T = 10{,}000^\circ$.

In principle, from the dependence of continuous emission on frequency, it is possible to determine the temperature of the electron gas. However, the results will be unreliable because the radiation intensity in the continuous spectrum is in general small and because the observed distribution is distorted by interstellar absorption. In dense nebulae ($N_e \gtrsim 10^4$), collisions of the second kind begin to quench the excited $2s$ state. In this case, the number of two-quantum transitions is diminished, and consequently the magnitudes of the Balmer and Paschen discontinuities are increased. Therefore, the density may be determined from the value of these discontinuities, although with no great degree of reliability. Moreover, account must be taken of the recombination emission from helium.

Figure 3 shows the theoretical distribution with respect to H_δ of the radiation intensity of nebulae in the continuous spectrum at various temperatures and densities of the electron gas [57]. Attention is drawn to the fact that at low temperature the intensity of the continuous spectrum depends essentially on the density of the electron gas.

The dependence of the Balmer (D_B) and Paschen (D_P) discontinuities on the density and temperature is shown in Fig. 4.

Let us stress again that the emission from interstellar hydrogen, in both the

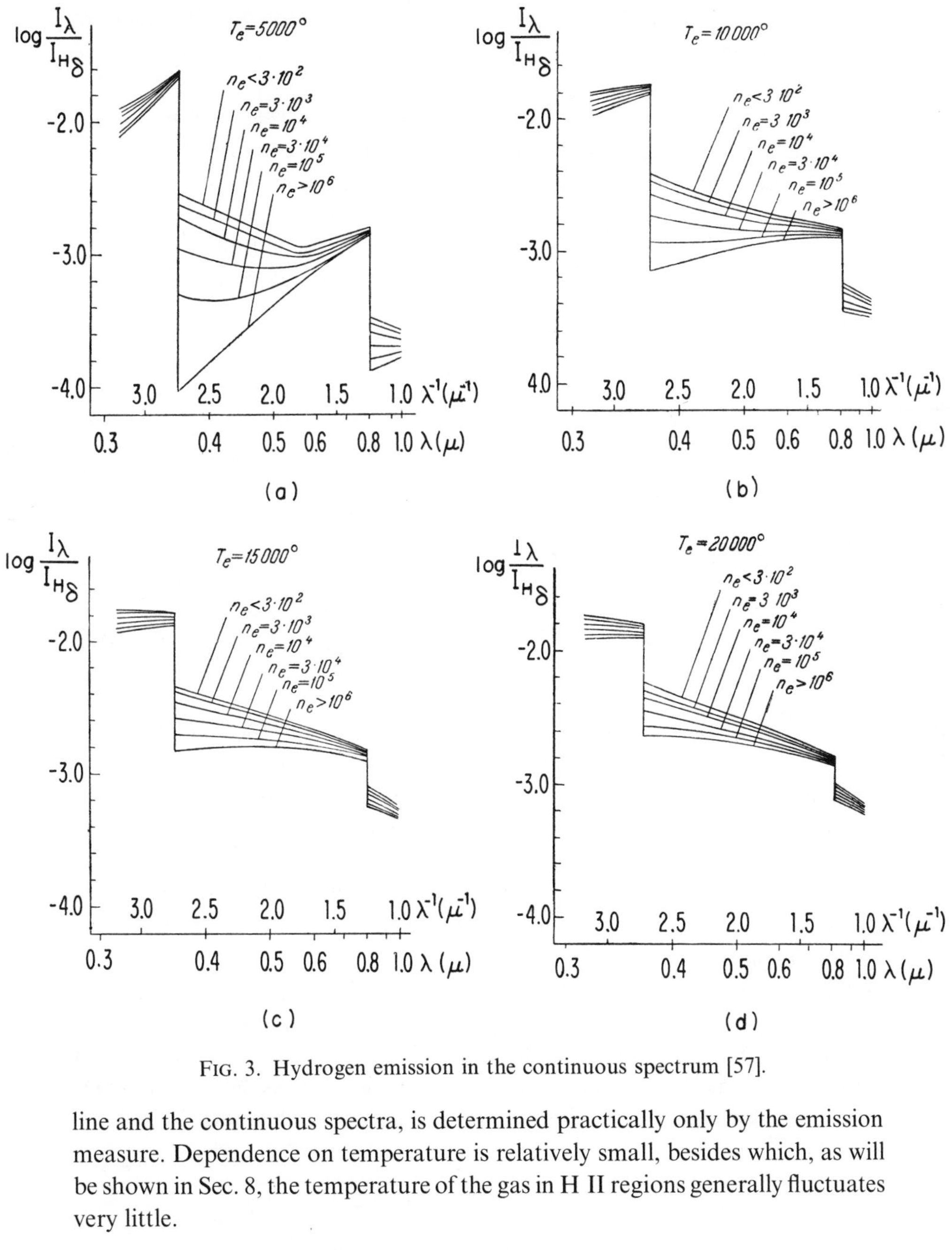

FIG. 3. Hydrogen emission in the continuous spectrum [57].

line and the continuous spectra, is determined practically only by the emission measure. Dependence on temperature is relatively small, besides which, as will be shown in Sec. 8, the temperature of the gas in H II regions generally fluctuates very little.

Helium Emission

Although the present section, as well as all the first chapter, is devoted to interstellar hydrogen, it is pertinent to consider here some observations on helium emission, since the methods for its calculation are the same as in the case of hydrogen. This problem is presented in greater detail in an article by Seaton [58].

Neutral helium atoms give rise to lines from the two series $ns \rightarrow 2p$ and $nd \rightarrow 2p$. In calculating their emission, it is possible to use fully the "hydrogenic" values for recombination coefficients and spontaneous transitions,

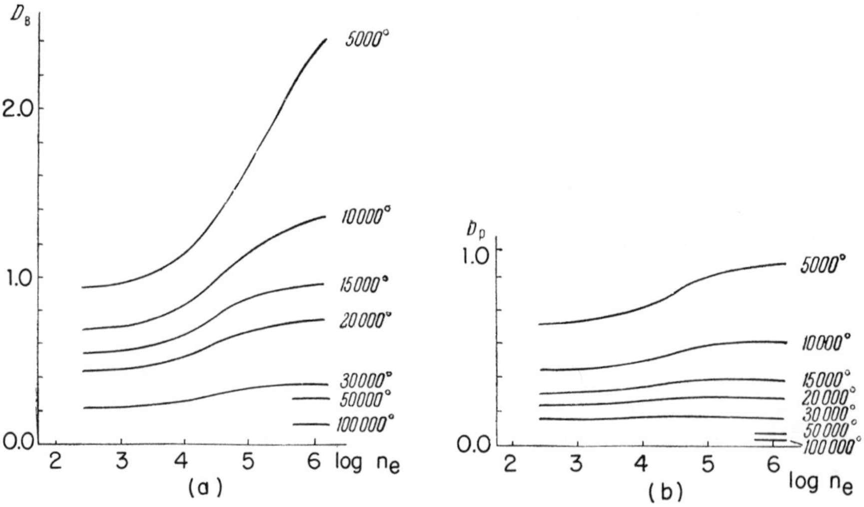

FIG. 4. Balmer and Paschen discontinuities under various conditions [57].

especially for the second series, since in this case the orbits corresponding to the upper and lower levels are located beyond the orbit of the $1s$ state and the second electron is almost completely screened from the nuclear charge. Therefore, in this case the emission coefficient of neutral helium is identical with the hydrogen emission coefficient, the only difference being the need to consider azimuthal quantum number degeneracy.

Let us introduce the effective recombination coefficient $\beta(n, n')$ for every line in such a way that, for example, $\beta_H(nd, 2p)N_p N_e$ is the number of quanta radiated by hydrogen atoms in the line $(nd \rightarrow 2p)$ per cubic centimeter per second. Then the corresponding effective recombination coefficients in the He lines at the same temperature, in both the singlet and doublet series, are determined by the relations [58]:

$$\begin{aligned} \beta_{He}(n^1D, 2^1P) &= \tfrac{1}{4}\beta_H(nd, 2p), \\ \beta_{He}(n^3D, 2^3P) &= \tfrac{3}{4}\beta_H(nd, 2p). \end{aligned} \tag{2.35}$$

For $\beta_H(nd, 2p)$, we have the expression

$$\beta_H(nd, 2p) = \frac{A_{nd2p}N_{nd}}{N_e N_p} = \frac{5A_{nd2p}h^3}{(2\pi mkT)^{3/2}}\, b_{nd}e^{\chi_n/kT}. \tag{2.36}$$

Here b_{nd} is found from Table 2.2. For example, $\beta_H(3d, 2p) = 5.27 \times 10^{-14}$ cm^3/sec when $T = 10{,}000°$.

Calculation of the recombination emission spectrum of singly ionized helium (recombination with He^{++}) proceeds precisely as for hydrogen, except that $Z = 2$ must be substituted in all the expressions for recombination coefficients, probability of spontaneous transitions, and so forth. Tables of values of b_n and b_{nl} calculated for hydrogen are applicable to singly ionized helium, but at one-fourth the temperature (when the value of the exponent $e^{-\chi_n/kT}$ remains as before). In other words, the column of values for $b_n e^{\chi_n/kT}$ in Table 2.1 calculated for hydrogen when $T = 2{,}500°$ determines the corresponding value of the identical parameter for ionized helium when $T = 10{,}000°$.

Table 2.4 gives the relative intensities of helium lines [58] when $T = 15{,}000°$ (the intensity of the transition line $3\,^3D \rightarrow 2\,^3P$, $\lambda = 5876$ Å is assumed to be 100).

Table 2.4

n	3	4	5	6
$I(n\,^3D, 2\,^3P)$	100	41	21	12
$I(n\,^1D, 2\,^1P)$	29	12	6	4

The relative line intensities of singly ionized helium when $T = 10{,}000°$ (the intensity of the transition line $4 \rightarrow 3$, $\lambda = 4686$ Å is taken as 100) are given in Table 2.5.

Table 2.5

n	5	6	7	8	9	10
$I(n, 4)$	21.9	14.6	2.9	7.3	5.4	4.2
$I(n, 5)$	—	6.8	5.0	3.7	2.8	2.1

We note that the value of the parameter β_{He} for the helium line $\lambda = 4686$ Å when $T = 10{,}000°$ is $\beta_{He}(4.3) = 2.08 \times 10^{-13}$. From what has previously been said about the β parameters, it follows that from the ratio of line intensities for helium and hydrogen the relative number of atoms and ions of these elements can be found. Actually, from the definition of line intensities as quantities proportional to $h\nu\beta N$, we obtain

$$\begin{aligned} \frac{I(4861)}{I(5876)} &= \frac{5876}{4861} \cdot \frac{\beta_H(4.2)N_p}{\beta_{He}(3\,^3D, 2\,^3P)N_{He^+}}, \\ \frac{I(4861)}{I(4686)} &= \frac{4686}{4861} \cdot \frac{\beta_H(4.2)N_p}{\beta_{He}(4.3)N_{He^{++}}}. \end{aligned} \tag{2.37}$$

Substituting numerical values, we find

$$\frac{N_{\mathrm{He}^+}}{N_p} = 1.00\frac{I(5876)}{I(4861)}, \qquad \frac{N_{\mathrm{He}^{++}}}{N_p} = 0.134\frac{I(4686)}{I(4861)}. \tag{2.38}$$

The relations allow the relative abundance of helium to be determined, together with its degree of ionization, from the observed lines of its emission spectrum. In particular, in planetary nebulae $N_{\mathrm{He}} \approx 0.2N_p$ on the average. In typical diffuse nebulae, helium lines are weak or are not observed at all. So far, helium has been discovered only in the Orion and Lagoon nebulae.

In the calculation of the radiation from neutral hydrogen it is necessary to take into account the presence of the metastable state $2\,^3S$. The direct transition $2\,^3S \to 1\,^2S$ is strongly forbidden, while the two-quantum transition $2\,^3S \to 2q \to 1\,^1S$ has a low probability ($A \approx 2.2 \times 10^{-5}$ sec^{-1}). The primary process of transferring helium atoms from this state is the transition to the state $2\,^1S$ by electron collisions, followed by the two-quantum transition to the $1\,^1S$ state. The coefficient of excitation by electron collision is estimated under various assumptions [58a] to be $q(2\,^3S \to 2\,^1S) \approx (0.4 - 2.5) \times 10^{-7}$ sec^{-1}. Considering that approximately half of the recombinations of helium result in the captured electron in the $2\,^3S$ state, we find the population of this level for $T = 10{,}000°$ to be

$$(0.4 \text{ to } 2.5) \times 10^{-7} N_{\mathrm{He}}(2\,^3S)N_e \approx 10^{-13} N_{\mathrm{He}} + N_e \tag{2.39}$$

or, approximately, $N_{\mathrm{He}}(2\,^3S) \approx 10^{-5} N_{\mathrm{He}+}$. Nebulae, for such a population, appear optically thick in absorption, in particular for the lines λ10,830 ($2\,^3S \to 2\,^3P$) and λ3889 ($2\,^3S \to 3\,^3P$). In fact, the latter line was observed in the Orion nebula in absorption [142] with an optical thickness at the center of the line of about 40. An analysis of the helium spectrum, taking this effect into account, was made by Pottasch [586]. In particular, quanta of the line λ3889 must be broken down into the quanta $\lambda = 43$ cm ($3\,^3P \to 3\,^3S$), λ7065 ($3\,^3S \to 2\,^3P$), and λ10,830 ($2\,^3P \to 2\,^3S$); this is similar to the case of the large optical thickness of lines of the Balmer series where H_β quanta are broken down into H_α and P_α.

But the ionization of the helium level $2\,^3S$ by L_α quanta may greatly reduce its population. For instance, if the helium concentration in the center of the Orion nebula is 2×10^3 cm^{-3}, the population of $2\,^3S$ is, as follows from (2.39), equal to 2×10^{-2} cm^{-3}. Observational data on λ3889 in this case give approximately 2×10^{-4} cm^{-3}. The difference of two orders of magnitude may be explained by ionization by L_α quanta, which may be accumulated in this region (see Sec. 3). This effect has been seen also in planetary nebulae by O'Dell [58c]. He found that there is a large accumulation of L_α quanta. This may be explained by the assumption that the planetary nebula is surrounded by an H I region, which is equivalent to supposing that $\tau_0 \gg 10^4$ in Eq. (2.39).

3. The Propagation of L_c and L_α Radiation in the Interstellar Medium

In the preceding section we treated emission in the continuous spectrum up to the Lyman limit and in the subordinate hydrogen lines where it is characterized by the fact that it can emerge freely from nebulae. This means that so far processes could be investigated in any element of volume, regardless of the properties and structure of the whole nebula. Quanta of the Lyman series must be treated differently. The optical thickness of all nebulae for lines of the Lyman series is much greater than unity. Therefore, Lyman quanta emitted in one element of volume are absorbed by a different volume element inside the same nebula. During the subsequent cascade transitions, all quanta of the Lyman series, except L_α quanta, are broken down. Quanta of the Balmer, Paschen, and other series formed in this way escape from the nebula, while the L_α quanta are repeatedly scattered and retained. To analyze the radiation field in the Lyman continuum (L_c) and in the L_α line, it is essential to study the transfer problem taking into account the multiple scattering in the whole nebula. The first research in this direction was carried out by V. A. Ambartsumyan [59] and Zanstra [60] in 1933–1934. At the present time, methods have been worked out that permit the various aspects of this problem to be studied with the required accuracy.

Here we shall confine ourselves to a series of questions that have a direct bearing on the physics of the interstellar medium, but only with that degree of completeness necessary for understanding the basic properties of the interstellar gas. A more complete account will be found either in the original works (certain of these are cited below) or in a book by V. V. Sobolev [61]. In order not to complicate the mathematical side of the problem, we shall consider only scattering in a one-dimensional medium. The angular distribution of L_α and L_c emission, for the calculation of which a three-dimensional problem must be solved, is of no interest since it cannot be observed. The solution of the one-dimensional problem yields an answer correct up to some numerical factor, of approximately unity, which is not essential for further analysis.

The Propagation of L_c *Radiation*

We begin with solution of the problem of calculating the L_c radiation field in a one-dimensional medium of infinite optical thickness.

The scattering process in the Lyman continuum leads to the ionization of hydrogen atoms, with resulting recombination to the ground state, followed by the emission of a "scattered" L_c quantum. Obviously, the probability of "survival" of an L_c quantum during a single scattering event (designated in the limit as Λ) is equal to the ratio of the probability of direct recombination to the ground state to the total probability of recombination to all levels, that is,

$$\Lambda = \frac{\alpha_1}{\alpha_2} = 0.39 \tag{3.1}$$

when $T = 10{,}000^\circ$. This figure depends only slightly on temperature; when $T = 20{,}000^\circ$, $\Lambda = 0.29$.

Essentially, the scattered radiation in the Lyman continuum can be practically considered monochromatic [62]. In fact, since the mean energy of the recombination electron is of the order of $0.5(\frac{3}{2}kT)$, the quantum radiated by the recombination has a frequency $\nu = \nu_c + 0.5(\frac{3}{2}kT/h)\ 1.078\nu_c$ when $T = 10{,}000^\circ$. The factor 0.5 allows for the fact that recombination of slow electrons is more probable. The absorption coefficient at this frequency is approximately 15 percent less than the absorption coefficient at the limit of the Lyman series. Therefore all the scattered L_c quanta can be considered to have a frequency $\bar{\nu} \approx 1.05\nu = 3.45 \times 10^{15}$ Hz and the absorption coefficient at this frequency is approximately $\bar{a}_1(\bar{\nu}) \approx 0.85a_1(\nu_c) = 5.35 \times 10^{-18}\ \mathrm{cm}^2$. We define the optical thickness correspondingly as $d\tau = \bar{a}_1(\bar{\nu})N_1 dr$.

Primary L_c radiation at frequency ν from stars is attenuated over the optical depth τ by a factor $\exp[-\tau(\bar{\nu}/\nu)^3]$. Therefore, the number of stellar L_c quanta absorbed per unit volume at this depth is

$$N_1 W \int_{\nu_c}^{\infty} F_\nu a_1(\nu) \exp\left[-\tau\left(\frac{\bar{\nu}}{\nu}\right)^3\right]\frac{d\nu}{h\nu}$$
$$= \frac{N_1 W \bar{a}_1(\bar{\nu})}{h\bar{\nu}} \int_{\nu_c}^{\infty} F_\nu \left(\frac{\bar{\nu}}{\nu}\right)^4 \exp\left[-\tau\left(\frac{\bar{\nu}}{\nu}\right)^3\right] d\nu. \quad (3.2)$$

We consider here that within the framework of a one-dimensional problem the flux of diluted emission is equal to WF_ν and not to πWF_ν, and also that $a_1(\nu) \sim \nu^3$. The probability of re-emission of an absorbed quantum is $\Lambda/2$, while its frequency will then be equal to $\bar{\nu}$, independent of the original frequency. In this way, the ratio of the emission coefficient at optical depth τ to the absorption coefficient per unit volume $\kappa = N_1\bar{a}_1(\bar{\nu})$ is equal to

$$g(\tau) = \frac{\varepsilon(\tau)}{\chi} = W\frac{\Lambda}{2}\int_{\nu_c}^{\infty} F_\nu \left(\frac{\bar{\nu}}{\nu}\right)^4 \exp\left[-\tau\left(\frac{\bar{\nu}}{\nu}\right)^3\right] d\nu. \quad (3.3)$$

The dilution coefficient W also changes with optical depth, although more slowly (because of the dependence on distance from the star). The quantity $g(\tau)$ is called the distribution function of the sources of escaping radiation.

As is known, the integral equation for determining the source function of multiscattered radiation $B(\tau)$ in the one-dimensional problem in the stationary case has the form [61]

$$B(\tau) = \frac{\Lambda}{2}\int_0^{\infty} B(\tau')e^{-|\tau-\tau'|}dr' + g(\tau). \quad (3.4)$$

The physical significance of this equation is very simple: it gives the radiation found in a volume element at depth τ due to absorption and subsequent re-emission of quanta that have arrived in this element from other elements of the medium, and due to $g(\tau)$, the single scattering of direct radiation from a star. It should be noted that the probability of re-emission is $\Lambda/2$.

Equation (3.4) has an explicit solution for an arbitrary function $g(\tau)$, namely,

$$B(\tau) = \frac{\Lambda}{2k}\left\{\left[\int_0^\tau g(\tau')e^{k\tau'}d\tau + \frac{1-k}{1+k}\int_0^\infty g(\tau')e^{-k\tau'}d\tau'\right]e^{-k\tau} + e^{k\tau}\int_\tau^\infty g(\tau')e^{-k\tau'}d\tau'\right\} + g(\tau). \quad (3.5)$$

Here $k = (1 - \Lambda)^{1/2}$. The solution of (3.5) assumes a particularly simple form if $g(\tau) = \frac{1}{2}\Lambda F e^{-\tau}$, which corresponds to the case of a flux of external monochromatic radiation falling on the medium in the absence of internal sources. Equation (3.5) then takes the form

$$\mathrm{B}(\tau) = F(1 - \sqrt{1 - \Lambda})e^{-\tau\sqrt{1-\Lambda}}. \quad (3.6)$$

This equation will be used subsequently.

It is apparent that the solution of the problem of the scattering of L_c emission requires substitution into Eq. (3.5) of Eq. (3.3), and we take $\Lambda = 0.39$ and $k = 0.78$. If we neglect the dependence of the dilution coefficient on r, that is, on the curvature of the layers, the general solution has the form

$$B_c(\tau) = \frac{\Lambda^2 W}{4k}\int_{\nu_c}^\infty F_\nu\left(\frac{\bar{\nu}}{\nu}\right)^4\left[\frac{e^{-\tau(\bar{\nu}/\nu)^3} - e^{-k\tau}}{k - (\bar{\nu}/\nu)^3} + \frac{e^{-\tau(\bar{\nu}/\nu)^3} + \dfrac{1-k}{1+k}e^{-k\tau}}{k + (\bar{\nu}/\nu)^3}\right. d\nu + g(\tau). \quad (3.7)$$

At large optical depths, the fundamental contribution to the integral is given by the high-frequency terms. Assuming, therefore, in the integral (3.7) that $(\bar{\nu}/\nu)^3 \ll k = 0.78$, we obtain an approximate equation correct when $\tau \gg 1$:

$$B_c(\tau) = \frac{\Lambda \cdot W}{2(1 - \Lambda)}\int_{\nu_c}^\infty F_\nu\left(\frac{\bar{\nu}}{\nu}\right)^4 e^{-\tau(\bar{\nu}/\nu)^3}d\nu. \quad (3.8)$$

To evaluate this integral we set $F_\nu = F_{\bar{\nu}}(\bar{\nu}/\nu)^n$. Hence, we find

$$B_c(\tau) = 0.32WF_{\bar{\nu}}\bar{\nu}\int_{\nu_c}^\infty\left(\frac{\bar{\nu}}{\nu}\right)^{4+n}e^{-\tau(\bar{\nu}/\nu)^3}d\left(\frac{\nu}{\bar{\nu}}\right) \approx \frac{0.11\Gamma(n+3)/3}{\tau^{(n+3)/3}}WF_{\bar{\nu}}\bar{\nu}. \quad (3.9)$$

The index $(n + 3)/3$ for class B stars is approximately 2–3; for class O stars it is approximately equal to $\frac{4}{3}$. The density of L_c emission falls relatively slowly with optical depth, since the shorter-wavelength radiation penetrates to considerable depths. Considerations for a three-dimensional medium lead to a change of k

to $[3(1 - \Lambda)]^{1/2} = 1.35$. This lowers $B(\tau)$ in Eq. (3.9) by approximately a factor of three.

The more detailed problems concerning the scattering of L_c radiation, among them the three-dimensional case, were resolved by R. E. Gershberg [62] by a somewhat different method. The characteristic feature of the distribution of L_c radiation should be noted in the transitional region between H I and H II. The initial L_c radiation in this case becomes bluer because of a decrease in the absorption coefficient with frequency according to the rule $a_1(\nu) \sim \nu^{-3}$. However, the scattered L_c radiation remains practically monochromatic.

As noted in the preceding section, the role of scattered L_c radiation in the luminosity of interstellar hydrogen is taken into account in an elementary way (by neglecting transitions to the ground state). However, in a number of cases (for example, in the study of the transition layer between the H II and H I zones), an explicit expression is required for the flux of diffuse L_c radiation. Remember that from the known source function $B_c(\tau)$ the flux of $H_c(\tau)$ radiation is found from the equation

$$H_c(\tau) = \int_0^\tau B_c(\tau')e^{-(\tau-\tau')}d\tau' - \int_\tau^\infty B_c(\tau')e^{-(\tau'-\tau)}d\tau'$$

$$\approx \frac{0.22}{\tau^{(n+6)/3}}\Gamma\left(\frac{n+6}{3}\right)WF_{\bar{\nu}}\bar{\nu}. \quad (3.10)$$

The radiation density (within the framework of a one-dimensional problem) is numerically equal to $2B_c(\tau)/c$. As was already noted, the extrapolation of these expressions to a three-dimensional problem only produces numerical factors of the order of unity.

The Propagation of L_α *Radiation*

Let us now proceed to a study of the diffusion of radiation in the line L_α. This problem may be investigated by various methods. We limit ourselves here to an account of the fundamental aspects of the simplest method, a study of the integral transfer equation.

Under the conditions of a highly rarefied interstellar gas, the profiles of spectral lines (L_α among them) are almost completely determined by the Doppler effect. If turbulent motion is discounted, the distribution of atomic velocities may be considered to be Maxwellian, depending on the gas temperature.

In this case, it is well known that the atomic absorption coefficient for the L_α line is equal to

$$k_\alpha(\nu) = \frac{\pi e^2}{mc}\frac{f_{2p1s}}{\sqrt{\pi}\Delta\nu_D}e^{-[(\nu-\nu_\alpha)/\Delta\nu_D]^2}, \quad (3.11)$$

where

$$\Delta\nu_D = \frac{\nu_\alpha}{c}\left(\frac{2kT}{m_H}\right)^{1/2} = \frac{(2kT/m_H)^{1/2}}{\lambda_\alpha},$$

$f_{2p1s} = 0.4162$ is the oscillator strength of the line L_α, and $\nu_\alpha = 2.42 \times 10^{15}$ sec^{-1} is the frequency of the center of the line. Substituting numerical values, we obtain

$$k_\alpha(\nu) = \frac{5.9 \times 10^{-12}}{T^{1/2}} e^{-[(\nu-\nu_\alpha)/\Delta\nu_D]^2} \text{ cm}^2, \quad \Delta\nu_D = 1.05 \times 10^9 T^{1/2} \text{ Hz.} \tag{3.12}$$

In particular, at the center of the L_α line at $T = 10{,}000^\circ$ the absorption coefficient is 9300 times that at the limit of the Lyman series.

Comparing Eq. (3.11) with the corresponding expression for the absorption coefficient in the wings of the line, where the profile is determined by attenuation of the emission

$$k_\alpha(\nu) = \frac{e^2}{mc} \frac{f_{2p1s} A_{2p1s}}{4\pi^2(\nu - \nu_\alpha)^2}, \tag{3.13}$$

we find that the Doppler profile (when $T = 10{,}000^\circ$) changes into Lorentzian wings when $[(\nu - \nu_\alpha)/\Delta\nu_D]^2 = 10.5$, that is, at a distance at which the absorption coefficient is $e^{10.5} = 4 \times 10^4$ times less than in the center of the line. When $T = 100^\circ$ this change occurs for $[(\nu - \nu_\alpha)/\Delta\nu_D]^2 = 8.0$ and the absorption coefficient is 3.1×10^3 times less than its value at the center of the line.

We assume in this section that by τ we mean the optical depth in the center of the L_α line, namely,

$$d\tau = k_\alpha(\nu_\alpha) N_1 dr = \frac{5.9 \times 10^{-12}}{T^{1/2}} N_1 dr. \tag{3.14}$$

If the optical thickness of the H II zone in the Lyman continuum is of the order of unity, then the optical thickness of this zone in the center of the L_α line is about 10^4 and in the Lorentzian wings it is less than unity. Consequently, L_α quanta scattered in the Lorentzian wings leave zone H II quite freely.

Therefore, in a study of the transfer of emission in the L_α line, account must be taken both of the complete annihilation of these quanta (for example, during their absorption by cosmic dust particles) and their shift to the far edges of the wings of the line, after which these quanta emerge from a given H II region. In both cases, L_α quanta appear as "lost" and do not participate in further scattering of this emission.

Many mechanisms are known for the annihilation or shifting of L_α quanta, some of which are examined in [6]. We shall consider here only those that prove to be the most applicable from the point of view of conditions in the interstellar medium.

As in the analysis of the scattering of L_c radiation, we introduce the parameter Λ as the average probability of "survival" in a single act of scattering. Let N_a be the number of atoms or dust particles per unit volume capable of "intercepting" L_α quanta and $a(\nu_\alpha)$ be their absorption coefficient at this frequency. Then for the parameter Λ we have

$$\frac{1-\Lambda}{\Lambda} = 2^{1/2}\frac{a(\nu_\alpha)N_a}{k_\alpha(\nu_\alpha)N_1}. \tag{3.15}$$

The factor $2^{1/2}$ embodies the fact that the absorption coefficient averaged over the profile is less than in the center of the line. Apparently, the most important mechanism for the annihilation of L_α quanta is their absorption by cosmic dust particles or their ionization of atoms and ions at potentials lower than $h\nu_\alpha =$ 10.2 eV.

For example, in the case of absorption of L_α quanta by particles of cosmic dust, it is possible to take $a(\nu_\alpha) \approx 8 \times 10^{-9}$ cm^2, assuming the mean radius of the particles equal to 5×10^{-5} cm and the effective absorption cross section of the particle to be approximately equal to its geometric cross section (the wavelength of an L_α quantum is considerably less than the particle dimensions). It is true that in the visible region of the spectrum dust particles usually scatter rather than absorb light; in this case they have no effect on the probability of quantum survival. However, L_α quanta whose energy is $\sim$ 10 eV in all probability eject electrons from dust particles, changing the structures of their constituent molecules, and therefore are not scattered, but undergo pure absorption. Therefore, Eq. (3.15) yields

$$\frac{1-\Lambda}{\Lambda} \approx 2 \times 10^3 T^{1/2}\frac{N_g}{N_1} = \begin{cases} 10^{-5} \text{ (H II)}, \\ 10^{-9} \text{ (H I)}. \end{cases} \tag{3.16}$$

Here N_g is the number of dust particles per unit volume. The numerical values in Eq. (3.16) are $N_g \approx 0.5 \times 10^{-12}$ cm^3, $N_1 \approx 10^{-2}$ cm^{-3} in the H II region and $N_1 \approx 10$ cm^{-3} in the H I zones, $T = 10{,}000°$ in H II and $T = 100°$ in the H I regions.

For the absorption of L_α quanta by atoms with ionization potentials below 10 eV, we can take $a(\nu_\alpha) \approx 6 \times 10^{-18}$ cm^2. Then

$$\frac{1-\Lambda}{\Lambda} \approx 10^{-6} T^{1/2}\frac{N_a}{N_1}. \tag{3.17}$$

Neutral atoms of sodium, magnesium, aluminum, silicon, and so forth can absorb L_α quanta. The total concentration of these elements relative to hydrogen is approximately 10^{-4}. If it is considered that in H II zones the degree of ionization of these elements is approximately equal to the degree of ionization of hydrogen, then for the H II zone in this case $1 - \Lambda = 10^{-8}$, which is considerably less than for absorption by dust particles. The consequences of this effect in an H I zone are not great, since here N_1 is larger by one to three orders of magnitude and N_a on the contrary is less (the degree of ionization as above). We note that L_α quanta can be absorbed by hydrogen atoms in the second state as well. In this case, N_a must be taken as $N_{2s} + N_{2p}$. Under the usual conditions of the interstellar medium, the role of this effect is negligible [$N_{2s} \approx 10^{-14} N_e N_p$ from Eq. (2.16), $N_{2p} \ll N_{2s}$] and, consequently, $1 - \Lambda \approx 10^{-18} N_e N_p / N_1$.

If $N_{2p} > N_{2s}$, the annihilation of L_α quanta is possible during the transition of a hydrogen atom from the $2p$ state to the $2s$ state, either by the emission of a radio quantum or by an electron collision and the subsequent two-quantum transition $2s \to 1s$, which also limits the population of the $2p$ sublevel.

Annihilation of L_α quanta is also possible if a hydrogen atom in the $2p$ state absorbs an H_β quantum, is transferred to the $4s$ or $4d$ sublevel, and then returns to the ground state through the chain $4s \to 3p \to 2s \to 1s$ or $4d \to 3p \to 2s \to 1s$ (the transition $2s \to 1s$ being a two-quantum process). Here, an H_β quantum, as well as an L_α quantum, is lost, but a new H_α quantum emerges. If this mechanism is present, the observed Balmer decrement must be more abrupt than that derived from Table 2.3. Since in a number of planetary nebulae there was in fact observed such an increased abruptness of the Balmer decrement, Pottasch [53] proposed that in these cases appreciable absorption of H_β quanta took place. For this to happen, the optical thickness for H_β must be comparable with unity. Hence we find the population of the $2p$ sublevels to be

$$N_{2p} \approx \frac{\tau(H_\beta)}{\dfrac{\pi e^2}{mc}\dfrac{f_{42}}{\Delta\nu_D} 2s_0 \times 1\ \mathrm{pc}} \approx \frac{2.7 \times 10^{-6}}{2s_0}\tau(H_\beta)\ \mathrm{cm}^{-3}. \tag{3.18}$$

Here the oscillator strength $f_{42} = 0.122$, the Doppler width of the H_β line $\Delta\nu_D = 2.6 \times 10^{10}\ \mathrm{sec}^{-1}$, and $2s_0$ is the extent of the H II region expressed in parsecs. In this way, for $\tau \approx 1$ and $2s_0 \approx 1$ pc, the population of the $2p$ sublevel must be about $10^{-6}\ \mathrm{cm}^3$, which can occur only if $N_e \gtrsim 10^6$ [see Eq. (3.32)]. However, even if the population of this sublevel is close to the quoted value, the absorption of L_α quanta by this mechanism cannot lead to a curtailment of the density of L_α radiation. In fact, two out of every three recombinations result in L_α quanta. At the same time, three Balmer quanta are also created, but of these only one-tenth are allocated to H_β. Thus, simultaneously with the formation of a single H_β quantum, six or seven new L_α quanta appear, while at the same time by this mechanism a single L_α quantum is annihilated and a single H_β quantum disappears.

There is an effect that can ionize a helium atom to the $2\,^3S$ level (metastable triplet state). Since in the given case the absorption coefficient falls inversely as the first power of the frequency and its value at the ionization threshold is equal to $2.8 \times 10^{-18}\ \mathrm{cm}^2$, we have $a_{He}(\nu_\alpha) \approx 2.8 \times 10^{-18}(4.8/10.2) \approx 1.3 \times 10^{-18}\ \mathrm{cm}^2$. For the parameter Λ we find

$$\frac{1 - \Lambda}{\Lambda} = 3 \times 10^{-7} T^{1/2} \frac{N_{He}(2\,^3S)}{N_1}, \tag{3.19}$$

where $N_{He}(2\,^3S)$ is the population of the metastable triplet state of helium.

This quantity must be large, since in the Orion nebula absorption is observed in the line He $\lambda3889$ ($2\,^3S \rightarrow 3\,^3P$) with an equivalent width of 0.2 Å. Thus, the concentration $N_{He}(2\,^3S) \approx 2 \times 10^{-4}$ cm^{-3} can be found in the central portion of the nebula. Consequently, $1 - \Lambda \approx 3 \times 10^{-5} \times 2 \times 10^{-4} \times 10 \approx 6 \times 10^{-8}$.

Now we shall consider the possibility that, as the result of a single scattering event, an L_α quantum will be immediately transferred to a far wing where $\tau < 1$, that is, to approximately $3\Delta\nu_D$. As the result of this, the quantum emerges from the medium without further scattering. This is possible, for example, in the case when the L_α quantum will be absorbed and re-emitted in another direction by a hydrogen atom traveling at a speed greater than $3(2kT/m_H)^{1/2} \approx 0.4T^{1/2}$ km/sec. The parameter Λ in this case is determined by the simple relation

$$\frac{1 - \Lambda}{\Lambda} \approx \frac{N_f}{N_1}, \tag{3.20}$$

which follows at once from Eq. (3.15); N_f is the number of fast hydrogen atoms per unit volume. A high efficiency of this mechanism demands an excess of fast hydrogen atoms. Such atoms can result, for example, from the recombination of protons from corpuscular fluxes emitted by stars that ionize H II regions [63]. For the present, we have no data allowing us to obtain even an approximate idea of the excess of fast hydrogen atoms above a Maxwellian distribution.

The other possibility of the escape of L_α quanta to beyond the boundaries of the Doppler profile is connected with the scattering of these quanta by free electrons (the Compton effect). The relative change of frequency of an L_α quantum during scattering by electrons is about $2h\nu_\alpha/mc^2 \approx 4 \times 10^{-5}$, which corresponds to a Doppler shift of 12 km/sec. Hence it follows that, although not every scattering by free electrons leads to the escape of a quantum beyond the limits of the Doppler thermal profile, a considerable portion of the total number can escape. A rough estimate of the parameter Λ in this case is given by

$$\frac{1 - \Lambda}{\Lambda} \approx \frac{\sigma_0 N_e}{k_\alpha N_1} 2^{1/2} \approx 1.5 \times 10^{-13} T^{1/2} \frac{N_e}{N_1}. \tag{3.21}$$

Here $\sigma_0 = 6.65 \times 10^{-25}$ cm^2, the effective Thomson cross section. Assuming $N_e \approx 10^3 N_1$ and $T \approx 10^4$, we find that $1 - \Lambda \approx 1.5 \times 10^{-8}$ in H II zones. In H I regions the role of this effect is small.

The escape of L_α quanta from the Doppler profile limits in H I regions can occur because of Raman scattering by H_2 molecules. In this case the quanta are displaced to frequencies that correspond to infrared transitions between rotational levels.

If there is a velocity gradient, L_α quanta can pass over into the wings of a line. This effect is discussed below.

In this way, the value of the parameter $1 - \Lambda$ for L_α lines in H II zones under all conditions is about 10^{-5}–10^{-7}, and in H I zones it must be appreciably less, apparently 10^{-8}–10^{-9}.

During a study of the scattering of L_α quanta, account must be taken also of incoherence, that is, the diffusion of frequency in the wings of the line profile. As a result of the incoherence of scattering, a quantum may be transferred to the wings of the line, after which it emerges from the region without further absorption or scattering.

Usually the problem of incoherent scattering is solved by assuming complete redistribution in frequency. Corresponding to this assumption, it is considered that the probability of the emission of quanta of frequency ν_2 does not depend on the frequency of the absorbed quantum ν_1, and is proportional to the absorption coefficient at frequency ν_2, that is, to $k_\alpha(\nu_2)$. Actually this is not entirely correct and the probability of re-emission of a quantum at frequency ν_2 depends on the frequency ν_1. However, it has been shown [61] that the solution of the problem of incoherent scattering on the assumption of complete redistribution in frequency and the accurate numerical solution taking into account some concrete relations between frequencies ν_1 and ν_2 are both close to the limits of accuracy that is needed in the problem of the diffusion of L_α emission in interstellar media. Therefore, we may limit ourselves to the assumption of complete redistribution in frequency.

We shall return to the consideration of an integral equation describing the transfer of radiation in the line L_α, similar to Eq. (3.4), but taking into account the incoherence of scattering [65, 61]. For future convenience, we shall use the dimensionless variable $x = (\nu - \nu_\alpha)/\Delta\nu_D$, so that the absorption coefficient (3.12) has the form $k_\alpha(x) = k_\alpha(\nu_\alpha)e^{-x^2}$. The source function for emission at frequency ν or dimensionless variable x, according to the assumption of complete redistribution in frequency, is written in the form $B_\alpha(\tau)e^{-x^2}/\pi^{1/2}$. The distribution function of sources of L_α quanta formed during the recombination of hydrogen atoms is denoted by $g_\alpha(\tau)e^{-x^2}/\pi^{1/2}$, where

$$g_\alpha(\tau) = \frac{\Lambda h\nu_\alpha}{2N_1 k_\alpha(\nu_\alpha)}(1 - X)(\alpha_t - \alpha_1)N_e N_p. \tag{3.22}$$

Here, we recall that $1 - X$ is that fraction of the cascade transitions and recombinations that terminates in the $2p$ state. Numerically, for $T \approx 10{,}000°$,

$$g_\alpha(\tau) \approx 2.2 \times 10^{-11} \frac{N_e N_p}{N_1} \text{ erg/cm}^2 \text{ sec.} \tag{3.23}$$

We notice that $g_\alpha(\tau)$ changes very slowly with optical depth in the L_α line. Actually, for the whole extent of the H II region, the value $N_e N_p/N_1$ is changed approximately by an order of magnitude, while the optical depth at the center of the L_α line changes by four orders of magnitude.

The integral transfer equation in the one-dimensional problem for the source function in the L_α line under conditions cited above has the form

$$B_\alpha(\tau) = \frac{\Lambda}{2}\int_0^{\tau_0} K(|\tau - \tau'|)B_\alpha(\tau')d\tau' + g_\alpha(\tau), \tag{3.24}$$

where

$$K(\tau) = \frac{1}{\pi^{1/2}}\int_{-\infty}^{+\infty} \exp(-2x^2 - \tau e^{-x^2})dx. \tag{3.25}$$

The physical meaning of the kernel of the integral equation $K(\tau)$ is clear. An L_α quantum of frequency x, emitted at an optical depth τ' with a relative probability of e^{-x^2}, has the probability of reaching depth τ without absorption of $\exp(-|\tau - \tau'|e^{-x^2})$ (since τe^{-x^2} is the optical thickness for a quantum of frequency x), after which it may be absorbed at this depth with a relative probability e^{-x^2}. Integration with respect to x in Eq. (3.25), with a normalization factor $1/\pi^{1/2}$, signifies that absorbed quanta of all frequencies can be re-emitted at any other frequency.

The peculiar feature of the kernel of Eq. (3.25) is that it decreases with an increase in τ considerably more slowly than the exponential kernel of Eq. (3.4). For large values of τ we have

$$K(\tau) = \frac{1}{\tau^2(\pi \ln \tau)^{1/2}}\left(1 + \frac{0.211}{\ln \tau} + \cdots\right). \tag{3.26}$$

This feature is connected with the fact that quanta re-emitted in the wings of the L_α line can travel considerably farther without subsequent scattering, since the absorption coefficient in the wings is small.

An approximate solution of Eq. (3.24) is readily found by recognizing the weak dependence of $g_\alpha(\tau)$ on τ noted above. The $B_\alpha(\tau)$ terms must then change slowly with optical depth. Since $K(\tau)$ decreases more slowly with τ than the exponential in Eq. (3.4), we may take $B_\alpha(\tau')$ in Eq. (3.24) outside the integral sign, while yet retaining sufficient accuracy when $\tau' = \tau$. We then obtain [66]

$$B_\alpha(\tau) = \frac{g_\alpha(\tau)}{1 - \Lambda + \frac{1}{2}\Lambda L(\tau) + \frac{1}{2}\Lambda L(\tau_0 - \tau)}, \tag{3.27}$$

where

$$L(\tau) = \int_0^{\tau} K(\tau')d\tau' = \frac{1}{\pi^{1/2}}\int_{-\infty}^{+\infty} e^{-x^2 - \tau e^{-x^2}}dx = \frac{1}{\tau(\pi \ln \tau)^{1/2}}\left(1 - \frac{0.2886}{\ln \tau} + \cdots\right). \tag{3.28}$$

Equation (3.24) and the identity (3.28), strictly speaking, are valid for $\tau \gg 1$ and $\tau_0 - \tau \gg 1$, that is, not in the vicinity of the boundaries of the region.

Retaining in (3.28) only the single function $L(\tau)$ in which we assume $\tau = \tau_0/2$, where τ_0 is the thickness of the whole H II region, and substituting the numerical values, we finally obtain for B_α when $T \approx 10{,}000^\circ$

$$B_\alpha \approx \frac{2.2 \times 10^{-11} N_e N_p / N_1}{1 - \Lambda + 2\Lambda/\tau_0[\pi \ln (\tau_0/2)]} \frac{\text{erg}}{\text{cm}^2\ \text{sec}}. \tag{3.29}$$

If the emission nebula is not enclosed by a H I region, we may assume $\tau_0 \approx 10^4$. Hence, $2/\tau_0/\pi^{1/2} \ln (\tau_0/2) \approx 4 \times 10^{-5}$. Consequently, the escape of L_α quanta in the wings of the line here has a higher probability than their annihilation (in isolated H II zones there is, as a rule, little dust). Substituting τ_0 and Λ in Eq. (3.29), we find

$$B_\alpha \approx 5 \times 10^{-7} \frac{N_e N_p}{N_1} \frac{\text{erg}}{\text{cm}^2\ \text{sec}}. \tag{3.30}$$

From the source function B_α, the population of the $2p$ level is immediately found. Considering only the diffusion of L_α quanta, we have the equilibrium condition

$$N_{2p} A_{2p1s} h\nu_\alpha = 2N_1 k_\alpha(\nu_\alpha) B_\alpha. \tag{3.31}$$

Actually, the weak dependence of B_α on τ implies that the emission here is almost isotropic. In this case, the L_α emission intensity is numerically equal to B_α. The factor 2, within the framework of a one-dimensional problem, takes into account the emission proceeding from both sides. It follows from Eq. (3.31) that

$$N_{2p} \approx \frac{2N_1 k_\alpha(\nu_\alpha)}{A_{2p1s} h\nu_\alpha} B_\alpha \approx 6 \times 10^{-18} N_e N_p\ \text{cm}^{-3}. \tag{3.32}$$

As is to be expected, the $2p$ sublevel is overpopulated relative to all levels with $n \geqslant 3$, but the population of the $2s$ sublevel is greater by three to four orders of magnitude.

The source function also determines the radiation pressure resulting from L_α emission. We present here, without derivation, the corresponding formula [66, 61]:

$$P(\tau) = -\frac{N_1 k_\alpha(\nu_\alpha)}{c\pi^{1/2}} \frac{dB_\alpha}{d\tau} \left[\int_0^\tau L(\tau')d\tau' - \tau \ln \tau \right] \approx -\frac{2 + \ln \tau}{\pi(\ln \tau)^{1/2}} \frac{N_1 k_\alpha(\nu_\alpha)}{c} \frac{dB_\alpha(\tau)}{d\tau}. \tag{3.33}$$

The approximate numerical factor given in Eq. (3.33) is correct when $\tau \gg 1$. The radiation pressure inside the H II region is negligibly small ($P(\tau) \sim 10^{-32} N_e N_p$ dyne/cm^2), but at the boundary of the H II zones, where $dB_\alpha/d\tau$ and N_1 change sharply, the pressure can be greater.

In the literature the phrase "accumulation of L_α quanta" is often used, which may create the impression that the density of L_α radiation considerably exceeds the density of the L_c radiation from which it is formed. In fact, the values of these quantities are comparable. The number of L_α quanta per unit volume exceeds the number of L_c quanta in the same volume by only a few times. In order to demonstrate this, let us compare the source functions B_α and B_c for regions inside the H II zones. Equation (3.9) is inapplicable here (during its derivation the condition $\tau \gg 1$ was used). However, an approximate equation for B_c when $\tau \lesssim 1$ can be obtained by analogy with Eqs. (3.27) and (3.22):

$$B_c(\tau) \approx \frac{g_c(\tau)}{1-\Lambda} \approx \frac{\Lambda h\bar{\nu}_c \alpha_1(T)}{2(1-\Lambda)\bar{a}_1} \frac{N_e N_p}{N_1}. \qquad (3.9')$$

We consider that here $1 - \Lambda \gg L(\tau)$, since $\Lambda = 0.39$, and that the L_c quanta are re-emitted immediately upon recombination at the ground level. Dividing Eq. (3.27), in which we set $\Lambda = 1$ and $\tau = \tau_0/2$, by Eq. (3.9′), we obtain, for $\Lambda = 0.39$,

$$\frac{B_\alpha}{B_c} = \frac{2.3\bar{a}_1}{k_\alpha(\nu_\alpha)L(\tau_0/2)} \approx 2\left(\ln\frac{\tau_0}{2}\right)^{1/2}, \qquad (3.34)$$

since $a_1/k_\alpha(\nu_\alpha) \approx 1/\tau_0$, if the optical thickness of the H II zone for L_c emission is about unity. In this case, the intensities of the L_α and L_c emissions are close to B_α and B_c, respectively. When $\tau_0 \approx 10^4$, it follows from Eq. (3.34) that the ratio of the L_α and L_c emission densities is about six.

The absence of appreciable accumulation is easily explained. Although L_α quanta experience considerably more scattering than L_c quanta before they emerge from the medium or are annihilated, they traverse a much shorter path between two scatterings, so that their lifetimes in the medium appear comparable.

If there is a velocity gradient in the medium (in a homogeneous, expanding, spherical nebula such a gradient occurs because of the curvature of the layers), then, depending on the motion of the quanta, it gradually moves into the wings of the line owing to a shift of the center of the L_α line in a medium moving with a velocity gradient. The solution of the problem of the scattering of L_α quanta, taking into account the velocity gradient, was given by V. V. Sobolev [66]. The source function for this case can also be written down in the form (3.27), where now the function $L(\tau)$ (for $\tau \gg 1$) must have the form

$$L(\tau) \approx \left(\frac{m_H}{2\pi kT}\right)^{1/2} \frac{dv}{d\tau} = \frac{(m_H/2\pi kT)}{N_1 k_\alpha(\nu_\alpha)} \frac{dv}{dr}, \qquad (3.35)$$

where dv/dr is the velocity gradient. If the macroscopic velocity v has the same order of magnitude as the thermal velocity of atoms $(2kT/m_H)^{1/2}$, then the loss of L_α quanta from a nebula because of a velocity gradient is approximately as effective as incoherent scattering [$L(\tau) \sim 1/\tau$].

V. V. Sobolev also examined the scattering of L_α emission in a three-dimensional, plane medium. Equation (3.27) is correct, and here there is a correction factor of $\frac{1}{2}$ in Eq. (3.28) and of $\frac{1}{3}$ in Eq. (3.35).

We have considered here the correct solution, as previously noted, for the case when the source function changes but little with optical depth. This condition is known to be realized completely in practically all H II regions, but close to the H II zone boundary, or in H I regions, a change of source function (B_α and g_α) with optical depth is more pronounced. However, even there, as a first approximation, at any rate, this simplification can be employed.

In fact, as seen from Eq. (3.9), the flux of L_c emission is damped in the transition layer considerably more slowly, not exponentially. Consequently, g_α also changes more slowly with optical depth.

The most general solutions of Eq. (3.24), applicable in the case of a marked dependence of g on τ (particularly in the case of the exponential character of this quantity), were investigated by V. V. Ivanov. If $1 - \Lambda$ is not too small compared with $L(\tau)$, then the solution of Eq. (3.27) can almost always be used. In the case of elastic scattering ($\Lambda = 1$), the accuracy of the solution of Eq. (3.27) becomes inadequate.

A complete study of the scattering of L_α quanta in the intermediate layer between H II and H I zones, and in the H I region, is rather complex (see Yada's work [67]). However, an elementary, qualitative account of these phenomena may be obtained from the following considerations. L_α quanta are formed in an H II region and then, on diffusion to its boundaries, pass over to the wings of the line, which are displaced from the center by a distance of about $3\Delta\nu_D \approx 3 \times 10^{11}$ Hz (see page 36). They form, as it were, two lines of L_α radiation, separated from each other by a distance of $6\Delta\nu_D \approx 6 \times 10^{11}$ Hz. The radiation proceeding into the H I region in these lines is already beyond the Doppler profile in this region, the width of which is ten times less than the width of the Doppler profile in the H II region (that is, equal to 10^{10} cy/sec). In this way, the overwhelming majority of L_α quanta in H I regions are scattered only in the Lorentz wings of this line. The same L_α quanta that fall into the Doppler portion of the H I profile will be rapidly returned reversibly to the H II region where they pass into the wings.

The Lorentz profile in the H I zone is determined only by the damping of the radiation, and therefore in this case the scattering of L_α quanta is coherent. Thus, Eq. (3.6) may be used to calculate the intensity of L_α radiation in the Lorentz limits of this line in H I regions, remembering that τ is the optical thickness in these wings (at a distance of 3×10^{11} cy/sec from its center).

It then follows that the intensity of L_α radiation is

$$\begin{aligned} I_\alpha &\sim \exp\left[-(1-\Lambda)^{1/2}\tau\right] \\ &= \exp\left\{-\left[\frac{a(\nu_\alpha)N_a}{k_\alpha(\nu_\alpha + 3\times 10^{11})N_1}\right]^{1/2} k_\alpha(\nu_\alpha + 3\times 10^{11})N_1 r\right\} \\ &= \exp\left\{-r\left[a(\nu_\alpha)k_\alpha(\nu_\alpha + 3\times 10^{11})N_a N_1\right]^{1/2}\right\}, \qquad (3.36) \end{aligned}$$

where r (pc) is the distance from the H II zone boundary. Corresponding to

Eq. (3.13), we have $k_\alpha(\nu_\alpha + 3 \times 10^{11}) \approx 1.9 \times 10^{-18}$ cm^2. Assuming that annihilation of L_α quanta is connected with absorption by dust ($a \approx 8 \times 10^{-9}$ cm^2, $N_g \approx 10^{-12}$ cm^3), we find

$$I_\alpha \sim \exp(-rN_1^{1/2}/3). \tag{3.37}$$

In this way, for $N_1 \approx 10$, practically all the L_α radiation will be absorbed at a distance of about 1 pc. Some uncertainty in the values of N_1 or N_g causes no appreciable error in Eq. (3.37), since these values enter into Eq. (3.36) under the square-root sign. Consideration of a three-dimensional medium leads to the appearance in Eq. (3.36) of the factor $3^{1/2}$, so that the true thickness of the absorption zone for L_α radiation is about 0.6 pc.

The L_α quanta absorbed by dust are transformed into radiation in the infrared. Since energy re-emitted in L_α constitutes a large portion of the total luminosity of a star, it follows that H II zones are surrounded by envelopes whose luminosity in the infrared is comparable to the bolometric luminosity of the star.

Much effort has been spent in studying the scattering of L_α radiation. Besides those previously given, there may be noted a series of papers by Zanstra [68, 69], Koelbloed [70], Miyamoto [71], V. V. Ivanov [77–79], Unno [72–74], V. S. Safronov [75], Field [76], and others. In several of these investigations, the scattering of L_α radiation was studied by a somewhat different method, that of solving the diffusion equation for the frequency displacement of a quantum. In particular, in paper [76] the diffusion equation is solved for the displacement in frequency of an L_α quantum during time t, assuming homogeneity of the radiation field. In other words, scattering of L_α quanta is studied here in an infinite medium where they are unable to escape from the medium, but where on the other hand the line profile broadens with time owing to incoherent scattering. The corresponding diffusion equation has an exact solution in two cases: (*a*) when a certain number of L_α quanta occur in the medium at a particular moment in time and then their further increase ceases; (*b*) when the formation of new L_α quanta proceeds in a medium at a constant rate. In both cases the line profile broadens according to the rule $(\ln t)^{1/2}$ and the central intensity is changed by $(\ln t)^{-1/2}$ in the first case and by $t(\ln t)^{-1/2}$ in the second. For the given problem, that is, when L_α quanta cannot leave the medium, the line profile remains in the form of a plateau. The introduction of the effect of atomic recoil during scattering leads to a small profile gradient near the center proportional to h/kT. We recall that if L_α quanta can escape from the medium (a finite medium), a dip is formed in the center of the line, which increases with increased optical thickness of the medium.

4. Radio Emission from Interstellar Hydrogen

In recent years, radio-astronomical investigations have formed probably the most powerful tool for studying the interstellar medium. One reason for this is the opportunity to observe regions of nonionized hydrogen that do not emit in the optical region of the spectrum. Moreover, radio waves are practically unabsorbed by cosmic dust, so that emission from very distant objects can be observed at radio frequencies. For this reason, data on the radio emission of H II regions substantially supplement the results obtained from optical observations.

The Transfer Equation for Thermal Radio Emission

Radio emission from the interstellar medium, often called cosmic radio emission, can be divided according to physical criteria into two components, the thermal component, which is dominant in the centimeter and partly in the decimeter range, and the nonthermal, emitted mainly in the meter band. Only the thermal component will be considered here; nonthermal radio emission from the interstellar medium is discussed in Chapter 4.

First of all, we shall consider the definition of thermal radio emission. In the preceding paragraphs, it was repeatedly noticed that the velocity distribution of electrons in interstellar space is described by the Maxwell equation. In reality, the relaxation time of the electron gas is of the same order of magnitude as the time required to travel the electron's free path [80];

$$\frac{(kT)^2}{e^4\bar{v}L}\frac{1}{N_e} \approx \frac{1.5 \times 10^4}{N_e}\ \text{sec}. \tag{4.1}$$

This is known to be less than all the remaining time scales for the interstellar medium. For example, the time required to establish ionization equilibrium is $1/\alpha_t N_e \approx 10^{13}/N_e$ sec. In Eq. (4.1) $\bar{v} = (8kT/\pi m)^{1/2}$ is the thermal velocity, and $L = 2 \ln [3(kT)/2(\pi N_e)^{1/2}e^3] \approx 40$, the so-called Coulomb logarithm, accounts for long-range interactions. Physically, L is the logarithm of the ratio of maximum and minimum values of the impact parameter between ions and electrons.

It is well known that emission takes place during the deceleration of free electrons in the field of protons and atomic nuclei. The fact that these emitting electrons are distributed according to thermodynamic equilibrium permits the given emission to be called thermal. We stress that this emission is still not in thermodynamic equilibrium, since for the latter, the medium would have to be optically thick.

If the electron distribution mentioned above does not satisfy Maxwell's equation (for example, in the case of cosmic-ray electrons), then their emission will be nonthermal. Another instance of nonthermal emission is the collective vibration of plasma electrons, which is measured, not by the emission intensity of the individual electrons, but by their amplitudes.

A characteristic of black-body radiation is the applicability of Kirchhoff's law.

Interstellar hydrogen produces both continuous and line radio emission. As in other cases, the problem of calculating radio-emission intensity requires the solution of the transfer equation; this solution is obtained in the present case by an elementary procedure. Let us write the transfer equation for radio emission in the following way [15]:

$$\frac{dI_\nu}{dr} = -\kappa(\nu)I_\nu + \varepsilon_t(\nu) + \varepsilon_n(\nu)_1 \tag{4.2}$$

where d/dr is the derivative along the line of sight, I_ν is the radio-emission intensity at frequency ν (measured either in erg/cm^2 sec ster Hz or in W/m^2 ster Hz), $\kappa(\nu)$ is the absorption coefficient per unit volume, and $\varepsilon_t(\nu)$ and $\varepsilon_n(\nu)$ are the thermal and nonthermal coefficients of radio emission, respectively. As a rule, it is more convenient to rewrite this equation, introducing the optical depth τ:

$$d\tau_\nu = \kappa(\nu)dr. \tag{4.3}$$

The name "optical depth" is not very suitable for a radio band, but it is universally adopted. Instead of Eq. (4.2) we have

$$\frac{dI_\nu}{d\tau_\nu} = -I_\nu + \frac{\varepsilon_t(\nu)}{\kappa(\nu)} + \frac{\varepsilon_n(\nu)}{\kappa(\nu)} = -I_\nu + B_\nu(T) + \frac{\varepsilon_n(\nu)}{\kappa(\nu)}. \tag{4.4}$$

Here, according to Kirchhoffs law, instead of $\varepsilon_t(\nu)/\kappa(\nu)$, the universal temperature function $B_\nu(T)$ is used, which in the radio band ($h\nu \ll kT$) is determined by the Rayleigh–Jeans equation

$$B_\nu(T) = \frac{2kT}{c^2}\nu^2. \tag{4.5}$$

The essential simplification that allows an immediate solution of the transfer equation in the radio band consists of the fact that here, in contrast to the problem of optical radiation transfer in stellar atmospheres, the kinetic temperature of the gas does not at all depend on the radio-emission field, and therefore, the function $B_\nu(T)$ in Eq. (4.4) may be regarded as given; the same applies to the function $\varepsilon_n(\nu)/\kappa(\nu)$.

The following problem occurs most frequently. Consider a radio-emission source whose optical thickness along the line of sight at some frequency ν is equal to $\tau_0(\nu)$, while this source is observed against a radio-emission background of intensity $I_\nu^{(f)}$ emitted from regions situated far from the source. We can neglect emission and absorption in front of the source. We shall consider that the gas temperature of the source is a function of the optical depth. The solution of the transfer equation is written for this problem in the form

$$I_\nu = \int_0^{\tau_0(\nu)} \left[B_\nu[T(\tau')] + \frac{\varepsilon_n(\nu)}{\kappa(\nu)}\right] e^{-\tau'}d\tau' + I_\nu^{(f)}e^{-\tau_0(\nu)}. \tag{4.6}$$

We neglect for the present the term $\varepsilon_n(\nu)/\kappa(\nu)$ and replace $B_\nu(T)$ in Eq. (4.5) by

$$I_\nu = \frac{2k\nu^2}{c^2}\int_0^{\tau_0(\nu)} T(\tau')e^{-\tau'}d\tau' + I_\nu^{(f)}e^{-\tau_0(\nu)}. \tag{4.7}$$

Hence it follows that the intensity is readily expressed in terms of brightness temperature defined by the expressions

$$T_b(\nu) = \frac{c^2}{2k\nu^2}I_\nu; \qquad T_b^{(f)}(\nu) = \frac{c^2}{2k\nu^2}I_\nu^{(f)}. \tag{4.8}$$

The brightness temperature is the temperature of a black body of the same dimensions as the object observed, emitting at a given frequency the same amount of radiation. Equation (4.7) is rewritten in terms of brightness temperature in the form

$$T_b(\nu) = \int_0^{\tau_0(\nu)} T(\tau')e^{-\tau'}d\tau' + T_b^{(f)}e^{-\tau_0(\nu)}. \tag{4.9}$$

Only H II zones yield thermal radio emission in the continuous spectrum, where, as is known, the temperature of the electron gas is practically constant and equal to $T = 10{,}000^\circ$. In this case Eq. (4.9) is rewritten in the form

$$T_b(\nu) = T(1 - e^{-\tau_0(\nu)}) + T_b^{(f)}e^{-\tau_0(\nu)}. \tag{4.10}$$

Thus, the intensity of a thermal radio-emission source for a known background brightness temperature is determined only by its optical thickness $\tau_0(\nu)$. In particular, if $\tau_0(\nu) \gg 1$,

$$T_b(\nu) = T \approx 10{,}000^\circ; \tag{4.11}$$

at the opposite extreme $[\tau_0(\nu) \ll 1]$,

$$T_b(\nu) = T\tau_0(\nu) + T_b^{(f)}(\nu). \tag{4.12}$$

Continuous radio emission from the H II zone itself $[T_b(\nu) - T_b^{(f)}]$ cannot have a brightness temperature exceeding $T \approx 10{,}000^\circ$. In the meter band the value of $T_b^{(f)} > 10{,}000^\circ$; here the radio emission of the Galaxy possesses a nonthermal character. In this case an H II zone with large optical thickness is observed in absorption.

Let us now consider the term $\varepsilon_n(\nu)/\kappa(\nu)$ in Eq. (4.6), which describes the nonthermal component of radio emission arising directly inside an H II region. If the optical thickness $\tau_0(\nu) \leqslant 1$, we can write

$$\int_0^{\tau_0(\nu)} \frac{\varepsilon_n(\nu)}{\kappa(\nu)}e^{-\tau'}d\tau' \approx \frac{\varepsilon_n(\nu)}{\kappa(\nu)}\tau_0(\nu) \approx \frac{2\nu^2}{c^2}kT_b^{(n)}(\nu), \tag{4.13}$$

where $T_b^{(n)}(\nu)$ is the brightness temperature of the nonthermal emission source. If $\tau_0(\nu) \gg 1$, the nonthermal component will be greatly decreased. However, the nonthermal component in thermal radio-emitting sources is in general small $[\varepsilon_T(\nu) \gg \varepsilon_n(\nu)]$. Substituting for ε_n/κ in Eq. (4.6) the right-hand side of Eq. (4.13) divided by $\tau_0(\nu)$, we obtain in place of Eq. (4.10) the relation

$$T_b(\nu) - T_b^{(f)} = \left[T - T_b^f + \frac{T_b^{(n)}}{\tau_0(\nu)}\right](1 - e^{-\tau_0(\nu)}). \tag{4.14}$$

Equation (4.14) allows the determination of the optical thickness $\tau_0(\nu)$ of the source from the observed radio emission. The term $T_b^{(f)}$ is eliminated by an observation in an adjacent region, while $T_b^{(n)}$ has a dependence on frequency that is characteristic of nonthermal radio emission.

Optical Thickness of Radio Emission

We now introduce a formula for optical thickness $\tau_0(\nu)$. By definition,

$$\tau_0(\nu) = \int_0^l \kappa(\nu)dr, \tag{4.15}$$

where l is the dimension of the H II region along the line of sight ($l \leqslant 2s_0$). Absorption in the continuous spectrum in the radio region is caused by free-free electron transitions. The atomic absorption coefficient for this case is given by Eq. (I.34). Allowing also for induced emission, we have for the absorption coefficient per unit volume

$$\kappa(\nu) = k(\nu)\frac{h\nu}{kT} N_e N_p = \frac{16\pi^2 e^6 \bar{g}}{c(6\pi mkT)^{3/2}} \frac{N_e N_p}{\nu^2}. \tag{4.16}$$

Equation (4.16) was obtained with the help of a quantum-mechanical calculation, although it can also be derived by classical methods. This is obvious, since the Planck constant does not appear. From the classical point of view, the absorption process of radio waves can be described as the production by the wave of oscillations of electrons, which then transmit their acquired energy during collisions with protons. Equation (I.33) determines the value of the Gaunt factor in the radio region. Substituting Eq. (4.16) into Eq. (4.15), and using Eq. (I.33), we obtain for the optical depth [81]

$$\tau_0(\nu) = \frac{16\pi^2 e^6 \bar{g}}{c(6\pi mkT)^{3/2}} \frac{1}{\nu^2} \int_0^{l_0} N_e N_p ds = \frac{3 \times 10^{16}}{\nu^2 T^{3/2}} \left[17.7 + \ln \frac{T^{3/2}}{\nu}\right] \mathrm{ME}$$
$$= \frac{3.2 \times 10^{11}}{\nu^2} \left[1 - 0.21 \log\left(\frac{\nu}{10^9}\right)\right] \mathrm{ME}, \tag{4.17}$$

for $T = 10^4$. It is very significant, then, that thermal radio emission, as well as optical emission, is defined in terms of the emission measure. The brightness temperature of an optically thin nebula (neglecting background and nonthermal emission), is

$$T_b = T\tau_0(\nu) = 10^4\tau_0(\nu)_0 = \frac{3.2 \times 10^{15}}{\nu^2}\left[1 - 0.21 \log\left(\frac{\nu}{10^9}\right)\right]\text{ME}. \quad (4.18)$$

Consequently, the intensity of radio emission is

$$I_\nu = \frac{2kT_b}{c^2}\nu^2$$

$$\approx 1.0 \times 10^{-21}\left[1 - 0.21 \log\left(\frac{\nu}{10^9}\right)\right]\text{ME erg/cm}^2\text{ sec ster Hz}. \quad (4.19)$$

Thus, the intensity of radio emission of an optically thin H II region, within the accuracy of a logarithmic factor, does not depend on frequency and is proportional to its optical emission;

$$\frac{I_\nu}{I(H_\beta)} \approx 3.3 \times 10^{-14}\left[1 - 0.21 \log\left(\frac{\nu}{10^9}\right)\right]. \quad (4.20)$$

This relation is used for determining the total absorption by cosmic dust particles, which weakens the H_β emission, but leaves radio emission completely unaffected.

From the relations (4.10) and (4.17), it follows that, for $\nu \ll 0.6\ \text{ME}^{1/2}$ MHz, the brightness temperature is almost constant and equal to 10,000°, while for $\nu \gg 0.6\ \text{ME}^{1/2}$ MHz it quickly falls off. The emission intensity behaves in the opposite way. Therefore, from the location of the break in the spectrum, the emission measure of the source may be determined immediately. In Fig. 5, as

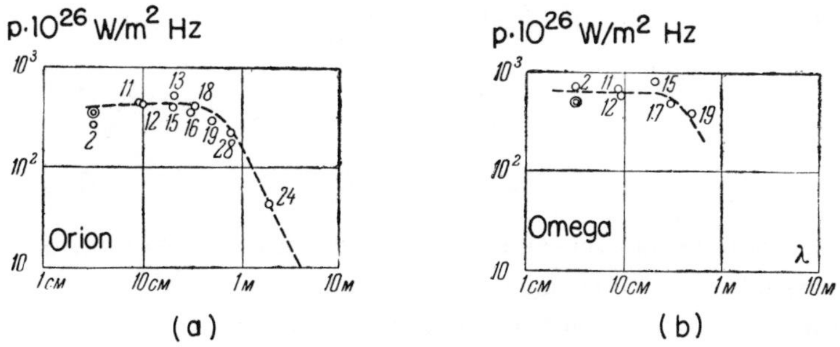

FIG. 5. Radio spectra of (*a*) the Orion nebula and (*b*) the Omega nebula [82].

an illustration, radio-emission spectra [82] are given over the intensity range of the Orion and Omega nebulae. In the Orion nebula, a sharp drop in the spectrum begins at $\lambda \approx 60$ cm, while in the Omega nebula the drop is at $\lambda \approx 40$ cm. Hence, their emission measures are found to be equal to 5×10^5 and 10^6 respectively. When the dimensions of the nebula are known, its density and mass can be immediately found from the emission measure.

The direct determination of the brightness temperature of a nebula is possible only when the resolving power of the radio telescope permits the construction of the isophot distribution in two dimensions. However, at the present time, this is impossible for the majority of nebulae. In this case, measurements give not the intensity but the flux F_ν of radio emission from the whole nebula and the background,

$$F_\nu = \int I_\nu d\Omega, \tag{4.21}$$

where the integral is taken over the solid angle through which the source is observed. The unit of flux F_ν is erg/cm^2 sec Hz or W/m^2 Hz. Denoting the radio-emission background flux arriving in the same solid angle as $F_\nu^{(f)}$, we can write [83]:

$$F_\nu - F_\nu^{(f)} = \frac{2k\nu^2}{c^2}(T - T_b^{(f)}) \int [1 - e^{-\tau_0(\nu,\theta,\phi)}] d\Omega. \tag{4.22}$$

For the sake of brevity, the non-thermal-emission term is here omitted. The optical depth in the integral (4.22) depends on the shape and angular dimensions of the nebula. We shall suppose that the nebula is spherical and that its angular radius is equal to θ_0. Then, calling $\tau_0(\nu)$ the optical diameter, we have

$$\tau(\nu, \theta) = \tau_0(\nu)(1 - \theta^2/\theta_0^2)^{1/2} \tag{4.23}$$

and $d\Omega = 2\pi\theta d\theta$. Substituting Eq. (4.23) into (4.22), we find

$$F_\nu - F_\nu^{(f)} = \frac{4\pi k\nu^2}{c^2}(T - T_b^{(f)})\theta_0^2\left[\frac{1}{2} + \frac{e^{-\tau_0(\nu)}(\tau_0(\nu) + 1) - 1}{[\tau_0(\nu)]^2}\right]. \tag{4.24}$$

In particular, for large or small values of $\tau_0(\nu)$ we have

$$F_\nu - F_\nu^{(f)} = \frac{4\pi k\nu^2}{3c^2}(T - T_b^{(f)})\theta_0^2\tau_0(\nu) \qquad [\tau_0(\nu) \ll 1], \tag{4.25}$$

$$F_\nu - F_\nu^{(f)} = \frac{2\pi k\nu^2}{c^2}(T - T_b^{(f)})\theta_0^2 \qquad [\tau_0(\nu) \gg 1]. \tag{4.26}$$

These equations are convenient to use when the flux of radio emission from the source is studied at two frequencies considerably separated from one another. For example, if at frequency ν_1 the optical thickness is $\tau_0(\nu_1) \gg 1$, while at frequency $\nu_2 \tau_0(\nu_2) \ll 1$, comparison of Eqs. (4.25) and (4.26) at a known background temperature immediately yields $\tau_0(\nu_2)$ and consequently the emission measure. In the general case for an arbitrary value of τ_0, Eq. (4.24) must be supplemented by the obvious relation

$$\frac{\tau_0(\nu_1)}{\tau_0(\nu_2)} = \left(\frac{\nu_2}{\nu_1}\right)^2 \frac{17.7 + \ln (T^{3/2}/\nu_1)}{17.7 + \ln (T^{3/2}/\nu_2)}. \tag{4.27}$$

The simultaneous solution of Eqs. (4.27) and (4.24), which may be achieved by a graphical method, permits the determination, not only of the emission measure, but of the temperature of the nebula as well. This was done for a series of nebulae by Wade [83] (Rosette nebula, $T = 8{,}600°$, $N_e = 17$), Yu N. Pariiskii [84] (Orion nebula, $T = 11{,}730°$), and others.

The hypothesis concerning the sphericity and homogeneity of a nebula is gross. Therefore, Pariiskii [84], while using this method, proposed to determine the quantity $\tau_0(\nu, \theta, \phi)$ directly from observations of high resolving power. These observations are made at frequencies such that $\tau_0(\nu, \theta, \phi) \ll 1$ over the

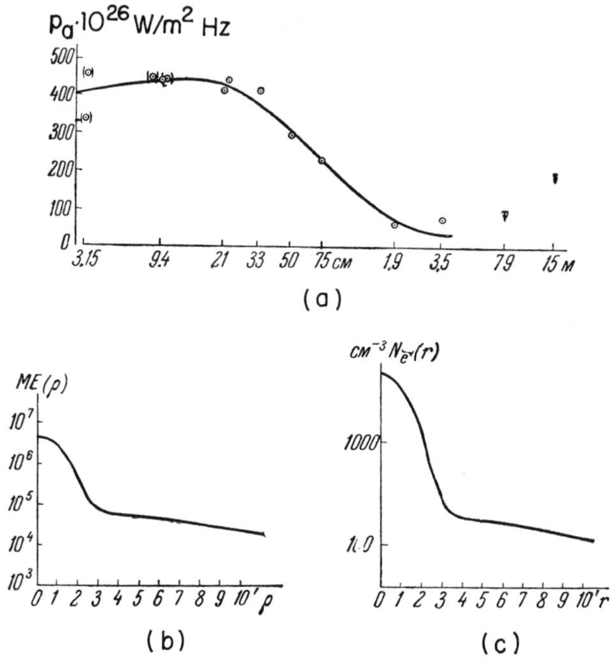

FIG. 6. Orion nebula [84]: (*a*) observed and calculated spectra; (*b*) emission measure; (*c*) electron density.

whole nebula, so that $T_b \sim \tau_0$. Therefore, finding the true distribution of optical thickness in the nebula from the observed distribution of brightness temperatures, we may also calculate the value of the integral in Eq. (4.22) for those cases when $\tau_0(\nu, \theta, \phi)$ is comparable with or greater than unity for a portion or for the whole of the nebula, bearing in mind that τ_0 is inversely proportional to ν^2 throughout. Similarly, integrated fluxes from nebulae are found at any frequency, and comparison of them with observations permits a determination of the temperature without any assumption concerning sphericity or homogeneity. The distribution of the emission measure (optical thickness) and the electron density in the Orion nebula as a function of the distance from its center is shown in Fig. 6. In this diagram, the theoretical dependence of the integrated flux of radio emission on wavelength calculated from this distribution is given. It is obvious that this generalized method is applicable if it is possible to study a nebula with high resolving power even at a single frequency.

The emission measure and the density and mass of a nebula can be determined from observations at a single frequency if we know the absolute value of the flux of radio emission and the angular radius θ_0 of the nebula. Actually, assuming the nebula to be optically thin and considering that the background flux and its brightness temperature are already eliminated, Eq. (4.22) can be rewritten for a spherical nebula in the form

$$F_\nu = \frac{2k\nu^2}{c^2} T \int \tau(\nu, \theta) d\Omega = \frac{\pi k \nu^2}{c^2} T\tau_0(\nu)\theta_0^2. \tag{4.28}$$

Substituting here Eq. (4.17) and the numerical values for the remaining constants, we find

$$F_\nu = 1.6 \times 10^{-21}\theta_0^2 \left[1 - 0.21 \log\left(\frac{\nu}{10^9}\right)\right] \text{ME erg/cm}^2 \text{ sec Hz}. \tag{4.29}$$

The emission measure determined in this way is connected with the distance r from the nebula by the relation ME $\approx N_e^2 r 2\theta_0$. Therefore, from Eq. (4.29) the density and mass of the nebula are at once determined. The resulting formula may be written in the convenient form [85]

$$N_e \approx \frac{60(F_\nu \times 10^{26})^{1/2}}{\phi^{3/2} r^{1/2}}; \qquad \frac{M}{M_\odot} = 4 \times 10^{-6} \phi^{3/2} r^{5/2} (F_\nu \times 10^{26})^{1/2}. \tag{4.30}$$

Here $\phi = 2\theta_0$ is the diameter of the nebula in degrees; F_ν is measured in watts per square meter per cycle per second and r in parsecs. These or similar equations were used in a series of papers on the determination of the physical characteristics of nebulae [86, 85]. They give only upper limits for mass and density, since the assumption concerning homogeneity and sphericity is inadequate. With an increase of density above a certain limit, the temperature also increases and their effect is partially compensated [83], since $\tau \approx N_e^2 T^{-3/2}$.

The $\lambda = 21$ cm *Radio Line*

We turn to the radio-emission line spectrum of interstellar hydrogen. In principle, it consists of many lines, but as yet only one has been observed, the line $\lambda = 21.11$ cm ($\nu = 1420.4$ MHz), which arises from the transition between the hyperfine-structure sublevels of the ground state of a hydrogen atom (see Appendix I). The importance of the study of interstellar hydrogen in this line consists first of all in the fact that, as yet, this is the sole procedure for direct observation of neutral hydrogen in interstellar space.

The feasibility of observing this line was indicated for the first time by van de Hulst in 1945 [87]. The idea was then worked out in detail by I. S. Shklovskii

[88], who calculated the transition probability and estimated the emission intensity. In 1951 the line $\lambda = 21$ cm was discovered almost simultaneously by Ewen and Purcell [89], Muller and Oort [90], and Christiansen and Hindman [91]. The study of interstellar hydrogen in the line $\lambda = 21$ cm gave very valuable information on the structure of our Galaxy and its rotation, on the distribution of interstellar gas, and on the central portion of the Galaxy. The following section will be devoted to these questions, but here we shall consider only the theoretical aspects of this problem.

The transfer equation for radio emission at $\lambda = 21$ cm can also be written in the form (4.4), except that now $\kappa(\nu)$ becomes the absorption coefficient in this line, and the continuous emission coefficient close to $\lambda = 21$ cm is substituted for $\varepsilon_n(\nu)$ (both for thermal and nonthermal). The Kirchhoff–Planck function, as before, may be written approximately as the Rayleigh–Jeans relation, and the spin temperature T_s may be introduced.

The solution of the equation so obtained, written in terms of the brightness temperature, is

$$T_b(\nu) = \int_0^{\tau(\nu)} \left[T_s(\tau') + \frac{c^2}{2k\nu^2} \frac{\varepsilon_n(\nu)}{\kappa(\nu)} \right] e^{-\tau'} d\tau'. \tag{4.31}$$

Here the term that takes into account background emission is omitted, and by $\tau(\nu)$ we mean the optical thickness along the line of sight in the Galaxy. We shall further consider that the value $\varepsilon_n(\nu)/\kappa(\nu)$ is constant along the line of sight and that the total optical thickness of the Galaxy in a given direction is less than unity. Then, replacing ε_n/κ by the expression on the right-hand side of Eq. (4.13), we obtain

$$T_b(\nu) = \int_0^{\tau(\nu)} T_s(\tau') e^{-\tau'} d\tau' + T_b^{(n)} \frac{1 - e^{-\tau(\nu)}}{\tau(\nu)}. \tag{4.32}$$

Without significant error, Eq. (4.32) may be considered correct for large optical thicknesses.

If the spin temperature is constant along the line of sight, we have

$$T_b(\nu) = T_s(1 - e^{-\tau(\nu)}) + T_b^{(n)} \frac{1 - e^{-\tau(\nu)}}{\tau_\nu}. \tag{4.33}$$

Thus, the intensity of radio emission is also here characterized by the optical thickness. The value of $T_b^{(n)}$ may readily be determined from observations in the spectral region adjacent to the line. It is easy to see that $T_b(\nu) \to T_b^{(n)}$ when $\tau(\nu) \to 0$.

The spin temperature or excitation temperature characterizes the distribution of hydrogen atoms in the hyperfine-structure sublevels. In fact, if this distribution is described in the form of a Boltzmann equation,

$$\frac{N_1}{N_0} = \frac{\tilde{\omega}_1}{\tilde{\omega}_0} e^{-h\nu_{10}/kT_s} = 3e^{-h\nu_{10}/kT_s}, \tag{4.34}$$

we can obtain by the use of Eq. (4.34) and Einstein's coefficients Planck's equation for that same temperature T_s and consequently, in the limit of low frequencies, also Eq. (4.5). Here $\tilde{\omega} = 2F + 1$ is the statistical weight, N_0 and N_1 are the number of atoms in the ground and excited levels per unit volume respectively, and ν_{10} is the frequency of the line $\lambda = 21$ cm.

The connection between spin and kinetic temperature may be found [92] from the equilibrium equation

$$N_1[N_H q_{10} + A_{10} + B_{10}\rho(\nu_{10})] = N_0[N_H q_{01} + B_{01}\rho(\nu_{10})]. \qquad (4.35)$$

Here $N_H = N_1 + N_0$ is the total number of hydrogen atoms, A_{10}, B_{10}, and B_{01} are the Einstein coefficients, q_{01} and q_{10} are the probabilities of excitation and de-excitation of levels during atomic collisions, and $\rho(\nu_{10})$ is the density of interstellar emission at frequency ν_{10}. There exists between coefficients q_{10} and q_{01} a connection of the type (I.45), but here the kinetic temperature of the gas T_H must be used;

$$\frac{q_{01}}{q_{10}} = 3e^{-h\nu_{10}/kT_H}. \qquad (4.36)$$

The Einstein coefficients are related by the known expressions

$$3A_{10} = \frac{8\pi h\nu_{10}^3}{c^3}\, 3B_{10} = \frac{8\pi h\nu_{10}^3}{c^3}\, B_{01} = 1.8 \times 10^{-29} B_{01}. \qquad (4.37)$$

Taking into account Eq. (I.13), we find that $B_{01} = 4.75 \times 10^{14}$ cm^3/erg sec^2. The emission density is written in the form of the Rayleigh–Jeans equation at a certain "radiation" temperature T_R;

$$\rho(\nu_{01}) = \frac{8\pi k T_R}{c^3}\, \nu_{01}^2. \qquad (4.38)$$

Substituting Eqs. (4.34), (4.36), and (4.37) in Eq. (4.35), we find, remembering that $h\nu$ is considerably less than T_s, T_H, and T_R,

$$T_s = \frac{yN_H T_H + T_R}{1 + yN_H}; \qquad y = \frac{h\nu_{01} q_{10}}{kT_H A_{10}}. \qquad (4.39)$$

If $yN_H \gg 1$, the spin temperature is almost equal to the kinetic temperature of the gas. This means that transitions between two sublevels are caused chiefly by collisions and the population of the sublevels is determined from Boltzmann's equation. Using the data in Appendix I for the values of A_{10} and q_{10}, we find that the spin and kinetic temperatures will be equal if the inequality

$$1600 N_H > T_H^{1/2} \qquad (4.40)$$

is satisfied. In H I regions this condition is realized ($T_H \approx 100°$, $N_H > 0.1$ cm^{-3}), but in more rarefied regions it may be violated. In particular, this condition is clearly not met in the intergalactic medium, since in this case the frequency of collision is insufficient to maintain the population of an excited sublevel. In connection with this, Wouthuysen [93] and Field [76] considered the possible excitation of a sublevel of hyperfine structure from the transition $F = 0 \rightarrow 2p \rightarrow F = 1$ during absorption and emission of L_α quanta. In this case the population of the excited sublevel is determined by the density of the L_α emission.

It remains to determine the dependence of optical depth on frequency. The absorption coefficient $k_0(\nu)$ per atom is connected with the Einstein coefficient B_{01} by the well-known equation

$$\int k_0'(\nu)d\nu = \frac{h\nu_{01}}{c}\frac{h\nu_{01}}{kT_H}B_{01} = \frac{1.02 \times 10^{-14}}{T_H}\,\text{cm}^2/\text{sec}. \tag{4.41}$$

We recall that a prime on the absorption coefficient means that induced emission is taken into account.

The profile of the spectral line $\lambda = 21.11$ cm is determined only by Doppler shifts (because of the smallness of the damping constant). However, we must consider here not only the thermal motion of atoms, but the macroscopic motion of the gas as well, since as a rule the latter predominates. Therefore, we write the velocity-distribution function in a general form, calling $f(V)dV$ the mean probability that an atom will have a radial velocity between V and $V + dV$. According to the Doppler relation we have

$$\frac{\Delta\nu}{\nu_{01}} = \frac{\nu - \nu_{01}}{\nu_{01}} = \frac{V}{c}, \qquad \Delta\nu = \frac{V}{21\ \text{cm}}. \tag{4.42}$$

From Eqs. (4.41) and (4.42) it is easy to obtain the absorption coefficient per frequency interval:

$$k_0'(\nu) = \frac{(h\nu_{01})^2}{ckT_H}B_{01}f(V)\frac{dV}{d\nu} = \frac{h^2\nu_{01}}{kT_H}B_{01}f\left(c\frac{\Delta\nu}{\nu_{01}}\right)$$
$$= \frac{2.14 \times 10^{-13}}{T_H}f(21\Delta\nu)\ \text{cm}^2, \tag{4.43}$$

where $\Delta\nu = \nu - \nu_{01}$ is the frequency difference in cycles per second. It follows from Eq. (4.34) that when $h\nu \ll kT_H$ a quarter of all the hydrogen atoms are found in the ground sublevel. Therefore, the optical thickness will be

$$\tau_0(\nu) = \frac{1}{4}\int k_0'(\nu)N_H dr = 5.35 \times 10^{-14}\int \frac{N_H f(V)}{T_H}dr. \tag{4.44}$$

If the gas temperature is considered to be constant and taken outside the integral sign, the remaining expression will represent the number of hydrogen atoms in a column of unit cross section along the line of sight having a radial velocity in the unit velocity range (1 cm/sec) about V. Calling this number $\mathfrak{N}(V)$, we can rewrite Eq. (4.44) in the form

$$\tau(\nu) = \frac{\mathfrak{N}(V)}{1.87 \times 10^{13} T_{\mathrm{H}}}, \qquad V = 21(\nu - \nu_{01}) \text{ cm/sec.} \tag{4.45}$$

This equation, obtained by van de Hulst, Muller, and Oort [94], plays an important part in the analysis of the spatial distribution of interstellar hydrogen.

Thus, from Eq. (4.33), the optical depth at different frequencies can be determined by measuring the profile of the line $\lambda = 21.11$ cm. The number of hydrogen atoms in a column of unit cross section along the line of sight with certain radial velocities can then be determined from Eq. (4.45). We shall consider in the following section a method for studying the spatial distribution of interstellar hydrogen using the formula for differential galactic rotation in conjunction with Eq. (4.45), but we shall first look at a number of general considerations.

If over the whole profile the optical thickness is much less than unity, the emission intensity is

$$I_\nu = \frac{2k\nu_{01}^2}{c^2}[T_{\mathrm{H}}\tau(\nu) + T_n]. \tag{4.46}$$

Excluding the continuous spectrum (T_n) and integrating over the complete line profile, we obtain

$$\int (I_\nu - I_n)d\nu = \frac{2kT_{\mathrm{H}}}{c^2}\nu_{01}^2 \int \tau(\nu)d\nu$$

$$= \frac{2kT_{\mathrm{H}}}{21^2}\frac{d\nu/dV}{1.87 \times 10^{13}T_{\mathrm{H}}}\int \mathfrak{N}(V)dV = 3 \times 10^{-33}\mathfrak{N}, \tag{4.47}$$

where $\mathfrak{N}$ is the total number of hydrogen atoms along the line of sight. In this way, from the total emission in the line $\lambda = 21$ cm, the total number of atoms in a column of unit cross section may be determined, and, if its extent is known, also its mean density.

On the other hand, if in the center portion of the line profile the optical thickness is much greater than unity, we can determine from the measured brightness temperature, in agreement with Eq. (4.33), the spin temperature, which in the majority of cases, as has already been noted, is equal to the kinetic temperature of the gas. We shall now estimate the number of hydrogen atoms in a column of unit cross section that has an optical thickness of approximately unity (on the assumption that there is no systematic motion in a given direction). Assuming a velocity scatter of 7 km/sec, we find that $\mathfrak{N}(V)$ in Eq. (4.45) constitutes approximately the fraction $\frac{1}{7} \times 10^5 \approx 1.4 \times 10^{-6}$ of all the atoms in the column. Therefore, the total number of atoms in a column of unit optical thickness in the center of the line is

$$\mathfrak{N}(\tau = 1) = 1.87 \times 10^{13} \times 7 \times 10^5 T_{\mathrm{H}} \approx 1.3 \times 10^{19} T_{\mathrm{H}}. \tag{4.48}$$

If $T \approx 100°$ and $N_H \approx 0.5$ cm^{-3}, the length of such a column is 800 pc. In directions toward the center and partly toward the anticenter of the Galaxy, where there is no rotational line broadening, interstellar hydrogen is indeed opaque at the line $\lambda = 21$ cm.

A study of the line profile of $\lambda = 21$ cm toward the center of the Galaxy showed a brightness temperature of 125°. Approximately the same temperature was obtained from measurements in the direction of the anticenter, where $\tau \approx 1$. Evidently this value defines the kinetic temperature of interstellar gas in H I zones. Later it will be shown that the equilibrium temperature of regions of nonionized hydrogen is even below 20°. It is natural to suppose that the kinetic and, consequently, the spin temperatures of neutral hydrogen atoms are different in various parts of the Galaxy and in various cosmic regions. Thus the observed temperature is somehow averaged. In order to clarify the nature of this averaging, let us imagine that along the line of sight we find separate clouds with temperatures of $T_1, T_2, T_3, \ldots, T_n$. Let the number of atoms in a column of unit cross section along each of these regions be $\mathfrak{N}_1, \mathfrak{N}_2, \mathfrak{N}_3, \ldots, \mathfrak{N}_n$. Then from Eq. (4.45) it follows that the total optical thickness of this system of clouds "strung" along the line of sight is

$$\tau = \frac{1}{1.87 \times 10^{13}} \sum_{i=1}^{n} \frac{\mathfrak{N}_i}{T_i}. \tag{4.49}$$

Comparing Eq. (4.49) with Eq. (4.45), we may conclude that the harmonic mean of the temperature, which fixes the optical thickness of hydrogen at $\lambda = 21$ cm, may be determined from the expression

$$\overline{\left(\frac{1}{T_H}\right)} = \frac{1}{\mathfrak{N}} \sum_{i=1}^{n} \frac{\mathfrak{N}_i}{T_i} = \frac{\Sigma \mathfrak{N}_i / T_i}{\Sigma \mathfrak{N}_i}. \tag{4.50}$$

In this way, the observed temperature of neutral hydrogen in the radio band is a harmonic mean of the temperatures of separate regions. This was first noticed by Kahn [17].

Absorption at the line $\lambda = 21$ cm may be observed against a background of strong sources whose brightness temperature close to this hydrogen line is greater than the kinetic temperature of the interstellar gas in H I regions. We consider the following problem for the analysis of this effect. Suppose that at a certain distance from the observer a strong source of continuous emission is observed with a brightness temperature T_a. The optical depth up to this source we denote by $\tau(\nu)$ and the optical thickness of interstellar hydrogen behind the source by $\tau'(\nu)$. It is then obvious that in the direction of the source the brightness temperature inside the profile of the line $\lambda = 21$ cm will be equal to

$$T_b = T_H(1 - e^{-\tau(\nu) - \tau'(\nu)}) + T_a e^{-\tau(\nu)}. \tag{4.51}$$

It is assumed here that the nonthermal radio emission in this direction is concentrated only in the source, and that its dimensions are small in comparison with the antenna pattern. If the observation is made in a direction adjacent to that of the source, then, assuming that T_H and the total optical thickness are here the same as in the direction of the source, we have

$$T'_b = T_H(1 - e^{-\tau(\nu) - \tau'(\nu)}). \tag{4.52}$$

Using Eq. (4.52), we can rewrite Eq. (4.51) in the form

$$\Delta T = T_b - T_a = T'_b - T_a(1 - e^{-\tau(\nu)}). \tag{4.53}$$

Here T_a may be determined from an observation of the source at a frequency close to the line, T'_b from observations at the line frequency of the portion of sky close to the source, and T_b from observations of the source itself in the line. Hence Eq. (4.53) provides a determination of optical depth up to the source at various frequencies;

$$\tau(\nu) = -\ln\left(1 + \frac{\Delta T - T'_b}{T_a}\right). \tag{4.54}$$

An analysis of the absorption line $\lambda = 21$ cm also allows us to find the density and velocity distribution of neutral interstellar hydrogen. It is curious to note that if there is no neutral hydrogen behind the source, that is, if $\tau'(\nu) = 0$, then the profile of the absorption line

$$\Delta T = T_b - T_a = (T_H - T_a)(1 - e^{-\tau(\nu)}) \tag{4.55}$$

must be a mirror image of the profile of the same line beyond the source, $T'_b = T_H(1 - e^{-\tau(\nu)})$ (recall that $T_H < T_a$). However, in actual conditions behind sources often distributed close to the plane of the Galaxy neutral hydrogen almost always exists.

Other Radio Lines

It is easier to study separate clouds of interstellar gas from line absorption at $\lambda = 21$ cm than from their emission lines. Since strong sources of continuous radiation are found relatively far from the Sun, their lines of sight intercept several clouds, and the total contour represents, as a rule, the superposition of line profiles formed in the individual clouds. However, it is possible, with some degree of certainty, to resolve this contour into a sum of Gaussian profiles, and to take each Gaussian to represent an absorption line in an individual cloud. Thus, we assume that

$$\tau(\nu) = \tau_0 \exp\left[-\left(\frac{\nu - \nu_0}{\Delta\nu_D}\right)^2\right], \tag{4.56}$$

where τ_0 is the optical depth at the center of the line, and $\Delta\nu_D$ is the Doppler half-width. Then, substituting Eq. (4.56) in Eq. (4.45), and integrating with respect to frequency, we obtain for the total absorption in the cloud the expression

$$\int \tau(\nu)d\nu = \tau_0\pi^{1/2}\Delta\nu_D = \frac{\int N_H dr}{3.9 \times 10^{14}T}, \tag{4.57}$$

where $\int N_H dr$ is the number of atoms in a column of unit cross section through the cloud. On the other hand, if the velocity dispersion is strictly thermal, the temperature of the gas is

$$T_H = \frac{m_H\lambda_{01}^2}{2k}(\Delta\nu_D)^2 = 2.6 \times 10^6(\Delta\nu_D)^2\ {}^\circ\text{K}, \tag{4.58}$$

where the frequency is in cycles per second. Thus, by determining from observations the total absorption and the width of the individual profiles, we can both find the density of the gas and set an upper limit on its temperature, since the profile is extended, not only thermally, but by macroscopic motions as well.

Shuter and Vershur [94a] studied by this method some contours of absorption lines against the background of the sources Cassiopeia A, Cygnus A, and Taurus A. In four observed contours they resolved more than forty Gaussian curves for which they found τ_0 and $\Delta\nu_D$, and consequently the density and temperature of these clouds. The number of clouds with a given τ_0 decreases exponentially with optical depth (the maximum recorded number was at $\tau_0 = 1.6$). The upper limit of the gas temperature was from 20 to 120°K, and the average value of the concentration of hydrogen atoms was $N_H \approx 3$ cm^{-3} (the maximum value was about 17 cm^{-3}). The reliability of these results is limited somewhat by the arbitrary separation of the Gaussian from the general contours. In addition, certain selection effects were present, such as the fact that clouds with higher temperatures have smaller optical thicknesses, and therefore are more difficult to observe. The authors emphasize, however, that the temperature can be determined considerably more precisely from absorption lines of weak clouds ($\tau_0 \ll 1$), where the brightness temperature $T_a\tau(\nu) \sim T_a T_H^{-3/2}$ is large, than from emission lines for which the brightness temperature is given by $T_H\tau \sim T_H^{-1/2}$. In clouds with large optical thickness ($\tau_0 \gg 1$) the situation is reversed.

It is known that the excited states of a hydrogen atom possess fine structure whose transitions between the sublevels give lines in the radio band. The highest probability occurs for the transitions $2\,^2P_{3/2} \to 2\,^2S_{1/2}$. The wavelength of this line is $\lambda = 3.03$ cm ($\nu = 9847$ MHz). Moreover, since the sublevel $2P_{3/2}$ is populated, the chance of observing this line is considerably greater relative to the probability of observing lines of other fine-structure transitions, for example, $2\,^2S_{1/2} \to 2\,^2P_{1/2}$, where the Einstein coefficient is considerably less. Using the population of the $2\,^2P_{3/2}$ sublevel computed in the previous section, Eq. (3.32), we find the emission coefficient and intensity in the line $\lambda = 3.03$ cm to be

$$\varepsilon_{3.03} = \frac{h\nu_{2p2s}}{4\pi} A_{2p2s} N_{2p} = 2.1 \times 10^{-41} N_e N_p \text{ erg/cm}^3 \text{ sec ster},$$

$$I_{3.03} = \int \varepsilon_{3.03} dr \approx 6 \times 10^{-23} \text{ME erg/cm}^3 \text{ sec ster.} \tag{4.59}$$

The study of a line of such an intensity is as yet beyond the technical limits of the observer. However, if this line could be discovered, it would be possible to compare the theoretical calculations of the diffusion of L_α quanta with observed data. Note that, if a nebula is sufficiently bright, we must consider not only spontaneous transitions, but transitions induced by the action of the thermal emission from the nebula itself; in this case the line intensity already approaches the limits of observation, since here the saturation of the $2\,{}^2P_{3/2}$ level can create a maser effect [94b].

The transition in the hyperfine structure of a deuterium atom gives a spectral line of wavelength $\lambda = 91.6$ cm ($\nu_D = 327.38$ MHz). The probability of spontaneous transition in this line [15] is

$$A^{D}_{10} = 5.2 \times 10^{-17} \text{ sec}^{-1}. \tag{4.60}$$

In this band of wavelengths, the nonthermal emission of the Galaxy has a brightness temperature far greater than the kinetic temperature of the interstellar gas. Therefore, applying Eq. (4.33) to the deuterium line, we have

$$T^D_b = T_n(\nu_D) \frac{1 - e^{-\tau(\nu)}}{\tau(\nu)} = T_n(\nu_D) \left[1 - \frac{1}{2} \tau(\nu_D) \right]. \tag{4.61}$$

Consequently, the deuterium line $\lambda_D = 91.6$ cm must be observed in absorption. One might discover this line against the background of the spectrum of the bright nonthermal source Sagittarius A at the center of the Galaxy, since here $T_a \gg T_H$ and the effect will be appreciable. In this case,

$$T^D_b = T_a e^{-\tau(\nu)} [1 - \tau(\nu_D)]. \tag{4.62}$$

The optical thickness at the deuterium line is not difficult to evaluate by analogy with Eqs. (4.43) and (4.48). We obtain successively the quantities (a unit frequency interval corresponds to a velocity interval of 91.6 cm/sec):

$$B^D_{01} = 7.1 \times 10^{14} \text{ cm}^3/\text{erg sec}^2; \qquad k'_0(\nu_D) = \frac{7.4 \times 10^{-14}}{T_H} f(91.6\Delta\nu),$$

$$\tau_D(\nu) = \frac{1.85 \times 10^{-14}}{T_H} \int N_D f(91.6\Delta\nu) dr = \frac{1.85 \times 10^{-14}}{T_H} \mathfrak{N}_D(V). \tag{4.63}$$

When the velocity dispersion of the interstellar gas is about 7 km/sec, the total number of deuterium atoms in a column of unit cross section is $\mathfrak{N}_D \approx 7 \times 10^5 \mathfrak{N}_D(V)$. Thus the optical depth at the center of the line $\lambda_D = 91.6$ cm is

$$\tau_D \approx 2.7 \times 10^{-20} \mathfrak{N}_D / T_H. \tag{4.64}$$

The optical thickness along the diameter of the Galaxy, 7.5×10^{22} cm, at the deuterium line is

$$\tau_D \approx 2 \times 10^3 N_D / T_H. \tag{4.65}$$

If $T_H \approx 100^\circ$ and $N_D \approx 2 \times 10^{-4}$ cm^{-3}, then $\tau_D \approx 5 \times 10^{-3}$. Attempts up to the present time to discover this line have not been successful.

The interesting possibility of observing transitions between the high levels of a hydrogen atom was noted by Wild [95] and N. S. Kardashev [96]. It follows from Eq. (2.4) that the number of atoms at levels with large n values is relatively large. The highest probabilities for spontaneous emission from these levels arise from transitions between neighboring quantum numbers. It follows from Eq. (I.2) that transitions of the type $n \to n - 1$ fall into the radio band $\lambda > 1$ mm, if $n > 28$. On the other hand, the principal quantum number is restricted by the condition $n < 600$ (if $\lambda < 10$ m). The population of these levels may be calculated from Eq. (2.4), taking $b_n = 1$. The absorption coefficient per atom per unit frequency is easily determined by the formula

$$k'_n(\nu) = \frac{\pi e^2}{mc} f_{n,n+1} \frac{h\nu}{kT} f(V) \frac{dV}{d\nu}, \tag{4.66}$$

where $f(V)$ is a function of the velocity distribution that describes the line profile. The absorption coefficient, calculated per unit volume, is

$$\kappa_n(\nu) = \frac{\pi e^2}{mc} \frac{h^3 n^2}{(2\pi mkT)^{3/2}} \frac{n}{6} \frac{h\nu}{kT} N_e N_p f(V) \frac{dV}{d\nu}. \tag{4.67}$$

We note that here $e^{\chi_n/kT} \approx 1$, and the oscillator strength of this transition equals $n/6$ (see Appendix I). Substituting numerical values and using the expression $\nu_{n(n-1)} = 6.58 \times 10^{-5} n^{-3}$, we obtain

$$\kappa_n(\nu) = 5.8 \times 10^{-12} \frac{N_e N_p}{T^{5/2}} f(V) \frac{dV}{d\nu}. \tag{4.68}$$

The profiles of these lines can be very complex. First, in addition to thermal motion, the macroscopic velocity of the H II region also affects the profile, and second, when $n > 60$ the effectiveness of electron collisions is comparable with that of spontaneous transitions, so that the line profiles will be distorted by intermolecular effects such as the Stark effect. However, for $n < 60$ ($\lambda < 30$ cm), the role of collisional broadening in the center of the line is small. If we take for the central portion of the line profile a Gaussian distribution,

$$f(V)\frac{dV}{dv} = \frac{1}{\pi^{1/2}\Delta\nu_D} e^{-[(\nu-\nu_0)/\Delta\nu_D]^2}, \tag{4.69}$$

where $\Delta\nu_D$ is the Doppler half-width, the optical depth in the center of the line in question ($T = 10{,}000°$) will be

$$\tau(\nu_{n(n-1)}) \approx 10^{-3}\frac{\text{ME}}{\Delta\nu_D}. \tag{4.70}$$

Comparing Eq. (4.70) with the optical depth in the neighboring region of the continuous spectrum, Eq. (4.17), we obtain

$$\frac{\tau(\nu_{n(n-1)})}{\tau(\nu)} \approx \frac{3.1 \times 10^{-14}\nu^2}{\Delta\nu_D} \approx \frac{10^{-3}\nu}{\overline{(V^2)}^{1/2}}, \tag{4.71}$$

where $\overline{(V^2)}^{1/2}$ is the velocity dispersion. Substituting $\overline{(V^2)}^{1/2} \approx 12$ km/sec, we obtain, for $\nu > 1200$ MHz ($\lambda < 25$ cm), an optical thickness in these lines of more than 1 percent of the optical thickness in the continuous spectrum. If the latter is less than unity (which is almost always true in this region), such transitions can be observed. The radiative transition $105 \rightarrow 104$ ($\nu = 5463$ MHz) was actually observed in the Orion and Omega nebulae by Dravski and Dravski [96a], and the transition $91 \rightarrow 90$ in the Omega nebula by Borozdich and Sorochenko [96b]. The brightness temperature in excess of the continuous spectrum constituted, according to Eq. (4.71), about 4 percent. The average of several experimental values [96a] gives about 7 percent in the Omega nebula, and about 6.5 percent in the Orion nebula, with a mean error of ± 3 percent.

In recent years some new lines of highly excited hydrogen and helium have been discovered [96c, d, e, f]. The widths of all these lines were unexpectedly small and correspond to temperatures of from 4000° to 9500°. Moreover, there is no trace of broadening of these lines by the Stark effect. Therefore there are some difficulties in the theoretical interpretation of these observations. One explanation is that the current theory of Stark broadening is incorrect because it does not take into account the perturbation of both levels [96g]. Another explanation [96h] refers this effect to the overpopulation of higher levels ($b_{n+1}/b_n = 1.0007$ when $n > 100$). This overpopulation corresponds to the negative excitation temperature $T_{ex} = -360°$K. If the optical depth is not small, there is a maser effect. It is necessary to note that in this range of frequencies the optical depth of large nebulae is big enough, in contrast to the optical Balmer lines, where there is also overpopulation ($b_n/b_2 \gg 1$) but the optical depth is vanishingly small. The maser effect leads to decrease in the widths of lines and therefore may explain the observational data.

5. The Distribution of Interstellar Hydrogen

Methods described in the previous sections can be used to study individual diffuse nebulae and gas complexes, as well as the distribution of interstellar hydrogen in the Galaxy as a whole. The morphology of diffuse nebulae will be given in Sec. 10. We consider here the distribution of interstellar hydrogen in the Galaxy in general.

Ionized Hydrogen

The distribution of H II regions of interstellar hydrogen can be obtained both by photographing the sky through light filters that isolate the emission lines of nebulae against the background of the night sky, and by measuring the thermal radio-emission flux from these regions.

Of the hydrogen lines, the most conspicuous is the H_α line, the intensity of which is connected with the emission measure by the relation (see Sec. 2)

$$I(H_\alpha) = 2.62 \times 3 \times 10^{-8}\ \text{ME} = 7.9 \times 10^{-8}\ \text{ME erg/cm}^2\ \text{sec ster}. \quad (5.1)$$

A study of the distribution of interstellar hydrogen from this line was started by G. A. Shain and V. F. Gaze [97–102]. It allowed many new emission nebulae to be detected, and revealed characteristic aspects of their structure and distribution and, at least approximately, their density and mass from their emission measure. An unusual number of emission nebulae are found in the constellation Cygnus (Fig. 7).

Owing to interstellar absorption, only relatively nearby nebulae can be observed by optical methods. Moreover, absorption distorts the estimate of the emission measure, since its value is not always known. Radio observations are free from these disadvantages. The distribution of ionized hydrogen in the Galaxy has been most thoroughly studied by Westerhout from observations of thermal radio emission at 21.6 cm, close to the hydrogen line $\lambda = 21$ cm but not coincident with it [103]. The procedure is as follows. Isophots were constructed of the intensity distribution of radio emission. They were corrected for nonthermal components (if the character of the spectrum is known, these can be determined from observations at wavelengths longer than $\lambda = 3$ m, as will be shown below), so that it is then easy to determine the emission measure in each direction. If we assume that in all H II regions the electron density ($\overline{N}_e$) is approximately the same, and that, on the average, ionized-hydrogen regions occupy a certain fraction of the entire region, then, knowing the dimensions of the system, it is easy to calculate $\alpha\overline{N}_e^2$. The distribution of $\alpha\overline{N}_e^2$ as a function of distance from the center of the Galaxy is then calculated from the distribution of the isophots along the Milky Way. For example, a maximum in the thermal radio emission is observed in the direction of galactic longitude $l \approx 26^\circ$, and not in the direction of the center. This means that in the central portion of the

Galaxy (for $R < 2$ kpc) there is almost no ionized hydrogen and a maximum density occurs approximately at a distance $R_{\odot} \sin 26^{\circ} \approx 10 \times 0.44 \approx 4.4$ kpc, where $R_{\odot}[=10$ kpc$]$ is the solar distance from the galactic center. (In works prior to 1964 the old scale of the Galaxy was used, for which $R_{\odot} = 8.2$ kpc.

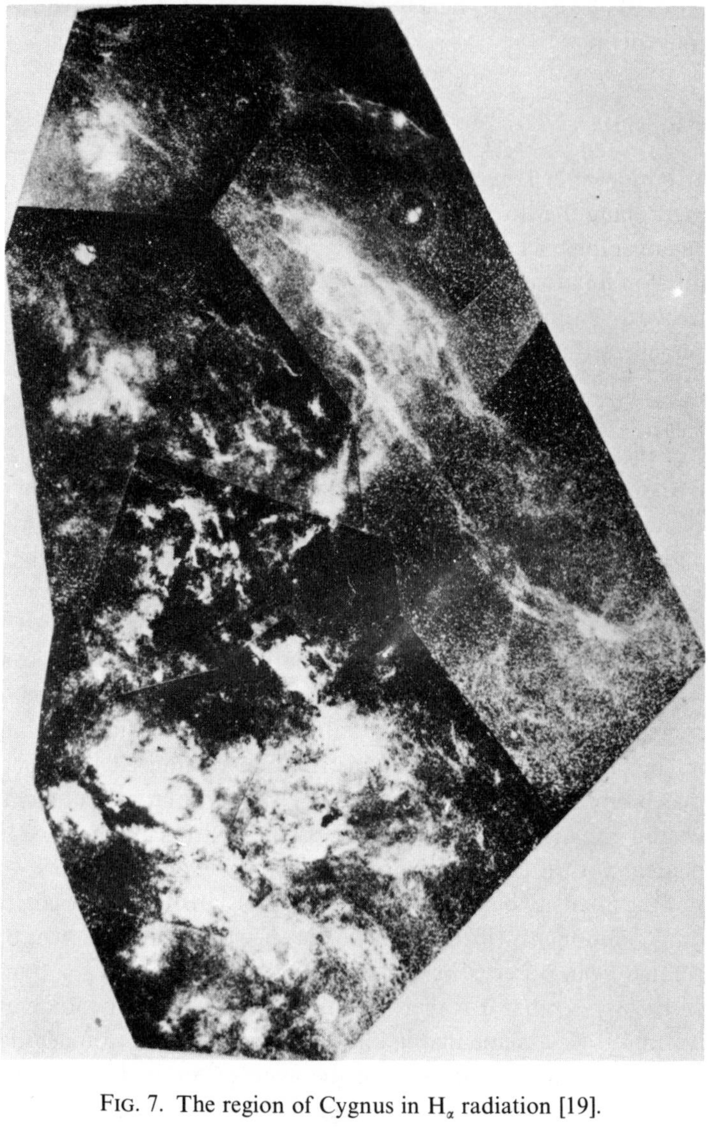

FIG. 7. The region of Cygnus in H_{α} radiation [19].

Here all data and figures correspond to the new scale.) The distribution of ionized hydrogen obtained by Westerhout in the galactic plane is presented in Fig. 8. There are a number of arguments, to be sure, that allow us to suppose that the maximum of radiation is produced not by a ring, but by a local density increase beyond the center of the Galaxy [104]. Hydrogen is ionized in approximately 5 percent of all the regions near the galactic plane. The total mass of ionized hydrogen amounts to $6 \times 10^7\ M_\odot$, that is, about 0.1 percent of the entire mass of the Galaxy. The distribution of ionized hydrogen in the Andromeda Nebula has an annular character, it is true, while the maximum densities of H I and H II approximately coincide.

On the other hand, in a radio survey of the Galaxy [103a] at wavelength $\lambda = 20.8$ cm, the significance of this ring was not confirmed. However, southern

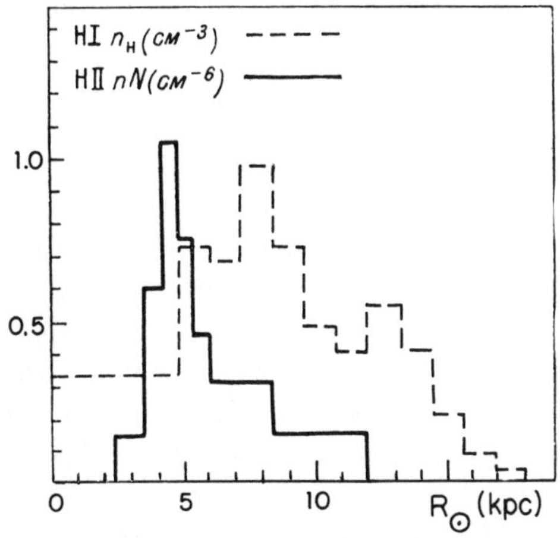

FIG. 8. The distribution of neutral and ionized hydrogen in the Galaxy from radio data [103]. Since the density of ionized hydrogen is determined from the degree of emission, the graph gives the quantity nN, where N is the concentration in the central region and n is the mean density in the galactic plane.

parts of the Milky Way during this study showed complicated chaotic structure, which added difficulties to the interpretation. The external arms with a half-width of 3° were better developed.

We recall that the observations of radio emission in the continuous spectrum include both thermal emission, considered in Sec. 4, and nonthermal emission, considered in Sec. 18. Therefore, it is first of all necessary in an analysis of the galactic radio emission to separate the thermal from the nonthermal component. This may be done by means of a comparison of observations at two different frequencies, since the dependence of the brightness temperature on frequency is different for each of these forms of radiation. In fact, the observed brightness temperature is the sum of the two components, or

$$T_b(\lambda) = T_b^{\text{nonthermal}}(\lambda) + T_b^{\text{thermal}}(\lambda). \tag{5.2}$$

For an optically thin layer, we have $T_b^{\text{thermal}}(\lambda) \approx \lambda^2$. As will be shown in Sec. 18 [see Eq. (18.11)], the relation $T_b^{\text{nonthermal}}(\lambda) \approx \lambda^{2+\alpha}$ holds over a wide range of frequencies for nonthermal radio emission. With this in mind, we obtain from Eq. (5.2)

$$T_b^{\text{thermal}}(\lambda_1) = \frac{T_b(\lambda_1) - (\lambda_1/\lambda_2)^{\alpha+2} T_b(\lambda_2)}{1 - (\lambda_1/\lambda_2)^2}. \tag{5.3}$$

It is easy now, with the help of this expression, to isolate from the continuous spectrum in the centimeter and decimeter ranges the nonthermal radiation, which has a high intensity in the meter range. Until the present time the observations used for this purpose were those of Mills [16] made at $\lambda = 3.5$ m. There is a certain ambiguity in the choice of spectral index α (in [103] α was taken as $\alpha = 0.7$, in [103a], $\alpha = 0.6$, and in [103b] $\alpha = 0.4$).

Neutral Hydrogen

The distribution of neutral hydrogen is studied by emission or by absorption of its radio line $\lambda = 21$ cm. Here, the most important results are obtained for studying the distribution of interstellar hydrogen, partly because neutral hydrogen is a basic constituent of the interstellar medium, and, more important, because measurement of the profile of this line allows a detailed study of the distribution of interstellar gas velocity. There are not at the present time many measurements of the line $\lambda = 21$ cm arising in separate H I regions and we shall not consider them here.

The following method for studying the distribution of neutral hydrogen was worked out by van de Hulst, Muller, and Oort [94]. Consider the profile of the radio line $\lambda = 21$ cm in any direction in the Galaxy. If at a certain frequency or over the total profile the optical thickness is less than unity, then from Eqs. (4.33) and (4.45) the optical thickness of the Galaxy may be determined in a given direction, and hence the total number of atoms in a column of unit cross section along the line of sight is

$$\mathfrak{N}(V) = 1.83 \times 10^{13} T_{\mathrm{H}} \tau(\nu), \tag{5.4}$$

where the velocity V is related to a shift $\Delta\nu$ of the line center by the relation $V = 21\Delta\nu$.

If the observed profile is freed from the effect of terrestrial and of solar motion, absorption of continuous emission close to $\lambda = 21$ cm in the terrestrial atmosphere, and so forth, the corrected line profile depends on the following atomic and gaseous motions: (*a*) thermal atomic motions, (*b*) disordered motion of regions of interstellar gas and motion inside the regions, (*c*) differential rotation of the Galaxy, (*d*) systematic deviation of the gas motion from purely circular rotation in the galactic plane. The number of atoms $\mathfrak{N}(V)$ with a velocity in the range from V to $V + 1$ cm/sec depends on all these forms of motion.

If systematic motion of the interstellar gas is disregarded (items *c* and *d*), interstellar hydrogen is opaque in the line $\lambda = 21$ cm (Sec. 4). In this case

$\tau(\nu) > 1$, and we can deduce from an analysis of the profile that the harmonic mean of the spin temperature of the gas is approximately equal to the kinetic temperature. From observations in the direction of the center, $\bar{T}_H \approx 125$. In the direction of the galactic anticenter, $\tau \approx 1$, and the value $\bar{T}_H \approx 120$ has been obtained for the temperature.

Taking into account systematic motion, for example galactic rotation, "extends" the permissible velocity range, and hence diminishes $\mathfrak{N}(V)$ many times, and brightens, as it were, the interstellar gas. The velocity of galactic rotation is approximately 30 times the velocity of the disordered motion, so that, other conditions being equal, $\mathfrak{N}(V)$, and consequently $\tau(\nu)$, may be reduced 30 times. This "brightening" in $\lambda = 21$ cm allows us to detect the distribution of interstellar hydrogen in the Galaxy at great distances from the Sun.

In order to obtain in a neat form the dependence of $\tau(\nu)$ on systematic velocities, the observed profile must be freed from the smearing effect of thermal and macroscopic motion of the gas in these regions, and above all, from the disordered motion of the regions themselves. The method for doing so is as follows [94]: let $a(\nu)$ be the profile of line $\lambda = 21$ cm determined only by systematic motion, and $g(\nu)$ the observed profile, distorted by the effect of disordered and thermal motion. The velocity distribution of these motions may be written in the form of a normalized function $f(V)$. Obviously, the observed profile can be obtained by superimposing distributions $a(\nu')$ and $f[(\nu - \nu')\ 21\ \text{cm}]$, whereby emission from all parts ν' of the profile is shifted by disordered motion to frequency ν, accumulating radiation at this latter frequency. Thus

$$g(\nu) = \int_0^{\infty} f[21(\nu - \nu')] \frac{dV}{d\nu} a(\nu')d\nu'. \tag{5.5}$$

It is more convenient in this equation to change to the dimensionless frequency $x = (\nu - \nu_{01})/\Delta\nu_D$:

$$g(x) = \int_{-\infty}^{+\infty} f(x - x')a(x')dx'. \tag{5.6}$$

If the velocity-distribution function is described by Maxwell's equation (thermal motion), then

$$f(x - x') = \frac{1}{\pi^{1/2}} e^{-(x-x')^2}. \tag{5.7}$$

Observations of interstellar regions show that their velocity distributions are better described by a simple exponential function [105]:

$$f(x - x') = \tfrac{1}{2}e^{-|x-x'|}. \tag{5.8}$$

Solution of the integral equation (5.6) with the kernel (5.8) leads at once to

$$a(x) = g(x) - \frac{d^2g}{dx^2}, \tag{5.9}$$

which is easily seen by direct substitution. For function (5.7) the solution of (5.6) is written in the series form

$$a(x) = g(x) - \frac{d^2g}{dx^2} + \frac{1}{2!}\frac{d^4g}{dx^4} - \frac{1}{3!}\frac{d^6g}{dx^6} + \cdots. \tag{5.10}$$

Terminating the series at the nth derivative, we obtain the so-called Eddington approximation of nth order. Usually, only the first two terms are used in this method. This is obviously valid if the Doppler width of the disordered motion is much less than the structural features of the profile. However, as Ollongren and van de Hulst showed [106], this method can also be used when the Doppler width of the function $f[(\nu - \nu')/21\text{ cm}]$ is large compared with the profile itself $a(x)$. To do this, some numerical coefficient that depends on the size of the frequency interval in which the profile fluctuations occur must be placed in front of the second derivative. Thus, designating the observed profile by $\tau'(\nu)$ and the line profile freed from the effect of disordered and thermal motion by $\tau(\nu)$, we have

$$\tau(\nu) = \tau'(\nu) - (\Delta\nu_D)^2 \frac{d^2\tau'(\nu)}{d\nu^2}, \tag{5.11}$$

where $\Delta\nu_D$ is selected during the analysis of the profiles. Substituting $\tau(\nu)$ obtained in this way in Eq. (5.4), we obtain the dependence on systematic velocities alone of the number of atoms along the line of sight.

The quantity $\mathfrak{N}(V)$ must now be expressed in the form of a function depending on the distance to the center of the Galaxy in galactic coordinates. Let us suppose at first that the deviation of the gas motion from circular symmetry can be neglected. Let the observer be in the galactic plane at a distance R_0 from the center. Then it follows, from the well-known Oort equations of the theory of galactic rotation, that the relative radial velocity of an element of gaseous mass, located at a distance R from the center and at a galactic elevation z, is determined by the equation

$$V_r = R_0[\omega(R, z) - \omega(R_0, 0)] \sin l \cos b. \tag{5.12}$$

Here $\omega(R, z)$ is the angular velocity of rotation at a distance R and at an elevation z (the vector ω is directed toward the south galactic pole, so that the rotation in Fig. 10 is in a clockwise direction); l and b are the galactic longitude and latitude. The distance from this element of mass is

$$r = R_0 \cos l \pm (R^2 - R_0^2 \sin^2 l)^{1/2}. \tag{5.13}$$

The plus or minus sign is chosen according to the position of the element relative to the tangent point of the line of sight. This equation is readily obtained by considering the triangle: observer, galactic center, and the given point. If

Eq. (5.12) is now differentiated with respect to r, then, considering l = const., $z = 0$, and $b = 0$, we obtain the velocity gradient along the line of sight:

$$\frac{dV_r}{dr} = (R^2 - R_0^2 \sin^2 l)^{1/2} \frac{R_0}{R} \frac{d\omega}{dR} \sin l. \tag{5.14}$$

Let us return now to Eq. (5.4). By definition, $\mathfrak{N}(V)$ is the number of atoms along the line of sight whose velocities lie in the range V to $V + 1$ cm/sec. When $\mathfrak{N}(V)$ is free from effects due to disordered and thermal motion, atoms with such velocities can be found only close to those points along the line of sight where the projection of the linear velocity is V. If we assume a monotonic dependence for $\omega(R)$, then for $R > R_0$ there is one such point, while for $R < R_0$ there are two such points with the given radial velocity, one behind the point of intersection and the other in front of it. The extension of these regions along the line of sight is obviously equal to $(dr/dV) \times 1$ cm/sec. Therefore, the density of neutral hydrogen is $N_H = \mathfrak{N}(V)dV_r/dr$. Substituting Eq. (5.4) here and expressing the velocity gradient in kilometers per second per kiloparsec, we obtain

$$N_H = 6.0 \times 10^{-4} T_H \tau(\nu) \frac{dV_r}{dr}. \tag{5.15}$$

The combined application of Eqs. (5.12) to (5.15) permits the determination of the density of neutral hydrogen at any point for which observations are possible from the measured profile of the line $\lambda = 21$ cm for a known law of galactic rotation.

A known difficulty arises from the presence of two regions with identical radial velocities when $R < R_0$. Here the optical thicknesses of the two regions are simply additive. They may be separated by successive approximations by assuming some initial distribution of gas in height above the galactic plane and by remembering that close and distant regions in the same line of sight have different heights [107].

Rates of galactic rotation may also be determined from the profile of the radio line $\lambda = 21$ cm [108]. Indeed, it is easy to see that the maximum rate along the line of sight is reached at the point where the line of sight touches the locus of galactic rotation (if $\omega(R)$ does not increase with R). Thus, line profiles at $\lambda = 21$ cm freed from disordered motion observed in the longitude range from 270° to 90° must fall sharply either on the high-frequency side (observations counter to the galactic rotation), or on the low-frequency side (observations along the galactic rotation). Since the distance from the point of contact is known ($R = R_0 \sin l$), then, converting the frequency of the break in the profile into velocity $V_r = 21\Delta\nu$, we find at once from Eq. (5.12) the angular velocity at this distance for $b = 0$, $z = 0$. The change of the galactic rate of rotation with distance from the center is shown in Fig. 9 [109]. We point out that here the linear rate of rotation is given. The angular rate falls monotonically with an increase in R. This curve is drawn as the upper limit from various observations, since if the point of contact occurs in a region low in hydrogen (between spiral arms) the velocity obtained by the break in the profile in this case relates to different, although neighboring, points along the line of sight, and therefore leads to lower values. The presence of a maximum for $R \leqslant 1$ kpc will be considered below. New investigations [109a, b] show that the velocity distribution is different in the northern and southern hemispheres of the Galaxy. It was found that in the plane of the Galaxy there is some flow of gas with velocities $\sim$ 10 km/sec and dimensions $\sim$ 100 pc.

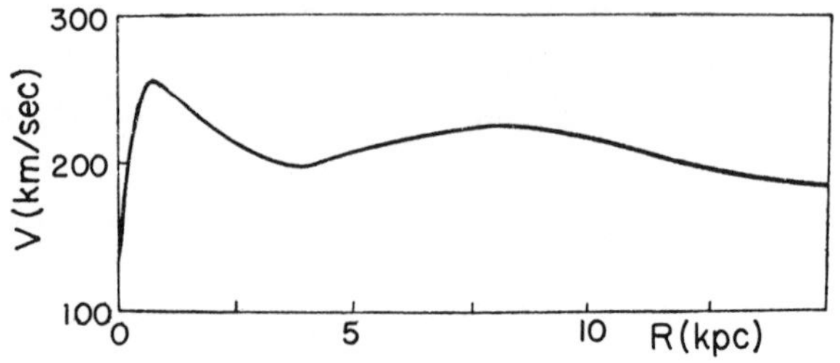

FIG. 9. Linear rate of rotation of the Galaxy from radio data [109].

However, we must note that the actual line profiles do not indicate an abrupt fall to zero immediately after the point of maximum velocity. This is connected partly with the inaccuracy introduced during the elimination of disordered velocities, and partly with the fact that the true gas motion is not entirely circular. Moreover, in this method for obtaining the velocity at a given distance, only one point of the profile is used, which also increases the probability of error. In papers [108a, b], a "method" is used to determine the angular rate of rotation of interstellar gas from an analysis of the total line profile, which therefore possesses a somewhat higher accuracy. The results agree closely with those of Fig. 9, although the curve is somewhat lower.

We note in passing that, using the dependence $V(R)$, we can determine the potential and mass distribution in the Galaxy [110]. This distribution is determined, of course, by the stars, which constitute 98 percent of the entire galactic mass. Interstellar hydrogen acts here as a potential indicator.

The Distribution of Hydrogen in the Galaxy

It is well known that the interstellar medium is highly concentrated in the galactic plane. Radio observations have been used to define more accurately the thickness of the hydrogen layer Δz (up to half density), which gradually increases from the central region outward and near the Sun reaches 350–500 pc. In the portion of the Galaxy within the Sun's orbit the gas layer is extremely flat, with an average deviation (where the density is a maximum) from the galactic plane of less than 50 pc. However, in the exterior portion considerable variations are observed; for example, in the region of the anticenter the layer bends toward the north, and in the region farthest beyond the center, to the south [111]. The average distance of the layer from the galactic plane $\bar{z}$ along its diameter, where deviations are especially great, is given in Table 5.1.

Table 5.1

R	Core	0.1	3	5	8	10	12	14
Northern hemisphere								
$\bar{z}$	0	0	0	0	+50	+250	+600	—
Δz	8(H II)	80(H I)	120	220	600	800	1200	—
Southern hemisphere								
$\bar{z}$	—	—	—	—	0	−100	−350	−600
Δz	—	—	—	—	400	900	1000	2000

All the observations from [112, 113] are recorded here, as well as values for Δz.

The total mass of the neutral hydrogen layer in the Galaxy, according to radio observations, is $\sim 1.4 \times 10^9$ solar masses, which amounts to less than 2 percent of the galactic mass. The mass of ionized hydrogen is 20 times less. At the boundaries of the layer, hydrogen is not uniformly distributed and forms a spiral structure, as shown in Fig. 10 [114]. Although the total picture seems complex, the whole may be considered to be annular formations of spiral structure, which were discovered earlier in our own and other galaxies from the distribution of hot stars and diffuse nebulae. The gas density in the arms is several times that between them. The width of the arms is about 400 pc, and their thickness agrees with the thickness of the gas layer. Probably more than 20 percent of the mass of all the interstellar hydrogen is concentrated in the arms. The densest ring, with a mean concentration $N_H \approx 1\ \text{cm}^{-3}$, has a radius of about 8 kpc. A second arm passes near the Sun ($R = 10$ kpc), and a third occurs at a distance of ~ 13 kpc from the center. Let us recall that the maximum density of ionized hydrogen is found at a distance of ~ 4.4 kpc.

In Fig. 11(a), on a larger scale than in Fig. 10, is shown the outline of the spiral arms in the vicinity of the Sun, as compiled by Bok [115]. The Sun is near the internal portions of the arm that passes through Cygnus and Carina. This arm has a projection in the direction of the Orion nebula and is therefore usually called the Orion arm. The Perseus arm is located beyond it, including in particular the well-known binary cluster h and χ Persei. The Sagittarius arm passes near the galactic center. In addition to the principal arms, there occur isolated projections and cross connections, such as the accumulation of gas proceeding from the Sun to the Sagittarius arm. Also the arm itself has a complex structure. However, it must be noted that the structure of nearby objects is determined with low accuracy from the $\lambda = 21$ cm line, since in this case systematic veloc-

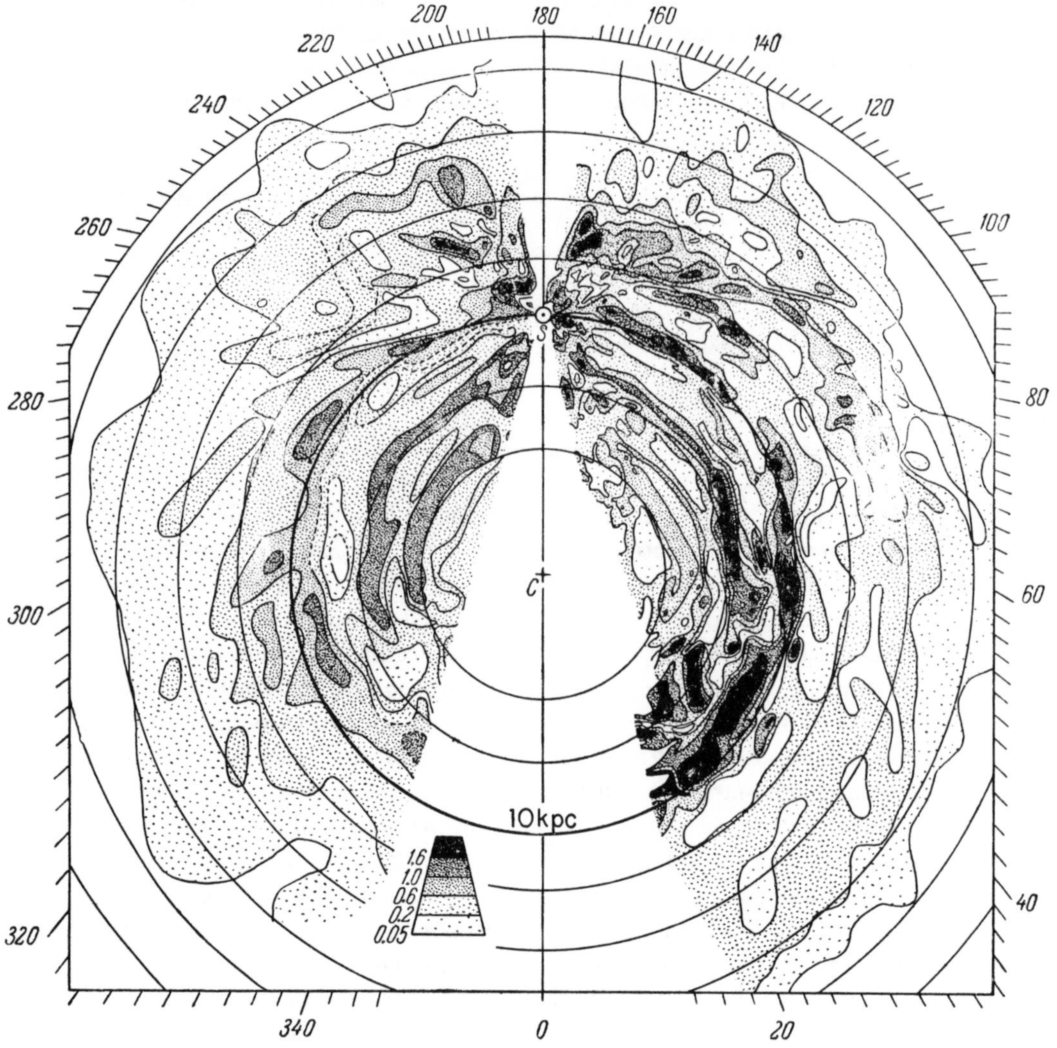

FIG. 10. Distribution of neutral hydrogen in the Galaxy [114].

ities are random in order and generally close to zero in the direction toward the center and anticenter. Therefore, Fig. 10, and especially Fig. 11(*a*), are as yet more or less rough outlines.

The spiral branches can be traced not only from 21-cm data, but by other methods, such as nonthermal galactic radio emission (see Sec. 19), the distribution of hot stars, thermal radio emission from H II regions surrounding hot stars, and so forth. Figure 11(*b*) shows a scheme of spiral arms constructed by N. S. Kardashev from all published radio data (the lines broken with points).

The positions of the spiral arms determined from long-period Cepheids [115a] in the neighborhood of the Sun are indicated by hatching in this figure. For comparison are shown logarithmic spirals (solid and dashed lines) that best fit the radio data.

The density distributions obtained depend on the assumed law of galactic rotation. In earlier investigations, it was assumed that gas motion in the Galaxy was circular [116]. However, it has been shown [114, 117] that the hydrogen distribution is considerably more symmetric if systematic radial gas motion is also considered possible. In Kerr's work [114], it is assumed that gas and stars (among them the Sun) travel radially from the galactic center at a velocity of about 7 km/sec, and that this velocity slowly decreases with distance from the center. Such a velocity distribution makes the northern and southern portions of the Galaxy (right and left of the empty sector in Fig. 10) symmetric. In the work of I. L. Genkin [117], Eq. (5.12), the so-called K term is added to Eq. (5.12). The treatment of radio data with different values of K showed that a good spiral structure is obtained for $K = -2$ km/sec kpc. With other values of K the picture remains nonsymmetric. The contraction of the Galaxy in the region of the Sun agrees, to a first approximation, with the general spread of gas and of stars when the expanding velocity diminishes with recession from the center. On the other hand, Brace [117a] was able to observe neither a contraction nor an expansion, and came to the conclusion that the assumption of circular motion of the interstellar gas between the 3-km arm and the Sun is indeed valid. The method of this investigation consists in constructing a chart that shows the dependence of the intensity of radiation on latitude and velocity at a given longitude close to the center. These charts were then compared with similar charts constructed on the basis of observed data. Although this comparison, too, is unable to give a definitive answer, Kerr's model [114] certainly disagrees with the observations.

In general, the accuracy of recent observations, and in particular the indeterminacy of data reduction and the irregular structure of the arms themselves, prevent a clarification of the details of the motion of interstellar gas.

It is interesting to notice that observations in the direction of the galactic center yield a certain increase in intensity corresponding to an increase in temperature of some tens of degrees for the range of frequencies corresponding to velocities from +4.5 to +6 km/sec over a very large range of galactic latitudes. This is interpreted as the presence of clouds of neutral hydrogen close to the solar system that are receding from the Sun with velocities of 5–6 km/sec in the direction of the galactic center.

A study of the immediate surroundings of the Sun has shown that hydrogen forms, as a first approximation, a plane-parallel layer with positive fluctuations [118]. If the average density of the layer is eliminated, dense gas condensations are very noticeable in Scorpius and Ophiuchus, in Gemini, Taurus, and Orion, and in certain weaker condensations. They are all situated in a great circle inclined at 20–30° to the galactic plane and forming a local gas system of average concentration ~ 0.5 cm^{-3}. We shall return to this system to study the distribution of dust (Sec. 12) and interstellar polarization of light (Sec. 14).

As is well known, an investigation of spectral laws of interstellar gas and absorption by interstellar dust shows that all interstellar media have a patchy structure (see Secs. 7 and 12). Therefore, subsystems of interstellar gas are often thought of as aggregates of interstellar gas clouds. Statistical methods are usually required for an analysis of their characteristics, although, as is seen from

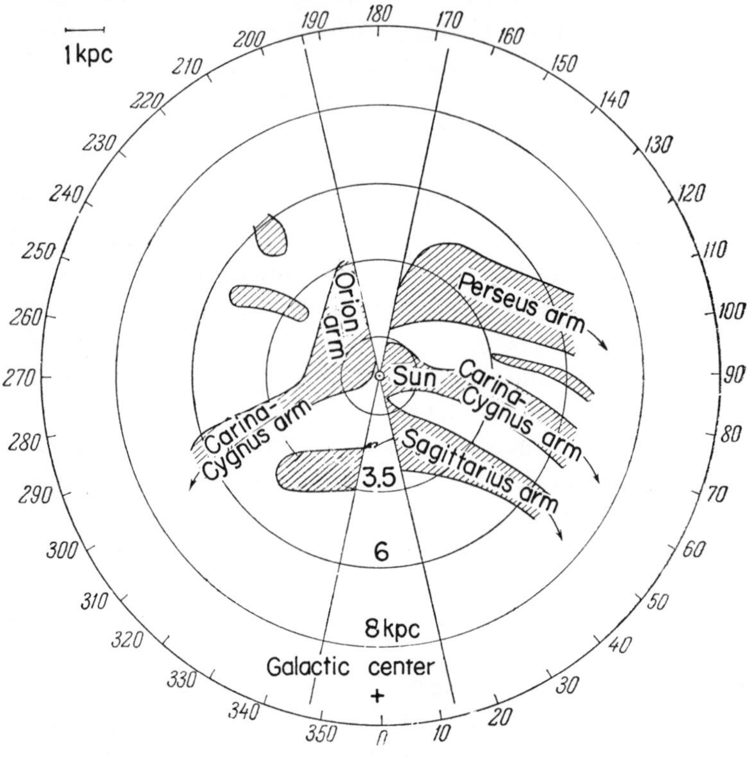

FIG. 11(*a*). Schematic chart of the spiral arms of the Galaxy in the vicinity of the Sun [115].

an application of these methods to interstellar absorption, the assumption of a random distribution of the cloud parameters and their positions, on which these methods are based, is in reality not valid. Nevertheless, statistical analysis does allow an estimate of a series of parameters for interstellar gas and dust clouds. Obviously, a statistical analysis of the cloud structure is also possible from observations at the 21-cm line. One of the first works in this direction was that of Takakubo [118a] (see also [118b]). The method consists essentially in calculating, and then comparing with observations, the line profiles, and hence the cloud parameters, based on a given distribution of the clouds in altitude, their dimensions, their average density, and the velocity distribution function of the motion of the clouds themselves, as well as the internal motions (turbulent and thermal) of the gas. A similar analysis of observations at medium galactic latitudes ($10^\circ \leqslant |b| \leqslant 25^\circ$) led to the following cloud parameters: average

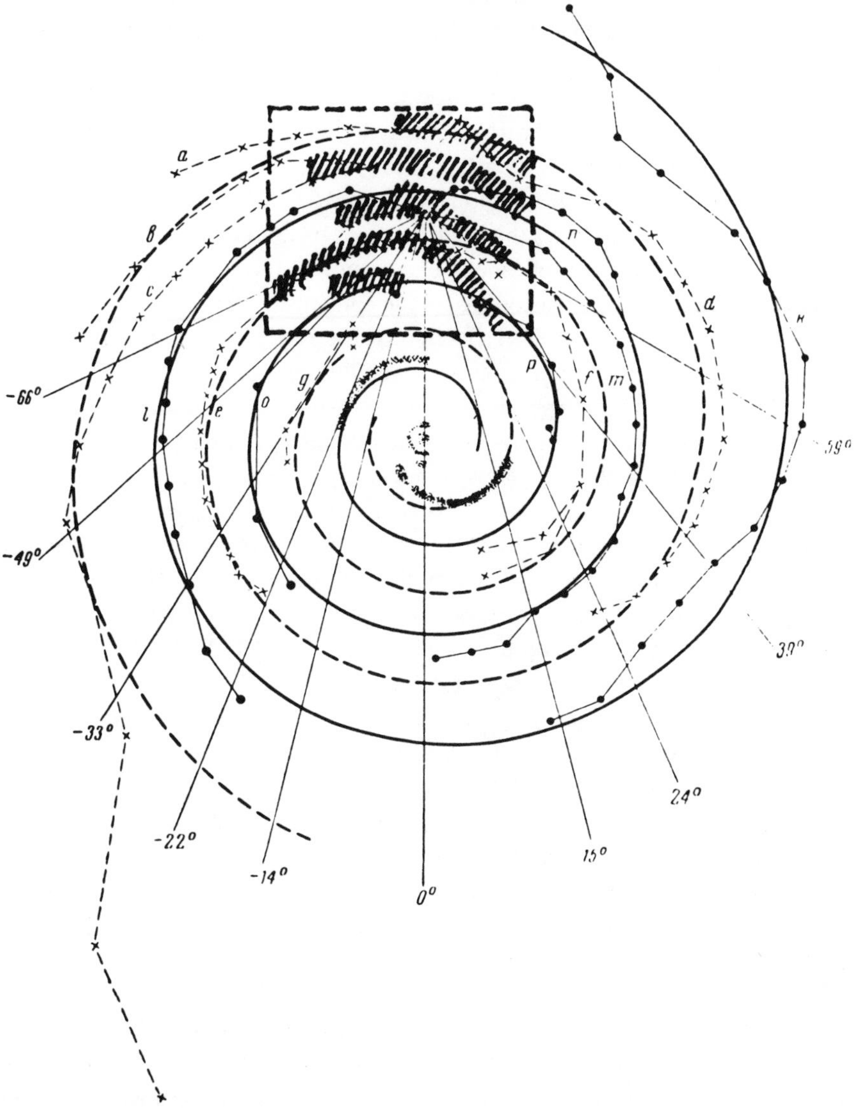

FIG. 11(*b*). Chart of the spiral arms from all published radio data; hatching shows positions determined from long-period Cepheids.

dimension, 3.5 pc; mass, 50 solar masses; average hydrogen density in a single cloud, 14 cm^{-3}; the space average close to the galactic plane, 0.7 cm^{-3}. The clouds occupy about 5 percent of interstellar space and the line of sight in 1 kpc pierces on the average 10–11 clouds [118a]. These results are in good agreement with independent determinations of cloud parameters from investigations of interstellar lines and interstellar absorption.

From recent investigations of interstellar hydrogen clouds we must note the determination of their mass-distribution function $\rho(M)$, M being the mass of a cloud. Statistical analysis of observations of interstellar absorption gives the relation $\rho(M) \sim M^{-\beta}$, where $\beta \approx 2$ for $M < 5 \times 10^3 M_\odot$ and $\beta = 1$ to 1.5 for $M > 5 \times 10^3 M_\odot$ [118c]. The theoretical discussion of the dynamics of an interstellar cloud by Field and Saslaw [118d] on the basis of previous ideas of Oort and Spitzer gives the function $\rho(M) \sim M^{-3/2}$ in accordance with observations. This idea includes the calculation of fusion of clouds by impact and the consequent dissipation of kinetic energy, and analysis of the formation of massive complexes by this fusion. After the condensation of stars in these complexes and formation of their H II regions, the rest of the interstellar matter flies away in the form of small clouds, driven by the pressure of the H II region of nearly formed stars.

It follows from measurements of radial velocities that gas in the vicinity of the Sun moves toward the galactic plane with a velocity of about 6 km/sec [118]. This means that the mass of gas that crosses 1 pc^2 close to the galactic plane per year is approximately equal to 7×10^{-8} solar masses. At the present time the ultimate fate of this gas is not clear, since the flux in the plane of the Galaxy has an almost null divergence. Possibly the movement to the plane is temporary and connected with the formation of a local system.

Observations at high galactic latitudes show that the line $\lambda = 21$ cm in these directions has broad wings corresponding to a mean velocity dispersion of not less than 30 km/sec. Hence, it follows that at high latitudes there are many fast clouds of small mass [18:18]. The existence of these regions also follows from observations of interstellar line absorption [119c]. This question will be considered in more detail in Sec. 19. Close to the galactic plane there also exist a certain number of regions with velocities $\sim$30–40 km/sec. Generally speaking, the number of these regions is greater than is given by the distribution functions (5.8) (see Sec. 7).

Very interesting results were obtained by studies of the galactic center using radio methods [119, 120–123, 281]. Figure 12 gives the isophots of a radio study of this region at a wavelength of $\lambda = 10$ cm [281], that is, in the continuous spectrum. The central region is quite conspicuous, as is the so-called Sagittarius A source, consisting of two portions, one of which is brighter. It is possible that this portion of the observed Sagittarius A source is situated at the galactic center itself. Observations at wavelength $\lambda = 21.6$ cm (in the continuous spectrum close to the hydrogen line) [86] also disclosed here a central source of dimensions 80×35 pc with brightness corresponding to an emission measure of 3.6×10^5. More detailed studies of the central source at wavelengths of 3.2 and 9.4 cm were made by Yu. N. Pariiskii [121, 122] and revealed a certain eccentric, small (diameter about 8 pc), but very dense source of radio emission. Its emission measure is ME $\sim 2 \times 10^6$. However, this source is most probably nonthermal [123].

From Fig. 12 it is clear that on both sides of the center, at a distance of about 100 pc, there are two maxima, which appear as the cross section of a ring, or two opposing arms, that provide thermal radio emission. Their thickness is

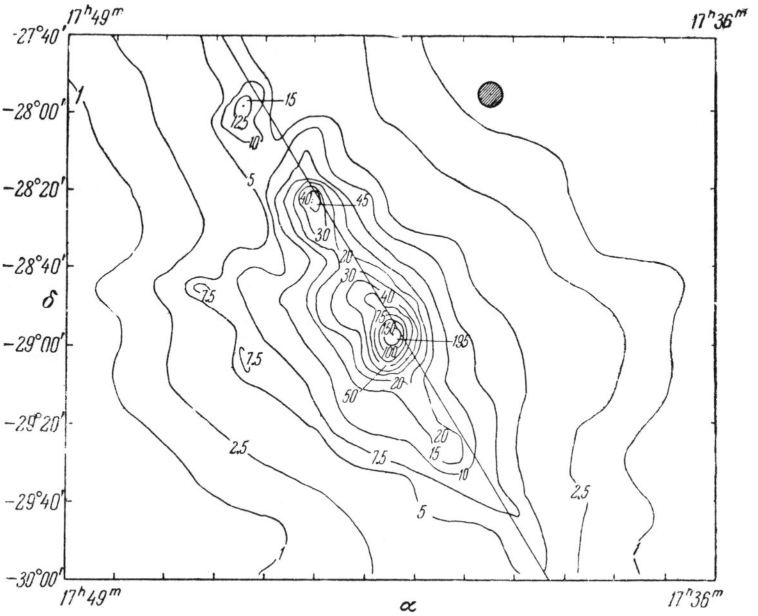

FIG. 12. Isophots of the region of the center of the Galaxy at a wavelength of 10 cm [281].

~ 20 pc, their electron concentration $\sim 20 \text{ cm}^{-3}$, and their mass about 10^6 solar masses [120].

Observations of the line $\lambda = 21$ cm, both in emission and in absorption, against the background of the central sources essentially reinforce this picture. In the region $R < 3$ pc the profiles of the $\lambda = 21$ cm line show broad wings both in emission and in absorption, indicating velocities of up to 250 km/sec. Evidently, some formations (about eight) resembling fragments of arms are found that rotate about the center at high velocity. They correspond to the maximum rate of rotation in Fig. 9. A high circular velocity is indicated from the presence of an intense concentration of mass in the galactic center. The period for a single revolution at a distance of $R \approx 0.5$ kpc is only 12×10^6 yr, and at the Sun's distance it is 250×10^6 yr. More detailed observations show in the central region ($R < 800$ pc) the presence of a thin, rapidly rotating gas disk with sharp boundaries. In addition, in line profiles from the central source an absorption line is visible belonging to gas moving with a radial velocity of

53 km/sec [124]. It was possible to follow the transition of this absorption line into an emission line visible on both sides of the center. The dimensions of this formation and the dependence of its radial velocity on longitude show that it represents a spiral arm of complex structure found at a distance of 3 kpc and receding from the galactic center at a velocity of about 50 km/sec. Examination at high resolving power [125] shows that the $\lambda = 21$ cm line is visible in absorption at the maximum on the left-hand side of Fig. 12, while it does not occur in the spectrum of the right-hand maximum. Evidently, here the arm is broken up. The 3-kpc arm rotates about the center of the Galaxy with a velocity of approximately 50 km/sec [125a]. The general circular velocity at this distance, on the other hand, is approximately 250 km/sec. It can be assumed that material in the arm was thrown out from the central regions, and therefore retained a relatively low angular momentum. It should be noted that the OH absorption line, which will be discussed in more detail in Sec. 9, shows a complex structure with rapid radial motion at various points. The OH line is more sensitive than the 21-cm line and reflects the motion of finer fluctuations in the gas [125a] since here the absorption coefficient is considerably larger.

The central bulge is not visible in H_α because of strong absorption. Certain weak emission areas close to the direction of the galactic center are situated much closer to us. The H_α lines from these nebulae indicate a motion of approximately 10–15 km/sec. However, using a Fabry standard with high resolution, we see in directions close to the center emission of weak narrow lines shifted approximately 135 km/sec [126]. Such weak lines can be discovered because they do not coincide with the H_α lines of the geocorona.

Neutral hydrogen exists not only in galaxies, but in intergalactic space as well. Although its density there is undoubtedly small, its total mass can be great because of the huge expanse of the medium. An examination of this question and some observations were given by Field [127]. Hydrogen is easily observed in absorption in the direction of a powerful source, such as the radio galaxy Cygnus A. Owing to the expansion of the metagalaxy, at a given frequency only those atoms found at a certain distance can absorb. Therefore, the distance to the source does not enter the expression for optical depth. The latter has the form, analogous to Eq. (5.15),

$$\tau(\nu) = \frac{1.7 \times 10^3}{dV_r/dr} \frac{N_H}{T_s} = 1.7 \times 10^4 \frac{N_H}{T_s}. \tag{5.16}$$

Here $dV_r/dr \approx 0.1$ km/sec kpc is Hubble's constant. The spin temperature T_s, in contrast with galactic conditions, is not equal to the gas temperature. Excitation of hydrogen takes place in intergalactic space not by collisions, but by absorption of L_α quanta. If quanta are ejected from galaxies (for this to be possible the degree of ionization of intergalactic hydrogen must be less than 13 percent), calculations show that $T_s = 32°$ [128]. If the same L_α quanta are formed primarily by recombinations in the gas itself (for this the gas must be strongly ionized), the number of these quanta per unit volume is larger, and T_s is approximately the temperature of the gas itself, although to calculate this accurately is difficult. The gas temperature is also unknown, but it is hardly less than 10,000°.

A search for absorption lines in the spectrum of the radio galaxy Cygnus A led to a negative result [129]. This gives for an upper limit when $T_s = 32°$ (low hydrogen ionization) the value $N_H < 2.6 \times 10^{-5}$ cm^{-3} [129a]. This work also gives an upper estimate of $N_H < 2.1 \times 10^{-5}$ cm^{-3} obtained from attempts to measure the 21-cm line in emission from regions near the southern galactic pole. An increase in observational accuracy could eliminate the hypothesis of neutral intergalactic hydrogen—the absence of the line when highly sensitive measurements are made will mean that hydrogen is ionized and that its temperature is fairly high.

We comment briefly here on a method for estimating the concentration of intergalactic magnesium ions proposed by I. S. Shklovskii [129b]. If intergalactic hydrogen is neutral, then magnesium must be found in the first ionization state. Mg II has a strong resonance doublet $\lambda\lambda$ 2796–2803. Emission from Mg II, observed in distant quasistellar sources, is shifted by the red shift into the visual part of the spectrum. Intergalactic magnesium ions must absorb the continuous spectrum of the source over the whole spectral band between the laboratory and the shifted wavelengths. Such a depression has not yet been observed. It therefore follows that if the intergalactic gas is an H I region with normal chemical composition, the upper limit of the hydrogen concentration is $N_H < 2 \times 10^{-7}$ cm^{-3}. If, on the other hand, the intergalactic gas is ionized, its concentration cannot be determined by this method.

2. The Physical State of the Interstellar Gas

6. Interstellar Radiation and Ionization of the Interstellar Gas

The Spectral Density of Radiation in Interstellar Space

The physical state of the interstellar gas is determined primarily by its interaction with radiation. Therefore, to study the interstellar medium we must know in the first place the density and spectral composition of the total stellar emission.

The spectral density of radiation close to hot, bright stars (inside the H II zone) is described by Eq. (1.11) for the total spectrum. In H I regions, usually distributed far from hot stars, the mean intensity of interstellar radiation for $\lambda > 912$ Å can be determined by the summation of expressions of the type (1.11) for the emission that arrives from all the stars surrounding the region. Such a summation can be carried out by direct calculation [130] for the immediate neighborhood of the Sun. However, for studies of the physical state of the interstellar medium, the mean intensity of radiation must be known at various points in the Galaxy. Ultraviolet radiation in the region 912 Å $< \lambda <$ 2000 Å is of special interest, since this radiation ionizes the majority of elements of the interstellar gas in H I regions.

Since observational data on this region of stellar spectra are still insufficient, we shall use models of hot stars from theoretical calculations. Moreover, the method described below for calculating the mean intensity of emission is sufficiently general that it is easy to allow for possible changes of data in ultraviolet stellar radiation [131].

The intensity of stellar emission in H I regions is determined by two factors: (1) the distribution of stars in space, and (2) absorption by cosmic dust particles. Both are associated with extreme heterogeneity. This is particularly true of cosmic dust. If, however, only the mean values are of interest, some idealization can be introduced into the problem. It is often assumed that the dependence of the mean density of cosmic dust $\rho_g(z)$ and the mean number of stars $n_*(z)$

in the Galaxy at the height z are approximated by the expressions

$$\rho_g(z) = \rho_g(0)e^{-|z|/\beta_g}; \quad n_*(z) = n_*(0)e^{-|z|/\beta_*}, \tag{6.1}$$

where the value of the parameter β_g for cosmic dust is approximately 100–120 pc, and on the average β_* for stars of spectral classes O and B is approximately 50 pc, for stars of class A approximately 75 pc, and for red giants approximately 400 pc.

The absorption coefficient κ_g of light per unit volume by cosmic dust particles is proportional to their density. Therefore, if we assume that the dust in the galactical plane is distributed homogeneously, the optical depth along the path from an observer in the galactic plane to a point lying at a distance r on the line of sight along galactic latitude b is determined by the expression [132]

$$\tau_g = \int_0^r \kappa_g dr = \frac{\kappa_{0g}\beta_g}{|\sin b|}(1 - e^{-r|\sin b|/\beta_g}), \tag{6.2}$$

which is readily obtained from the first of Eqs. (6.1) by noting that $\kappa_g \sim \rho_g$ and $r|\sin b| = |z|$. Here κ_{0g} is the absorption coefficient in the galactic plane. The heterogeneous distribution of cosmic dust causes κ_{0g} to depend very strongly on the galactic longitude. However, as we shall see below, final expressions for the intensity of interstellar radiation depend only logarithmically on κ_{0g}. Therefore, the assumption that κ_{0g} is equal to some mean value is not in grave error. We shall therefore disregard the dependence of κ_g on frequency. In the spectral region close to the ultraviolet this dependence is relatively weak.

The average intensity of radiation from all stars distributed inside the solid angle $d\Omega$ with its axis directed toward latitude b and with its vertex at a point situated in the galactic plane is

$$\bar{I}_\nu(b)d\Omega = d\Omega \int_0^\infty I_\nu n_*(r|\sin b|)r^2 dr = \tfrac{1}{4}R_*^2 F_\nu n_*(0)d\Omega \int_0^\infty \exp\left[-\frac{r|\sin b|}{\beta_*} - \frac{\kappa_{0g}\beta_g}{|\sin b|}(1 - e^{-r|\sin b|/\beta_g})\right]dr. \tag{6.3}$$

Here, Eq. (1.1) is substituted in place of I_ν, and use is made of Eqs. (6.1) and (6.2). The integral in Eq. (6.3) is readily calculated for the case $\beta_g = 2\beta_*$. We obtain [131]

$$\bar{I}_\nu(b)d\Omega = \frac{R_*^2 F_\nu n_*(0)}{4\kappa_{0g}}\left[1 - \frac{\sin b}{\kappa_{0g}\beta_g}(1 - e^{-\kappa_{0g}\beta_g/|\sin b|})\right]d\Omega. \tag{6.4}$$

The condition $\beta_g \approx 2\beta_*$ is fulfilled as a first approximation for class O and B stars, and somewhat less well for class A stars. Since the ionization of elements in an H I region is produced by a radiation field with λ from 912 to approximately 2000 Å, the use of Eq. (6.4) for calculating ionization gives correct results for stars of the classes mentioned. It is not possible to use this formula for spectral studies of interstellar radiation in the visible and near-ultraviolet regions. The best results are given in this case by the approximation $\beta_g \approx \beta_*$, for which the integral (6.3) is also readily calculated:

$$\bar{I}_\nu(b)d\Omega = \frac{R_*^2 F_\nu n_*(0)}{4\kappa_{0g}}(1 - e^{-\kappa_{0g}\beta_g/|\sin b|})d\Omega. \tag{6.5}$$

Equations (6.4) and (6.5) immediately allow a determination of the anisotropy in interstellar radiation. Thus, comparing the average intensity of radiation from stars at $b = 0$ and $b = 90°$, we obtain from Eq. (6.4)

$$\frac{\bar{I}_\nu(90°)}{\bar{I}_\nu(0°)} = 1 - \frac{1 - e^{-\kappa_{0g}\beta_g}}{\kappa_{0g}\beta_g} \approx \tfrac{1}{2}\kappa_{0g}\beta_g, \tag{6.6}$$

since $\kappa_{0g}\beta_g \ll 1$. The anisotropy from Eq. (6.5) for $\kappa_{0g}\beta_g \ll 1$ is half this value. To avoid misunderstanding, we stress that $\bar{I}_\nu(b)$ is the radiation, averaged over 4π, from stars found in a given direction.

We can find from Eq. (6.4) the average intensity of interstellar emission ($d\Omega = \cos b\, db\, dl$) averaged over all directions.

$$\bar{\bar{I}}_\nu = \int \bar{I}_\nu(b)\frac{d\Omega}{4\pi} = \frac{R_*^2 F_\nu n_*(0)}{4\kappa_{0g}}\left[1 - \frac{1}{2\kappa_{0g}\beta_g} + \kappa_{0g}\beta_g \int_{\kappa_{0g}\beta_g}^{\infty} e^{-y}\frac{dy}{y^3}\right]$$

$$\approx \tfrac{1}{8}R_*^2 F_\nu n_*(0)\beta_g[\tfrac{3}{2} + \mathrm{Ei}(\kappa_{0g}\beta_g)]. \tag{6.7}$$

The latter equality is correct for $\kappa_{0g}\beta_g \ll 1$. In Eq. (6.5), the quantity $2[1 + \mathrm{Ei}(\kappa_{0g}\beta_g)]$ replaces the quantity in brackets in Eq. (6.7). Thus, the total stellar radiation from a certain spectral class close to the galactic plane can be represented in the form of diluted radiation from a single star with a mean dilution coefficient of

$$\overline{W}(0) = \tfrac{1}{8}[\tfrac{3}{2} + \mathrm{Ei}(\kappa_{0g}\beta_g)]R_*^2 n_*(0)\beta_g. \tag{6.8}$$

In calculating $\overline{W}$ we must choose a value for κ_{0g}. This value is unknown for the ultraviolet regions of the spectrum. However, considering the weak dependence of $\overline{W}$ on κ_{0g}, we assume that $\kappa_{0g} \approx 2$ kpc^{-1}; then $\kappa_{0g}\beta_g \approx 0.2$, $\mathrm{Ei}(0.2) \approx 1.22$, and the numerical coefficient in Eq. (6.8) is equal to 0.32. Changes by a factor of 2 in κ_{0g} increase or decrease Eq. (6.8) by not more than 20–30 percent.

We also estimate the mean dilution coefficient outside the galactic plane. For the sake of simplicity we shall restrict ourselves in the first place to the case when the observer is located above the galactic center at heights exceeding β_*, and in the second place to the assumption that the radiation results from only half the layer of stars. In other words, we shall consider that the radiation from stars found above the galactic plane is not reduced by absorption, while the radiation from stars found below the galactic plane is completely absorbed.

Therefore, we have for the mean dilution coefficient at height z_0

$$\overline{W}(z_0) = \tfrac{1}{4}R_*^2 n_*(0) \int_0^\infty e^{-z/\beta_*} dz \int_0^{R_0} \frac{2\pi r dr}{r^2 + (z - z_0)^2}. \tag{6.9}$$

Here $R_0 \approx 15$ kpc, the galactic radius. When $z_0 \gg \beta_*$, we obtain

$$\overline{W}(z_0) = \tfrac{1}{4}R_*^2 n_*(0)\beta_* \ln \frac{R_0^2 + z_0^2}{z_0^2}. \tag{6.10}$$

The numerical value of $\overline{W}(0)$ runs from 2×10^{-21} for stars of class O5 to 2×10^{-13} for stars of classes F to M. Correspondingly, $\overline{W}$(9 kpc) runs from 5×10^{-22} to 4×10^{-14}. Stars in spectral class B, for which $\overline{W}(0) \approx 4 \times 10^{-17}$ and $\overline{W}$(9 kpc) $\approx 2 \times 10^{-17}$, give the largest contribution to the total radiation. Further calculations require values of the quantity F_ν for stars of various spectral classes over all intervals of their spectra. As was already remarked in Sec. 1 (see also [64]), there is sharp disagreement between the theoretical value of F_ν and the data observed from rockets for B stars in the interval $1600 < \lambda < 2400$ Å. Therefore, the theoretical values of F_ν (the calculations of model stellar atmospheres) cannot be used here, while the observed values of F_ν for the total ultraviolet region of stars from different spectral classes are still not available. Earlier calculations [130, 131], which did not allow for this discrepancy, gave a very exaggerated value of the total ultraviolet stellar radiation. This discrepancy was then allowed for [131a, b]. Results of the most recent calculations are given in Table 6.1. Here $u_\lambda = (4\pi/\lambda^2)\bar{\bar{I}}_\nu \times 10^{-8}$ is the spectral density of the total

Table 6.1

λ (Å)	912	1000	1250	1500	1750	2000	2250	2500
$u_\lambda \times 10^{20}$ (erg/cm^3 Å)	100	250	350	370	370	430	2100	3100
$\bar{\bar{I}}_\nu \times 10^{20}$ (erg/cm^2 sec Hz)	0.07	0.20	0.43	0.66	0.93	1.4	8.4	15.4

stellar radiation per angstrom wavelength interval. In paper [131a] a value of $\bar{\bar{I}}_\nu$ was obtained that differed from the data in Table 6.1 by a few tens of percent (after excluding the spectral region $\lambda > 2000$ Å, where the difference was approximately three times as large, possibly because of ambiguities associated with the radiation from cooler stars). We note that calculations based on black-body curves for the radiation from individual stars yield a value for the total intensity that is from four to eight times that presented in Table 6.1, while the calculated values from models of hot stars [28–31] are approximately 50 times the results in Table 6.1 (for $\lambda = 1000$ Å).

The greatest contribution to interstellar radiation in the visible portion of the spectrum is given by stars of the later A subclasses and the red giants. The density and intensity of interstellar radiation over the total spectrum were calculated in papers [130, 131] (see also [1]). The total density (integrated over the spectrum) of interstellar radiation close to the galactic plane at an elevation of 9 kpc is equal to

$$u(0) \approx 3 \times 10^{-13}\ \mathrm{erg/cm^3}; \quad u(9\ \mathrm{kpc}) \approx 6 \times 10^{-13}\ \mathrm{erg/cm^3}. \quad (6.11)$$

Rather high values, obtained by direct calculation of the radiation flux from stars visible on the Earth, were used previously. Uncertainty is introduced here by the chance proximity of several hot stars, in particular Sirius.

In the previous calculations scattered interstellar emission was not allowed for. Therefore, if we denote the extinction coefficient by κ_{0g}, the values obtained for u_λ are low. On the other hand, if for κ_{0g} we substitute the true absorption coefficient, we obtain too high a value for the density of interstellar radiation. The difference between them is small, since in the first place the spectral density depends on κ_{0g} only logarithmically, and in the second place dust particles absorb preferentially in the ultraviolet spectral region and do not scatter light, that is, the extinction coefficient is close to the true absorption coefficient. The method for calculating the intensity of scattered interstellar radiation is described in Sec. 13.

Near H II zones, but outside them, the intensity of ultraviolet interstellar radiation, as V. V. Ivanov has remarked, is evidently greater after allowance is made for the continuous emission originating in H II zones from two-quantum transitions and the L_α radiation penetrating the H I regions. Apparently, the values in Table 6.1 or in Eq. (6.11) will be increased because of this by a factor less than 1.5, since in accordance with Fig. 1 the intensity of stellar L_c radiation, which supplies L_α and dual-photon quanta, is appreciably less than the radiation from the stars at $\lambda > 912$ Å.

The foregoing values for the intensity and the density of interstellar radiation are required for solving the following problems: (1) determination of the degree of ionization of various elements; (2) calculation of the mean energy imparted to an electron during each ionization; (3) elucidation of the role of light pressure; (4) calculation of the energy dissipated by relativistic electrons in the reverse Compton effect. The first two problems will now be considered.

Ionization of the Elements

The degree of ionization of elements is different in H II regions, where there are quanta with $\lambda < 912$ Å, from that in H I regions, where there are no such quanta. Therefore, a study of the ionization of elements must be carried out for H II and H I regions separately.

We recall that hydrogen in an H II zone is almost completely ionized ($N_p/N_1 \sim 10^3$). The next element to occur, helium, also forms two ionization zones, He II and He III. The areas in which hydrogen is ionized but helium is not can be called He I zones. The radius of helium ionization zones can also be calculated from Eq. (1.3), but we must now use the number of quanta emitted from a star with $\lambda \leqslant 504$ Å for the He II zone and with $\lambda < 227$ Å for the He III zone [133]. The probability of recombination for singly ionized helium atoms hardly differs from the corresponding hydrogen value, but for doubly ionized helium it is four times as much. Using models of the photospheres of hot stars, one can calculate [134] that, for example, a B0.5 and an O7.5 star have He II zones whose radii are, respectively, one-third and two-thirds the radii of the corresponding hydrogen ionization zones. The He III zones here are small. On the other hand, stars of the Wolf–Rayet type (O5) have an He III zone whose radius is more than 40 percent less than the radius of the hydrogen ionization zone, while the radii of the He II and H II zones are almost the same. This difference decreases with an increase in stellar temperature.

Let us note, however, that quanta with $\lambda < 504$ Å, which ionize helium, are absorbed by hydrogen. In fact, since the absorption coefficient of hydrogen is $a_1 \sim \nu^{-3}$, close to $\lambda \approx 504$ Å it is $a_{\mathrm{H}} = 6.3 \times 10^{-18}(504/912)^3 \approx 1.1 \times 10^{-18}$ cm^2, while the absorption coefficient of helium close to the ionization threshold is $a_{\mathrm{He}} \approx 7.6 \times 10^{-18}$ cm^2, which is not much greater than a_{H}. However, the total absorption by hydrogen in an He II zone, in spite of its relatively great abundance, is not large, since the degree of ionization of hydrogen in this region is substantially greater than the degree of ionization of helium.

This appears reasonable if we remember that the volume of the ionization zone is determined by the number of ionizing quanta emitted from a star. Since the H II volumes are much greater than the He II volumes for class B and late O stars, only a small number of $\mathrm{L_c}$ quanta are absorbed inside an He II zone. The degree of helium ionization inside He II and He III zones may be estimated by using the same simple relations that were employed in Sec. 1 for determining the degree of hydrogen ionization.

The other chemical elements do not usually form their own ionization zones. The fact is that when an element is of low abundance its absorption beyond the limit of the series does not essentially reduce the flux of ionizing radiation. Under these conditions the width of the transition zone is comparable with the radius of the ionization zone. It is true that in very dense nebulae the relative width of the transition zone may be reduced. For example, in nebula NGC 6523, an O^{++} region was detected surrounding the H II zone [135]. In the majority of cases, however, sharply defined ionization zones of even such relatively abundant elements as oxygen and nitrogen are not observed.

The degree of ionization of any element N^+/N is found from the equation

$$\frac{N^+N_{\mathrm{e}}}{N} = 4\pi \int_{\nu_i}^{\infty} a_i(\nu)\bar{I}_\nu \frac{d\nu}{h\nu} \Bigg/ \sum_{n=i}^{\infty} \alpha_n(T), \tag{6.12}$$

which has already been used for calculating the degree of ionization of hydrogen [see Eq. (1.14)]. Here $h\nu_i$ is the ionization potential and the subscript i refers to the principal quantum number of the ground state. In calculating the sum of the recombination coefficients, hydrogenlike values of $\alpha_n(T)$ can be used only for excited states. The corresponding recombination coefficient for the ground state must be calculated separately. However, at temperatures of the order of 10,000°, the use of hydrogenlike values does not lead to appreciable errors. For example, for elements in the second period of the periodic table, the probability of recombination to the ground state (in units of 10^{-13} cm^3/sec) is 1.7, 0.7, 0.8, and 0.7 for C, N, O, and F, respectively, while for a hydrogenlike atom the probability of recombination is 0.74. In the third period, sodium constitutes an anomaly, since its recombination coefficient to the ground state is 2×10^{-16} cm^3/sec, while for a hydrogenlike atom the corresponding recombination coefficient is 6×10^{-15} cm^3/sec. Several such anomalously small recombination coefficients are known, corresponding to small values of the absorption coefficient from the ground state. However, in these cases the probability of recombination to excited levels remains close to the probability of recombination of hydrogenlike atoms, especially for azimuthal quantum numbers $l \geqslant 3$ ($n \geqslant 4$). An analysis of the departure of absorption coefficients from excited levels for $l < 3$ and for small values of n from the corresponding values for hydrogenlike atoms [136] showed that, as a rule, the absorption coefficients, and hence the recombination coefficients, are somewhat less than for hydrogen, although the opposite is found. In any case, substitution for the sum in the denominator of Eq. (6.12) of $\alpha_t - \alpha_1$ for elements of the second period, $\alpha_t - \alpha_1 - \alpha_2$ for elements of the third period, and so forth and use of the hydrogenic values of these quantities can cause an error of not more than half an order of magnitude when $T \approx 10{,}000°$. Usually the uncertainty in the $\bar{I}_\nu$ values is greater. In calculating multiple ionization, the dependence of $\alpha(T)$ on Z must be taken into account. However, it must be emphasized that at low temperatures the role of the ground level is increased, so that the use of hydrogenlike values for the recombination coefficients can lead here to considerable errors.

Absorption coefficients for ionization of electrons from the $2p$ state were calculated by Seaton [18:121]. They can be expressed in the approximate form

$$a_i(\nu) = a_i(\nu_i)\left[\gamma\left(\frac{\nu_i}{\nu}\right)^s + (1 - \gamma)\left(\frac{\nu_i}{\nu}\right)^{s+1}\right], \tag{6.13}$$

where the values of the absorption coefficient at the ionization limit $a_i(\nu_i)$ and the parameters γ and s are given in Table 6.2.

There are no similar estimates for ionization from the states $3s$, $3p$, $4s$, and so forth. Values for the recombination coefficients are given for certain elements and ions in [1]. Recombination coefficients for these elements can be found there also.

The foregoing description is sufficient to calculate, or at least to estimate, the degree of ionization of various elements. Detailed calculations have up to now not been published, and therefore we are limited here only to general considerations.

If the ionization potential of an element is greater than 13.6 eV, the element can be ionized only in the H II region. The degree of ionization of a given element can be estimated very approximately by comparing it with the degree of ionization of hydrogen. For example, for the ratio of the degree of ionization of elements from the second period to the degree of ionization of hydrogen we obtain approximately from Eq. (6.12)

$$\frac{N^+}{N} \approx \frac{N_p}{N_1} \frac{a_i(\nu_i)}{a_1(\nu_c)} \frac{F_{\nu_i}}{F_{\nu_c}} \frac{1}{Z^2}, \tag{6.14}$$

where we replace the integral in the numerator of Eq. (6.12) by the product $a_i F_{\nu_i}$ close to the ionization limit. We also take into account here what was

Table 6.2

Element	$a_i(\nu_i) \times 10^{18}$	s	γ	Element	$a_i(\nu_i) \times 10^{18}$	s	γ
C II	4.3	2	1.67	O II	8.1	2	2.45
C I	10.5	1	1.9	O I	2.5	1	4.0
N II	6.8	2	2.06	Ne I	5.0	1	4.3
O III	3.4	2	1.30	Na II	8.0	2	4.2
N I	8.9	1	3.1				

previously said concerning recombination coefficients. For single-stage ionization, $Z = 1$; for double-stage, $Z = 2$; and so forth. It follows from Table 6.2 that the ratio of $a_i(\nu_i)$ to $a_1(\nu_c)$ exceeds unity by not more than twice (with the exception of O1 stars). The ratio F_{ν_i}/F_{ν_c} changes much more than this. In B1.5–B0.5 stars it is approximately $(\nu_c/\nu_i)^{5 \text{ to } 6}$. In very hot O9–O5 stars, this ratio is approximately $(\nu_c/\nu_i)^{0.4 \text{ to } 2}$. Thus, the degree of ionization of heavy elements in H II regions is less than the degree of ionization of hydrogen, especially in the case of double ionization, but it is still fairly high because $N_p/N_1 \approx 10^3$. It must be understood that Eq. (6.14) provides only an order-of-magnitude estimate. The degree of ionization for a number of the most abundant elements (O, N, Ne) was calculated under the assumption that stellar radiation beyond the Lyman limit can be described by a Planck function at some temperature T_0 [135a]. A critical value of T_{0c} was found at which, for the dilution coefficient $W = 10^{-17}$ and electron concentration $N_e = 10\ \text{cm}^{-3}$, the degree of ionization for some particular state was equal to 0.5 ($N^{++}/N^+ = 0.5$). A rough estimate of this quantity can be obtained from the relation $T_{0c} \approx \chi/8k$, where χ is the corresponding ionization potential. The connection between T_0 and the spectral class of the star remains, however, unclear.

Ionization is also possible by means of electron collisions at high temperatures. The degree of ionization in this case can be obtained as a first approximation from Eq. (1.21). The critical electron temperature for which $N^+/N = 0.5$ is determined by the relation $T_{0c} = \chi/11k$.

In H I regions only elements with ionization potentials of less than 13.6 eV are ionized, but, because of the low electron pressure, their degree of ionization here is greater than in H II zones. The fact is that, if in H II zones hydrogen supplies free electrons, then free electrons are produced in H I zones only by the ionization of carbon, sodium, magnesium, and other elements present in insignificant quantities. Therefore, in an H I zone there are at least three orders of magnitude fewer free electrons than in an H II zone, which brings about an increase in the degree of ionization in H I. Actually, this increase is not so marked, since the recombination coefficient increases at the same time because of a fall in temperature. We now estimate the degree of ionization of carbon in H I regions. The value of the integral in the numerator of Eq. (6.12) is readily calculated with the aid of Tables 6.1 and 6.2. The total recombination coefficient of carbon at low temperatures is [137]

$$\alpha_C(T) = \frac{1.26 \times 10^{-10}}{T^{1/2}}[1 - 0.19 \log T] \text{ cm}^3/\text{sec}. \tag{6.15}$$

We are still able to calculate from this equation temperatures in the H I region. Finally, Eq. (6.12) for carbon in interstellar space takes the form

$$\frac{N_C^+ N_e}{N_C} = \frac{0.072 T^{1/2}}{1 - 0.19 \log T}, \tag{6.16}$$

which when $T \approx 100°$ and $N_e \approx 10^{-3}$ cm^{-3} gives $N_C^+/N_C \approx 10^3$.

The values of the quantity N^+N_e/N for certain other elements were calculated in a series of papers [35, 138, 139]. A short selection of these calculations is given in Table 6.3.

Table 6.3

		Element					
		C	Na	Al	K	Ca	Ca^+
$\frac{N^+N_e}{N}$	$T = 100°$	1	0.3	100	8	130	0.002
	$T = 1000°$	5	1.5	400	40	650	.01

The anomalously weak ionization of sodium is explained by the fact that its absorption coefficient in the ground state, where all the atoms are found, is very small ($\sim 10^{-19}$ cm^2), while at the same time the recombination coefficients to the excited levels retain their hydrogenlike values. Potassium behaves similarly. It must also be noted that the degree of ionization of aluminum can be still greater because of autoionization. This effect takes place because the Al atom first makes a transition to the $3s3p^2$ level, which at 3424 cm^{-1} is above the ionization limit. The probability of this transition during absorption of a quantum is greater than the probability of direct ionization. Autoionization may increase the degree of ionization of aluminum by a factor of three [139].

Energy Acquired by Electrons During Ionization

It is well known that the heating of interstellar gas is due primarily to the following mechanism. During ionization, the electron acquires a kinetic energy equal to $h(\nu - \nu_i)$, where ν is the frequency of the absorbed quantum. On the other hand, during recombination, together with the free electron, kinetic energy of the order of kT is lost. If $h(\nu - \nu_i) > kT$, a constant thermal flow occurs from the radiation to the electron cloud. Therefore, for an analysis of the temperature of the interstellar gas it is important to determine $\bar{\varepsilon}$, the mean energy acquired by an electron during ionization. It equals, obviously, the ratio of energy absorbed during ionization to the number of absorbed quanta, that is,

$$\bar{\varepsilon} = h(\bar{\nu} - \nu_i) = h\left\{\frac{\int_{\nu_i}^{\infty} a_i(\nu)\bar{I}_\nu d\nu}{\int_{\nu_i}^{\infty} a_i(\nu)\bar{I}_\nu \frac{d\nu}{\nu}} - \nu_i\right\}. \tag{6.17}$$

These integrals are easy to evaluate for a known distribution of energy in the spectrum of interstellar radiation. If we approximate the data in Table 6.1 by the dependence $\bar{I}_\nu \sim \nu^{-3}$ (which is approximately satisfied for 1000 Å $< \lambda <$ 2000 Å), and use Eq. (6.13), we then have

$$\bar{\varepsilon} = h\nu_i\left\{\frac{\frac{\gamma}{s+2}\left[1 - \left(\frac{\nu_i}{\nu_c}\right)^{s+2}\right] + \frac{1-\gamma}{s+3}\left[1 - \left(\frac{\nu_i}{\nu_c}\right)^{s+3}\right]}{\frac{\gamma}{s+3}\left[1 - \left(\frac{\nu_i}{\nu_c}\right)^{s+3}\right] + \frac{1-\gamma}{s+4}\left[1 - \left(\frac{\nu_i}{\nu_c}\right)^{s+4}\right]} - 1\right\}. \tag{6.18}$$

In particular, for carbon we have $s = 1$, $\gamma = 1.9$, and $\nu_i/\nu_c = 0.83$. It follows from Eq. (6.18) that $\bar{\varepsilon} = 0.1\, h\nu_i = 1.6 \times 10^{-12}$ erg $= 1$ eV. We recall that carbon appears as a source of free electrons in H I regions.

In a similar manner we can also calculate the value of $\bar{\varepsilon}$ for an H II zone. Here the energy of the radiation is transmitted by the electron cloud during the ionization of both hydrogen and helium (in He II and He III zones). Therefore, strictly speaking, we should determine $\bar{\varepsilon}_{\mathrm{H}}$, $\bar{\varepsilon}_{\mathrm{He\,I}}$, and $\bar{\varepsilon}_{\mathrm{He\,II}}$, while during the calculation of $\bar{\varepsilon}_{\mathrm{H}}$ outside the helium ionization zones account must be taken of the absence of radiation with $\lambda <$ 504 Å. Such calculations have been made by V. I. Pronik [140]. The results are given in Table 6.4.

The value of $\bar{\varepsilon}$ has a relatively weak dependence on stellar temperatures. Data from this table will be used for the calculation of temperature in the interstellar gas.

Table 6.4

Spectral class	ε_H (10^{12} erg)	$\varepsilon_{He\,I}$ (10^{12} erg)	$\varepsilon_{He\,II}$ (10^{12} erg)
Wolf-Rayet	5.3	7.9	9.7
Wolf-Rayet	5.2	7.9	7.7
Wolf-Rayet	5.1	7.9	6.7
O5V	4.8	5.3	—
O7.5V	4.7	5.0	—
O9V	4.2	—	—
O9.5V	4.2	—	—
B0V	4.2	—	—
B0.5V	3.9	—	—
B1V	3.6	—	—
B1.5V	3.6	—	—

7. The Formation of Spectral Lines in the Interstellar Gas

Observational Conditions for Absorption Lines

Interstellar gas was discovered for the first time by Hartman in 1904 from absorption lines. Since then, lines have been observed from only five atoms (neutral Na, K, Ca, Fe, and ionized Ca II, Ti II, and Fe II) and two molecules (neutral and ionized CH and the cyanogen molecule CN). Such paucity in the number of detected absorption lines is connected with the specific peculiarities of their observation.

First of all, we must bear in mind that, because of the low gas density and the considerable dilution of the interstellar radiation field, practically all the atoms and ions are found in the ground state. If the ground state is split into sufficient close sublevels, then not only the ground state, but the sublevels as well, can be populated. In this way we can observe only those lines that are connected with transitions from the ground state. Meanwhile, for the majority of chemical elements, these series fall into the ultraviolet region of the spectrum.

Absorption lines can be observed only against the background of the spectrum of distant stars with high luminosities, because otherwise the number of atoms in the gas layer between the star and the observer is insufficient for the formation of noticeable absorption. Moreover, the interstellar line must not blend into the stellar lines formed in the stellar atmosphere. Therefore, it is best for the interstellar lines to be observed in spectra from distant O and B stars, since almost all the elements enumerated above are strongly ionized in the atmospheres of these stars and consequently do not yield the corresponding stellar lines.

In the future, when it becomes possible to observe the ultraviolet spectral region (912 Å $< \lambda <$ 2000 Å) with a higher resolution, the number of observed interstellar absorption lines will increase sharply. Radiation at wavelengths

$\lambda < 912$ Å is absorbed by interstellar hydrogen, and therefore will not be observed.

Naturally, there can be observed, or will be observed in the future, only elements of sufficient abundance. We shall consider, although somewhat arbitrarily, as sufficiently abundant elements the number of whose atoms is not less than one ten-millionth (10^{-7}) of the number of hydrogen atoms. Elements of the periodic table up to nickel are included, with the exception of Li, Be, B, Sc, and V.

The following limitations are connected with the state of ionization. The elements He, N, O, Ne, Cl, and A occur in their neutral state in H I regions in sufficient abundance. The remaining elements can be in neutral and singly ionized states and Ca and Ti in doubly ionized states. Consequently, in H I regions only lines from these ionization states can be formed. All elements are ionized in H II zones, many multiply ionized, while there are almost no neutral atoms (except He and Ne). Moreover, observational conditions for absorption lines in H II zones are unfavorable. First, only a relatively small number of interstellar clouds, about 5 percent, are ionized, and second, the resonance lines for the majority of ions fall in the spectral region $\lambda < 912$ Å. Rocket observation of interstellar lines of atoms and ions O I, S II, Al III, C II in the wavelength range 1260–1720 Å in the spectra of two stars, δ and π Scorpii, have recently been made [140a].

In the future we shall limit ourselves to lines that are formed in H I regions and have wavelengths $\lambda > 912$ Å.

Table 7.1 lists elements whose concentrations are greater than 10^{-7} times the hydrogen concentration. The wavelengths of the primary lines arising from the ground state are also indicated. Where wavelengths differ by only a few angstroms, only a mean value for the wavelength of the multiplet is given. The terms and transitions are quoted, together with the statistical weights multiplied by the value of the oscillator strengths (where these are known). Wavelengths of the observed lines are set in italics.

We consider now the data from this table.

(1) *The ground state with a single s electron.* Many lines in the visible spectrum are observed here. Obviously just as many Mg II lines must occur in the ultraviolet region of the spectrum. These lines are actually quite strong in spectra from distant galaxies, where the red shift transfers them to the visible part of the spectrum. It must be stressed that with all such elements it is best to observe a resonance doublet, since the following line has a much smaller oscillator strength. The complexity of the configurations of the ions of Ti, Mn, Fe, Co, and Ni explains the large number of lines comparable in strength in their spectra.

(2) *The s^2 configuration.* Lines of Mg I must occur in the ultraviolet spectrum; Ti I and Mn I do not occur, since these atoms are ionized; Fe and Ca are also ionized, but are more abundant, so that lines from their neutral atoms are seen.

Table 7.1

Element	Wavelength (Å)	Transitions	Terms	$\tilde{\omega}f$
Na I	*5896*	$3s$–$3p$	^2S–^2P	0.65
	5890	$3s$–$3p$	^2S–^2P	1.31
	3302	$3s$–$4p$	^2S–^2P	0.019
Mg II	2803	$3s$–$3p$	^2S–^2P	0.60
	2796	$3s$–$3p$	^2S–^2P	1.20
K I	*7699*	$4s$–$4p$	^2S–^2P	0.66
	7665	$4s$–$4p$	^2S–^2P	1.32
Ca II	*3968*	$4s$–$4p$	^2S–^2P	0.67
	3934	$4s$–$4p$	^2S–^2P	1.33
Ti II	*3384*	$4s$–$4p$	^4F–^4G	1.4
	3349	$4s$–$4p$	^4F–^4G	2.6
	3229	$4s$–$4p$	^4F–^4F	0.3
	3242	$4s$–$4p$	^4F–^4F	1.0
	3235	$4s$–$4p$	^4F–^4G	2.3
	3073	$4s$–$4p$	^4F–^4D	
Mn II	2606	$4s$–$4p$	^7S–^7P	
	2592	$4s$–$4p$	^7S–^7P	
	2574	$4s$–$4p$	^7S–^7P	
Fe II	2600	$4s$–$4p$	a^6D–z^6D	
	2587	$4s$–$4p$	a^6D–z^6D	
	2384	$4s$–$4p$	a^6D–z^6F	
	2374	$4s$–$4p$	a^6D–z^6F	
	2344	$4s$–$4p$	a^6D–z^6P	
Co II	2059	$3d$–$4p$	^3F–^3G	
	2012	$3d$–$4p$	^3F–^3G	
Ni II	1752	$3d$–$4p$	^2D–^2F	
	1742	$3d$–$4p$	^2D–^2D	
Mg I	2852	$3s^2$–$3s3p$	^1S–^1P	1.68
Al II	1671	$3s^2$–$3s3p$	^1S–^1P	
Ca I	*4227*	$4s^2$–$4s4p$	^1S–^1P	1.6
Ti I	5174	$4s^2$–$4s4p$	^3F–^3F	
	5064	$4s^2$–$4s4p$	^3F–^3D	
	4681	$4s^2$–$4s4p$	^3F–^3G	
Mn I	4031	$4s^2$–$4s4p$	^6S–^6P	3.7
Fe I	*3860*	$4s^2$–$4s4p$	^5D–^5D	0.07
	3720	$4s^2$–$4s4p$	^5D–^5F	0.11
	3440	$4s^2$–$4s4p$	^5D–^5P	
Co I	3527	$4s^2$–$4s4p$	^4F–^4F	
Ni I	3392	$4s^2$–$4s4p$	^3F–^3F	
	3369	$4s^2$–$4s4p$	^3F–^3D	0.5

Table 7.1—*cont.*

Element	Wavelength (Å)	Transitions	Terms	$\tilde{\omega}f$
C II	1335	$2p$–$2s2p^2$	^2P–^2D	
	1037	$2p$–$2s2p^2$	^2P–^2S	
Al I	3961	$3p$–$4s$	^2P–^2S	
	3944	$3p$–$4s$	^2P–^2S	0.066
	3092	$3p$–$4s$	^2P–^2D	
Si II	1817	$3p$–$3s3p^2$	^2P–^2D	
	1530	$3p$–$4s$	^2P–^2S	
	1263	$3p$–$3d$	^2P–^2D	
C I	1657	$2p^2$–$2p3s$	^3P–^3P	0.19
	1561	$2p^2$–$2s2p^3$	^3P–^3D	
Si I	2516	$3p^2$–$3p4s$	^3P–^3P	
N I	1200	$2p^3$–$2p^23s$	^4S–^4P	
	1135	$2p^3$–$2s2p^4$	^4S–^4P	
S II	1259	$3p^3$–$3s3p^4$	^4S–^4P	
O I	1302	$2p^4$–$2p^33s$	^3P–^3S	
	1039	$2p^4$–$2p^34s$	^3P–^3S	
	1025	$2p^4$–$2p^33d$	^3P–^3D	
S I	1807	$3p^4$–$3p^34s$	^3P–^3S	
	1473	$3p^4$–$3p^34s$	^3P–^3D	
	1388	$3p^4$–$3s3p^5$	^3P–^3P	

(3) *The p^1 configuration.* Here, almost all the lines, with the exception of Al I, fall into the ultraviolet region. The lines of C II and Si II must be extremely intense since both of these elements occur in H I zones in the singly ionized state.

The Al I doublet requires special consideration. Up to the present it has not been observed, although the line $\lambda = 3302$ Na I from an element of similar abundance, but with lower oscillator strength, is observed. It is evidently due to the fact that Na is more weakly ionized than Al, which is apparent from Table 6.3 [139]. Moreover, it is possible that the favorable conditions for observing sodium are explained by its unusually low degree of ionization.

(4) *The p^2 configurations.* These are not very favorable for observation of lines. Carbon and silicon are on the whole ionized.

(5) *The p^3 configurations.* The N I line must be extremely intense.

(6) *The p^4 configurations.* Oxygen lines must be very strong. Moreover, they must be observed against a background of the intense spectra from hot stars before the Lyman continuum. Therefore, conditions for observing O I absorption lines in the ultraviolet region are extremely favorable. Sulfur is ionized in H I regions.

(7) *The p^5 and p^6 configurations.* Lines of Cl I must be weak because of the

relatively low abundance of this element. Moreover, in H I zones chlorine is ionized. Lines of Ne I and A I are located too far in the ultraviolet region, so that data on these configurations are not included in the table.

Thus, we see that all elements with lines in the visible region of the spectrum and found in ionization states favorable for observation have been detected. A study of the most intense ultraviolet lines will no doubt soon be made.

Analysis of Absorption Lines

We shall examine the data that can be derived from observations of interstellar absorption lines. The absorption-line profile is determined by atomic motion, and the intensity (equivalent width), by the number of absorbing atoms in a column from the Sun to the star of 1-cm^2 cross section against the background of the spectrum in which the line is observed.

A detailed study of interstellar-line profiles presents a very difficult observational problem, requiring large telescopes. Of several works devoted to this problem, that of Beals [141] must first be mentioned, together with the extensive studies of Adams [142] obtained with high-dispersion spectra of interstellar lines from 300 class O and B stars, and the work of Münch [143], who studied Ca II and Na I lines in the spectra of 112 stars. It was shown that these lines usually consisted of several (up to seven) lines of different intensities. This indicates that interstellar gas is not a uniform medium with smooth density gradients, but consists of discrete clouds traveling with different velocities. The radial velocities of these clouds are determined from the displacements of the components, while from the number of components one can estimate the dimensions of the space occupied by the clouds and the average number found along the line of sight. It is necessary to remark, however, that the dispersion used up to the present time enables one to distinguish separate components belonging to clouds whose relative velocities are larger than 5–7 km/sec, or comparable with the dispersion in velocity. Therefore, only those components are split that belong to a small number of rapidly moving clouds.

It is also possible in an investigation using very high dispersion to analyze each separate profile, and, in particular, to determine the average square of the velocity of the gas in the separate clouds. By comparing the average values of the squares of the velocities according to interstellar lines of calcium or sodium with those values obtained from the hydrogen $\lambda = 21$ cm line, one can estimate the thermal and turbulent velocities. An analogous method as applied to the study of radio lines of molecules and hydrogen will be described in Sec. 9. Here, of course, the accuracy of this method is not high because of the specific differences in radio and optical observations.

As an example, let us consider the results of Takakubo [143a] for the values of the dispersion of internal velocities and optical depths for some components of interstellar lines in the spectrum of the star ε Orionis. Values of these quantities are presented in Table 7.2 for clouds also observed in the line $\lambda = 21$ cm. Since the thermal velocity of calcium atoms is much less than that of hydrogen atoms, we can assume that the velocity dispersion of the calcium atoms corresponds to the turbulent velocities, and here is of the order of 2.5–3.5 km/sec.

Table 7.2

Optical observations			Radio observations		
Optical depth	Radial velocity (km/sec)	Velocity dispersion (km/sec)	Optical depth	Radial velocity (km/sec)	Velocity dispersion (km/sec)
0.30	−13.7	3.5	0.026	−14.6	5.2
.66	−5.8	2.3	.132	−4.2	3.5
.98	+9.6	2.4	.287	+9.4	7.9

The excess velocity dispersion of hydrogen atoms above that of calcium atoms defines the thermal motion. These data are sufficiently accurate, however, for temperature calculations.

The measurement of interstellar absorption contours is difficult in the majority of cases, and therefore usually only the equivalent widths and the curve of growth, well known in astrophysics, are obtained.

Let I_0 be the intensity of stellar radiation in the spectral region near the absorption line in question. Calling N the number of absorbing atoms per unit volume, we have for the observed intensity inside the line profile

$$I_\nu = I_0 \exp(-\tau_\nu) = I_0 \exp\left[-\int_0^{r_0} a_1(\nu) N dr\right], \tag{7.1}$$

where r_0 is the distance to the star. Contrary to the case of stellar atmospheres, it is not necessary here to consider diffusion of radiation, because the flux of quanta scattered by the medium in a narrow cone along the line of sight is negligibly small in comparison with the radiation flux from the star. Equivalent widths can be calculated either on a wavelength scale or on a frequency scale. In the present case the wavelength scale is generally used. The equivalent width is the quantity

$$W_\lambda = \int_0^\infty \frac{I_0 - I_\lambda}{I_0} d\lambda = \int_0^\infty (1 - e^{-\tau(\lambda)}) d\lambda. \tag{7.2}$$

Neglecting the natural width of the line, which is valid for all interstellar lines if $\tau(\lambda_0) < 10^3$, we have for the absorption coefficient the expression

$$a_1(\lambda) = \frac{\pi e^2}{mc} \frac{f \lambda_0^2}{\Delta\lambda_D} f\left(\frac{\lambda - \lambda_0}{\lambda_0} c\right). \tag{7.3}$$

Here $\Delta\lambda_D$ is the Doppler half-width of the line, λ_0 is the wavelength at the center of the line, and $f(V)$ is the normalized dimensionless velocity-distribution function. If the line profile depends only on thermal atomic motion, then

$$f\left(\frac{\lambda - \lambda_0}{\lambda_0} c\right) = \frac{1}{\pi^{1/2}} \exp\left[-\left(\frac{\lambda - \lambda_0}{\Delta\lambda_D}\right)^2\right].$$

The Doppler half-width is equal to $\Delta\lambda_D = \lambda_0(2kT/m_a c^2)^{1/2}$, where m_a is the mass of the atom.

In the majority of cases the line profile formed in H I areas is determined not by thermal velocities (the temperature is low in H I), but by macroscopic motion. It is also possible in this case to use the Gaussian distribution function, substituting for $(2kT/m_a)^{1/2}$ the velocity dispersion, but, as has already been stated, it is better here to introduce Blaauw's [105] distribution function

$$f\left(\frac{\lambda - \lambda_0}{\lambda_0} c\right) = \tfrac{1}{2} \exp\left(-\frac{|\lambda - \lambda_0|}{\Delta\lambda_D}\right), \quad \Delta\lambda_D = \frac{\lambda_0}{c} \bar{V}, \tag{7.4}$$

where $\bar{V} \approx$ 7–8 km/sec is the mean cloud velocity. Defining the optical thickness τ_0 at the center of the line as

$$\tau_0 = \frac{\pi e^2 f \lambda_0^2}{mc\Delta\lambda_D} f(0) \int N dr = \frac{\pi e^2 f \lambda_0^2}{mc\Delta\lambda_D} f(0)\mathfrak{N}, \tag{7.5}$$

we put the expression (7.2) into the form

$$W_\lambda(\tau_0) = \int_0^\infty \left\{1 - \exp\left[-\tau_0 \frac{f\left(\frac{\lambda - \lambda_0}{\lambda_0} c\right)}{f(0)}\right]\right\} d\lambda$$
$$= \Delta\lambda_D \int_{-\infty}^{+\infty} (1 - e^{-\tau_0 f(x)/f(0)}) dx, \tag{7.6}$$

where the variable $x = (\lambda - \lambda_0)/\Delta\lambda_D$ is introduced in place of λ. For the distribution (7.4) this integral is expressed by the tabular function Ei(x) [143]:

$$W_\lambda(\tau_0) = 2\Delta\lambda_D[\gamma + \ln \tau_0 + \mathrm{Ei}(\tau_0)], \tag{7.7}$$

where $\gamma = 1.781$.

An explicit expression for Eq. (7.6) may be obtained for a single case. An analysis of the velocities of the separate components of the interstellar absorption lines catalogued in [142] shows that the number of clouds with high radial velocities is comparatively large, larger even than results based on the distribution function (7.4). If we adopt a distribution function falling still more slowly at high velocities [144],

$$f(V)dV = \frac{1}{2 \ln \eta} \frac{dV}{V}, \tag{7.8}$$

where η is some normalization constant characterizing the ratio of the number of high velocities to that of thermal velocities, the dependence of equivalent width on optical depth at the center of the line is determined by the expression [145]

$$W_\lambda(\tau_0) = 2\Delta\lambda_D \left\{\frac{\tau_0}{\eta} e^{\tau_0/\eta}[\mathrm{Ei}(\tau_0/\eta) - \mathrm{Ei}(\tau_0)]\right\}. \tag{7.9}$$

For a Maxwellian velocity distribution the integral in Eq. (7.6) can be calculated only by a numerical method [35].

In this case, the expression of $W_\lambda(\tau_0)$ for small values of τ_0 may be written as the series

$$W_\lambda(\tau_0) = \pi^{1/2}\Delta\lambda_D \sum_{n=1}^{\infty} \frac{(-1)^{n+1}\tau_0^n}{n!n^{1/2}} \left[f(x) = \frac{1}{\pi^{1/2}} e^{-x^2}\right]. \tag{7.10}$$

For large values of τ_0 we obtain the corresponding asymptotic expansion

$$W_\lambda(\tau_0) = 2\Delta\lambda_D(\ln \tau_0)^{1/2}\left(1 + \frac{0.2886}{\ln \tau_0} - \frac{0.1355}{(\ln \tau_0)^2} + \cdots\right). \tag{7.11}$$

In all cases, the equivalent width is proportional to τ_0 for $\tau_0 \ll 1$. When $\tau_0 \gg 1$, the increase in equivalent width with an increase in the number of atoms in the column along the line of sight is logarithmic, like the curves of growth in stellar atmospheres.

Thus, the single observed quantity, the equivalent width, depends on two parameters that characterize the interstellar gas: the velocity range of $\Delta\lambda_D$ and τ_0, which is proportional to the number of atoms along the line of sight. Only at low values of τ_0 does the equivalent width not depend on $\Delta\lambda_D$. Therefore, in order to determine both parameters, observations must be combined from at least two lines. To avoid introducing still another unknown, the relative abundance of the elements, Wilson and Merrill proposed the use of two lines from the same element [146]. The most convenient lines in this respect are the close doublets Na I and Ca II. In the future, we may use for this purpose the lines Mg II, C II, O I, and others. The method of investigation, identical in every case, may be illustrated with Na I and Ca II as an example.

Let τ_0 be the optical depth in the center of the weaker component. The optical depth at the center of the stronger component will then be twice as great, since the corresponding oscillator strengths are in the ratio 2:1. We obtain from Eq. (7.7) the ratio of the equivalent widths of lines from the two components, the so-called doublet ratio DR,

$$\mathrm{DR} = \frac{W(2\tau_0)}{W(\tau)} = \frac{\gamma + \ln 2\tau_0 + \mathrm{Ei}(2\tau_0)}{\gamma + \ln \tau_0 + \mathrm{Ei}(\tau_0)}. \tag{7.12}$$

The DR may be obtained similarly for the case of Maxwellian distribution. In this way, knowing the doublet ratio, we obtain τ_0 immediately, after which we find $\Delta\lambda_D$ from Eq. (7.7), and the number of atoms $\mathfrak{N}_a$ from Eq. (7.5).

The doublet-ratio method has, however, definite limitations. If $\tau_0 < 0.1$, then the DR ≈ 2 and hardly depends on τ_0. Then, considering Eqs. (7.10) and (7.5), we find

$$W_\lambda(\tau_0) = 2\Delta\lambda_D\tau_0 = \frac{\pi e^2}{mc} f\lambda_0^2\mathfrak{N}. \tag{7.13}$$

Consequently, the number of absorbing atoms can be found for weak lines, but the degree of velocity dispersion cannot be measured. On the other hand, if $\tau_0 > 3$, then the DR is almost unity and again hardly depends on τ_0. It follows from Eq. (7.11) that here the number of absorbing atoms is not determined, but in return $\Delta\lambda_D$ may be estimated as in the case of curves of growth in stellar atmospheres.

Both parameters $\mathfrak{N}$ and $\Delta\lambda_D$ may be found by the doublet-ratio method only when $0.3 < \tau_0 < 3$. Then the doublet ratio itself has the values $1.2 \lesssim \text{DR} \lesssim 1.8$.

When τ_0 in the Na I lines is large, instead of the strong line $\lambda = 5890$ Å another line $\lambda = 3302$ Å may be used if it is observed. Since the ratio of oscillator strengths is now 21.8, if we designate by τ_0 the optical thickness at the line $\lambda = 3302$ Å, we find for the ratio of the equivalent widths of the line the expression

$$\frac{W(5896)}{W(3302)} = \frac{W(21.8\tau_0)}{W(\tau_0)} = \frac{\gamma + \ln(21.8\tau_0) + \text{Ei}(21.8\tau_0)}{\gamma + \ln\tau_0 + \text{Ei}(\tau_0)}. \tag{7.14}$$

Further calculations are similar to those above.

Table 7.3 gives the values of DR and $W_\lambda/\Delta\lambda_D$ as functions of τ_0 for the distribution functions $\pi^{-1/2}e^{-x^2}$ [35] and $\frac{1}{2}e^{-|x|}$ [143]. It follows from the table that the difference in the doublet ratio for the different distribution functions is small. Therefore, data from observations of τ_0 depend little upon the choice of distribution function. A determination of the Doppler half-width is sensitive, however, to the choice of the form of the function $f(V)$.

We introduce here also simple approximate equations that allow us to change immediately from equivalent width measurements to number of atoms and mean velocities [143]. For example, for the D_1 and D_2 sodium doublet at large values of τ_0 and for the distribution function (7.4) we have

$$\begin{aligned} \log\mathfrak{N} &= 12.88 + \frac{0.301\,W(D_1)}{W(D_2) - W(D_1)} + \log[W(D_2) - W(D_1)], \\ \bar{V} &= 36.7[W(D_2) - W(D_1)] \text{ km/sec.} \end{aligned} \tag{7.15}$$

Here equivalent widths are expressed in angstroms.

From the doublet-ratio method, Strömgren [35] estimated for the first time the gas density in interstellar clouds and in the intercloud medium. The calculation technique is quite simple. The number of atoms in a given ionization state is found from the equivalent width. From this the number of atoms per unit volume is obtained by dividing by the stellar distance. From a consideration of ionization equilibrium, the total number of atoms of all ionization stages per unit volume is determined. At this stage of the calculation it is necessary to

Table 7.3

$\log \tau_0$	$\frac{1}{\pi^{1/2}} e^{-x^2}$		$\frac{1}{2} e^{-\lvert x \rvert}$	
	DR	$\log \frac{W_\lambda}{\Delta\lambda_D}$	DR	$\log \frac{W}{\Delta\lambda_D}$
−1.0	1.91	−0.76	1.95	−0.71
−0.6	1.84	−.39	1.88	−.33
−.4	1.77	−.21	1.83	−.14
−.2	1.66	−.04	1.75	+.04
.0	1.53	+.11	1.66	+.20
+.2	1.39	+.24	1.55	+.35
+.4	1.29	+.34	1.44	+.48
+.6	1.21	+.42	1.35	+.60
+1.0	1.13	+.52	1.24	+.76
+2.0	1.07	+.66	1.13	+1.02

know the number of free electrons, and therefore the calculation is somewhat arbitrary, although this is also true in the determination of when we divide $\mathfrak{N}$ by r_0. It is true that the free-electron density can also be estimated from the ratio of the intensities of the lines of ionized and neutral calcium. As the result of this work it has been shown that absorption lines originate in H I regions. The density inside the regions is about 10 cm^{-3}. The density of the intercloud medium is determined from the weak lines. Strömgren's estimate in this case gives $N_\mathrm{H} \approx 0.1\ \mathrm{cm}^{-3}$.

Another detailed study of interstellar lines by the doublet method, made by Münch [143], was devoted to the analysis of velocities. In this work, the spiral arms, which appear as the systematic, galactic-longitude-dependent mixture of the main components (Fig. 13), were investigated by means of interstellar-absorption lines. It is clear from Fig. 13 that the great majority of clouds are found at certain distances, that is, in the arms. The velocity dispersion of the clouds proves to be variable—in the Orion arm, closest to the center of the Galaxy, the velocity range is 20–50 percent greater. In the earlier work of Ientsch and Unsold [147], an increase in velocity dispersion with distance from the Sun was noted. We observe that an analysis of the λ = 21 cm line also leads to an increase in the velocity dispersion of clouds of interstellar gas as one approaches the galactic center. According to [107], the root-mean-square velocity of the clouds is given by

$$(\overline{v^2})^{1/2} = \left(\frac{756}{R} - 56\right)^{1/2} \text{ km/sec,} \tag{7.16}$$

where R (kpc) is the distance from the galactic center. A comparison of the observed interstellar absorption lines with hydrogen lines also allows a determination of relative abundances of the corresponding elements. This can be done by comparing optical depths in the centers of the lines if these are small,

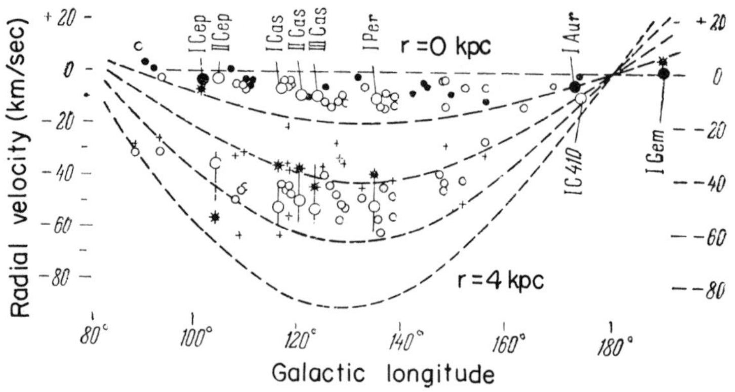

FIG. 13. Radial velocities of interstellar clouds at various galactic longitudes [18:21]. The broken curves correspond to different distances from the Sun. The clouds belong to the Orion and Perseus arms.

or by comparing equivalent widths if the optical depths are large; Eq. (7.5) or Eq. (7.6) is used, respectively. Such an investigation was carried out, in particular by Howard, Wentzel, and Magee [147a]. They determined by means of the doublet method optical depths in interstellar lines of calcium and sodium in 39 stars with $|b| > 15°$, and then compared them with the equivalent widths (7.6) for the hydrogen line $\lambda = 21$ cm in the same directions. The abundance of sodium relative to hydrogen was found from this study to be the same as the solar value, $N_{\mathrm{Na}}/N_{\mathrm{H}} \approx 2 \times 10^{-6}$, while the value for the relative calcium abundance was 60 times smaller than the normal value. The treatment of ionization was, however, very uncertain. There are also some differences in the average velocities, since measurements of 21-cm radiation give the velocity of the whole cloud, while optical absorption lines measure the velocity of a column of gas whose cross section is equal to the visual surface of the star.

Forbidden Lines

In addition to the emission spectra of hydrogen and helium, and the absorption lines of other elements, emission spectra of a number of elements can be formed under the conditions of the interstellar medium by means of electron collisions. Similar spectra are also observed in nebulae. The explanation of these spectra and the conditions for their formation were given by Bowen in 1927 [148, 149]. He showed that if any atom or ion possesses excited levels close to the ground state, that atom or ion can be excited on collision with thermal electrons, so that from reverse transitions to the ground state quanta are emitted that are observed as an emission spectrum.

Since the energy of electrons in interstellar space, even in H II zones, is relatively small, only comparatively low levels can be excited by electron collision. Therefore, lines can occur in emission only from the infrared to the near ultraviolet spectral region. In particular, the permitted lines listed in Table 7.1 can be excited by electron collisions. Only elements of low abundance have such lines, so that they are not observed. The more abundant elements (C, N, O, and so forth) and their ions possess terms from which downward permissible transitions are possible, but which are displaced from the ground state and therefore cannot be excited by electron collisions. However, these elements still have terms with the same principal quantum number, but with a different l value. These terms are metastable low-lying states and usually split into several levels.

Transitions between these terms in many cases give quanta in the visible or infrared region of the spectrum. It is not possible to observe absorption in these lines, since the absorption coefficient for forbidden transitions is very small. However, in the conditions of interstellar space at low gas densities and strong dilution of emission, the metastability of these levels does not prevent emission. In other words, collisions of the second kind, photoexcitation, and ionization rarely occur and the excited atom has a lifetime sufficient for a spontaneous downward transition.

The wavelengths (Å) and probabilities of spontaneous transitions (sec^{-1}) between the low-lying levels of atoms and ions of the more abundant elements are given in Table 7.4. The observed lines are given in italics. The table consists of data from Garstang [150–152], Pasternak [153], Naqvi and Talwar [154], and Czysak and Krueger [154a]. In case of disagreement, the latest data are given. The transition probabilities were calculated by the authors. The data in the table, generally speaking, have various degrees of accuracy.

We shall now consider electron configurations having low-lying electron levels.

(1) *Configurations with a single s electron* have only a split ground term with fine and hyperfine structure. Transitions between these states can occur only with abundant hydrogen, since the probability of these transitions is very small. The next levels are situated relatively far above the ground state, and downward transitions from them are permitted. Atoms and ions of Na I, Mg II, Ca II, and so forth that form this configuration have insufficient abundances to be seen in emission. However, we know already that all of them are detected or will

Table 7.4

1. Configuration p^1
Transition $^2P_{3/2}$–$^2P_{1/2}$

Ion	λ	A
C II	1,562,500	2.36×10^{-6}
N III	573,100	4.75×10^{-5}
O IV	258,700	5.19×10^{-4}
A II	892,800	1.26×10^{-5}
Si II	348,400	2.13×10^{-4}
S IV	105,300	7.73×10^{-3}

2. Configuration p^2

	$^3P_1$–$^3P_0$		$^3P_2$–$^3P_0$		$^3P_2$–$^3P_1$	
	λ	A	λ	A	λ	A
C I	6,100,000	0.78×10^{-7}	2,300,000	0.2×10^{-13}	3,690,000	0.27×10^{-6}
N II	2,037,000	2.1×10^{-6}	761,600	1.3×10^{-12}	1,217,000	7.4×10^{-6}
O III	881,800	2.6×10^{-5}	326,000	3.5×10^{-11}	517,000	0.97×10^{-4}
Ne V	241,600	1.3×10^{-3}	89,930	5.2×10^{-9}	143,300	4.5×10^{-3}
Si I	1,296,000	8.2×10^{-6}	447,800	1.5×10^{-9}	684,000	4.2×10^{-5}
S III	336,500	4.7×10^{-4}	120,100	4.7×10^{-8}	186,800	2.4×10^{-3}
Cl IV	203,700	2.1×10^{-3}	74,570	2.8×10^{-7}	118,000	8.2×10^{-3}

	$^1D_2$–$^3P_0$		$^1D_2$–$^3P_1$		$^1D_2$–$^3P_2$	
	λ	A	λ	A	λ	A
C I	9809	5.5×10^{-8}	9823	7.8×10^{-5}	9850	2.3×10^{-4}
N II	6527	4.2×10^{-7}	*6548*	1.0×10^{-3}	*6584*	3.0×10^{-3}
O III	4932	1.9×10^{-6}	*4959*	7.1×10^{-3}	*5007*	2.1×10^{-2}
Ne V	3301	1.9×10^{-5}	*3346*	0.14	*3426*	0.38
Si I	1588	2.6×10^{-6}	16,068	$.48 \times 10^{-3}$	16,455	2.8×10^{-3}
S III	8831	0.9×10^{-5}	*9070*	2.5×10^{-2}	*9532*	6.4×10^{-2}
Cl IV	7262	2.2×10^{-5}	7531	8.0×10^{-2}	8046	0.20

	$^1S_0$–$^3P_1$		$^1S_0$–$^3P_2$		$^1S_0$–$^1D_2$	
	λ	A	λ	A	λ	A
C I	4621	2.6×10^{-3}	4627	1.9×10^{-5}	8727	0.50
N II	3063	0.034	3071	1.6×10^{-4}	5755	1.08
O III	2322	0.23	2332	7.1×10^{-4}	4363	1.6
Ne V	1575	4.2	1593	6.8×10^{-3}	2972	2.6
Si I	6527	0.036	6590	0.0046	10,991	3.3
S III	3721	0.85	3797	0.016	*6312*	2.5
Cl IV	3118	2.6	3203	0.038	*5323*	3.2

Table 7.4—*cont.*

3. Configuration p^3

	$^2D_{5/2}$–$^4S_{3/2}$		$^2D_{3/2}$–$^4S_{3/2}$		$^2D_{3/2}$–$^2D_{5/2}$	
	λ	A	λ	A	λ	A
N I	*5201*	7.0×10^{-6}	*5198*	1.6×10^{-5}	1.2×10^6	5.3×10^{-9}
O II	*3729*	4.1×10^{-5}	*3726*	1.3×10^{-4}	4.76×10^6	1.5×10^{-7}
Ne IV	2442		2440		4.0×10^6	2.5×10^{-7}
S II	*6717*	4.6×10^{-4}	*6731*	1.8×10^{-3}	3.2×10^6	3.3×10^{-7}
Cl III	5517	1.0×10^{-3}	*5538*	7.0×10^{-3}	4.5×10^6	3.2×10^{-6}
Ar IV	*4711*	9.6×10^{-3}	*4740*	0.077	0.8×10^6	2.3×10^{-5}

	$^2P_{1/2}$–$^4S_{3/2}$		$^2P_{3/2}$–$^4S_{3/2}$		$^2P_{1/2}$–$^2D_{5/2}$		$^2P_{1/2}$–$^2P_{3/2}$	
	λ	A	λ	A	λ	A	λ	A
N I	3466	2.7×10^{-3}	3466	6.7×10^{-3}	10,398	0.031	10,407	0.048
O II	2471	2.3×10^{-2}	2471	5.7×10^{-2}	*7319*	.061	7330	.101
Ne IV	1609		1608		4417	.21	4720	.53
S II	*4076*	0.13	*4068*	0.34	10,372	.086	10,339	.20
Cl III	3353	0.37	*3343*	0.96	8550	.11	8502	.35
Ar IV	2869	0.97	2854	2.5	7332	.12	7263	.68

	$^2P_{3/2}$–$^2D_{5/2}$		$^2P_{3/2}$–$^2D_{3/2}$		$^2P_{3/2}$–$^2P_{1/2}$	
	λ	A	λ	A	λ	A
N I	10,398	0.054	10,407	0.025		
O II	7319	.115	*7330*	.061	10^8	6×10^{-11}
Ne IV	4714	.40	4720	.44	10^7	9×10^{-9}
S II	10,321	.21	10,287	.17	2×10^6	1.0×10^{-6}
Cl III	8424	.36	8434	.39	10^6	7.6×10^{-6}
Ar IV	7236	.67	7169	.91	55×10^5	5.2×10^{-5}

4. Configuration p^4

	$^3P_1$–$^3P_2$		$^3P_0$–$^3P_2$		$^3P_0$–$^3P_1$	
	λ	A	λ	A	λ	A
O I	632×10^3	8.9×10^{-5}	442×10^3	1.0×10^{-10}	1470×10^3	1.7×10^{-5}
Ne III	154×10^3	1.1×10^{-3}	108×10^3	2×10^{-8}	361×10^3	6.1×10^{-3}
Na IV	90.4×10^3	3×10^{-2}	63.5×10^3	2.4×10^{-6}	213×10^3	5.6×10^{-3}
S I	252×10^3	1.4×10^{-3}	174×10^3	7.1×10^{-8}	565×10^3	3.0×10^{-4}
Cl II	143×10^3	7.5×10^{-3}	100×10^3	4.8×10^{-7}	334×10^3	1.4×10^{-3}
Ar III	90×10^3	3.1×10^{-2}	63.7×10^3	2.7×10^{-6}	218×10^3	5.1×10^{-3}

Table 7.4—*cont.*

	$^1D_2$–$^3P_2$		$^1D_2$–$^3P_1$		$^1D_2$–$^3P_0$	
	λ	A	λ	A	λ	A
O I	*6300*	6.9×10^{-3}	*6364*	2.2×10^{-3}	6392	1.1×10^{-6}
Ne III	*3869*	0.20	*3968*	0.06	4012	1.2×10^{-5}
Na IV	3245	.71	3366	.21	3420	4.6×10^{-5}
S I	10,820	.027	11,306	8.0×10^{-3}	11,540	5.0×10^{-6}
Cl II	8580	.10	9125	0.29	9382	1.2×10^{-5}
Ar III	*7136*	.32	7751	.083	8036	2.9×10^{-5}

	$^1S_0$–$^3P_2$		$^1S_0$–$^3P_1$		$^1S_0$–$^1D_2$	
	λ	A	λ	A	λ	A
O I	2958	0.7×10^{-4}	2972	0.078	5577	1.28
Ne III	1794	5.1×10^{-3}	1815	2.2	3343	2.8
Na IV	1497	0.022	1523	8.3	2804	5.6
S I	4507	7.3×10^{-3}	4589	0.35	7725	1.8
Cl II	3583	1.8×10^{-2}	3675	1.34	6153	2.3
Ar III	3005	4.2×10^{-2}	3109	4.02	5191	3.1

5. Configuration p^5

	$^3P_{1/2}$–$^2P_{3/2}$	
	λ	A
Ne II	*127,900*	8.61×10^{-3}
Na III	73,300	4.57×10^{-2}
Mg IV	44,930	0.199
Cl I	113,500	1.23×10^{-2}
Ar III	69,840	5.29×10^{-2}
Ca IV	32,100	0.545

6. Configuration *Fe*

Transition	λ
$^6D_{7/2}$–$^6P_{9/2}$	26,000
$^6D_{5/2}$–$^6D_{9/2}$	15,000

be observed in absorption. The same applies to atoms or ions with s^2 configurations.

(2) *Configurations with a single p electron* have a doubly split ground term (see Table 7.4). These transitions give lines in the far-infrared region, which are as yet unobserved.

(3) *Configurations with two p electrons*. The structures of the terms for this case are shown in Fig. 14. The ground term is split, so that transitions to it give

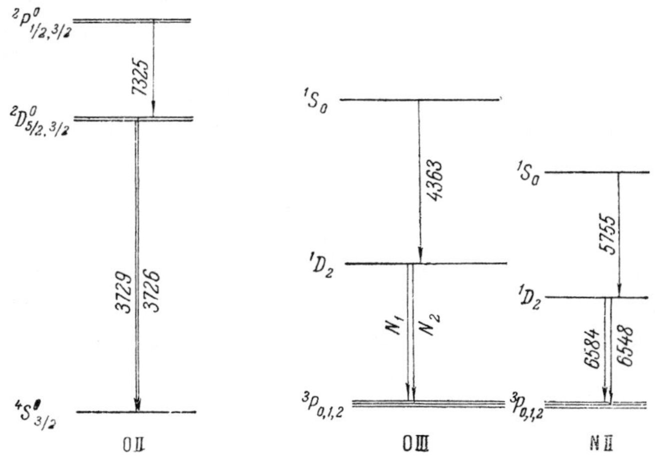

FIG. 14. Lower terms of ions O II, O III, N II.

triplets, of which one or two components are usually observed. We note that these three terms have the same principal quantum number, which explains their proximity. Therefore, from the possible combinations ^1S, ^1P, ^1D, ^3S, ^3P, ^3D, composed of orbital quantum numbers ($L = 0, 1, 2$) and spins ($S = 0, 1$) for the two p electrons, only $^1S_0$, $^1D_2$, $^3P_{0,1,2}$ satisfy Pauli's principle. Indeed, in the ^3D configuration, for example, L must equal 2 and S equal 1, that is, the two electrons must have identical azimuthal quantum numbers and parallel spins, which is impossible.

The relative positions of the terms are determined by Hund's rule: of terms formed from equivalent electrons, the lowest is that of maximum multiplicity (here ^3P), while the lowest level is that with the smallest internal quantum number J (here $^3P_0$). All the transitions for atoms and ions in this configuration of interest to us are given in Table 7.4. The $D \rightarrow P$ transitions in this configuration are called "nebular," the $S \rightarrow D$ "auroral," and the $S \rightarrow P$ "transauroral." Transitions between the levels of the ground term give lines in the far-infrared spectral region, which as yet are beyond the limits of possible observation. The transauroral transitions have a lower probability than the auroral; moreover, they fall as a rule in the ultraviolet spectral region. Lines from the auroral transitions of diffuse nebulae, except the Orion nebula and other very bright nebulae, are difficult to observe. This is owing in part to the low surface brightness of the nebulae, but mainly to the relatively low electron temperature, which makes the probability of exciting the third level small. The most intense are the nebular lines, which occur in almost all nebulae.

(4) *Configurations with three p electrons.* Of the possible combinations, the $^2P_{1/2,\,3/2}$, $^2D_{3/2,\,5/2}$, and $^4S_{3/2}$ levels (Fig. 14) satisfy Pauli's principle. Here the excited terms are split, so that all the transitions to the ground term have a doublet character. The probability of the auroral transitions $P \rightarrow D$ is much greater than the probability of the nebular transitions $D \rightarrow S$. In important ions the terms are situated relatively low, so that many lines in the nebular transitions are actually observed.

(5) *Configurations with p^4 and p^5 electrons.* Since the p^6 configuration constitutes a filled shell, the lower terms of the p^4 and p^5 configurations are equivalent to those already discussed for the p^2 and p^1 configurations, respectively (the sums of the quantities L, S, J of complementary configurations are equal to zero, that is, the quantum numbers of these configurations have identical absolute values). The sole difference consists in the fact that, in conformity with Hund's rule, the order of levels in the ground term is inverted; the lowest level now has the highest internal quantum number (for example, in the p^4 configuration the lowest level is $^3P_2$).

Population of Levels

In order to calculate the intensity of the forbidden emission lines, it is necessary to determine first the population of the excited levels, just as in the calculation of the recombination spectrum. Here the equilibrium equation is also the starting point for the solution of this problem. In this case, we must consider excitation and de-excitation from electron collisions, and spontaneous transitions. Recombinations and absorption from excited levels may be neglected owing to the low probability of these processes. The role of recombinations might be questioned, if only because this process would give, roughly speaking, emission lines that would be weaker than hydrogen lines the smaller the abundance of this element. At any rate, observations show that emission in forbidden lines is usually more intense than hydrogen emission.

To avoid introducing mathematical complications, we shall study a system of three levels in increasing order of energy. In other words, we shall consider, for example, the lower terms of the p^2 or p^4 system. Hence, the equilibrium equations for the second and third terms are:

$$\begin{aligned}(N_1 q_{12} + N_3 q_{32}) N_e + N_3 A_{32} &= N_2 A_{21} + N_2 N_e (q_{21} + q_{23}),\\ (N_1 q_{13} + N_2 q_{23}) N_e &= N_3 (A_{31} + A_{32}) + N_3 N_e (q_{32} + q_{31}).\end{aligned} \tag{7.17}$$

In accordance with the symbols in Appendix I, the quantity q_{ik} $(i < k)$ is the excitation coefficient and q_{ki} is the de-excitation coefficient. Equation (I.45) is the connection between them; it may be written in the form

$$q_{ik} = \frac{\tilde{\omega}_k}{\tilde{\omega}_i} q_{ki} e^{-h\nu_{ik}/kT}. \tag{7.18}$$

The solution of the system of equations (7.17) may be presented in the form

$$\begin{aligned} N_2 &= \frac{N_1 N_e \{q_{12}[A_{31}+A_{32}+N_e(q_{31}+q_{32})]+q_{13}(N_e q_{32}+A_{32})\}}{N_e q(N_e q_{32}+A_{32})+[A_{21}+N_e(q_{21}+q_{23})][A_{31}+A_{32}+N_e(q_{31}+q_{32})]}, \\ N_3 &= \frac{N_e N_1 q_{13}+N_e N_2 q_{23}}{A_{31}+A_{32}+N_e(q_{32}+q_{31})}. \end{aligned} \tag{7.19}$$

These rather cumbersome expressions can be simplified. We recall that the $3 \to 1$ transition is the transauroral and the $2 \to 1$ the nebular. Usually $h\nu_{31}$ is considerably greater than kT, and therefore the probability of excitation of the third level by electron collision is relatively small, that is, $q_{13} \ll q_{12}$. Moreover, even if the probability of spontaneous nebular transitions is not very high, the $2 \to 1$ transition from electron collisions may be neglected in comparison with spontaneous transitions. Hence, omitting the relatively small terms in Eqs. (7.19), we obtain

$$\frac{N_2}{N_1} = \frac{q_{12}N_e}{A_{21}+N_e q_{21}}; \quad \frac{N_3}{N_2} = \frac{q_{13}}{q_{12}} \frac{A_{21}+N_e q_{21}}{A_{31}+A_{32}+N_e(q_{31}+q_{32})}. \tag{7.20}$$

These relations become very simple when second-order collisions may in general be neglected, that is, if $N_e q_{21} \ll A_{21}$. Then,

$$\frac{N_2}{N_1} = \frac{q_{12}N_e}{A_{21}}, \quad \frac{N_3}{N_1} = \frac{q_{13}N_e}{A_{31}+A_{32}}. \tag{7.21}$$

The criteria of applicability of Eqs. (7.20) or (7.21) are easily formulated by introducing the critical electron concentration

$$N_e^{(k)} = \frac{A_{ki}}{q_{ki}}. \tag{7.22}$$

If the number of electrons N_e is much less than $N_e^{(k)}$, then for a given level the de-excitation by electron collisions is unimportant. In the opposite case it must be considered.

After the population of the various levels is determined, the emission coefficients ε_{ki} are found from the obvious formula

$$\varepsilon_{ki} = \frac{h\nu_{ik}}{4\pi} A_{ki} N_k. \tag{7.23}$$

In the particular case of Eqs. (7.21) we obtain

$$\varepsilon_{21} = \frac{h\nu_{21}}{4\pi} q_{12}N_eN_1, \quad \varepsilon_{31} = \frac{h\nu_{31}}{4\pi} q_{13}N_eN_1 \frac{A_{31}}{A_{32} + A_{31}},$$
$$\varepsilon_{32} = \frac{h\nu_{32}}{4\pi} q_{13}N_eN_1 \frac{A_{32}}{A_{32} + A_{31}}. \tag{7.24}$$

The first of Eqs. (7.24) does not in general include the probability of spontaneous transition, while in the second and third these probabilities enter as ratios whose sum is equal to unity. In other words, Eqs. (7.24) give identical emission coefficients both for forbidden and for allowed transitions. This is readily understood. If collisions of the second kind are unimportant, each excitation by electron collision results in a downward spontaneous transition and consequently in the emission of a quantum regardless of the probability of this transition.

The difference between forbidden and allowed transitions arises when it is not possible to neglect collisions of the second kind. In dense nebulae, these collisions are able to "quench" excited metastable levels, but are not able to de-excite excited levels having allowed transitions. In this case the intensity of the forbidden lines is reduced relative to the intensity of the allowed lines.

If excitations and downward transitions are induced primarily by electron collisions, a Boltzmann distribution is established with the electron temperature as a parameter. In this case, the population of the excited metastable level and consequently the energy of emission are proportional to the gas density. Under the conditions of interstellar space, collisions of the second kind never quench allowed levels, and therefore the population is proportional to the number of excitations, that is, to the square of the density. Therefore, with an increase of density, the intensities of allowed lines are strongly enhanced.

In this way, the intensity of emission in forbidden lines depends first of all on the coefficients of excitation by electron collisions. The problem of calculating the excitation and de-excitation coefficients of metastable levels, which is very important for the physics of interstellar gas and of nebulae, has been discussed in many papers, beginning with the first researches of Menzel and co-workers [9]. The problem was investigated most thoroughly by Seaton (see review in [18]). Since this problem is basically physical in character, it is described in Appendix II.

The de-excitation coefficient is expressed in terms of the so-called "collision efficiency" Ω_{ki} from Eq. (II.5):

$$q_{ki} = \frac{h^2}{2\pi m^2}\left(\frac{m}{2\pi kT}\right)^{1/2} \frac{\Omega_{ki}}{\tilde{\omega}_k} = \frac{8.63 \times 10^{-6}}{\tilde{\omega}_k T^{1/2}} \Omega_{ki}. \tag{7.25}$$

Tables of values for Ω are also given in Appendix II. The excitation coefficient is readily calculated from Eqs. (7.18) and (7.25), taking into account the condition $\Omega_{ki} = \Omega_{ik}$.

Knowing the excitation coefficients, we find at once the intensities of the forbidden lines. Actually, since absorption in the forbidden lines is negligible, we have for the intensities

$$I_{ki} = \int \varepsilon_{ki} dr = \frac{h\nu_{ik}}{4\pi} A_{ki} \int N_k dr. \tag{7.26}$$

Let us now suppose that collisions of the second kind are unimportant, and the population of the level is determined by the first of Eqs. (7.21). Then,

$$I_{ki} = \frac{h\nu_{ik}}{4\pi} \int q_{ik} N_i N_e dr = \frac{h\nu_{ik}}{4\pi} \frac{q_{ik} N_i}{N_p} \mathrm{ME}. \tag{7.27}$$

In the latter equation, it was assumed that the temperature, and consequently q_{ik}, are constant along the line of sight and that the ratio of the number of ions of a given element in the ground level to the number of protons is also constant and given. In this case the emission intensity in a forbidden line, as well as for hydrogen, depends only on the temperature and the emission measure. However, in the absence of the recombination spectrum of hydrogen, the dependence of Eq. (7.27) on temperature is expressed here much more strongly. Therefore, the ratio of forbidden-line intensities between themselves, and their ratio to hydrogen lines, is much more sensitive to temperature changes. It is very significant, then, that I_{ki} is proportional to the degree of ionization of an element and its relative abundance. Consequently, both can be measured from observations on forbidden lines.

Finally, it must be remembered that, with the approach of the electron concentration to the critical point determined by Eq. (7.22), the population of the excited level, and consequently the emission coefficient, diminish in comparison with Eqs. (7.24). This means that the density may be found from the ratio of the intensities of definite lines belonging to a single ion, since the values of $N_e^{(k)}$ are different for different levels.

The Determination of Temperature and Electron Concentration from Observations of Forbidden Lines

Several methods based on the foregoing considerations have been worked out for determining temperature, density, and chemical composition from data on observations of forbidden lines. The most suitable methods for determining temperature will be described in the following section. We shall limit ourselves here to a few examples.

The most convenient way to determine the temperature of bright nebulae is by comparing the intensities of nebular and auroral lines. The first method of this kind was the procedure of V. A. Ambartsumyan [155] for determining temperature from the intensity ratio of the auroral line λ 4363 of doubly ionized oxygen to the nebular doublet $N_1 + N_2$ ($\lambda\lambda$ 5007, 4959) of the same ion O III. Substituting in Eq. (7.26) the total expression for the population (7.19) and using values for the coefficients Ω and A taken from tables in Appendix II, we find from simple calculation

$$\frac{I(4363)}{I(\mathrm{N}_1 + \mathrm{N}_2)} = 13e^{-33{,}000/T}\left\{\frac{1 + 2.8 \times 10^3 T^{1/2}/N_e}{1 + 2.3 \times 10^5 T^{1/2}/N_e}\right\}. \tag{7.28}$$

When densities are not too great ($N_e < 10^5$ cm^{-3}), collisions of the second kind may be neglected and Eq. (7.28) assumes the simple form

$$\log \frac{I(4363)}{I(\mathrm{N}_1 + \mathrm{N}_2)} = -0.80 - \frac{14{,}300}{T}. \tag{7.29}$$

Obviously, from Eq. (7.27) we can obtain

$$\frac{I(4363)}{I(\mathrm{N}_1 + \mathrm{N}_2)} = \frac{\nu_{23} q_{13}}{\nu_{12} q_{12}} = \frac{\nu_{23}\Omega_{13}}{\nu_{12}\Omega_{12}} e^{-h\nu_{23}/kT}, \tag{7.30}$$

where substitution of numerical values leads to Eq. (7.29).

With an increase in electron density there is first of all a decrease in the population of the second level, for which $N_e^{(k)} \approx 2640 T^{1/2} \approx 2.6 \times 10^5$ cm^{-3} at $T \approx 10{,}000^\circ$, and the intensity ratio of the auroral and nebular lines is increased. With an electron concentration greater than $N_e^{(k)} = 2.5 \times 10^5 T^{1/2} = 2.5 \times 10^7$ cm^{-3}, the intensity ratio of the lines again proves to be independent of density:

$$\log \frac{I(4363)}{I(\mathrm{N}_1 + \mathrm{N}_2)} = 1.120 - \frac{14{,}300}{T}. \tag{7.31}$$

The reason for this is that at very high densities transitions between levels arise principally from the effect of electron collisions, and therefore the populations of these levels are determined by the Boltzmann distribution at the temperature of the electron gas. Substituting in Eq. (7.19) $A_{ki} \ll q_{ki}$ and q_{ik}, we obtain

$$\frac{N_2}{N_1} = \frac{q_{12}}{q_{21}} = \frac{\tilde{\omega}_2}{\tilde{\omega}_1} e^{-h\nu_{21}/kT}, \quad \frac{N_3}{N_1} = \frac{q_{13}}{q_{31}} = \frac{\tilde{\omega}_3}{\tilde{\omega}_1} e^{-h\nu_{31}/kT}, \tag{7.32}$$

and consequently for the intensity ratio

$$\frac{I(4363)}{I(\mathrm{N}_1 + \mathrm{N}_2)} = \frac{\nu_{32} A_{32} \tilde{\omega}_3}{\nu_{21} A_{21} \tilde{\omega}_2} e^{-h\nu_{32}/kT}, \tag{7.33}$$

which on substitution of numerical values leads to Eq. (7.31). In this way, when $N_e \ll 2.5 \times 10^5$ cm^{-3} or $N_e \gg 2.5 \times 10^7$ cm^{-3}, the temperature of the electron gas is determined from the intensity ratio of lines λ 4363 and $N_1 + N_2$ in a well-defined way. When $2.5 \times 10^5 < N_e < 2.5 \times 10^7$ cm^{-3}, the density must be known in order to determine the temperature.

Relations similar to Eq. (7.28) may be obtained for other elements as well. For example, for singly ionized nitrogen the intensity ratio of the nebular and auroral lines is

$$\frac{I(5755)}{I(6548 + 6584)} = 61.5e^{-21,500/T} \frac{1 + 340T^{1/2}/N_e}{1 + 1.7 \times 10^5 T^{1/2}/N_e}. \tag{7.34}$$

The combined use of Eqs. (7.28) and (7.34) permits density and temperature to be measured simultaneously. However, this is possible only when N_e is comparatively large (from 10^4 to 10^7 cm^{-3} for $T = 10{,}000°$). In Fig. 15(*a*, *b*, *c*) graphs are given representing the dependence on N_e and T [57] of the intensity ratios of the auroral and nebular lines (the numbers are near the curves). From these curves it is easy to develop a graphical solution of Eqs. (7.30) and (7.34). These curves show well the regions where the influence of temperature and density of the intensity ratios is dominant.

The high accuracy of temperature determinations by these methods is explained by the fact that the energy interval between the terms is comparable with the kinetic energy of the electrons.

For a density determination, it is more convenient to compare mutually the line intensities of a nebular doublet, for example, in the case of O II, the intensity of lines λ 3729 and λ 3726. This method is proposed by Seaton and Osterbrock [156]. Here $^2D_{5/2}$ and $^2D_{3/2}$ may be chosen for the second and third levels. All the previously obtained relations for three-level atoms are used in this case. Since the energies of terms *2* and *3* are now almost the same (that is, $h\nu_{32} \ll kT$), the ratio of the intensities $I(3729)$ and $I(3726)$ depends very little on temperature and is determined generally from the value of the electron density.

However, for more accurate calculations, account must also be taken of the third split term, $^2P_{1/2,\,3/2}$. Thus, in this problem, calculation of the populations of five levels, $^4S_{3/2}$ and the excited levels $^2D_{3/2}$, $^2D_{5/2}$, $^2P_{1/2}$, and $^2P_{3/2}$ is required. The corresponding system is analogous to Eq. (7.17) [156]. The rest of the procedure is similar to the three-level problem. In Appendix II, values of Ω_{ik} are given for each of the excited transitions in this case.

The ratio $I(3729)/I(3726)$ obtained from the solution of this problem is

$$\frac{I(3729)}{I(3726)} = 1.5 \frac{1 + 0.33 \times 10^{-19,600/T} + \dfrac{2.3N_e}{100\sqrt{T}} \times (1 + 0.75 \times 10^{-19,600/T} + 0.14 \times 10^{-39,200/T})}{1 + 0.40 \times 10^{-19,600/T} + \dfrac{9.9N_e}{100\sqrt{T}} \times (1 + 0.84 \times 10^{-19,600/T} + 0.17 \times 10^{-39,200/T})}. \tag{7.35}$$

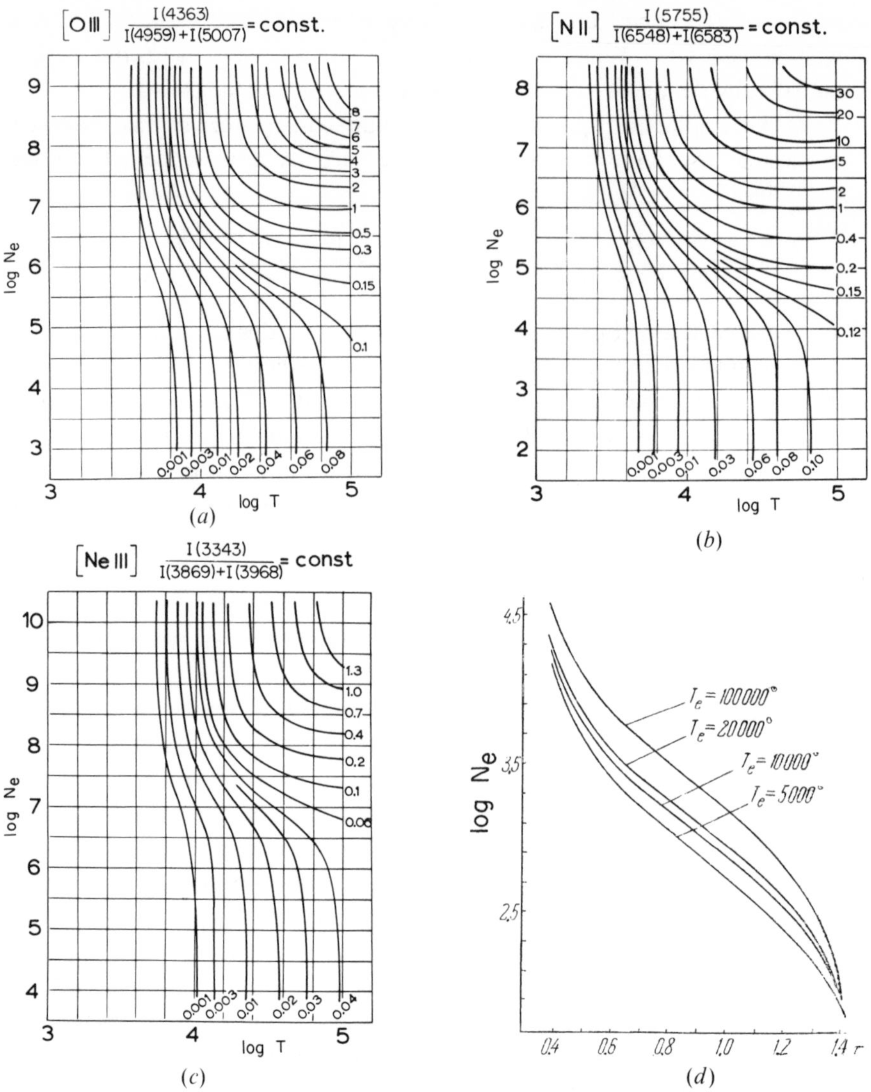

FIG. 15. Intensity ratios of auroral and nebular lines for (*a*) O III, (*b*) N II, (*c*) Ne III; (*d*) dependence of the ratio $r = I(3729)/I(3726)$ [O II] on N_e at various values of T [57].

We shall consider the limiting cases. Let $T \approx 10{,}000^\circ$ and the electron density be low ($N_e \ll 10T^{1/2} \approx 10^3 \text{ cm}^{-3}$). Then the population of the $^2D_{5/2}$ and $^2D_{1/2}$ sublevels is determined only from excitation by electron collisions from the ground state. Each excitation results in a downward spontaneous transition and the ratio of line intensities λ 3729 and λ 3726 must be proportional to the effective collision, that is,

$$\frac{I(3729)}{I(3726)} = \frac{\Omega(^2D_{5/2} - {}^4S_{3/2})}{\Omega(^2D_{3/2} - {}^4S_{3/2})} = 1.50. \tag{7.36}$$

The temperature terms in Eq. (7.35) allow for transitions to the ^2D level of atoms previously excited to the ^2P state. When $T \lesssim 10{,}000^\circ$, the effect of this process is small.

In the opposite case, when the electron density is high ($N_e \gg 10T^{1/2}$ cm^{-3}), de-excitation by electron collision of the $^2D_{5/2}$ and $^2D_{3/2}$ sublevels becomes effective and the population of these sublevels is determined from the Boltzmann distribution $N_3\tilde{\omega}_2 = N_2\tilde{\omega}_3$. The temperature does not enter into this relation as in the case of hydrogen hyperfine structure, since in this case the energy difference of the levels is small. Therefore, for the ratio of the intensities we have, using the data of Table 7.4,

$$\frac{I(3729)}{I(3726)} = \frac{\tilde{\omega}(^2D_{5/2})\,A(^2D_{5/2} - {}^4S_{3/2})}{\tilde{\omega}(^2D_{3/2})\,A(^2D_{3/2} - {}^4S_{3/2})} = 0.43. \tag{7.37}$$

Similarly, from Eq. (7.35), for large values of N_e, we obtain the somewhat smaller value 0.35. Observations of this ratio in dense nebulae give in fact the same value, independent of density, equal to 0.35, and not 0.43. In the opinion of the authors of this method [156], the calculations of the probability of spontaneous transitions may be in error, and in further calculations they used observational and not theoretical values.

Hence, we see that during changes of electron density from 100 cm^{-3} to, say, 10,000 cm^{-3} the intensity ratio of the components of the nebular doublet (O II) is appreciably altered, because, if the density of a nebula falls within these limits, it can be determined by this ratio. For ease of calculation, a graph is drawn, given in Fig. 15(*d*) [57].

This method may be used for the analysis of other pairs of lines of those ions that form a p^3 configuration, since here we need a single lower level and split excited states. However, in other cases the change in the ratio may be less and therefore the accuracy will be less. For example, for the O II ion, the ratio of the line doublet of the auroral transition $I(7320)/I(7330)$ over the entire density range varies between the limits 1.24 and 1.31.

Chemical composition can also be determined from the forbidden lines. For example, if the role of collisions of the second kind is negligible, then from the ratio of Eq. (7.27) applied to the intensity of hydrogen lines (Sec. 2) the relative concentration of the ion in question is immediately determined if we assume that the temperature is known, and, in the general case, the density. The next step in the determination of chemical composition is the conversion of the numbers of ions in a given state of ionization to the total number of atoms of the element. Unfortunately, it is not possible to make this conversion with sufficient accuracy, for a number of reasons. It is difficult to compute accurately the degree of ionization, because the energy distribution in stellar ultraviolet spectra is not well known. The degree of ionization and the electron density are different in various parts of a nebula, while it remains unclear how these values change from region to region. It is possible that the lines of different elements and ions originate in different regions.

Consequently, the chemical composition of nebulae and the interstellar medium cannot be determined with sufficient reliability. As the result of investigations into the chemical composition of nebulae, primarily planetary nebulae, Aller ([157] and private communication) suggests the values cited in Table 7.5

Table 7.5

Nebula	Element								
	H	He	C	N	O	Ne	S	Cl	Ar
Orion	1000	130	0.25	0.05	0.40	0.62	0.1	0.8×10^{-3}	4×10^{-3}
Planetary	1000	180	.41	.24	.63	.25	.06	2×10^{-3}	6×10^{-3}
B stars	1000	160	0.16	0.08	0.63	0.80	0.05	1.6×10^{-3}	8×10^{-3}

for the relative number of atoms in the Orion nebula and the average values in planetary nebulae; the chemical composition of class B stars is also given for comparison.

Further analysis may alter the figures in the last three columns, but the order of magnitude will hardly change.

8. The Temperature of the Interstellar Gas

The temperature of the interstellar gas is determined from the equilibrium between the energy obtained by the gas from stars and the energy dissipated by radiation. It is also possible to indicate a number of mechanisms for heating and cooling the interstellar gas. Cosmic rays transfer a portion of their energy to the interstellar gas during the ionization of its atoms. From the presence of currents in interstellar space, it is possible to generate a Joule effect. Gas motion damped because of viscosity, collisions of interstellar gas clouds between themselves, shells of novae and supernovae moving in interstellar space resulting in shock waves, all heat up the interstellar gas. On the other hand, the presence of cosmic dust and molecules in interstellar space enhances the rate of gas cooling. Thus the problem of calculating the gas temperature in interstellar space is fairly complex, chiefly because so many insufficiently known factors have to be considered.

As a first approximation, this problem can be simplified considerably. First, if a stationary or quasi-stationary condition is assumed, heating due to gas-dynamic processes, which will be considered in Chapter 5, need not be taken into account. Second, cosmic rays and interstellar dust influence the temperature only when the energy dissipated by the interstellar gas by radiation is small. Third, as we shall see below, the temperature dependence of the radiation of the interstellar gas itself is so strong that the addition of other rather inefficient heating and cooling mechanisms does not lead to a very marked change in the equilibrium temperature. In view of what has been said, we shall limit ourselves to a study of the energy balance of the interstellar gas, taking into account only interactions with the radiation. The energy-balance equation was obtained first by V. V. Sobolev [158] and also by Baker, Menzel, and Aller [9]. It was studied in more detail by Spitzer and Savedov [159–162], also by Seaton [163] and V. I. Pronik [140].

The considerations on which the energy-balance equation is based are very simple. During the ionization of an atom or ion by stellar radiation, the electrons acquire kinetic energy equal to the difference between the energy of the quantum and the ionization energy. The mean value of this energy, designated here by $\bar{\varepsilon}_0$, was calculated in Sec. 6. During the period between ionization and recombination, electrons lose this energy in the excitation of the forbidden lines of the most abundant elements, the excitation of hydrogen atoms, and emission from free-free transitions, and finally the remainder of their energy is radiated during recombination. If the role of collisions of the second kind is small, each excitation of a metastable or allowed level leads to the emission of a quantum and hence to a loss of energy from the electron gas. Collisions of the second kind return to the electron gas a portion of the energy expended during excitation.

The Energy Balance in H II *zones*

We shall first consider the equation of energy balance for H II zones and for diffuse nebulae. Here, as was shown in Sec. 6, the electrons obtain energy from stellar radiation during the ionization of hydrogen and helium. The energy acquired at each ionization is given in Table 6.4. It varies from 9.7×10^{-12} erg for He III zones around very hot stars to 3.6×10^{-12} erg around cooler class B1.5 stars. If the gas is found in stationary conditions, it may be assumed that this energy extends over a period between two recombinations. In other words, the electron acquires on the average the energy $\bar{\varepsilon}_0$ over the period $1/N_p \alpha t$.

We shall now consider energy losses during emission. The most effective is the cooling mechanism during excitation of the metastable levels by electron collision. As before, we denote the ground state by 1 and the kind of atom or ion by the superscript (a). Then the energy lost by one electron per second is

$$q_{1k} N_1^{(a)} \sum_{i<k} \frac{h\nu_{ik} A_{ki}}{A_{ki} + N_e q_{ki}}. \tag{8.1}$$

The term $N_e q_{ki}$ in Eq. (8.1) allows for the return of energy to the electron gas by collisions of the second kind. For comparison with the energy flux, we relate Eq. (8.1) to the interval of time between two electron recombinations. Then, for the average energy lost during excitation to the kth level of an atom or ion of species (a) we obtain

$$\varepsilon_k^{(a)} = \frac{N_1^{(a)}}{N_p} \frac{q_{1k}}{\alpha_t(T)} \sum_{i<k} \frac{h\nu_{ik} A_{ki}}{A_{ki} + N_e q_{ki}}, \tag{8.2}$$

where $\alpha_t(T)$ is the total recombination coefficient [Eq. (1.24)].

With the help of the data cited in the preceding section and in Appendix II for the coefficients q and A, values of $\varepsilon_k^{(a)}$ can be calculated from Eq. (8.2), since for a given level they depend only on the temperature, except for the factor $N_1^{(a)}/N_p$. Unfortunately, this factor, which characterizes degree of ionization and abundance of the element, is the most difficult to determine. After the calculation of $\varepsilon_k^{(a)}$, the energy lost by an electron during excitation of forbidden lines is found by direct summation over all atoms in all the ionization states and at all levels. Many calculations of this type have been made. For example, according to one of the most recent [135a], the sum $\sum_{a,k,i} q_{1k}^{(a)} h\nu_{ik} N_1^{(a)}/N_p$ over all ions in the H II zone, except hydrogen, is approximately constant and equal to 2×10^{-24} erg/cm^3 sec for $10^3 \lesssim T_e \lesssim 5 \times 10^3$. This sum increases slowly for higher temperatures, attaining a value of 5×10^{-24} erg/cm^3 sec for $T =$ 10,000°, 1×10^{-23} erg/cm^3 sec for $T =$ 15,000°, and 1.9×10^{-23} erg/cm^3 sec for $T =$ 20,000°. The complexity and cumbersome nature of these calculations and the uncertainty in the values of the factor $N_1^{(a)}/N_p$ lead us to think that instead of theoretical calculations it is preferable here to use a semiempirical method. It is easy to determine from observations on nebulae the ratio of the energy emitted in any forbidden line of the visible spectrum to the energy emitted in the H_β line. This ratio will evidently be equal to the ratio of the emission coefficient $\bar{\varepsilon}_a$ to the emission coefficient in H_β. The quantity in Eq. (8.2) is the emission coefficient in a forbidden line, but calculated for one electron and for a period of time between two ionizations (the usual emission coefficient is calculated for 1 cm^3/sec ster).

Such an emission coefficient can be determined formally for the H_β line. Using Eq. (2.17), we obtain

$$\varepsilon_\beta^{(H)} = \frac{\varepsilon(H_\beta)4\pi}{\alpha_t(T)N_e N_p} = \frac{h\nu_{42} 16 A_{42} h^3 b_4}{(2\pi mkT)^{3/2}\alpha_t(T)} e^{\chi_4/kT}. \tag{8.3}$$

We must emphasize that $\varepsilon_\beta^{(H)}$ *is not the energy dissipated by an electron during emission in the* H_β *line*. The chief form of energy connected with the hydrogen atoms is in the lines of the recombination spectrum and not the kinetic energy of the gas. Equation (8.3) is simply the result of the division of Eq. (2.17) by $N_e N_p \alpha_t$ and multiplication by 4π. The expression (8.3) depends only on the temperature and is readily computed. Using Eq. (I.24), we obtain

$$\varepsilon_\beta^{(\mathrm{H})} = \frac{1.1 \times 10^{-8} b_4 e^{\chi_4/kT}}{T\left[0.42 + \frac{1}{2}\ln\left(\frac{159{,}000}{T}\right) + 0.47\left(\frac{T}{159{,}000}\right)\right]}. \tag{8.4}$$

When $T = 10{,}000^\circ$, $\varepsilon_\beta^{(\mathrm{H})} \approx 3 \times 10^{-13}$ erg. When the temperature is decreased by 50 percent, $\varepsilon_\beta^{(\mathrm{H})}$ is increased by 25 percent, and when T is increased by 50 percent, $\varepsilon_\beta^{(\mathrm{H})}$ is decreased by 15 percent.

We can now write for the average energy loss during the excitation of forbidden lines

$$\bar{\varepsilon}_a = \varepsilon_\beta^{(\mathrm{H})} \frac{\Sigma I_a}{I(\mathrm{H}_\beta)} = \frac{3 \times 10^{-13}}{(T/10^4)^{0.3}} \frac{\Sigma I_a}{I(\mathrm{H}_\beta)} \text{ erg.} \tag{8.5}$$

Here we must sum over the intensities of all the forbidden lines. The well-known disadvantage is that we observe only lines in the visible and adjacent areas of the spectrum. However, the error from this is not very great. The probability of excitation of an ultraviolet line when $T \approx 10{,}000^\circ$ is small, so that the emission of these lines is unimportant. The probability of exciting infrared lines is much greater, but, with a few exceptions, the most abundant of the nebular ions do not have infrared nebular lines. Moreover, the energy carried away by each quantum in this case is less and the infrared lines on the average are more strongly quenched by collisions of the second kind. We note, incidentally, that in Eq. (8.5) the role of these collisions is automatically allowed for. Here we are considering only the energy that is actually removed from the nebula.

In this way, $\bar{\varepsilon}_a$ can be found quickly from observations of $\Sigma I_a/I(\mathrm{H}_\beta)$. The nebular lines of [O III] ($\mathrm{N}_1 + \mathrm{N}_2$), [O II] ($\lambda$ 3727), and particularly the near-infrared lines of [S III] (λ 9069–9095), usually offer the greatest contribution to this summation, the last contributing approximately as much as all lines in the visible spectrum.

It is not difficult to calculate the loss of energy by an electron during the excitation of hydrogen atoms. As has already been noted, the excitation of hydrogen by electron collisions is usually unimportant in estimating its luminosity. However, when the energy balance is being considered, this mechanism must be allowed for. Electron collisions excite emission in the unobserved L_α line much more frequently than in the observable Balmer spectrum.

We are considering here only losses of energy from excitations to the second level of a hydrogen atom. The average energy lost by an electron in this process during the time between ionizations (or recombinations) is

$$\bar{\varepsilon}_\mathrm{H} = \frac{h\nu_\alpha q_{12} N_1}{\alpha_t(T) N_\mathrm{p}}. \tag{8.6}$$

Using Table I.6, we find that, when $T = 10{,}000^\circ$, $\bar{\varepsilon}_\mathrm{H} = 4 \times 10^{-12} N_1/N_\mathrm{p}$ erg, and, since $N_1/N_\mathrm{p} \approx 10^{-3}$, in this case the losses on excitation of the L_α line by electron collision are small. However, when $T = 20{,}000^\circ$, we have $\bar{\varepsilon}_H = 2.4 \times 10^{-9} N_1/N_p$ erg, and here the losses in hydrogen excitation are comparable with the losses by excitation of the forbidden lines. With a further rise in temperature, these losses increase rapidly. However, the degree of ionization is increased with rising temperature, so that the increase of $\bar{\varepsilon}_H$ is not as significant. Thus the excitation of hydrogen by electron collisions need not be considered in the energy-balance equation if $T < 15{,}000^\circ$, but must be considered if $T > 15{,}000^\circ$. In the case of relatively large concentrations of neutral hydrogen atoms, calculation of these losses is necessary at much lower temperatures.

Similarly, the energy of the electron can be consumed in the excitation of allowed lines in the most abundant elements.

The energy losses due to emission during free-free transitions is determined by Eq. (2.25). In the period between two recombinations an electron dissipates the energy

$$\bar{\varepsilon}_{ff} \approx \frac{1.4 \times 10^{-27} T^{1/2}}{\alpha_t(T)} \approx 8.3 \times 10^{-17} T \text{ erg}. \tag{8.7}$$

During the calculation of Eq. (8.7), it was assumed that in the expression for $\alpha_t(T)$ in the brackets in Eq. (I.24) $T = 10{,}000^\circ$.

Electrons usually recombine after expending a considerable part of their energy. The average value of the kinetic energy lost by an electron during recombination to all levels in a hydrogen atom is obviously equal to

$$\begin{aligned} \beta_t(T) &= \sum_{n=1}^{\infty} \tfrac{1}{2} m \int_0^\infty v^3 \sigma_n^i(v) f(v) dv \\ &= 2\pi m \left(\frac{m}{2\pi kT}\right)^{3/2} \int_0^\infty \sum_{n=1}^{\infty} \sigma_n^{(i)}(v) v^5 e^{-mv^2/2kT} dv. \end{aligned} \tag{8.8}$$

With the help of Eq. (I.21), the integral in Eq. (8.8) is easily calculated. A corresponding calculation made by Seaton [164], in which he used more accurate values for the Gaunt factors, produced the approximate equation, similar to Eq. (I.24),

$$\begin{aligned} \beta_t(T) &= 5.20 \times 10^{-14} Z \left(\frac{\chi_1 Z^2}{kT}\right)^{1/2} \\ &\quad \times \left\{ -0.071 + \tfrac{1}{2} \ln \left(\frac{\chi_1 Z^2}{kT}\right) + 0.64 \left(\frac{kT}{\chi_1 Z^2}\right)^{1/3} \right\} kT. \end{aligned} \tag{8.9}$$

Dividing $\beta_t(T) N_e N_p$ by $\alpha_t(T) N_e N_p$, we obtain the average value for the energy $\bar{\varepsilon}_r$ lost by electrons during recombination over a period of time between two recombinations. This ratio for hydrogen is

$$\bar{\varepsilon}_r = \frac{\beta_t(T)}{\alpha_t(T)} = kT \frac{-0.071 \times \frac{1}{2} \ln(\chi_1/kT) + 0.64 (kT/\chi_1)^{1/3}}{0.43 + \frac{1}{2} \ln(\chi_1/kT) + 0.47 (kT/\chi_1)^{1/3}} \approx 0.8kT \tag{8.10}$$

over a wide temperature range, from 3,000° to 30,000°. We note that the average kinetic energy of an electron is $\frac{3}{2}kT$, that is, twice as great as Eq. (8.10). Thus, the lower the energy of an electron, the greater is the probability of its recombination. Therefore, the average energy of recombining electrons is less than the average kinetic energy of the electron gas.

All the data have now been assembled for establishing the equilibrium equation. It follows from Eqs. (8.5), (8.7), and (8.10) that $\bar{\varepsilon}_0 = \bar{\varepsilon}_a + \bar{\varepsilon}_{ff} + \bar{\varepsilon}_r$, or

$$\bar{\varepsilon}_0 = \frac{3 \times 10^{-13}}{(T/10^4)^{0.3}} \frac{\Sigma I_a}{I(H_\beta)} + 2 \times 10^{-16}T. \tag{8.11}$$

The excitation of hydrogen atoms by electron collisions is not taken into account here.

Three parameters enter Eq. (8.11). If two of them are known, the third is determined by solving the equation. In particular, it may be used to find the temperature if $\bar{\varepsilon}_0$, which is determined from the measurement of $\Sigma I_a/I(H_\beta)$, is known. Of even greater interest is the case when observations determine not only the sum $\Sigma I_a/I(H_\beta)$, but the temperature as well (for example, from the [O III] lines). Then the value of $\bar{\varepsilon}_0$ is found from Eq. (8.11), which up to now can be calculated only from models of the photosphere. The nomogram of Eq. (8.11), making its use easier, is given in Fig. 16 [140].

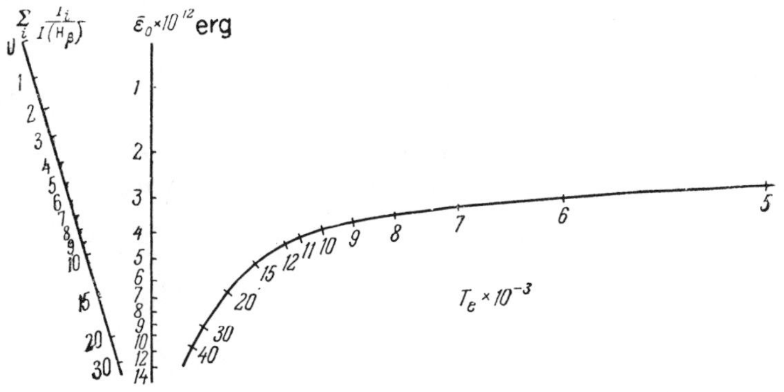

FIG. 16. Nomogram for the determination of temperature, given $\bar{\varepsilon}_0$ and the total relative intensity of forbidden lines [140].

We note that the values of $\bar{\varepsilon}_0$ calculated from Eq. (8.11) for temperatures obtained from observation are systematically somewhat lower than those calculated from model spectra of hot stars. It is possible that not all cooling mechanisms were taken into account here.

In more recent calculations of the temperature of H II zones, the excitations of sublevels of the ground state of the most abundant ions have been taken into account [164a], making use of the new models of high-temperature stars [164b]. It was shown that in rarefied nebulae the temperature is usually lower than in the normal, denser planetary and diffuse nebulae, but in any case it was higher than 5000°. These results were confirmed by observations of O'Dell [164c].

The Determination of Temperature from Observations

Temperatures in H II zones, including their subdivision into He I, He II, and He III zones, are obtained from the solution of Eq. (8.11) and are presented in Table 8.1. Account is also taken here of the stratification of ionization of elements that give forbidden lines [140].

Table 8.1

Spectral class	He I	He II	He III
Wolf-Rayet	8000	9700–10300	14000–16000
O5–O7.5	7600	8400–8800	—
O9–B0	7000	—	—

The temperature in an H II zone is on the average close to 10,000°. This value is usually taken as standard.

A method was described in the preceding section for determining the temperature of nebulae from the ratio of the intensities of auroral and nebular lines. However, this method is inapplicable to ordinary diffuse nebulae, and especially to the surrounding H II zones, since the third level, responsible for the auroral lines, is hardly excited, owing to the relatively low temperature. Here only the intensities of nebular lines of different ions can be compared. For example, Seaton [163] proposed to determine the temperature from the relative intensities of the lines O I, O II, N I, and N II. Since the ionization potentials of nitrogen and oxygen are close together, we have the equation

$$\frac{N_{\text{O II}}}{N_{\text{O I}}} = \frac{N_{\text{N II}}}{N_{\text{N I}}}. \tag{8.12}$$

Now in accordance with the equations in Sec. 7, we express the intensities of nebular lines in terms of the number of atoms and ions, and thus form the ratio

$$\frac{I[\text{N II}] \times I[\text{O I}]}{I[\text{O II}] \times I[\text{N I}]} = \frac{I(6548 + 6584)\, I(630 + 6364)}{I(3729 + 3726)\, I(5198 + 5102)}$$

$$= 0.146 \frac{1 + 3.4 \times 10^{-4} N_e}{1 + 10^{-3} N_e T^{-1/2}} (1 + 1.6 \times 10^{-2} N_e T^{-1/2}) \times 10^{-9350/T}. \tag{8.13}$$

Here the degree of ionization and the relative abundance of the elements are excluded by the assumption (8.12). The temperature can be determined from the observed relative intensities at a known electron concentration, from Eq. (8.13). The method is especially suitable for nebulae with weak ionization.

Another method, proposed by V. I. Pronik [140], is based upon comparison of the [O II] and [O III] nebular-line intensities. In diffuse nebulae and H II zones, oxygen occurs only in singly and doubly ionized states, since the ionization potentials of H I and O I are close together, while the ionization potential of O III is very high. Ions of O IV can also exist where helium is doubly ionized, that is, in very limited regions of an H II zone. Hence we have the obvious equality

$$\frac{N_{\mathrm{O\,II}} + N_{\mathrm{O\,III}}}{N_{\mathrm{p}}} = \frac{N_{\mathrm{O\,I}}}{N_{\mathrm{H\,I}}}. \tag{8.14}$$

Expressing the number of oxygen ions $N_{\mathrm{O\,II}}$ and $N_{\mathrm{O\,III}}$ in terms of the intensities of lines $\lambda = 3727$ Å and $N_1 + N_2$, respectively, by the use of Eq. (7.30), and the number of protons N_{p} in terms of the intensity of H_β from Eq. (2.19), the condition (8.14) can be written in the form

$$\frac{I(3727)\tilde{\omega}(\mathrm{O\,II})}{I(\mathrm{H}_\beta)\Omega(\mathrm{O\,II})} e^{h\nu(\mathrm{O\,II})/kT} + \frac{I(N_1 + N_2)}{I(\mathrm{H}_\beta)} \frac{\tilde{\omega}(\mathrm{O\,III})}{\Omega(\mathrm{O\,II})} e^{h\nu(\mathrm{O\,III})}$$

$$= \frac{N_{\mathrm{O\,I}}}{N_{\mathrm{H\,I}}} \frac{h\nu_{42}A_{42}h^3 16 b_4}{(2\pi mkT)^{3/2}} \frac{8.63 \times 10^{-6}}{T^{1/2}} e^{\chi_4/kT}. \tag{8.15}$$

At low densities, the coefficients $I(3727)/(H_\beta)$, $I(N_1 + N_2)/I(\mathrm{H}_\beta)$, and $N_{\mathrm{O\,I}}/N_{\mathrm{H\,I}}$ in this equation are functions of temperature only. If the intensities of lines $\lambda = 3727$ Å and $N_1 + N_2$ are determined from observations and used to obtain the relative abundance of oxygen, the electron temperature may be found from Eq. (8.15). A nomogram that permits a convenient solution of Eq. (8.15) is shown in Fig. 17. It was constructed for various values of $N_{\mathrm{O\,I}}/N_{\mathrm{H\,I}}$. The relative abundance of oxygen is generally taken as $N_{\mathrm{O\,I}}/N_{\mathrm{H\,I}} \approx 6 \times 10^{-4}$.

Possible errors in this value lead to only a small change in the value obtained for the temperature. At high densities, collisions of the second kind must be considered. A table of corrections that allows for this effect is given in [57].

The principal advantage of this method is that it permits an investigation of how temperature varies inside emission regions, and, in particular, of the dependence of temperature on the distance to the star. In addition, if we assume that the chemical composition of all nebulae is the same, we can determine from stellar spectra the dependence of the temperature on the density of nebulae. This analysis is of great interest, since the temperature of H II zones may vary because of differences in stellar spectra (different values of $\bar{\varepsilon}_0$), a change in the role of collisions of the second kind, the influence of the degree of ionization, and so forth.

Using this method, V. I. Pronik [135] discovered a number of cases of temperature changes with distance to the star. The electron temperature is somewhat higher, as a rule, in the vicinity of a star. This effect can be explained by the dependence of cooling efficiency on the degree of ionization of N and O.

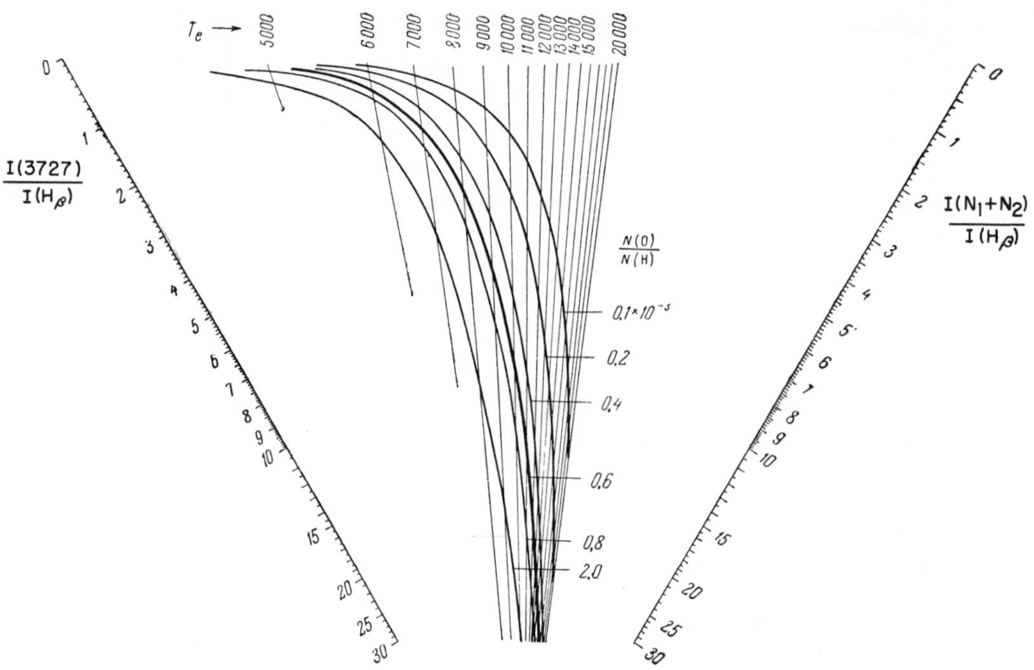

FIG. 17. Nomogram for determining T from the relative intensities of [O II] and [O III] [140].

A similar investigation of temperature was carried out for H II regions found in spiral arms and in the galactic center [135a]. This study was based on the observation of the pronounced difference in the ratio of intensities of the close lines H_α and [N II] λ 6584 in spiral arms, where it was approximately equal to 3, and in central parts of the Galaxy, where this ratio was diminished to approximately 0.1. This can be explained by the change of electron temperature in various regions, since the intensity of forbidden lines is very sensitive to temperature change. These authors [135a] found, on the basis of their calculations and in agreement with observations, that in the spiral arms $T = 6500°$, and in the galactic nucleus, T_e is from 20,000 to 30,000°. The authors of this work propose that such a pronounced difference in physical conditions is explained by corpuscular radiation from stars in the centers of galaxies. However, this may be connected with the flow of gas from central regions, which is observed in many galaxies.

Temperature in H I *Regions*

We now turn to the calculation of the temperature in the H I regions of non-ionized hydrogen. Original data are much less certain here. In H II zones, the abundance of ionizing quanta and free electrons assured the predominance of the chief processes (hydrogen ionization and cooling by excitation of abundant ions), so that the role of the remaining mechanisms for heating and cooling was negligible. In H I regions, there are not many free electrons, heating during ionization is less effective, and the number of ions is also less. Therefore, account must be taken here of many different mechanisms for heating and cooling.

Nevertheless, we shall start the temperature calculation in an H I region by confining ourselves to emission processes only, because here the values are less uncertain. Stellar emission heats the gas in H I zones as in H II zones, only now instead of ionization and recombination of hydrogen, we must consider ionization and recombination of those elements for which the ionization potential is less than 13.6 eV. The most abundant of these elements is carbon, so that we assume that the heating of gas in H I zones is due to the recombination and ionization of carbon atoms. The energy G_{ei} acquired per unit volume of gas per second by this process is

$$G_{ei} = \bar{\varepsilon}_0 \alpha_{\mathrm{C}}(T) N_{\mathrm{e}} N_{\mathrm{C\,II}}, \tag{8.16}$$

where $\bar{\varepsilon}_0$ is the average kinetic energy acquired by an electron during a single ionization [see Eqs. (6.18) and (6.14)], and $\alpha_{\mathrm{C}}(T)$ is the recombination coefficient (6.16) to all levels of the carbon atom. In agreement with the results of radio observations, the temperature in H I regions is low. This means that electron collisions can excite only the low levels, primarily the components of the fine structure of the ground term of p^1, p^2, and p^4 configurations. The reverse transitions give radiation in the far-infrared region of the spectrum, which up to now has not been observed. Of the most abundant elements with such levels, ionized carbon ranks first (the distance between the fine-structure levels is 8×10^{-3} eV or 92° on the temperature scale). The line $\lambda = 156\,\mu$ is emitted during this transition. In addition, ionized silicon and iron are relatively abundant. The distances between the fine-structure levels are 413° for Si II and 554° and 961° for Fe II. The wavelengths of the corresponding transitions are $\lambda = 34.6\,\mu$ (Si II) and $\lambda = 27$ and $16\,\mu$ (Fe II).

At low temperatures the probability of exciting the metastable level of an ion, other things being equal, is considerably greater than for an atom, since ions attract electrons. Therefore, the excitation of atoms in H I regions is usually neglected. However, very abundant neutral oxygen can play a considerable role in the cooling of the transition layer between H I and H II [162].

The energy lost by the excitation of metastable levels is calculated as follows. Neglecting collisions of the second kind, we have

$$L_{ei} = \sum h\nu_{ik} q_{ik} N_i N_{\mathrm{e}}. \tag{8.17}$$

The numerical value of L_{ei} depends on the choice of chemical composition. Seaton [18:121] calculated L_{ei} for the four above-mentioned levels of C II, Si II, and Fe II, for which he assumed for the relative concentrations of the elements $N_C:N_{Si}:N_{Fe} = 1:0.37:0.18$. Therefore, we have for the summation (8.17)

$$L_{ei} = \frac{3 \times 10^{-20}}{T^{1/2}}\left(e^{-92/T} + 10e^{-413/T} + 2.6e^{-554/T} + 3.4e^{-961/T_e}\right)N_eN_C \text{ erg/cm}^3 \text{ sec.} \quad (8.18)$$

Data for the effective cross sections (Ω parameters) for these levels are given in Appendix II.

We shall show that the role of collisions of the second kind in H I regions is small because of the low electron density. In fact, the critical electron concentration for C II, for example, is

$$N_e^{(k)} = \frac{A_{ik}}{q_{ki}} = \frac{2.36 \times 10^{-6}}{8.63 \times 10^{-6}} \frac{\tilde{\omega}_k T^{1/2}}{\Omega_{ki}} \approx 0.43T^{1/2} \text{ cm}^{-3}, \quad (8.19)$$

which is certainly a large number of free electrons per unit volume in H I regions.

Let us now compare this value of L_{ei} with the energy losses due to free-free transitions and recombinations. Equations (2.25) and (2.29) can be used approximately in this case, if N_p is changed to N_C. If $T \approx 100°$, the emission coefficient for free-free transitions will be $\sim 10^{-26}N_eN_C$ erg/cm³ sec and for recombination reaches the value $\sim 6 \times 10^{-25}N_eN_C$ erg/cm³ sec, where the main part of the radiation from recombination is associated with binding energy. At the same time, for $T \sim 100°$, losses by excitation of the forbidden levels are much greater, approximately $3 \times 10^{-21}N_eN_C$ erg/cm³ sec. Therefore, the calculation can be limited to L_{ei} alone.

Thus, when the temperature of the interstellar gas in a stationary state is determined only by its interaction with radiation, the energy-balance equation is written in the very simple form $L_{ei} = G_{ei}$ or, substituting numerical values,

$$\frac{2 \times 10^{-22}}{T^{1/2}}[1 - 0.19 \log T] = \frac{7 \times 10^{-20}}{T^{1/2}}(e^{-92/T} + 10e^{-413/T} + 2.6e^{-554/T} + 3.4e^{-961/T}). \quad (8.20)$$

It is significant that N_eN_C cancels, so that the equilibrium equation does not depend on density, and it depends only slightly on the choice of chemical composition if the last three terms on the right-hand side are small.

Thus the solution of Eq. (8.20) completely determines the temperature of an H I region while taking into account the reaction of the gas with radiation. As a result, we obtain

$$T \approx 15°. \quad (8.21)$$

This value for the equilibrium temperature in nonionized hydrogen regions seems very low. In fact, the temperatures, even for the harmonic mean from 21-cm radiation, was found to be 125°, while the arithmetic mean temperature must be still higher. Measurements of the 21-cm line in absorption from a massive cloud ($N_H \approx 20$, mass about $10^3 M_\odot$ with a dimension of 13 pc), against the background of the Cassiopeia A source, gave a value for the temperature of approximately 50° [165]. We shall therefore consider other mechanisms for heating and cooling gas in H I regions connected with the presence in the interstellar medium of cosmic rays, magnetic fields, dust particles, and molecules.

Cosmic-ray particles, by ionizing interstellar gas, transfer a portion of their energy to the gas. It is known (see, for example, [12]) that the transfer of energy from a relativistic proton, both by free and by bound electrons, is given by the expression

$$\frac{d\varepsilon}{dt} = \frac{2\pi e^4 N_H}{mc} L \approx 1.2 \times 10^{-20} N_H L \text{ erg/sec}, \tag{8.22}$$

where L is the Coulomb logarithm. Calling $N_{c.r.}$ the number of fast protons per unit volume, we have for the heating by cosmic-ray particles

$$G_{c.r.} = N_{c.r.} \frac{d\varepsilon}{dt} = 5 \times 10^{-19} N_H N_{c.r.} \text{ erg/cm}^3 \text{ sec}. \tag{8.23}$$

It is difficult to estimate $N_{c.r.}$. The fact is that the most effective particles are of relatively low energy, and do not reach the neighborhood of the Earth. Comparing Eq. (8.23) with Eq. (8.16) for $T = 15°$, we find for the condition under which heating by cosmic-ray particles becomes substantial

$$5 \times 10^{-19} N_H N_{c.r.} > 4 \times 10^{-23} N_e N_C. \tag{8.24}$$

In particular, for $N_C \approx 4 \times 10^{-4} N_H$ and $N_e \approx 10^{-3}$ cm^{-3}, it follows from (8.24) that heating by cosmic-ray particles becomes effective when $N_{c.r.} > 3 \times 10^{-11}$ cm^{-3}. Evidently, this condition is fulfilled in interstellar space. However, for $T > 15°$, the role of cosmic rays again becomes ineffective.

The Joule effect can also lead to heating of the gas. The energy released per unit volume during dissipation of a magnetic field of strength H is expressed as

$$G_H \approx \frac{\nu_m}{R^2} \frac{H^2}{8\pi}, \tag{8.25}$$

where ν_m is the coefficient of magnetic viscosity and R is a characteristic fluctuation dimension of the field. In H I clouds with a gas density $N_H \approx 10$ cm^{-3}, $\nu_m \approx 6 \times 10^{23}$ cm^2/sec. Putting $H \approx 3 \times 10^{-6}$ Oe and $R \approx 10$ pc, we find

$$G_H \approx 2.4 \times 10^{-28} \text{ erg/cm}^3 \text{ sec}, \tag{8.26}$$

so that for $N_C N_e < 6 \times 10^{-6}$ cm^{-6} the Joule heating becomes more effective than heating by radiation. Moreover, G_H increases with a decrease in the characteristic dimensions of the field. The magnetic viscosity ν_m is much less in H II regions, and therefore the Joule effect is ineffective there.

Photoemission of electrons at the surface of cosmic dust particles can also heat the gas. In principle, this mechanism is similar to the mechanism of heating from ionization and recombination. However, any reliable estimate is difficult to make here, since the emission properties of the surface of cosmic dust particles are completely unknown. If the particles are considered metallic, then from an estimate by Spitzer and Savedoff [161] that the number of dust particles per unit volume $\sim 10^{-12}$ cm^{-3}, we have

$$G_g \approx 10^{-27} \text{ erg/cm}^3 \text{ sec}. \tag{8.27}$$

However, if the particles are dielectric, which is more probable, their photoemission is negligibly small and $G_g \approx 0$.

Cosmic dust particles may also cool the gas. On collision with a particle, a hydrogen atom loses its own energy. In turn, the dust particle emits this energy in the infrared region of the spectrum. A rough estimate of this energy loss may be obtained by assuming that each hydrogen atom that collides with the surface of a dust particle gives up all of its own energy. On this hypothesis, we obtain

$$L_g \approx \tfrac{3}{2}kTN_{\mathrm{H}}\pi a^2 N_g \bar{v}_{\mathrm{H}} \approx \frac{3\sqrt{2\pi}(kT)^{3/2}}{m_{\mathrm{H}}^{1/2}} a^2 N_{\mathrm{H}} N_g$$
$$\approx 9.5 \times 10^{-12} T^{3/2} a^2 N_{\mathrm{H}} N_g \approx 2.4 \times 10^{-28} \text{ erg/cm}^3 \text{ sec}. \tag{8.28}$$

Here $\bar{v}_{\mathrm{H}}$ is the thermal velocity of hydrogen atoms, and a is the radius of the dust particle. The numerical value of L_g was obtained for $T = 100°$, $a \approx 5 \times 10^{-5}$ cm, $N_g \approx 10^{-12}$ cm^{-3}, and $N_{\mathrm{H}} \approx 10$ cm^{-3}. In general, L_g is small compared to L_{ei}, but when the density of dust particles is high appreciable cooling can occur.

As we shall see in the following section, it is extremely probable that in H I regions a significant concentration of H_2 molecules takes place. They also play a part in the cooling of interstellar gas.

During the collision of electrons and H_2 molecules the vibrational levels alone are excited, the lowest of which has an energy $\chi_m = 0.528$ eV. The effective cross section of this process is $\sigma_m \approx 7 \times 10^{-7}$ cm^2, and the probability of a spontaneous transition from the first vibrational level to the ground state is $A_m \approx 2 \times 10^{-7}$ sec^{-1}. Hence, the loss of energy by the electrons per unit volume is

$$L_{em} \approx N_e N_{\mathrm{H}_2} q_{em} \chi_m \approx N_e N_{\mathrm{H}_2} \chi_m \sigma_m \bar{v}_e e^{-\chi_m/kT}$$
$$\approx 3.7 \times 10^{-23} \sqrt{T} e^{-6130/T} N_e N_{\mathrm{H}_2} \text{ erg/cm}^3 \text{ sec}. \tag{8.29}$$

Collisions of the second kind have not played a part until now, since $N_e <$

$A_m/q_m \approx 4.8 \times 10^3 T^{-1/2}$. From Eq. (8.29) it is apparent that, after allowance for molecular collisions, the electron gas is drastically cooled only at temperatures $T \gtrsim 2000°$.

Collisions of hydrogen atoms with H_2 molecules and of the latter molecules among themselves are extremely effective for cooling interstellar gas in H I regions, since in this case the low-lying rotational levels are excited whose energy, determined by the rotational quantum number K, is

$$E_k = E_r K(K + 1), \; E_r = 1.2 \times 10^{-14} \text{ erg} = 0.00736 \text{ eV}. \tag{8.30}$$

Excitation by collision to the second rotational level, $K = 2$, is most probable. If we can neglect collisions of the second kind, the energy loss per unit volume by collisions between hydrogen atoms and molecules is

$$L_{H_2} = N_H N_{H_2} E_2 q_{02}, \tag{8.31}$$

where according to the calculations [166] the excitation coefficient q_{02} is equal to 3.1×10^{-16} cm^3/sec, 9×10^{-14} cm^3/sec, and 7×10^{-13} cm^3/sec at temperatures of 50, 100, and 150°, respectively. For example, when $T = 100°$, the energy loss by the excitation of H_2 is approximately $6.4 \times 10^{-27} N_H/N_{H_2}$ erg/cm^3 sec. With an increase in temperature, the cooling effect on the molecules is enhanced and may become predominant. Here, however, collisions of the second kind must be taken into account. The spontaneous-transition coefficient between different rotational levels K and K' of the para and ortho states of a hydrogen molecule is equal to zero unless $K' = K - 2$. In particular, for parahydrogen

$$A_{K,\,K-2} = 7.5 \times 10^{-13} \frac{K(K-1)(2K-1)^4}{2K+1} \text{ sec}^{-1}. \tag{8.32}$$

When $K = 2$ we have $A_{20} = 2.4 \times 10^{-11}$ sec^{-1}. Therefore, the critical concentration of hydrogen atoms is equal to

$$N_H^{(k)} = \frac{A_{20}}{q_{20} + q_{21}} \approx 1 \text{ cm}^{-3} \tag{8.33}$$

for $T = 100°$. Consequently, collisions of the second kind are effective in interstellar clouds and ortho- and parahydrogen molecules in low rotational states can be considered to have the Boltzmann distribution

$$N_{H_2}(K) = N_{H_2} \frac{(2K+1)e^{-E_K/kT}}{\sum_{K=0}^{\infty} (2K+1)e^{-E_K/kT}}. \tag{8.34}$$

It is true that the populations of levels with $K > 2$ can deviate from Boltzmann conditions, but the contribution of these levels to the summation in the denominator of Eq. (8.34) is small at the temperatures in which we are interested.

Radiative energy losses during rotational transitions in this case are equal to

$$L_{HH_2} = E_1 A_{10} N_{H_2}(1) + (E_2 - E_1) A_{21} N_{H_2}(2) + E_2 A_{20} N_{H_2}(2) + \sum_{K=3}^{\infty} (E_K - E_{K-2}) A_{K,K-2} N_{H_2}(K). \quad (8.35)$$

When temperatures are not too high ($kT \lesssim E_2$), radiation from the second level offers the greatest contribution, from which we obtain

$$L_{HH_2} \approx 8.7 \times 10^{-23} \frac{N_{H_2} e^{-515/T}}{1 + 3e^{-172/T} + 5e^{-515/T}} \frac{\text{erg}}{\text{cm}^3 \text{ sec}}. \quad (8.36)$$

In this way and by the effect of collisions of the second kind, cooling by H_2 molecules can be effective if the gas temperature is not too low and the number of molecules is sufficiently great.

Resuming the consideration of a method for calculating temperature in H I clouds, we can say that, under all conditions, nonionized hydrogen must rapidly cool to a temperature of about 100°. If heating by cosmic-ray particles or dissipation from interstellar magnetic fields is not significant, then the gas temperature must rapidly fall to about 15°. That such low temperatures evidently do not occur means that interstellar gas in H I regions does not usually reach a condition of thermal equilibrium. It will be shown in Chapter 5 that the heating of a gas during its motion must be sufficiently intense.

An observational determination of the gas temperature in H I regions is possible so far only from a study of the radio line $\lambda = 21$ cm. In the future, it will probably be possible to employ observations in the far-infrared region of the spectrum for this purpose.

We note here that, even in the simpler case of H II zones, a semiempirical method is used to calculate temperature. Naturally, in the case of H I clouds, observation of their infrared emission would allow a more reliable procedure to be worked out for determining temperature.

We add some comments about the observational determination of temperature in H I regions by the hydrogen line $\lambda = 21$ cm. When we use emission lines for this purpose it is necessary to remember that in this case the emission originates in both hot and cold regions. But in the case of absorption lines we measure temperature only in the corresponding H I cloud. This explains the apparent difference between the temperatures of emission ($\sim 125°$) and of absorption ($\leqslant 70°$) hydrogen lines. It was confirmed by Schuter and Verschur [166a] and by Marx [166b]. The observed temperature of H I clouds ranges from 25° to 120°K. The apparent difference of calculated and observed values may be explained by the heating of the interstellar cloud by shock waves (Sec. 21), by dissipation of Alfvén waves propagating in interstellar space along the interstellar magnetic field [166c], or by heating of an unknown number of superthermal particles [166d]. We also note the possibility of very effective heating by plasma waves.

Pikelner [166e] discussed the importance of superthermal particles based on Faraday rotation of the polarization of extragalactic radio sources in the galactic disk. The rotation measure requires that $\langle N_e \rangle \gtrsim 10^{-2}$ cm^{-3}. It was shown that H II regions cannot give a regular distribution of RM on the sky. But H I regions cannot give such high values of $\langle N_e \rangle$. We are forced to admit that hydrogen is partly ionized in H I regions in the spiral arms. There cannot be rotation in the halo because it explains also depolarization of the radio emission from the arms. Gas can be partly ionized only with superthermal particles.

Ionization and heating of the gas give us two equations for T, N_e, N_H, and q, where q is the loss of energy per atom in neutral hydrogen. Coulomb interaction with free electrons and ionization and excitation of He are also taken into account. The cooling rate $L(T)$ includes besides infrared transitions also optical ones and excitation of hydrogen. Between $T = 500°$ and $T = 5000°$, L is about constant; it increases sharply at $T > 5000°$. Field [166f] showed that there is thermal instability under these conditions. The temperature of the stable rarefied gas between clouds is $T \approx 5000°$. Then $N_e = 0.17 N_H$ and $q = 2.4 \times 10^{-24} N_e$. We have to choose a parameter that is dependent on q in our Galaxy. Let $N_H = 0.08$ cm^{-3}, $N_e = 1.4 \times 10^{-2}$ cm^{-3}, which do not contradict the Faraday rotation data. Then $q = 3.3 \times 10^{-26}$ erg/sec, and we can calculate the physical conditions at various densities. Table 8.2 explains the temperature of clouds ($T \lesssim 100°$) and the formation of the clouds themselves, since the equilibrium pressure has a minimum for intermediate densities. This model explains the peak brightness temperature at the 21-cm line; rarefied gas gives $T_b \approx 50°$ and the rest give clouds of different temperatures. Compression of the gas can take place only along the lines of magnetic force; it is a slow process and a good deal of the gas may be not in pressure equilibrium. Between the arms $N_H < 0.08$ cm^{-3} and the gas is stable, hence clouds are not formed.

An important point is the low-frequency absorption. Ellis and Hamilton [166g] and Alexander and Stone [166h] showed that there is absorption which

Table 8.2

T (°)	$10^{24} L(T)$	$10^2 N_e$	N_H	$N_H T$
10,000	5.90	1.04	0.032	320
5000	2.74	1.41	.085	425
2000	2.66	1.19	.095	190
1000	2.60	1.06	.107	107
500	2.40	0.99	.13	65
200	1.34	1.31	.36	73
100	0.55	2.56	2.0	197
80	.40	3.31	3.7	296
60	.25	5.03	9.8	590

corresponds to an emission measure ME $= N_e^2 l \approx 5$. The table "Physical conditions in interstellar gas of different densities" shows that the real ME $\approx$ 0.1. But $\tau \sim$ ME $\times T^{-3/2}$. In clouds $T \lesssim 100°$ and a thickness of several parsecs is enough to give the necessary absorption. The recombination spectrum of hydrogen is also increased in clouds.

The electron concentration in H I regions is a very important parameter, which is used in all calculations of ionization of elements, thermal emission of the gas, and so forth. Now recalculations are necessary. Superthermal particles also destroy interstellar grains, which is a reason why the average size of grains is about constant in the Galaxy.

9. Molecules in Interstellar Space

It is well known that, besides atomic absorption lines, absorption lines of the molecules CH, CH^+, and CN [167, 168, 168a], and of OH in the radio region [171a, b] are detected in interstellar space. Data on these observed lines are given in Table 9.1.

Table 9.1

Molecule	λ	Transition	f
CH	3143	$^2\Sigma^+ \leftarrow {}^2\Pi\frac{1}{2}$	
	3878.8	$^2\Sigma^- \leftarrow {}^2\Pi\frac{1}{2}(0, 0)$	
	3886.4	$^2\Sigma^- \leftarrow {}^2\Pi\frac{1}{2}(0, 0)$	0.002
	3890.2	$^2\Sigma^- \leftarrow {}^2\Pi\frac{1}{2}(0, 0)$	
	4300.3	$^2\Delta \leftarrow {}^2\Pi(0, 0)R(1)$	.002
CH^+	3745.3	$^1\Pi \leftarrow {}^1\Sigma^+(2, 0)R(0)$	
	3957.7	$^1\Pi \leftarrow {}^1\Sigma^+(1, 0)R(0)$	
	4232.6	$^1\Pi \leftarrow {}^1\Sigma^+(0, 0)R(0)$	.004
CN	3874.0	$^2\Sigma^+ \leftarrow {}^2\Sigma^+(0, 0)R(1)$	.07
	3874.6	$^2\Sigma^+ \leftarrow {}^2\Sigma^+(0, 0)R(0)$	
	3875.8	$^2\Sigma^+ \leftarrow {}^2\Sigma^+(0, 0)P(1)$	

A description of the terms of simple diatomic molecules and of the systematics of their spectra are given in Appendix III.

The Analysis of Interstellar Molecular Lines

All the observed interstellar molecular lines are weak and their equivalent widths are proportional to the number of molecules $\mathfrak{N}_m$ in a column of unit cross section along the line of sight between the observer and a star. Corresponding to Eq. (7.13), we have

$$W_\lambda = \frac{\pi e^2 \lambda^2}{mc^2} f \mathfrak{N}_m; \qquad W_\nu = \frac{\pi e^2}{mc} f \mathfrak{N}_m. \tag{9.1}$$

The equivalent widths W_λ of CH, CH^+, and CN lines listed in the table vary over a range from 0.003 to 0.014 Å [167, 142]. Substituting measured values in Eqs. (9.1) and using estimated oscillator strengths f (given in Table 9.1), Bates and Spitzer determined the mean densities of molecules along the line of sight to be [168]

$$N_{CH} \approx 3 \times 10^{-8}\ \text{cm}^{-3}, \qquad N_{CH^+} \approx 1.5 \times 10^{-8}\ \text{cm}^{-3},$$
$$N_{CN} \approx 10^{-9}\ \text{cm}^{-3}. \tag{9.2}$$

It may be assumed that the presence of molecules in interstellar space is connected with cosmic dust particles. In this case, correlation may be expected between the reddening of starlight (by absorption from dust) and the equivalent widths of molecular lines. For class O–B1 stars, such a correlation does in fact exist. Knowing the connection between absorption and gas density (when $N_H \approx 1\ \text{cm}^{-3}$ the reddening is equal approximately to $0^{m}.17\ \text{kpc}^{-1}$), we can determine the relative concentration of molecules. By this procedure, Bates and Spitzer obtained estimates of $N_{CH} \approx 1.6 \times 10^{-8} N_H$ and of $N_{CH^+} \approx 8 \times 10^{-9} N_H$. The mean radial velocities of interstellar gas from these lines are usually negative because of the galactic rotation in the region of the sky where observations are carried out and because of the so-called K effect.

In the case of the B2–B9 spectral classes, the molecular lines of CH^+ and, in part, of CN are observed from nonreddened stars. The radial velocities from these lines are usually positive ($\sim +5$ km/sec), that is, in this case the gas is traveling from the observer toward the star. In the opinion of Bates and Spitzer, the lines mentioned belong to molecules close to the star, which are formed in the gaseous envelopes surrounding stars. The number of molecules in a column of unit cross section is equal on the average to

$$\mathfrak{N}_{CH^+} \approx 1.9 \times 10^{13}\ \text{(B6–B9 stars in the Pleiades)}, \tag{9.3}$$

$$\mathfrak{N}_{CH} \approx 2.1 \times 10^{12}\ \text{(B2–B5)},$$
$$\mathfrak{N}_{CH^+} \approx 2.7 \times 10^{12}\ \text{(B2–B9)}. \tag{9.4}$$

Thus, it may be considered that there are at least two processes of molecule formation, that in interstellar areas and that in stellar neighborhoods.

The Formation and Dissociation of Molecules

The various processes of formation, ionization, and dissociation of molecules were analyzed by Bates and Spitzer [168] and by Herzberg [169]. An account is given below of the results obtained in these researches.

(1) *Radiative capture.* This process takes place in the following way. During the approach of two atoms, the total energy remains positive, that is, the electronic state of the system must be described by a potential curve corresponding to a positive energy. A spontaneous transition to a level with negative energy, with the emission of a quantum, binds the atom into a molecule. For example, in the case of a CH molecule (see Fig. 73), the approach is via the $^2\Sigma^-$ state curve, from which a spontaneous transition is possible to the $^2\Pi$ state. In the case of CH^+, the process is similar: approach in the $^1\Pi$ state and spontaneous transition to the $^1\Sigma$ state. The capture probability is proportional to the oscillator strength of downward transitions and depends very little on temperature. This process in interstellar space can take place only when the excited state of the molecule is formed from the ground states of the atoms in such a way that the downward transition from this state is allowed by selection rules. In the case of CH and CH^+ this is possible and the process of radiative recombination may be written as follows:

$$\begin{aligned} &C\,(^3P) + H\,(^2S) \rightarrow CH\,(^2\Sigma^-) \rightarrow CH\,(^2II) + h\nu, \\ &C + (^2P) + H\,(^2S) \rightarrow CH^+(^1II) \rightarrow CH^+(^1\Sigma) + h\nu. \end{aligned} \tag{9.5}$$

In the present case, it is true, there is one difficulty. The lowest state of carbon $^3P_0$ changes to the $^2\Pi_{1/2}$ state and not to the $^2\Sigma^-$ into which the $^3P_1$ and $^3P_2$ sublevels change. Let us remember that in interstellar space almost all the carbon atoms occur in the $^3P_0$ sublevel. However, it is possible that the approach does not proceed strictly adiabatically, so that a transition between sublevels can occur. Similarly, during the formation of CH^+ the upper sublevel $^2P_{3/2}$ is converted into the $^1\Pi$ state.

We quote below a series of reactions for the formation of the most important molecules from radiation capture [169]:

$$\begin{aligned} &H^+\,(^1S) + H\,(^2S) \rightarrow H_2^+\,(^2\Sigma_u^+) \rightarrow H_2^+\,(^2\Sigma_g^+) + h\nu, \\ &O\,(^3P) + H^+\,(^1S) \rightarrow OH^+\,(^3II) \rightarrow OH^+\,(^2\Sigma^-) + h\nu, \\ &O\,(^3P) + H\,(^2S) \rightarrow OH\,(^2\Sigma^-) \rightarrow OH\,(^2II) + h\nu, \\ &N^+\,(^3P) + H\,(^2S) \rightarrow NH^+\,(^2\Sigma^-) \rightarrow NH\,(^2II) + h\nu, \end{aligned} \tag{9.6}$$

and so on. The probability of these reactions is unknown in the majority of cases. For the reactions involving formation of CH and CH^+, the probability will be given below.

Quasi states are stable (that is, the atoms do not separate immediately) in molecules of OH^+, C_2, CO, CO^+, NO^+, N_2^+, O_2^+. In H_2^+ molecules, the $2\Sigma_u^+$ state is unstable, but the probability of downward spontaneous transition is small, so that if the temperature is not very low the downward transition can take place at the turning point (at small internuclear distances), before the particles go their separate ways. In other cases, quasi states are poorly investigated and their stability is only assumed.

In an H I region, all reactions can occur, with the exception of those in which H^+ participates. The formation of H_2^+ and O_2^+ is possible in the transition layer between H I and H II.

Certain molecules (H_2, NH, N_2) cannot be formed similarly, since a downward spontaneous transition is forbidden for them. In other cases (MgH, MgH^+), only molecular ground states can be formed from the ground states of the atoms, so that a spontaneous downward transition, which is necessary in order to transfer the molecules to a state of negative energy, is not possible.

Further examples of reactions for the formation of molecules may be quoted, for example, reactions with negative hydrogen ions: $C^+ + H^- = CH^+ + e$, $C^+ + H^- = CH + h\nu$, or exchange reactions: $OH + O = O_2 + H$, $CH + H = C + H_2$, and so forth. These reactions are insufficiently studied, but evidently their probability in the conditions of interstellar space is not high. However, in those cases when molecules are not formed through radiative capture (H_2, O_2), these reactions may be significant.

(2) *Photoionization and photorecombination.* The ionization potential of a molecule is usually 2 or 3 eV greater than the ionization potential of its constituent atoms. Direct calculations of the absorption coefficient (ionization) of molecules have not been made. In the case of a CH molecule, an estimate of its absorption coefficient may be obtained in the following manner. The electron cloud of CH is similar to the cloud of a nitrogen atom for which $a_1(\nu_i) \approx 3 \times 10^{-18}$ cm^2 for one outer electron. Because for CH the wave function is more diffuse, Bates and Spitzer estimated the absorption coefficient $a_\pi \approx 4 \times 10^{-18}$ cm^2 for the release of a single $p\pi$ electron and $a_\sigma \approx 6 \times 10^{-18}$ cm^2 for the release of two $p\sigma$ electrons. The energy required for the transition $CH^2(\Pi) \rightarrow CH^+(^1\Sigma^+)$ is 11.1 eV. From the data on the interstellar radiation field (Sec. 6), the probability of photoionization in interstellar space ($\sim 8 \times 10^{-12}$ sec^{-1}) is determined. The probability of recombination to the ground level is determined by Miln's relation; when $T = 100°$, we obtain $\alpha_{CH} \approx 2 \times 10^{-12}$ cm^3/sec. A rough estimate of recombination to excited levels gives a value for the total recombination probability of a CH^+ molecule: $\alpha_t \approx 7 \times 10^{-2}$ cm^3/sec. With an increase in temperature, this is reduced by $T^{-1/2}$.

(3) *Dissociative recombination.* If the $^1\Sigma^+$ curve of a CH^+ molecule is intersected by any unstable curve of a neutral CH molecule corresponding to excited levels of a carbon atom, the molecule can fall on this curve after recombination without emission, which leads to dissociation according to the formula $CH^+ + e \rightarrow C + H$. In individual cases this process is very probable (up to 10^{-17} cm^3/sec), but in the general case its rate is unknown. For the $^2\Sigma^+$ state, the probability of such a process is low.

(4) *Photodissociation.* This process consists of the absorption of a quantum,

after which the molecule changes to a state with a repulsive potential curve. Examples of such transitions are

$$\begin{aligned} &\mathrm{CH}\,(^2\mathrm{II}) + h\nu \rightarrow \mathrm{CH}\,(^2\Sigma^+) \rightarrow \mathrm{C}\,(^3P) + \mathrm{H}\,(^2S), \\ &\mathrm{CH}^+\,(^1\mathrm{II}) + h\nu \rightarrow \mathrm{CH}^+\,(^2\Sigma^+) \rightarrow \mathrm{C}^+\,(^2P) + \mathrm{H}\,(^2S). \end{aligned} \tag{9.7}$$

For these processes the energies of the quanta must be equal to 10 and 14 eV, respectively, and the oscillator strengths of the absorptions are equal to 3×10^{-2} and 10^{-2}. Hence it follows that the probability of photodissociation processes in H I regions is about 10^{-11} sec^{-1}. However, this is only a rough estimate. Dissociation is also possible with absorption of a quantum during transition to an attraction state, when, according to the Franck–Condon rule, the molecule falls on a potential curve at a point above the dissociation potential. For example, CH molecules possess such quasi states as $^2\Sigma^-$, $^2\Delta$, $^2\Sigma^+$, and CH^+ molecules have a $^1\Pi$ quasi state. The relative transition probability to states of negative energy (neglecting vibrational energies) but accompanied by dissociation, is about 9 percent for CH ($^2\Pi \rightarrow {}^2\Sigma^-$) and 0.5 percent for CH^+ ($^1\Sigma^+ \rightarrow {}^1\Pi$) of the total number of all transitions to corresponding states. Bates and Spitzer have estimated the dissociation rate through stable states: 5×10^{-12} sec^{-1} for CH and 5×10^{-13} sec^{-1} for CH^+, which is comparable with dissociation through unstable states.

CH *and* CH$^+$ *Molecules*

The foregoing data on various elementary processes allow us to construct equilibrium equations from which a theoretical determination of the relative concentrations of molecules in interstellar space is possible. Obviously, the greater the dissociation potential, the greater must be the concentration of molecules (taking into account the abundance of their constituent atoms). For example, hydrogen and nitrogen molecules, with the greatest dissociation potential, must be more abundant than CH, CH^+, CN, OH, and other molecules. These latter also are readily dissociated by exchange reactions, being converted, for example, into H_2 or N_2.

Under conditions of thermodynamic equilibrium, the degree of dissociation of diatomic molecules is determined by the equation

$$\frac{N_A N_B}{N_{AB}} = \frac{\tilde{\omega}_A \tilde{\omega}_B}{\tilde{\omega}_{AB}} \left(\frac{2\pi\mu}{h^2}\right)^{3/2} \frac{(kT)^{5/2} h^2}{8\pi^2 I} (1 - e^{-h\nu_0/kT}) e^{-U/kT}. \tag{9.8}$$

Here N_A, N_B, and N_{AB} are the number of atoms and molecules per unit volume, $\tilde{\omega}_A$, $\tilde{\omega}_B$, and $\tilde{\omega}_{AB}$ are the statistical weights of the ground levels, I is the moment of inertia, μ is the mass, ν_0 is the vibrational frequency of the atoms in the molecules, and U is the molecular dissociation energy.

However, in interstellar space, thermodynamic equilibrium certainly does not exist, so that here we must proceed as in the calculation of the degree of ionization of atoms, that is, we must construct an equilibrium equation in which the numbers of molecules formed and dissociated under the specific conditions of interstellar space are equated. Since quantitative data on the probability of the various molecular processes are up to the present imperfectly known, the solution of the corresponding equilibrium equations cannot be obtained with sufficient accuracy. Molecules of CH and CH^+, for which the probability of transitions is estimated, although very approximately, are an exception. Table 9.2 gives reactions, with corresponding estimates of probability [168 and 169].

Table 9.2

Reaction	Symbol	Probability	Comment
$C + H \rightarrow CH + h\nu$	γ_1	2×10^{-18}	If the population of the ^3P sublevel is proportional to the statistical weight
$C + H \rightarrow CH + h\nu$	γ_1	6×10^{-18}	If all C atoms are in the $^3P_0$ sublevel
$C^+ + H \rightarrow CH^+ + h\nu$	γ_2	2×10^{-18}	If the population of the ^2P sublevel is proportional to the statistical weight
$C^+ + H \rightarrow CH^+ + h\nu$	γ_2	0	If all C^+ ions are in the $^2P_{1/2}$ sublevel
$CH + h\nu \rightarrow CH^+ + e$	β_1	8×10^{-12}	
$CH^+ + e \rightarrow CH + h\nu$; $CH^+ + e \rightarrow CH' + h\nu$, $CH' \rightarrow CH + h\nu$	α_1	7×10^{-12}	
$CH' \rightarrow C + H$; $CH^+ + e \rightarrow C + H$	α_2	Unknown, but probably small	
$CH + h\nu \rightarrow C + H$	β_2	1.5×10^{-11}	
$CH^+ + h\nu \rightarrow C^+ + H$	β_3	5×10^{-13}	

Here a prime indicates dissociation through an intermediate state. With the symbols from the table, the equilibrium equations have the form

$$
\begin{aligned}
N_{CH}(\beta_1 + \beta_2) &= N_C N_H \gamma_1 + N_{CH^+} N_e \alpha_t, \\
N_{CH^+}[\beta_3 + (\alpha_1 + \alpha_2)N_e] &= \gamma_2 N_H N_{C^+} + N_{CH}\beta_1.
\end{aligned}
\tag{9.9}
$$

The solution of this system is written in the form

$$
\begin{aligned}
N_{CH} &= N_H \frac{\gamma_1[(\alpha_1 + \alpha_2)N_e + \beta_3]N_C + \gamma_2\alpha_1 N_e N_{C^+}}{\beta_3(\beta_1 + \beta_2) + [\beta_1\alpha_2 + \beta_2(\alpha_1 + \alpha_2)]N_e}, \\
N_{CH^+} &= N_H \frac{\gamma_1\beta_1 N_C + \gamma_2(\beta_1 + \beta_2)N_{C^+}}{\beta_3(\beta_1 + \beta_2) + [\beta_1\alpha_2 + \beta_2(\alpha_1 + \alpha_2)]N_e}.
\end{aligned}
\tag{9.10}
$$

This solution for the values of α, β, and γ introduced above, on the assumption that $N_e \approx 5 \times 10^{-4} N_H$, $N_C \approx 3 \times 10^{-4} N_H$, $N_C/N_{C^+} \approx 0.6 N_e$, and $T \approx 100^\circ$, was studied in detail by Bates and Spitzer [168]. Since values for the probabilities γ, γ_2, and α_2 are not unique, various cases were examined. We shall note only the essential: when $N_H \lesssim 20$, it is not possible under any assumption to explain the observations in interstellar space of the values for the number of CH and CH^+ molecules, since calculations give a concentration 500 to 1000 times smaller. The theoretical degree of ionization N_{CH^+}/N_{CH} is considerably more than the observed value. It might be possible to alter this result by considering processes of the type $H_2 + C \rightarrow CH + H$, or $C^+ + H^- \rightarrow CH^+ + e$ and by taking into account reactions on the surface of cosmic dust. However, so far we have not succeeded in obtaining agreement between the observed and the theoretical values for the concentrations of CH and CH^+, and particularly the degree of ionization of CH. The conclusion to be drawn is that atomic processes in interstellar gas clouds cannot explain the equilibrium state of CH and CH^+ molecules, provided the conditions there are normal and the reaction probabilities are correct.

This difficulty forces us to turn to a study of molecules in stellar neighborhoods. In the opinion of Bates and Spitzer, the abundance of molecules in stellar neighborhoods may be explained if we assume that they are formed in the cosmic clouds moving toward the star. Because of heating due to stellar radiation on the dust surface, evolution of methane CH_4 occurs, then hydrogen atoms in turn are dissociated from the CH_4 molecule, so that the time comes when these molecules occur in the form CH and then CH^+, after which they dissociate completely. In this way, the velocities of these molecules may be explained: their positive velocities (away from the observer) indicate that they form a cloud moving in the direction of the star. A numerical estimate shows that if the density of the dust is $\sim 1.4 \times 10^{-26}$ gm/cm^3, if 15 percent of it consists of CH_4 molecules, and if each one passes to the CH^+ state, their observed concentration can be explained. However, difficulties with CH remain, since CH molecules must rapidly be ionized near hot stars, and therefore there must be fewer of these molecules than of their ions, which is contrary to observation.

Strong lines of CN are observed in some stars immersed in emission nebulae [168a]. Analysis has shown that these lines are formed, apparently, in non-ionized gas adjacent to the H II zone. It is possible that radicals are formed on dust particles when a flux of short-wavelength radiation emanates from the expanding H II zone. However, such processes have not yet been thoroughly studied.

It is interesting to note that the CN molecule has a low sublevel with energy corresponding to $\lambda = 0.254$ cm above the ground state. The absorption line originating from this sublevel has been observed, in contradiction to the assumption that we may observe in interstellar space only such absorption lines as originate from the lowest level of the ground state. This discrepancy may be explained by excitation of this sublevel by relict black-body radiation for which $T = 3^\circ$ [169a, b].

Molecular Hydrogen

Hydrogen H_2 requires separate treatment. These molecules must be in great abundance, but up to now they have not been observed, since all their lines, which originate from the ground state, are in the ultraviolet spectral region.

Two hydrogen atoms give on approach the two molecular terms $^1\Sigma$ and $^3\Sigma$ (Fig. 18). The second triplet state is possible, because of Pauli's principle,

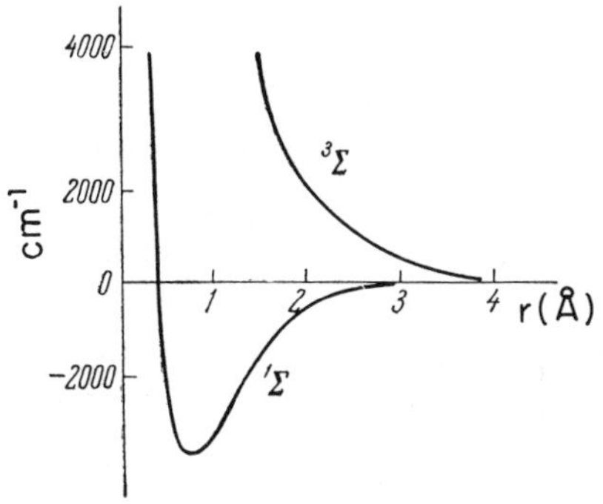

FIG. 18. The lower states of H_2 molecules.

only when the principal quantum number of one of the electrons is greater than unity (in addition, it must not differ from zero and the azimuthal quantum number). However, because of this, the energy of the molecule formed is increased (the second H level is very high), so that there is a dissociation potential larger than 4.5 eV. Thus only the term $^1\Sigma$ is stable, corresponding to an antiparallel spin distribution leading to an exchange attraction. Therefore, this molecule cannot be formed by radiative capture, since the spontaneous transition $^3\Sigma \rightarrow {}^1\Sigma$ is strongly forbidden by selection rules. (Nuclear spins also can be parallel and antiparallel. The corresponding molecules give para- and orthohydrogen. Parahydrogen corresponds to even values of the rotational quantum number K of the vibrational ground state, orthohydrogen to the odd values of K.) The formation of H_2 molecules by collisions with excited hydrogen atoms is possible, but the probability of such collisions is negligibly small.

The most probable mechanisms for the formation of H_2 molecules, and possibly of other molecules, are the surface reactions of cosmic dust particles. Hydrogen atoms can adhere to the surface of a dust particle on collision with it. Since to expel an atom from the surface requires energy of about 1 eV, which is much more than the kinetic energy of atoms in H I regions, collisions of atoms do not in general disintegrate dust particles. A hydrogen atom either rebounds from another atom or forms a molecule with it. During the formation of an H_2

molecule, 4.47 eV of energy are liberated, of which there is excess sufficient to separate the molecule from the surface and even for partial vaporization of the dust particle. Thus, an H_2 molecule can be evaporated or sublimed directly from the surface of a dust particle, since the "melting point" of the H_2 solid phase is only 14°.

The probability of formation of an H_2 molecule by a similar method was estimated by McCrea and McNally [170, 170a]. We call w the probability that a hydrogen atom that has collided with the surface of a dust particle forms an H_2 molecule. Obviously, the number of molecules formed is equal to $\frac{1}{2}w$ times the total number of collisions. The latter is determined as an integral over the product of the speed of a hydrogen atom, the surface area $4\pi a^2 N_g$ of dust particles per unit volume (N_g is the number of dust particles per unit volume and a is the radius of a particle), and the velocity-distribution function. Hence the number of H_2 molecules formed per cubic centimeter per second is

$$\gamma_{\mathrm{H}_2} N_{\mathrm{H}}^{(0)} N_{\mathrm{H}} = \frac{1}{2} w \left(\frac{8kT}{\pi m_{\mathrm{H}}}\right)^{1/2} 4\pi a^2 N_{\mathrm{H}} N_g . \tag{9.11}$$

Here N_{H} is the number of hydrogen atoms in the free state and $N_{\mathrm{H}}^{(0)}$ is the total number of hydrogen atoms, that is, $N_{\mathrm{H}}^{(0)} = N_{\mathrm{H}} + 2N_{\mathrm{H}_2}$. The numerical value of γ for H_2 molecules (γ_{H_2}) under the conditions $w = 0.5$, $N_g \approx 10^{-14} N_{\mathrm{H}}^{(0)}$, $a \approx 5 \times 10^{-5}$ cm^3 is approximately $10^{-18} T^{1/2} \approx 10^{-17}$ cm^3/sec. We note in passing that the rate of formation of CH molecules by a similar procedure is $\gamma_{\mathrm{CH}} \approx 0.5 \times 10^{-17}$ cm^3/sec. Comparing with the values of γ in the case of radiative capture, we find that the mechanism of molecule formation on dust is actually more effective, but it is still insufficient to eliminate a discrepancy of 500–1000 times between the theoretical and observed values of the concentration. The value of γ increases with density and temperature, and therefore it is possible that during the collision of clouds, when N_{H}, N_g, and T are simultaneously increased, conditions become more favorable for the formation of molecules.

We shall estimate the probability of dissociation of hydrogen molecules. The threshold energy for the dissociation of the H_2 molecule is 14.5 eV. Therefore, direct dissociation of these molecules is not possible in H I regions. The possibility has been suggested of dissociation during the forbidden transition $^1\Sigma \rightarrow {}^3\Sigma$, which takes place with the absorption of a quantum corresponding to $\lambda = 1500$ Å [170]. However, the oscillator strength of this transition is small ($f \approx 10^{-9}$) [170a]. Thus, H_2 molecules formed in H I regions are no longer destroyed after the cloud has moved away from the hot star. Then dissociation and ionization of molecules take place together with the ionization of hydrogen. The probability of the destruction of H_2 molecules can be found by determining the average probability of ionization of H I clouds.

If we assume that in the Galaxy one O-type star flares up per hundred years, on the average, and that this ionizes $10^4 M_\odot$, which constitutes 10^{-5} of the total mass of galactic gas, we obtain for the average probability that a particular cloud will be ionized the value $\beta \approx 3 \times 10^{-15}\ \text{sec}^{-1}$. Writing the balance equation

$$\gamma_{H_2} N_H^{(0)} N_H \approx \beta N_{H_2}, \tag{9.12}$$

we find for the relative abundance of molecules

$$\frac{N_{H_2}}{N_H} \approx \frac{\gamma_{H_2} N_H^{(0)}}{\beta} \approx 3 \times 10^{-3} N_H^{(0)}. \tag{9.13}$$

Inside interstellar clouds and H I regions, $N_H \approx 20$, and consequently $N_{H_2} \approx 0.06\ \text{cm}^{-3}$. With an increase in density, the relative concentration of H_2 molecules is raised. Therefore, in the limit $N_H^{(0)} \approx 2N_{H_2}$ and from Eq. (9.13), it follows that $N_H \approx \beta/2\gamma \approx 160\ \text{cm}^{-3}$. Thus, at a density of $N_H^{(0)} > 200\ \text{cm}^3$ in an H I region and in the presence of dust, all of the hydrogen can condense into molecules. The data of Bok [171] show that in H I regions where there is dense dust hydrogen in the free state is limited ($N_H \lesssim 100\ \text{cm}^{-3}$).

The time required for ionization of the cloud has been estimated [170b] from the probability of chance encounters of hot O and B stars with the cloud. The velocity dispersion of the clouds was taken to be 7.5 km/sec. These authors obtained for an average over the Galaxy the value $N_{H_2}/N_H \approx 0.1$–10, independent of the height above the galactic plane. However, such an estimate of β is unsatisfactory for the following reasons: (1) during the lifetime of an O star the cloud moves only about 5–10 pc, that is, its own diameter; (2) it is not possible to assume that the cloud is in steady motion over a long period of time because of the interaction with the magnetic field; (3) the probability of formation of an O star close to the cloud is larger than the probability of approach, as is shown by the estimate made above.

At any rate, under all circumstances the estimate of the interstellar content of H_2 molecules is extremely uncertain. Therefore, it would be very important to observe their radiation. The first resonance line of electronic transitions is at a wavelength $\lambda = 1030$ Å. Obviously, it will be possible to observe it in the near future. Observations of these lines are the most useful for an analysis of the abundance of H_2 molecules.

The infrared lines $\lambda\lambda$ 28.2 μ ($k = 2 \to 0$), 2.4 μ ($v = 1 \to 0$, $k = 1$), 2.22 μ ($v = 1 \to 0, j = 2 \to 0$), 2.12 μ ($v = 1 \to 0, k = 3 \to 1$), and others are radiated during vibrational and rotational transitions. The calculation of the intensity of these lines does not present any difficulties in principle. At sufficiently high temperatures ($\gtrsim 500°$K) the state distribution of H_2 molecules is Boltzmannian. At low temperatures the probability of exciting even the first level by collisions is small, but the following mechanism is known to play a role. Molecules are excited by ultraviolet radiation to the first electronic state, from where spontaneous transitions are possible not only to the ground state but to excited vibrational and rotational levels as well [170c]. However, further transitions producing infrared radiation probably have a small probability.

Molecular Radio Lines

Up to the present, molecules are observed only in the visible spectral region. However, as I. S. Shklovskii [15] showed, a number of molecules may be observed in the radio band. Separation of the terms with $\Lambda > 0$ (Λ splitting), caused by the interaction of the rotation of the nucleus with the electron orbital angular momentum, gives lines in the radio band. The most abundant molecules occurring in interstellar space with $\Lambda > 0$ may be CH, OH, SiH, and NO, for which the ground state is $^2\Pi$. The transition frequencies of the Λ splitting of the rotational ground state of these molecules were calculated by I. S. Shklovskii [15]:

Molecule	λ (cm)
OH	18.3
CH	9.45
SiH	12.5
NO	67

Transitions between the sublevels of the Λ splitting are allowed, but the probability of spontaneous transitions is low because the frequencies are low. Since the molecules are found in the densest regions, collisions between them are expected to be quite frequent, and therefore the populations of the excited sublevels of OH, SiH, and NO are determined by the Boltzmann distribution. For the higher-frequency line $\lambda = 9.45$ cm (CH), the population must be found by calculation of excitations caused by collisions with hydrogen atoms.

The lines of OH must be very intense because of their large abundance. Four lines of this radical, $\lambda\lambda$ 1612, 1665, 1667, and 1720 MHz have actually been observed, which correspond to transitions between the Λ sublevels that are due to hyperfine structure [171a–e]. The transition probabilities for the most intense of these lines are 2.66×10^{-11} sec^{-1} ($\nu = 1667.4$ MHz) and 2.47×10^{-11} sec^{-1} (1665.4 MHz). The theoretical ratio of the optical depths in the centers of these lines in the absence of saturation is 1.8.

If the profiles of the lines depend only on thermal motions, the ratio of the optical depth of the most intense OH line to that of $\lambda = 21$ cm is

$$\frac{\tau_{\mathrm{OH}}(1667.4)}{\tau_{\mathrm{H}}(1410.4)} = \frac{2.66 \times 10^{-11}}{2.85 \times 10^{-15}} \cdot \frac{5}{3} \left(\frac{\nu_{\mathrm{H}}}{\nu_{\mathrm{OH}}}\right)^3 \left(\frac{m_{\mathrm{OH}}}{m_{\mathrm{H}}}\right)^{1/2} \frac{N_{\mathrm{OH}}}{N_{\mathrm{H}}} = 4.0 \times 10^4 \frac{N_{\mathrm{OH}}}{N_{\mathrm{H}}}. \tag{9.14}$$

The relative abundance of OH molecules can be determined immediately in this way from observations of OH and H. It is true that the line profiles are usually influenced by turbulence or other macroscopic motions of the gas. However, we can make use of the fact that the Doppler widths due to thermal motions are inversely proportional to the square root of the mass of the molecule or atom, while the turbulent Doppler width is independent of the masses of the particles. Therefore, by determining from observations the half-widths of lines of OH and H, we can separate their thermal and turbulent parts [171e], and at the same time determine not only the relative abundances of OH molecules, but the temperature and turbulent velocities inside the cloud as well. If we denote by $\Delta\nu_{OH}$ and $\Delta\nu_H$ the observed widths of the corresponding lines, we obtain for the temperature and root-mean-square turbulent velocity the expressions

$$T = \frac{m_H m_{OH}}{2k(m_{OH} - m_H)} [(\lambda_H \Delta\nu_H)^2 - (\lambda_{OH}\Delta\nu_{OH})^2], \tag{9.15}$$

$$\overline{v^2}_{turb} = \frac{m_H}{m_H - m_{OH}} \left[(\lambda_H \Delta\nu_H)^2 - \frac{m_{OH}}{m_H}(\lambda_{OH}\Delta\nu_{OH})^2\right]. \tag{9.16}$$

Obviously, other lines may be treated in a similar fashion.

Lines of OH were observed in absorption against the background of the sources Cassiopeia A [171a, c, e] and the galactic center [171b, d]. In the former case, absorption lines were observed that were formed in two clouds with radial velocities of -0.1 and -1.5 km/sec. The method described above, based on OH and H lines, was used to determine the gas temperature in these clouds (120° and 90°, respectively) and turbulent velocities (0.24 to 0.27 km/sec). The ratio of abundances of the molecules was $N_{OH}/N_H \approx 10^{-7}$ [171e].

Complex random motions of masses of gas were observed in the direction of the galactic center, with velocities ranging from 80 to -170 km/sec. The optical depths in the strongest lines, estimated from residual intensities, approached unity. Detailed interpretation of these OH spectra was difficult because the relative intensity of all four lines showed partial saturation (the theoretical intensity ratios from a thin layer are 1:5:9:1, in order of increasing frequency, while the observed values are 1:2.2:2.7:1).

The possibility of observation of the H_2 line $\lambda = 28\ \mu$ has been discussed by Raich and Cood [171f] and by Spitzer *et al.* [171g]. They calculated the intensity of the line, its curve of growth, and its equivalent width under different assumptions about the concentration of molecular hydrogen. It was shown that in principle this line can be detected by observation in the direction of the galactic plane where its intensity may correspond to black-body radiation with $T = 50°$. Up to now there have been no observations.

But there is the possibility of evaluating the concentration of H_2 molecules from indirect observations, by the lack of atomic hydrogen in different interstellar objects [171h, i], which sometimes give high concentrations of H_2. However, this method is inconclusive.

More recently, there have been investigations of the origin of H_2 molecules [171j]. It was shown that the rate of formation of such molecules on cosmic dust particles is negligible if their temperature exceeds 8°K because of weak adhesion of the hydrogen atom to the surface of the dust particle. The possibility of escape of H_2 molecules from the atmosphere of late-type stars in their stellar wind has also been discussed [171k].

Very unexpected was the discovery of OH emission in the vicinity of some H II regions [171l, m]. These lines have high intensity (perhaps as high as several millions of degrees) but unusually small width, even as small as 400 Hz. Most of them are linearly or circularly polarized, with the degree of polarization up to 100 percent. They originate in small regions approximately 10^{15} cm across and usually change considerably in the course of a few days. They are apparently due to an interstellar maser effect, but up to now there is no satisfactory explanation of its mechanism [171n, o].

An attempt to observe the line $\lambda 3080$ of this radical OH was unsuccessful [171p]. This gives an upper limit for the concentration of OH of nine times less than is indicated by radioastronomical data. Perhaps this molecule is formed in very small regions, and its maser effect might be explained by some physical–chemical process.

Complex Molecules and Diffuse Lines

In principle, the spectra of triatomic molecules should be observed. The spectra of these molecules, consisting of H, C, O, and N, have been partially studied in laboratories. In the accessible visible and near-infrared regions, they have low oscillator strengths so that interstellar lines are weak and these molecules cannot be observed. However, it is possible that lines from these molecules will be discovered in the far ultraviolet. Electron transitions for a series of unsaturated molecules, such as NO_2, and free radicals lie in the accessible spectral region, but the number of such radicals that are known is small. Certain of them occur in comets and in planetary atmospheres but in interstellar space their concentration is small because of the low density and possible destruction by stellar ultraviolet radiation.

In addition to sharp atomic and molecular lines in interstellar space, there occur the so-far unidentified diffuse lines. Since the hypothesis of their molecular origin is the most probable, we shall discuss the data on these lines in the present section. A list of diffuse lines with an indication of their widths and intensities on an arbitrary scale is given in Table 9.3.

The diffuse character of these lines is difficult to explain. If they were from molecules composed of atoms heavier than Ne, smearing of the rotational structure might be expected, that is, blending of the rotational levels. However, such molecules have low abundances and cannot give noticeable lines in interstellar space.

The opinion has been expressed that the broadening of molecular lines may be the result of predissociation [169]. This process is as follows. The absorption

Table 9.3

λ (Å)	Intensity	$\Delta\lambda$ (Å)
4430.6	10	20
4760	—	Wide
4890	—	50–60
5780.55	3	2
5797.13	1	1.5
6180	—	50–60
6203.0	1	—
5270.0	1	—
6283.9	6	6
6613.9	2	4

of a light quantum transforms a molecule into an excited state. If the attraction curve of this excited state intersects the repulsion curve of the other state, then close to the point of intersection a spontaneous transition between them is possible. As a result, the molecule disintegrates after a very short time. It is easy to see that this produces broadening and smearing of the rotational levels. We consider a numerical estimate. The molecule exists in the excited state for 10^{-8} sec. During this time, it performs $10^{-8}/10^{-13} \approx 10^5$ vibrations and about $10^{-8}/10^{-11} \approx 10^3$ rotations. Since predissociation takes place after a few vibrations, not even a single rotation is possible after the molecule has attained the predissociated state. This means that rotation ceases to be quantized and the rotational levels are smeared. The probability of predissociation depends on the time during which a molecule stays near the point of intersection, that is, it is maximum when the turning point of the vibration is close to the point of intersection.

Unfortunately, there are several difficulties in explaining diffuse interstellar lines by the phenomenon of predissociation. First, in order that predissociation may smear a line in the visible region, the dissociation potential of the corresponding molecule must be less than the energy of a quantum (about 1.5 eV) and such diatomic molecules are unknown at the present time. In triatomic molecules, however, this is possible. For example, in the well-studied HCO molecule, the transition of the vibration (6.0) structure corresponds with the very diffuse line $\lambda = 7639$ Å. Other radicals containing hydrogen, such as HO_2, C_2H, HNO, HN_2, and so forth, can have similar wide lines. The dissociation energy of these radicals is very small, but they have been little studied. The fact that strong bands are observed means that the dissociation probability must be considerable—in fact, each quantum absorbed disrupts the molecule. It is highly improbable that there exists in interstellar space a mechanism for forming triatomic molecules with an efficiency sufficient to compensate for their destruction during predissociation. It may be correct to suggest that diffuse lines are formed during predissociation of still more complex molecules. The restoration of complex molecules is more probable, since here each collision with an atom leads to condensation and to the formation of molecular complexes. It is well known that there is a noticeable correlation between the reddening of stellar radiation (that is, absorption by dust) and the intensity of certain diffuse lines, in particular λ 4430, predicted earlier by Wilson [194b]. An example of such a correlation, taken from the work [194a], is shown in Fig. 19.

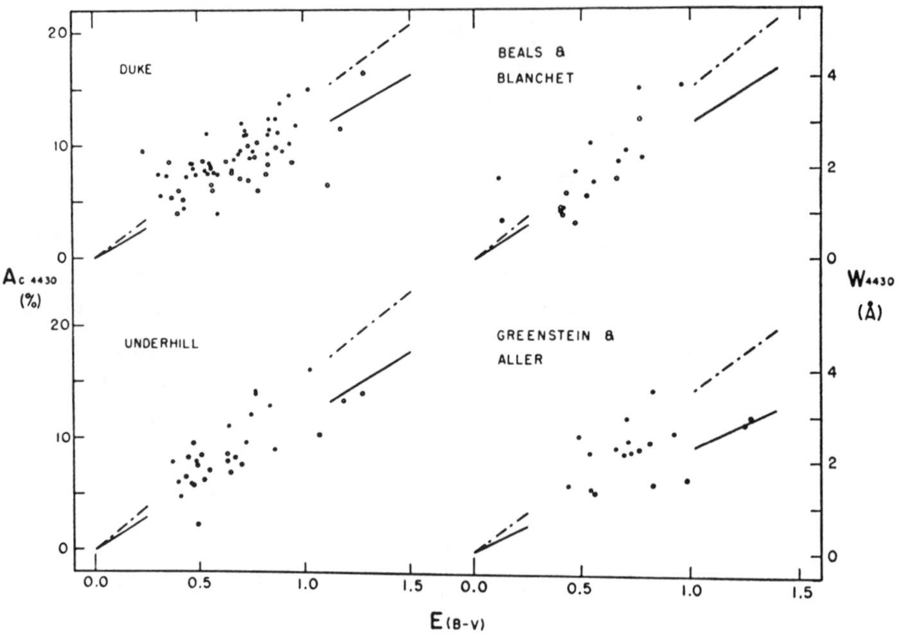

FIG. 19. Intensities of the 4430-Å band plotted against $E(B - V)$ [194a]: *solid circles*, observations in galactic longitude region $90° \leqslant l \leqslant 110°$; *open circles*, $30° \leqslant l \leqslant 70°$.

Therefore, we can assume that diffuse lines are formed during absorption by metallic atoms and ions embedded in dust particles. Here absorption in the separate lines is possible if the atoms have unfilled internal levels screened from the effect of the surrounding atoms. Such spectra are seen in the laboratory in crystals of rare-earth salts and in elements of the iron group. They are composed, especially at low temperatures, of a multitude of more or less sharp "lines." Several arguments are advanced against this hypothesis [169]: first, such transitions must be due to multiplets, and second, they are forbidden by Laporte's rule. Greenberg and Lichtenstein [168b] demonstrated the effect of impurities having resonance lines in the vicinity of λ 4430 on a complex refractive index. Using the Mee theory, extinction is calculated with respect to wavelength for spherical particles of "contaminated ice" of various sizes. An acceptable size distribution can explain the line λ 4430. Herbig has suggested [168c] that absorption in the band λ 4430 occurs as a result of a system of molecular lines from H_2 excited during absorption from the metastable level C $^3\Pi$ and broadened by intermolecular interactions if the absorbing molecules are adsorbed on dust particles. There is a serious difficulty here in maintaining the population of this level, whose energy is approximately equal to 10 eV.

Still other hypotheses have been proposed. In analogy with the predissociation phenomenon, it might be supposed that such lines arise during the preionization of negative ions, for example, during the transition $2s^22p^5$ $^2P \rightarrow 2s2p^6$ ^2S in the O^- ion, shifting an electron into an unstable orbit. These transitions are permissible and possess high values for the oscillator strengths f. When $f \approx 10^{-2}$, the concentration of oxygen ions must be about 10^{-6} N_H to excite the line $\lambda = 4430$ Å, and if $f \approx 1$, as, for example, in the above O^- transition, 10^{-8} N_H O^- ions are needed. However, it is not known whether the wavelengths in this case are suitable, since the energy of the quantum must be greater than the energy of affinity. For O^-, this energy exceeds 3 eV, which is greater than the energy of all diffuse lines. Moreover, here we must keep in mind the remarks made above about predissociation: each quantum absorbed disrupts a negative ion, so that a very effective mechanism for the formation of negative ions must exist. Thus, the question concerning the origin of diffuse absorption lines in interstellar space still remains open.

Rapid molecular growth can lead to the formation of molecular complexes that are neither dust particles nor molecules. This question will be discussed in the following chapter.

10. The Morphologic Features of Diffuse Nebulae

The brighter H II regions are called diffuse nebulae. In certain cases, these consist of ordinary clouds of interstellar gas with a density of 10–20 particles per cubic centimeter, ionized by radiation from one or several hot stars. An example of such a region is the "North America" nebula. In other cases nebulae consist of more compact formations expanding into the surrounding medium. Since the surface brightness of a nebula diminishes with expansion, the brightest are young dense nebulae, small in size. An example of such a young object is the well-known nebula in the constellation of Orion.

Despite their name, diffuse nebulae do not consist of amorphous, structureless formations. G. A. Shain [97, 98] recorded a number of characteristic peculiarities and details often found in nebulae and evidently connected with hydrodynamic and magnetohydrodynamic effects. Such a peculiarity is the elongated form of many large nebulae, resulting from their expansion in an external magnetic field, which restricts their transverse motion [172] (Fig. 20*a*). Another characteristic type of large nebulae has a comparatively symmetric ring-shaped form. They constitute nebulae with matter concentrated at the periphery (Fig. 20*b*). Their formation is also connected with expansion [173], accompanied by

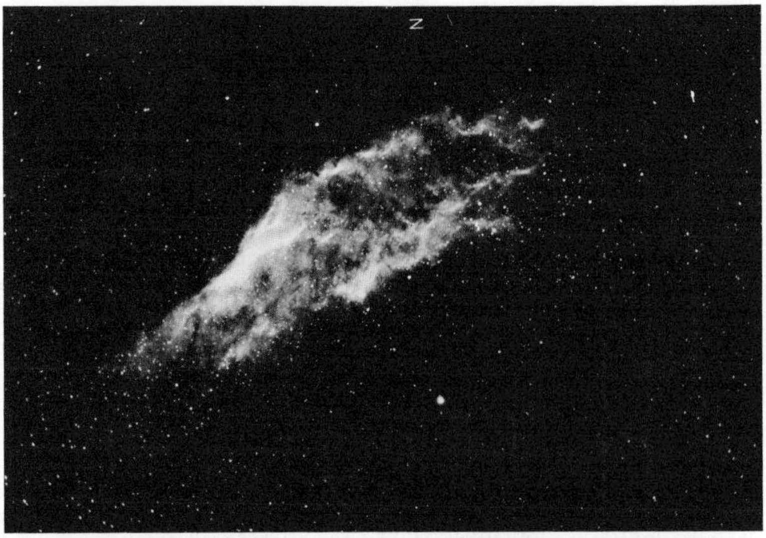

(a)

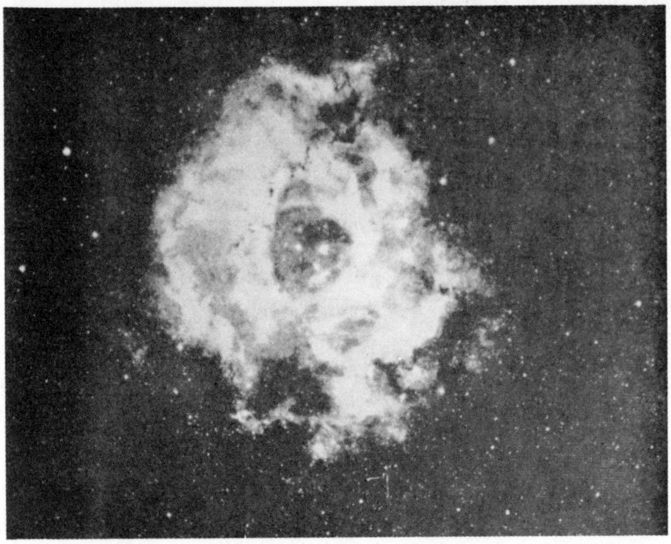

(b)

FIG. 20. (*a*) NGC 1499, the "Wave" or "California" nebula; (*b*) NGC 2237, 2238, and 2244 and S 162, the "Rosette" nebula. (Crimean Observatory.)

interaction with the surrounding medium. Depending on the degree of its expansion, a ring-shaped nebula often becomes elliptical.

Nebulae tend to associate with other diffuse bodies. For example, the region in the constellation Cygnus (see Fig. 7) shows many emission clouds in a small area. Maximum thermal radio emission occurs in this direction. The large number of nebulae is partially explained because the line of sight passes along one arm. However, R. N. Ikhsanov showed that a large fraction of the nebulae close to γ Cygnus (in the southern part of the mosaic, Fig. 7) form a single complex and are not spread out along the arm [226]. This conclusion follows from the fact that there are comparatively few hot stars in the area to ionize hydrogen. If the nebulae were distributed in a large volume, their ionization would demand many more stars than are observed. Further, with nebulae associated in a single complex, whose distance in this case must be about 1000–1500 pc and the mass about $2 \times 10^5 M_\odot$, there is still an insufficient number of hot stars grouped near the Cygnus VI association, and we must assume the presence of unobserved stars.

Even more frequently, associations of nebulae with dark clouds are observed. In the majority of cases, this is not a random projection and these genetically connected objects form a single complex. Evidently, this was originally a dense mass of dark matter, in which young stars emerged, among them massive hot giants, as the result of certain condensation processes. The ultraviolet radiation from these stars ionized and heated a portion of the gas and converted it into a nebula. The expansion of the hot gas, which will be discussed in more detail in Chapter 5, allowed the ionization of new mass, while the denser portions remained unionized. They have the form of dark projections with their edges turned to the star and condensed by the pressure of hot gas. Highly elongated projections are called "elephant trunks" (Fig. 21). In a later stage of development, the "trunks" disintegrate, forming separate dark clots, or globules, which are visible in the same photograph.

The parts of globules and projections exposed to a star are ionized, and the pressure in a thin surface layer is increased. A bright rim of expanding gas is

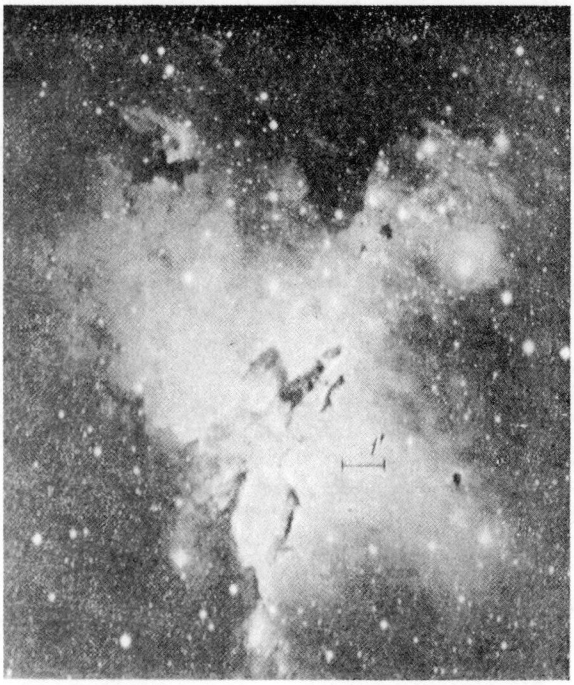

FIG. 21. A portion of nebula NGC 6611 with "elephant trunks," hoops, and globules. (Palomar Observatory.)

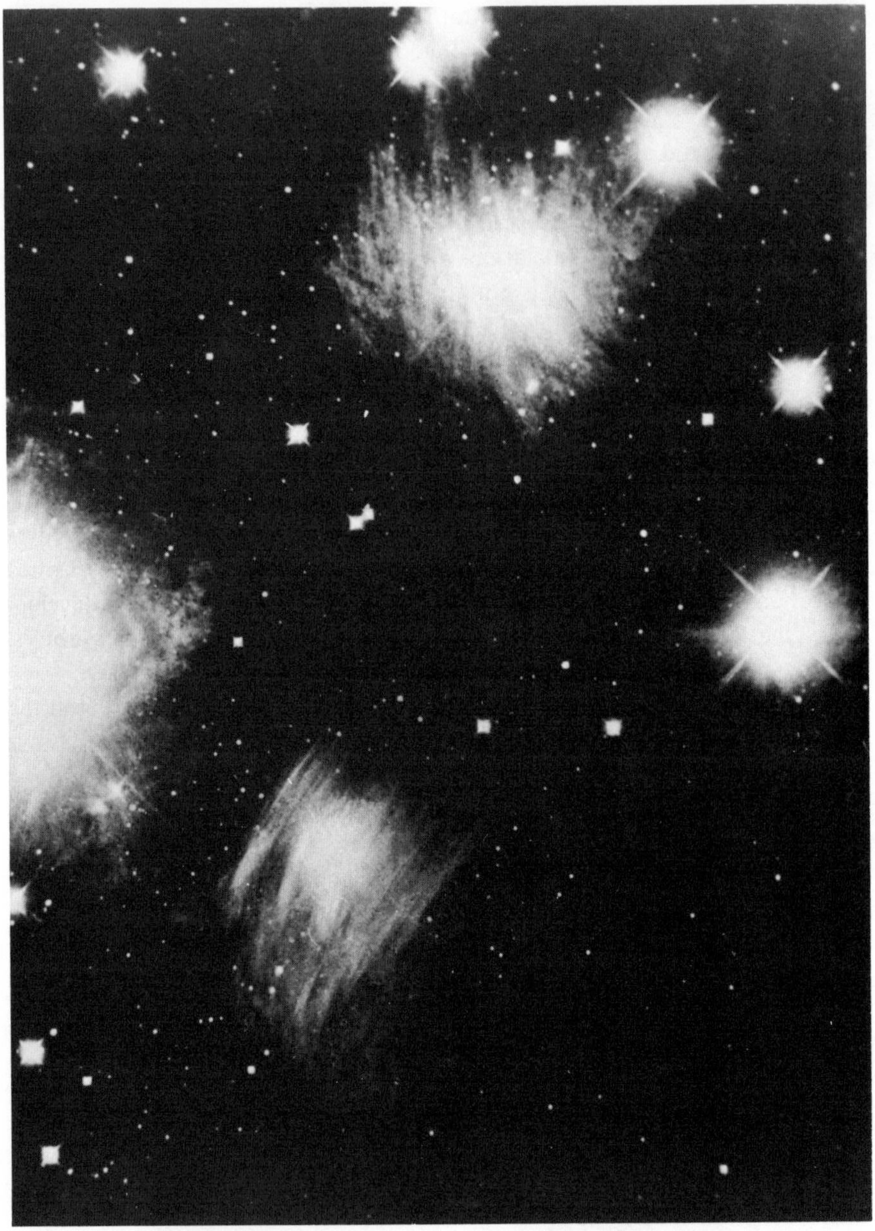

Fig. 22. Reflecting nebulae in the Pleiades; thin filaments are visible. (Palomar Observatory.)

formed, denser than an H II region, called a "hoop." In certain cases, the boundary between the bright and dark gas is almost flat or cylindrical; the hoop then has the shape of a long filament. Typical examples of such hoops are seen in Fig. 21.

There are usually few "filaments" of such origin in nebulae. At the same time in large nebulae there occur a large number of true filaments, frequently of a complex shape [97, 174]. The origin of these filaments is not clear; they are possibly connected with electromagnetic forces. A thin filamentary structure is also observed in dust formations—bright reflection nebulae, for example in the Pleiades (Fig. 22), and dark nebulae. No explanation has yet been given for these structures either.

Individual Nebulae

We shall consider a few individual nebulae that can serve as specific examples of different types. These nebulae are often referred to in the literature.

(1) *The Orion nebula* (NGC 1976 or M 42) is the brightest diffuse nebula in the northern sky (Fig. 23). It is part of a large complex of bright nebulae and hot stars forming several associations. Its distance is about 500 pc. A large ring

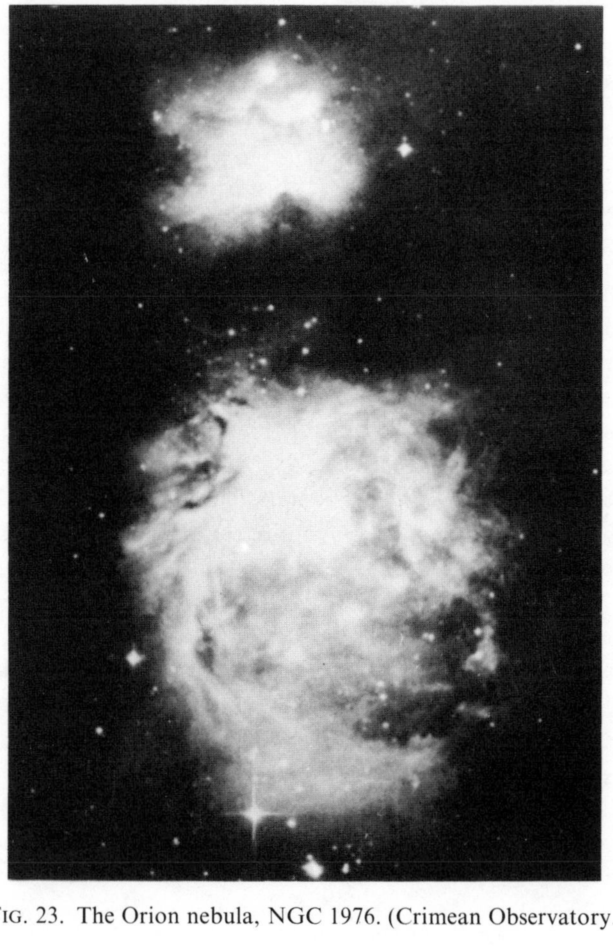

FIG. 23. The Orion nebula, NGC 1976. (Crimean Observatory.)

of luminous gas and many nebulae and filaments are visible in a photograph of this region taken in H_α radiation. The basic constituent of the complex is an envelope of neutral hydrogen, whose thickness is about 30 pc and whose internal radius is about 40 pc, expanding at the rate of 10 km/sec [175]. The external portion of the envelope is visible as a large luminous arc (Barnard's arc), measuring 60×180 pc^2. Inside the ring is found an expanding H II region. Menon, who observed the envelope in the 21-cm line, estimated the mass of the complex (H I and H II) to be 110,000 $M_\odot$. The associated stars in the complex are very young. Among them there are many class O stars whose age is measured in a few million years. At the same time, there are stars that are less massive, for example, class A0 stars, which have not yet completed their contraction stage, characterized by weak H_α radiation. Correlated with the youth of the complex are the "high-velocity" hot stars. The velocities of these stars are given in Table 10.1.

Table 10.1

Star	Spectral class	Velocity (km/sec)	Age (10^6 yr)
AE Aur	O9.5	106	2.7
μ Col	B0	123	2.2
53 Ari	B2	59	4.9

The directions of motion of these stars and their positions show that they emerged from the center of the ring. According to Blaauw [176], high-velocity stars are formed from a compact system of two or three massive stars after the most massive star finishes its evolution before the others and is converted into a supernova, throwing off a considerable portion of its mass. This reduces the mutual attraction and the rapidly rotating stars separate with high velocities. The nonthermal source of radio emission adjacent to the dark "Horsehead" nebula [177] evidently represents the envelope of a supernova whose expansion was halted by the dense nebula on one side.

All these facts indicate that the Orion complex consists of a region where stars, together with radiating nebulae, are born, a huge "factory," which has been operating for several million years. The most massive stars have already disappeared through evolution, while the less massive continue to excite the luminescence of nebulae, and the expanding H I ring gradually disperses into interstellar space, producing ordinary gas clouds. The material for the creation of all these stars and nebulae evidently consisted of the large mass of dense, cold gas, similar, for example, to the complex dark clouds in Taurus-Ophiuchus.

The Orion nebula is comparatively small. The diameter of the total nebula is about 1°. Its central portion, with a radius of about 4′ (0.58 pc), has a very high density—about 2×10^4 cm^{-3}, and its chaotic fluctuations in brightness suggest a "boiling kettle." At the periphery, the density determined at the line λ 3727 (O II) falls to 260 cm^{-3} and below [178]. It follows from this, incidentally, that the pressure gradient associated with the change in density must lead to its rapid dissolution [178a]. A velocity of 10 km/sec will thus be attained after the time $m_H vR/kT \approx 2 \times 10^{12}$ sec $\approx 3 \times 10^4$ yr. This young, dense nebula surrounds the multiple system of Orion–Trapezium, from which the name "Trapezium-type systems" arises. The hottest star, Trapezium θ^1 Orion of class O6, excites luminescence in the nebula. Apart from gas, the nebula contains much dust, about 1 percent by mass. This shows up through the reddening of stars in the Trapezium; absorption in the central portion of the nebula reaches 5–10 stellar magnitudes per parsec. In connection with this, the nebula is opaque in the visible spectral region and we see radiation from the front side of the nebula, since the longer the wavelength the thicker the effective radiating layer. Radiation from the remaining portion of the nebula is repeatedly scattered and is practically unabsorbed, so that the luminosity decrement of the nebula is altered little as a whole.

The nebula constitutes a powerful source of thermal radio emission, and this emission has a sharp maximum near θ^1 Orion. Various methods for determining density in the Orion nebula have led to somewhat diverse results [84, 179]. In particular, radio observations lead to systematically lower values. This may possibly be because the O II lines from which density is determined by the doublet method described above ($\lambda\lambda$ 3729 and 3726) are produced in the denser parts of the nebula, while radio observations give only a mean density [84].

Studies of line widths, and also variations of radial velocities and brightness fluctuations, indicate the presence of internal velocities of 7–8 km/sec. In discrete areas, bifurcated lines are observed, which may be explained on the hypothesis of internal motion of shock waves. The spectrum of the nebula and its chemical constitution were studied by Aller and Liller [180]. Deviations from the usual distribution of elements were not observed.

(2) *The Omega nebula* (NGC 6618 or M 17) is very similar to the Orion nebula. This is also a very dense, young nebula (Fig. 24), one of a group of bright nebulae in the constellation Sagittarius. The distance to the nebula is about 1700 pc, and the size of its center is about 4′, that is, 2 pc. Furthermore, two radio-emitting condensations were observed within the boundaries of the nebula. One of these yields up to 28 percent of the total flux, the other being much weaker [181a]. To the west of the nebula, there is an adjacent absorbing cloud. Radio observations show [181] that this cloud covers several smaller nebulae of half the size, including a nucleus that emits very intense radio waves. The electron concentration in the nucleus from radio observations is $N_e =$ 900 cm^{-3}. Optical observations in H_α give approximately the same value, $\sim$1100 cm^{-3}. The mass of the nebula is $\sim 350\ M_\odot$. The strong infrared lines of S III, whose absorption is not very great, shine through the extremely dark

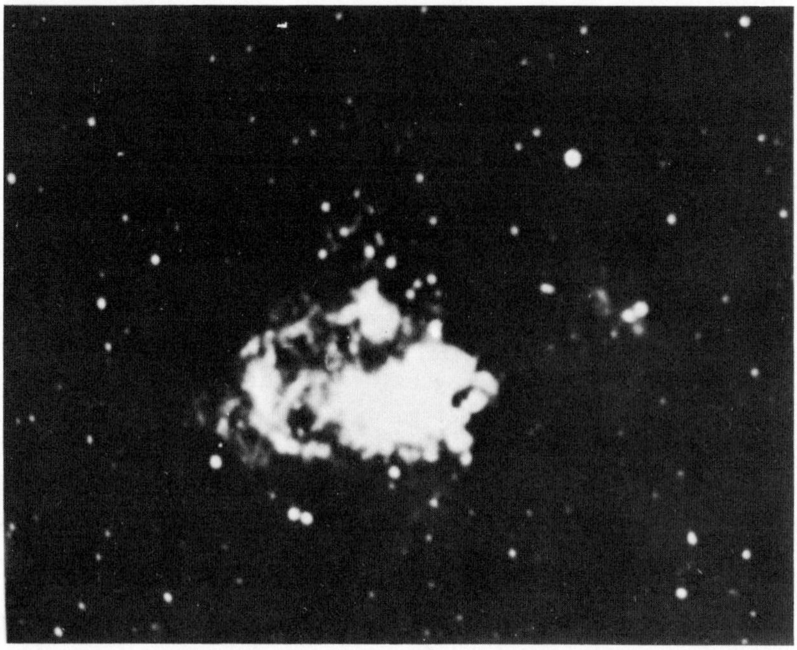

FIG. 24. The Omega nebula, NGC 6618. (Crimean Observatory.)

cloud to a certain extent. From a comparison of radio and optical observations, the total absorption up to the center amounts to $\sim 11^m$. Taking into account absorption up to the nebula, it may be shown that absorption in the dark cloud exceeds 8^m [182]. On the assumption that dust constitutes about 1 percent, the total mass of the cloud is equal approximately to 30 $M_\odot$ (density $N_H \approx 4 \times 10^3$ cm^{-3}) [183]. The absolute radio brightness in the Omega nebula is 30 times that in the Orion nebula. The known hot stars cannot provide such thermal radio emission. A star or a close system having a very large light source has been discovered in the center of the nebula from an infrared photograph. Possibly this star is the exciting center of the nebula. It is interesting that near the nebula, at a distance of 6 pc from the center, there is a class O5 star, receding from the center with a velocity of 200 km/sec. On the western side is a vast filamentary system and apparently a mixture of nonthermal radio emission. It is possible (R. E. Gershberg, [183b]) that one or several supernovae have recently erupted in the nebula. Velocities of separate regions in this nebula were obtained by means of interferometer measurements and found to lie in the range from -40 to 61 km/sec [183a]. The velocity of the central region of the nebula is 20.7 km/sec. It is thus possible that its filamentary structure is dispersing with velocities of approximately 50 km–sec.

(3) *The Lagoon nebula* (NGC 6523 or M 8) is situated in the constellation Sagittarius (Fig. 25). Its distance is about 1500 pc, its dimensions more than 10 pc. The nebula has a dense core and several envelopes whose brightnesses fall

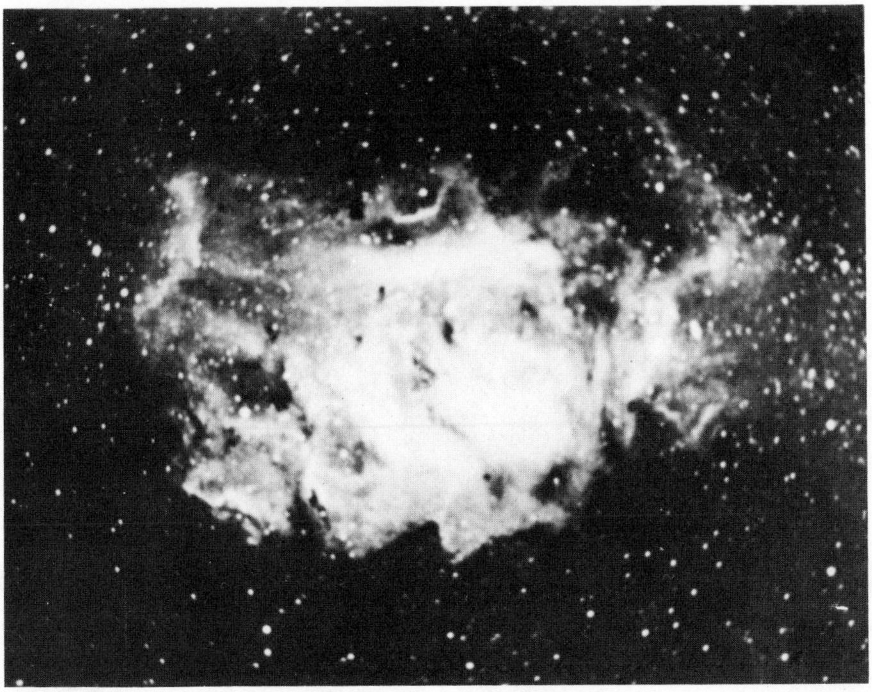

FIG. 25. The Lagoon nebula, NGC 6523. (Crimean Observatory.)

off with distance from the center [184, 185]. The envelopes are explained apparently by expansion of the nebula, dragging on the interstellar gas, with its subsequent ionization [135]. The density of the central core is more than 200 cm^{-3}, in the envelopes it falls to 70 cm^{-3} and 40 cm^{-3}, and on the periphery it is 20 cm^{-3} and less. The mass of the envelopes amounts to a hundred times the Sun's mass, while the total mass of the nebula is about 1500–2500 $M_{\odot}$.

The exciting stars are apparently the star HD 164,794 of class O5 and, to a lesser degree, HD 165,052 of class O7, found at the periphery of the nebula. Around the O5 star there is a zone where the luminescence of O III lines was discovered (about half the radius). In the remaining portion O II lines occur. This nebula is less dense than Orion and Omega, but is also comparatively young. This is an indication of the early class of the exciting star and the nature of other stars of the central cluster—the majority of these stars belong to type T Sagittarius that have not yet reached the main sequence—and of the presence of a small nucleus consisting of two nebulous stellar pinwheels with a strong continuous spectrum [186]. The density of the small nucleus is greater than 200 cm^{-3}. The temperature of the nebula is somewhat higher than 8000°. Dust in the nebula is comparatively scarce, and on the whole it is transparent, although the central portion is opaque. At the periphery, the nebula is bounded by dark clouds with characteristic hoops and filaments. In the directions where the nebula is brighter, the dark clouds are closer to the center, that is, the size of the Strömgren zones is less there.

Internal motions have also been observed in the Lagoon nebula [183a] similar to those in the Omega nebula. The velocity of disruption is here about 25 km/sec, while the average velocity for the whole nebula is approximately 8 km/sec.

(4) *The Trifid nebula* (NGC 6514 or M 20) is also located in the constellation Sagittarius (Fig. 26). The characteristics of this nebula are thin, dark fissures,

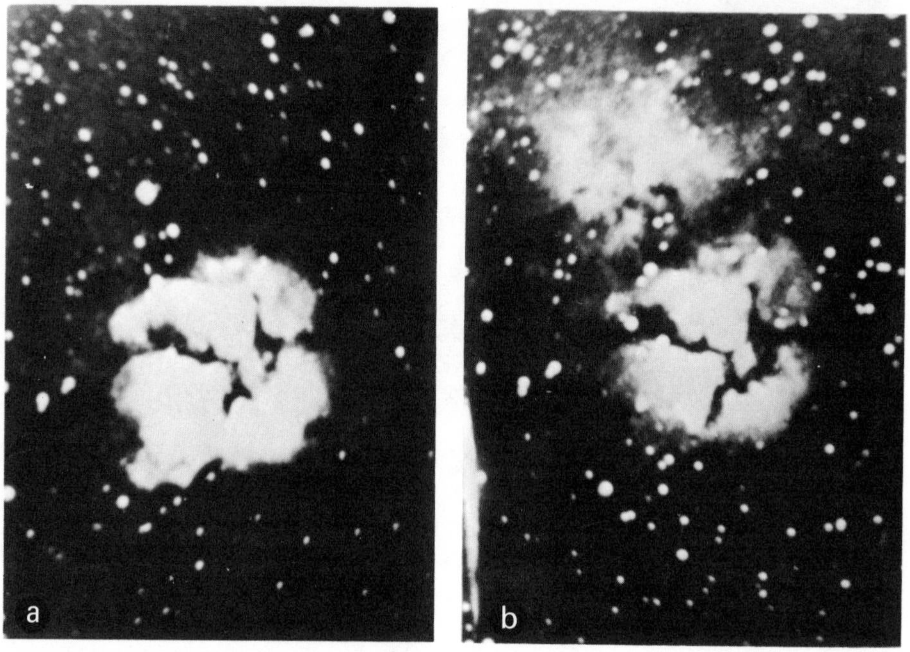

FIG. 26. The Trifid nebula, NGC 6514: (*a*) in H_α radiation; (*b*) in blue light, in which a reflecting nebula is visible. (Alma-Ata [20].)

as if the nebula were split into pieces. This is possibly cold material compressed on both sides by hot gas. To the northeast of the nebula, a reflecting nebula adjoins, visible in the continuous spectrum but invisible in H_α. In size and form it is similar to the emitting nebula. The reflecting nebula is illuminated by a supergiant of class A7.

(5) *The Wave or California nebula* (NGC 1499) is found in the constellation Perseus (Fig. 19). It is of interest because it seems to consist of distinct filaments with a weak background. The filaments are deflected, while the prominence is exposed to a rather distant hot star, ξ Persei [187]. The nebula is extremely elongated, which is especially noticeable on photographs of long exposure [336].

(6) *The North America nebula* (NGC 7000), together with the Pelican nebula (IC 5070) and other smaller objects (Fig. 27), is a typical example of an ionized

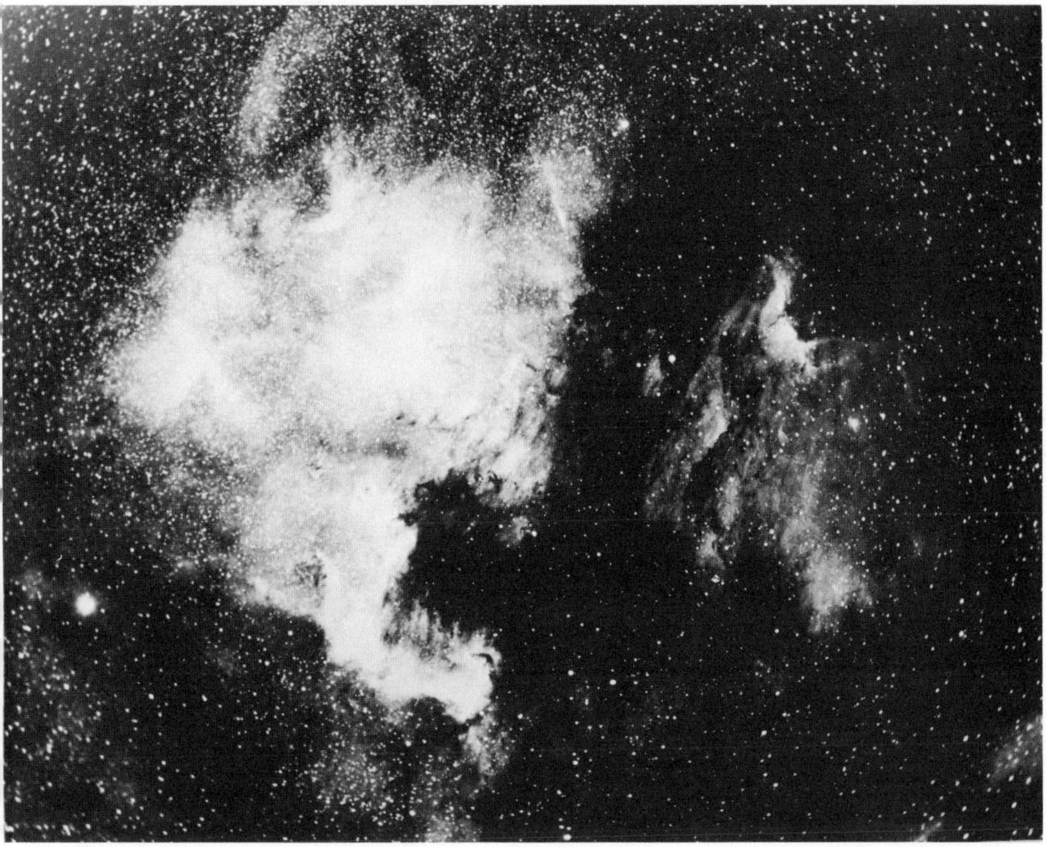

FIG. 27. The North America nebula, NGC 7000, and the Pelican nebula, IC 5070. (Crimean Observatory.)

cloud, an H II region with a concentration of $\sim 10\ \text{cm}^{-3}$. The whole group, whose distance is about 1000 pc, consists apparently of a single luminous cloud, separated by parts of projecting dark clouds, which form the Gulf of Mexico and a band between America and the Pelican. The principal portion of these clouds apparently belongs to the same complex as the luminous matter. The total diameter of the whole complex is approximately 50 pc. There is a single hot star of class O7 situated in the northern part of NGC 7000, which excites this portion. However, it cannot ionize the whole nebula. No other exciting stars in this object or in other nebulae of the complex, in the Pelican in particular, have been discovered. Apparently, one or several hot stars lie in the central portion of the complex behind the dark clouds of the Gulf of Mexico. In order to render a hot star invisible (weaker than 12^m), absorption by the cloud must be greater than 6^m. Hence, the density of the dust may be estimated to be $\rho \sim 10^{-23}\ \text{g/cm}^3$ and the mass of the total cloud, assuming that the gas is 100 times greater, is approximately 1000 $M_\odot$ [188]. In the same paper, it is shown that the degree of ionization of oxygen decreases with distance from the center. At the same time, line brightness in the filaments does not increase compared to the background, while in H_α emission the filaments are separated. Probably this is because most of the oxygen occurs in the O II state, since ionization is produced by radiation. Then, as follows from Eq. (6.12), the O II concentration diminishes with decrease in density, while the O III concentration scarcely changes. The continuous spectrum of the nebula is due primarily to the background of the distant stars of the Milky Way. In the nebula itself, this background is bright, but the cloud surrounding the nebula simultaneously obscures the background as well.

3. Interstellar Dust

11. Selective Interstellar Absorption and the Optics of Dust Particles

The existence of interstellar light absorption was long suspected, even by V. Ya. Struve, during the last century, but it was detected with certainty in 1930 from the well-known work of Trumpler [189].

The study of interstellar light absorption largely originated through methods of stellar astronomy; we shall not linger over the details of the problems here. We shall consider mainly what pertains to the physics of separate dust particles and dust clouds. The hypothesis of the absorption of stellar light by cosmic dust particles can now be considered as universally adopted, but certain difficulties may require further explanation.

The Dependence of Absorption on Wavelength

The most essential feature of interstellar absorption is its dependence on the wavelength of the light, the so-called selectivity. It is well known that interstellar absorption in the blue portion of the visible region of the spectrum is greater than in the red; consequently stars with appreciable absorption appear reddish. In Fig. 28 are reproduced microphotograms of two stars of the same spectral class B1. It is quite obvious that in one of them (26 Cephei) the blue part of the spectrum is reduced. Studies on stellar reddening have been carried out by various methods. The most effective is the method of photoelectric six-color photometry, developed by Stebbins, Whitford, and others [190]. In this method, stars of a single spectral class, but with different reddening, are compared in six spectral bands, close to the following wavelengths:

Band	U	B	V	G	R	I
λ_{eff} (Å)	3535	4220	4880	5715	7195	10310
$(1/\lambda)_{\text{eff}}(\mu^{-1})$	2.86	2.40	2.09	1.79	1.43	0.99

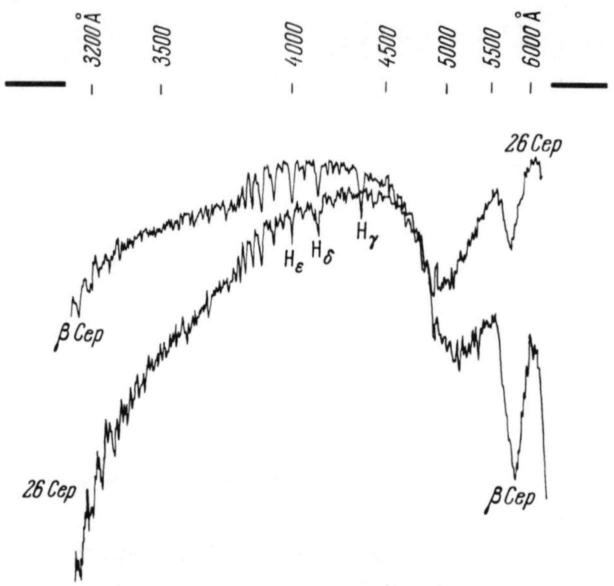

FIG. 28. Microphotograms of the spectra of two stars of a single spectral class B1, but with different degrees of reddening [237].

The mean curve obtained from the measurements, which describes the dependence of the interstellar absorption coefficient on wavelength over the range $1/\lambda = 3.0\ \mu^{-1}$ to $1/\lambda = 0.5\ \mu^{-1}$, is given in Fig. 29 [191]. Here absorption is expressed in stellar magnitudes relative to absorption at $\lambda = 1\ \mu$. Values of the absorption coefficient for $\lambda\lambda = 0.22\ \mu$ and $0.26\ \mu$ were estimated in [191a] from rocket data on six stars. These points lie within the limits of the errors on a continuation of the curve shown in Fig. 29.

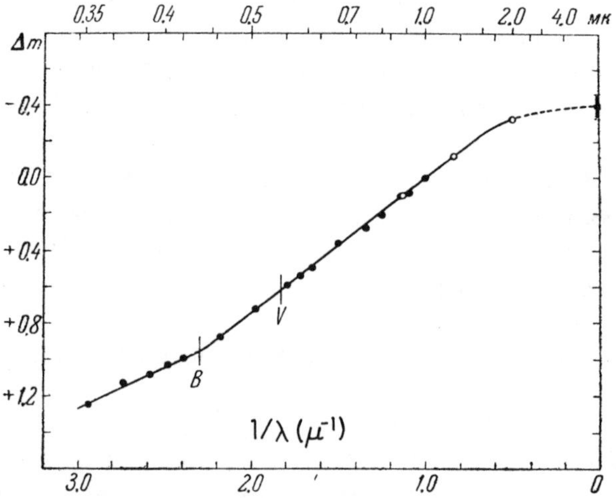

FIG. 29. The relation of absorption to wavelength [191].

As follows from Fig. 29, the interstellar absorption coefficient κ over the wavelength range 0.35 to 1 μ is to some degree of accuracy proportional to $1/\lambda$. We note that this method cannot determine the presence of neutral light absorption, which does not depend on wavelength, since it gives only the difference of absorption at various wavelengths. In this way, an empirically established dependence of the interstellar absorption coefficient $\kappa(\lambda)$ on the wavelength in a specified range is described by the formula

$$\kappa(\lambda) = C_1/\lambda + C_2. \tag{11.1}$$

The numerical value of the constant C_1 may be determined if the distance to the reddened star is known; the constant C_2, as previously stated, is not determined by this method. In Fig. 29 the absorption curve is extrapolated to $1/\lambda = 0$, where the linear dependence on $1/\lambda$ is upset. Extrapolation is inaccurate and therefore the region of possible relative values of κ when $\lambda \to \infty$ is indicated in the figure by a vertical line.

To estimate neutral absorption, it is possible to observe the decrease in radiation in the far-infrared region of the spectrum, where C_1/λ is small. For example, an attempt by V. I. Moroz [192] to discover the galactic center using emission of wavelengths up to 2.5 μ gave a negative result. Since there are many red and yellow stars in the central portion of the Galaxy, it might be anticipated that emission from the galactic center in this spectral region would be sufficiently intense. Since, within the limits of sensitivity of the apparatus, radiation at $\lambda \approx 2.5\ \mu$ has not been discovered from the galactic center, we must assume either that absorption from the center in the visible is greater than $8\text{–}10^{\text{m}}$ or that neutral absorption occurs whose value in this direction is about $1\text{–}2^{\text{m}}$.

At $\lambda \approx 4500$ Å, the selective interstellar absorption curve experiences a small jump. At the shorter wavelengths, the condition of Eq. (11.1) is well maintained, but with new values of C_1 and C_2. Apparently, in the ultraviolet region, during the increase in $1/\lambda$, the growth of the absorption coefficient must be slowed down and most probably suspended at a certain wavelength. Observations of stellar ultraviolet spectra made by O. A. Melnikov [193] confirmed this conclusion; however, there has been no immediate success in determining detailed absorption coefficients in the ultraviolet spectral region.

Integrating the absorption coefficient κ per unit volume along the line of sight to a star, we obtain the optical thickness τ_λ for interstellar absorption at a given wavelength. Frequently, instead of τ_λ, the value of absorption, expressed in stellar magnitudes,

$$A_\lambda = 1.086\ \tau_\lambda \tag{11.2}$$

is used. Absorption (accurate to the term C_2) in the six-color photometer bands is usually given in terms of the indices of the corresponding bands, for example, A_{U} or A_{B}.

Starlight is characterized by the difference in stellar magnitudes in two bands, for example, B − V or U − B. The null-point difference of these stellar magnitudes is chosen to correspond to nonreddened class A0 stars of the main sequence. The difference between the observed color index and normal light from nonreddened stars of the same spectral class is called the color excess, E. Obviously, E is the difference in absorption in two bands, for example $E_{B-V} = A_B - A_V$. The ratio R of selective absorption to the color excess for sufficiently narrow bands must be a constant, but in practice it depends on the starlight. The fact is that the intensity of stellar radiation changes within the limits of the band selected by each of the filters and the sensitivity of the collector of the six-color photometer, so that for the same absorption the values of A can be somewhat different. The values of

$$R = \frac{A_V}{A_B - A_V} = \frac{A_V}{E_{B-V}} \tag{11.3}$$

calculated from Fig. 29 and data on the energy distribution in stellar spectra vary over the range 3.2 for class O9 stars to 3.8 for M2 stars [194]. The determination of this value from observations verified the change of R with the spectrum, but, in addition, showed a weak dependence (especially for later stars) on the reddening value. Values of R obtained from observation differ somewhat from computed values. For example, for B0–B5 stars, from the results of observation we have $R = 3.0 \pm 0.2$, while data from Fig. 29 lead to a value of $R = 3.2$. The discrepancy may be due to inaccurate extrapolation of the absorption curve at $1/\lambda \to 0$ (which must be done from a computed value of R) and to a number of other reasons that so far have not been allowed for. Therefore, the mean value $R = 3.1$ was taken for O–B stars. The dependence of R on reddening and the spectral class of the star is shown in Fig. 30, where on the right-hand scale values are assumed, and on the left values are calculated for the parameter. It is well known that from Fig. 30, after the color excess has been determined from observations, the total absorption may be found at given wavelengths, which is essentially the fundamental method for the calculation of interstellar light absorption in stellar astronomy.

The dependence of κ on λ shown in Fig. 29 is valid for almost all the observed galactic regions. However, the following must be noted. Usually, what is determined from observations is not the curve κ_λ itself, but the value E_{U-B}/E_{B-V} or the ratio of two other color excesses. These ratios are sensitive to departures of the absorption curve from the dependence (11.1), and, in particular, to the value and location of the break in Fig. 29. Observations have shown [194a] that although in the first approximation these ratios are similar in various regions of the sky, there is nevertheless a noticeable dependence of these quantities on galactic longitude. The largest variations take place in the constellations Cygnus and Perseus, and particularly for the star Trapezium Orion. For example, Wampler [194a] obtained the dependence of the ratio E_{U-B}/E_{B-V} and Fernie and Marlborough [194c] the dependence of R on longitude shown in Tables 11.1 and 11.2. The last column of Table 11.1 gives the slope of the correlation

Table 11.1

Galactic longitude (deg)	$\frac{E_{U-B}}{E_{B-V}}$	Relative slope of correlation between intensity of line $\lambda 4430$ and E_{B-V}
62–102	0.88	1.0
102–122	.85	Large scatter of points
122–142	.80	1.27
152–212	.71	1.51
352–22	.76	1.41

Table 11.2

Galactic longitude (deg)	R
72–82	3.14
107–115	3.41
132–137	3.09
186–191	3.05
204–208	2.31
284–288	2.58
302–312	2.90
374–381	2.90

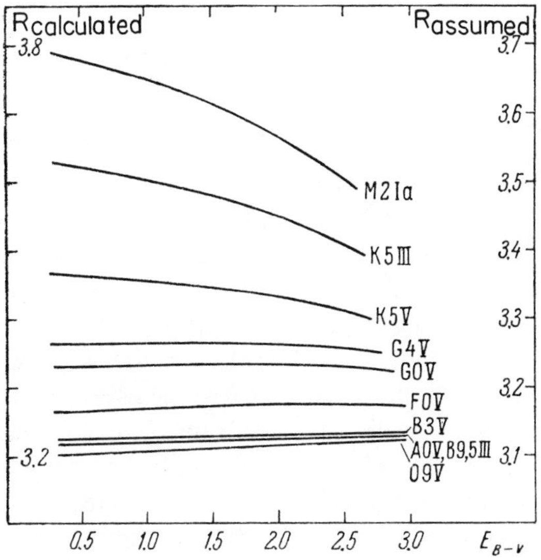

FIG. 30. Dependence of R on spectral class and on reddening [194].

between the intensity in the line $\lambda = 4430$ and the color index. The value of this slope in the constellation Cygnus was taken as standard. These data show most clearly the dependence of absorption properties of dust on the direction in the Galaxy. There are several explanations for this effect. The difference in the values of R and in the ratio E_{U-B}/E_{B-V} between the constellations Cygnus and Perseus is probably connected with absorption by oriented nonspherical particles. In Cygnus the line of sight goes along the arm and the direction of the magnetic field, while in Perseus it is perpendicular to it (see Sec. 14). In the Orion nebula the dimensions of the particles are probably different. There may be an influence of the variation in the composition of the particles, and, in particular, the difference in refractive index [194c]. Up to the present time attempts to separate these possibilities or to produce a somewhat concrete explanation have not been successful. New qualitatively reliable observations are needed (for example, in the far ultraviolet or infrared regions). However, as a first approximation to use in the study of physical properties, we may consider that the proportionality to $1/\lambda$ in the visible region of the spectrum of absorption by cosmic dust particles is universal.

This dependence allows us to deduce immediately that in absorption the particle size cannot be either much greater or much less than the wavelength of the visible spectral region. In fact, in the first case absorption would be neutral, and in the second case the absorption coefficient would be proportional to $1/\lambda^4$. It must be noted here that interstellar absorption cannot be absorption in the true meaning of the word, that is, the conversion of the energy of light waves into heat. Interstellar absorption can also be produced by the scattering of light on cosmic dust particles. Therefore, it is more correct to speak of interstellar absorption as extinction, although this term is rarely used in the literature. In dielectric particles small in comparison with the wavelength, the extinction coefficient is equal to the scattering coefficient, which, in agreement with Rayleigh's law, is inversely proportional to the fourth power of the wavelength.

The dependence $\kappa \sim 1/\lambda$ can obtain either during the scattering of light by particles with dimensions of the order of the wavelength, or by pure absorption by metallic particles with dimensions much less than a wavelength. Both possibilities are widely discussed in their application to interstellar light absorption. The most detailed studies of dielectric particles were carried out in 1946–1949 by van de Hulst [195, 196], and in the case of absorption by metallic dust particles by Schalen in 1936–1939 [197, 198]. Although at the present time it is difficult to reject completely any hypothesis, a whole series of similar data, which will be considered in Sec. 15, indicate the advantage of assuming that particles of cosmic dust are primarily dielectric, but with some admixture of metals. Therefore, we shall consider that interstellar absorption (extinction) is explained by scattering and by true absorption of light by cosmic dust particles, whose size is comparable with wavelengths in the visible spectrum.

Scattering of Light by Dielectric Particles

A study of the optical properties of particles with dimensions of the order of the wavelength of light offers no major difficulties. Maxwell's equations describe the propagation of an electromagnetic field, and determine the amplitude and phase of the light waves, both outside and inside the particles. However, on the mathematical side, the problem proves to be of comparative complexity, because of the necessity to match the internal and external solutions at the surface of the particles. The problem of light scattering by spherical particles was first solved completely by Mee in 1908. A detailed account of the present state of this problem is given in a book by van de Hulst [196]. The problem is solved exactly for particles of regular shapes—spheres, cylinders, disks, and so forth.

Cosmic dust particles are probably defectively shaped crystals. Their elongation (nonsphericity) is proved by the polarization of interstellar light. It is difficult to predict correctly the shape of particles that consist of many chemical components. Because of this, there is no need to study the optical properties of cosmic dust by the exact but cumbersome solution of Mee.

Study of the optics of these particles can be considerably simplified if it is assumed that their refractive index is almost unity. This assumption is very probable, since particles that are formed in a vacuum by adsorption of atoms on their surfaces have, as a rule, a very porous structure and consequently a low density (less than 1 g/cm^3).

The optics of particles with a refractive index close to unity and of sizes comparable with the wavelength of the light is rather simple [195, 196]. The simplicity of the case is explained because here light scattering may be calculated by considering the interference of beams passing through the particle (taking into account a phase change and reduction due to pure absorption) and beams refracted by the particle. Since the phase change on the passage of light through a medium of thickness h is

$$\delta = \frac{2\pi(n - 1)}{\lambda} h, \tag{11.4}$$

where n is the refractive index, then clearly in this particular problem the scattering properties of particles are determined by a single parameter δ. Therefore, results obtained in this way can be used for particles of different n values and of various sizes.

Consulting [196] for details of the solution of this problem, we shall limit ourselves here to certain observations necessary for further understanding.

The effective cross section for extinction of light by particles of refractive index close to unity is calculated from the formula (the symbol Re indicates the real part of the expression)

$$\sigma_e = 2\text{Re} \iint_S \{1 - \exp[-i\delta(x, y)]\}\, dx\, dy. \tag{11.5}$$

Here the integral is over the area S of the diametral cross section of a particle (coordinates x and y), $\delta(x, y)$ is the phase shift of a beam passing through a point with coordinates x and y, and the numerical factor 2 is due to Babinet's principle, which states that we must consider diffraction twice, once as the diffraction of the incoming light due to an "aperture" with the area of the particle and once as the diffraction of the refracted beam due to a "screen" of the same area. It is possible to write a term in the form of a simple exponential taking into account the phase shift, because when the refractive index is almost unity the light ray is hardly deflected from its original direction.

The expression in braces in Eq. (11.5) is proportional to the amplitude of the light scattered directly ahead. Generally speaking, in calculating the extinction coefficient, we must derive an expression for the amplitude of scattering in all directions of the radiation and then integrate the square of the amplitude coefficient with respect to the solid angle. However, it may be shown that the result of such an integration will agree with the real part of the expression for the amplitude of the radiation scattered straight ahead, which also is used in the expression (11.5).

For dielectric particles that have a real number for their refractive index, the extinction coefficient σ_e, numerically equal in this case to the coefficient of total scattering σ_S, has the form

$$\sigma_e = \sigma_s = 2 \iint_S (1 - \cos\delta)\, dx\, dy. \tag{11.6}$$

If the refractive index has an imaginary part different from zero, that is,

$$n = 1 + \xi - i\xi \tan\beta, \tag{11.7}$$

where $\xi \ll 1$ and β is the "loss angle," the relation (11.6) is then written in the form

$$\sigma_e = 2 \iint_S (1 - e^{-\delta \tan\beta} \cos\delta)\, dx\, dy, \qquad \delta = \frac{4\pi\xi}{\lambda} h(x, y), \tag{11.8}$$

where $h(x, y)$ is the chord along the line of sight. The pure absorption coefficient σ_a is determined from the integral

$$\sigma_a = \iint_S (1 - e^{-2\delta \tan\beta})\, dx\, dy, \tag{11.9}$$

since here the expression in parentheses represents the portion of the radiation passing through the particle as a beam with coordinates x and y. In calculating the integrals in Eqs. (11.6)–11.9), we must remember to take into account the dependence of the phase shift δ on the coordinates. Instead of the effective cross section σ, the parameter $Q = \sigma/S$ is often used; it is known as the *efficiency factor* for extinction, scattering, and absorption (Q_e, Q_s, and Q_a respectively).

Further calculations are guided by the geometry of the particles. For example, in the case of a spherical particle of radius a ($\delta = \pi a^2$), the phase change and the coordinates may be expressed in terms of the angle γ at which from the particle center the semichord is visible along the line of sight:

$$\delta = \frac{4\pi\xi}{\lambda} a \sin \gamma, \qquad \begin{aligned} x &= a \cos \gamma \sin \phi, \\ y &= a \cos \gamma \cos \phi, \end{aligned} \tag{11.10}$$

where ϕ is the azimuthal angle in the plane of the diametral cross section of the particle. The integrals (11.6)–(11.9) are readily calculated in this case. We give an expression for (11.8), the most complex of them,

$$\sigma_e = 2\pi a^2 \left[1 - 2e^{-\rho \tan \beta} \frac{\cos \beta}{\rho} \sin (\rho - \beta) - 2e^{-\rho \tan \beta} \left(\frac{\cos \beta}{\rho} \right)^2 \cos (\rho - 2\beta) - 2 \left(\frac{\cos \beta}{\rho} \right)^2 \cos 2\beta \right] \cdot \tag{11.11}$$

We note that in the present case it is easier to calculate the integral in (11.5) with $\delta = (4\pi\xi/\lambda)a(1 - \tan \beta) \sin \gamma$ and then to take the real part of the expression obtained. In Eq. (11.11), the designation

$$\rho = \frac{4\pi\xi a}{\lambda} \tag{11.12}$$

was introduced. If $\beta = 0$ (dielectric particles), we obtain

$$\sigma_e = 2\pi a^2 \left[1 - 2 \frac{\sin \rho}{\rho} + \frac{2}{\rho^2} (1 - \cos \rho) \right] \cdot \tag{11.13}$$

For very small particles ($\rho \ll 1$), extinctions are determined primarily by pure absorption and not by scattering. In this case as a first approximation

$$\sigma_e = \sigma_a \approx \frac{4\pi a^2}{3} \rho \tan \beta \approx \frac{16\pi^2 a^3}{3\lambda} \xi \tan \beta, \tag{11.14}$$

that is, the absorption coefficient here is in fact inversely proportional to the wavelength, in accordance with what was previously said about metallic particles. Van de Hulst calculated similarly the extinction of light for cylinders [196].

In the foregoing formulation of the problem, polarization of light is not taken into account, since total scattering and absorption are in fact allowed for only by the introduction of the phase shift δ, which is identical for both states of polarization. In the case of light scattering by spherical particles, it is in general not necessary to consider polarization because of the symmetry of such particles. Light scattering by nonspherical particles may lead to a change in polarization. Unfortunately, within the framework of the foregoing solution to the problem of light scattering by particles with refractive index close to unity, we cannot take polarization into account.

On the other hand, this means that in the present problem we are generally limited by the consideration of scalar wave propagation, for which the calculation of the phase shift is particularly simple. Similarly, Greenberg [199] arrived at an expression for the effective cross section of scattering of light by ellipsoidal particles. It is also described by Eq. (11.13), where instead of πa^2, the visible area of the geometric cross section, equal to $\pi a B = \pi a(a^2 \cos^2 \chi + b^2 \sin^2 \chi)^{1/2}$, must be substituted, and the parameter ρ is replaced by

$$\rho = \frac{4\pi\xi}{\lambda}\frac{ab}{B} = \frac{4\pi\xi}{\lambda}\frac{ab}{(a^2 \cos^2 \chi + b^2 \sin^2 \chi)^{1/2}}. \tag{11.15}$$

Here a is the equatorial radius, b is the polar semiaxis, and χ is the angle between the axis of symmetry and the direction of light propagation, so that the solution is correct both when $b < a$ (flattened particles) and when $b > a$ (elongated particles).

Using the same method, Greenberg calculated the extinction of light by particles bounded by cones or paraboloids of rotation.

In Fig. 31 curves are drawn for the extinction factor of dielectric particles for various values of the refractive index. It is quite apparent that a solution of the problem on the assumption that $n - 1 \ll 1$ gives a sufficiently accurate result up to $n \approx 1.5$.

It must be understood that the small fluctuations on this curve obtained during the exact resolution of the problem are quite unimportant in the study of interstellar absorption. Moreover, the larger fluctuations of the extinction curve introduced as the phase shift passes through a particular value are multiples of π, and do not have a large value since they are rounded off by averaging over the size-distribution function of the particles.

The difference between an exact and an approximate solution becomes appreciable only when $\rho \ll 1$, or when the actual extinction process occurs in accordance with Rayleigh's law, $\sigma_e \sim 1/\lambda^4$, while the solution of Eq. (11.11) when $\beta = 0$ gives $\sigma_e \sim 1/\lambda^2$. However, even in the presence of a small but finite imaginary part of the refractive index close to $\rho = 0$, pure absorption predominates over scattering, and in this case Eq. (11.14) describes the extinction curve correctly, at least qualitatively. Consideration of the true absorption (Fig. 32) also smooths fluctuations in the extinction coefficient. This is because light absorption by particles reduces the interference of passing and diffracted radiation.

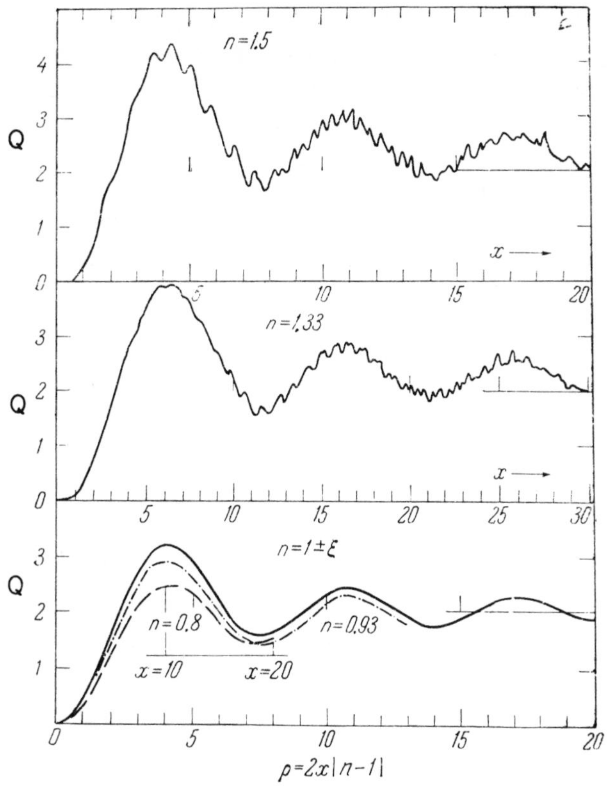

FIG. 31. Extinction factor Q for spherical dielectric dust particles with various refractive indices [196].

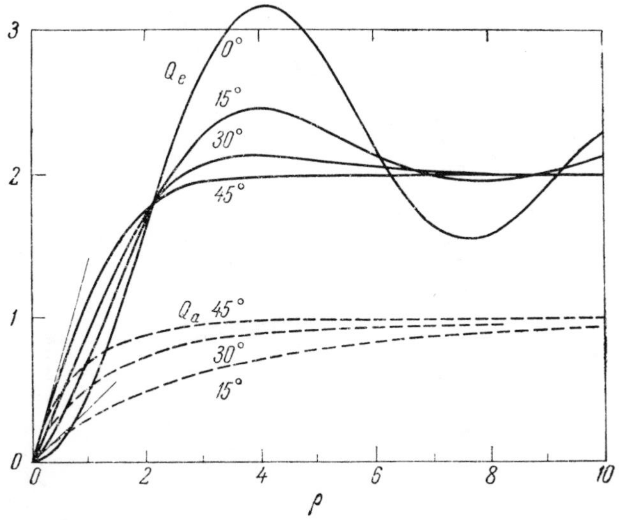

FIG. 32. Extinction and absorption factor when n is almost 1 [196]. The refractive index is equal to $1 + \xi - i\xi \tan \beta$.

The dependence of the effective cross section on the orientation of the axis of symmetry for ellipsoidal particles along and across the line of sight is shown in Fig. 33. The marked differences between σ_l and σ_p are explained not only by the difference in the visible areas of the geometric cross section, but by the phase shift at various orientations, which leads to a change in the location and shape of the fluctuations in the extinction curve.

Extinction curves for infinite cylinders are given in Fig. 34. The dependence on polarization, as already noted, can be obtained only from an exact solution

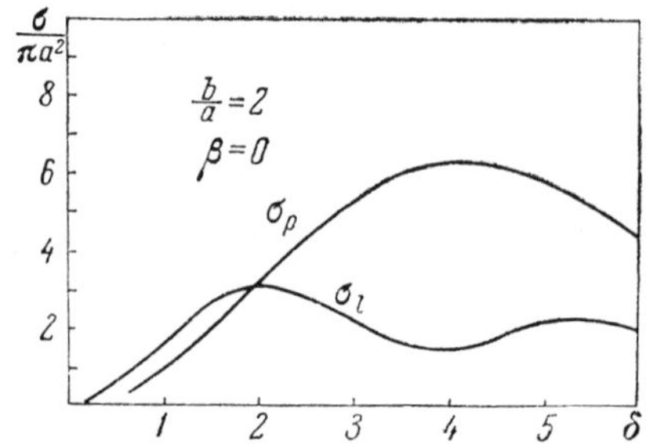

FIG. 33. The effective extinction cross section divided by πa^2 for elongated spheroids, when the orientations of the axis of symmetry of the particles are along (σ_l) and across (σ_p) the line of sight [265].

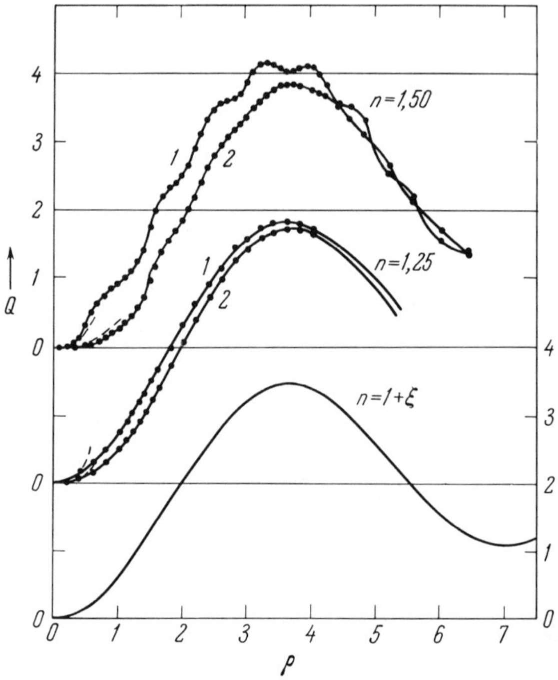

FIG. 34. The extinction factor of long cylinders for light with the electric vector (1) parallel and (2) perpendicular to the axis [196].

of the problem. As seen from Fig. 34, this dependence is unimportant for the determination of the extinction value, but for the study of interstellar polarization it is important. Extinction curves for cylindrical particles with considerable metallic impurities are shown in Fig. 35. It should be noted here that for large values of ρ the sign of the polarization is reversed. This is also shown in Fig. 34, when $n = 1.5$.

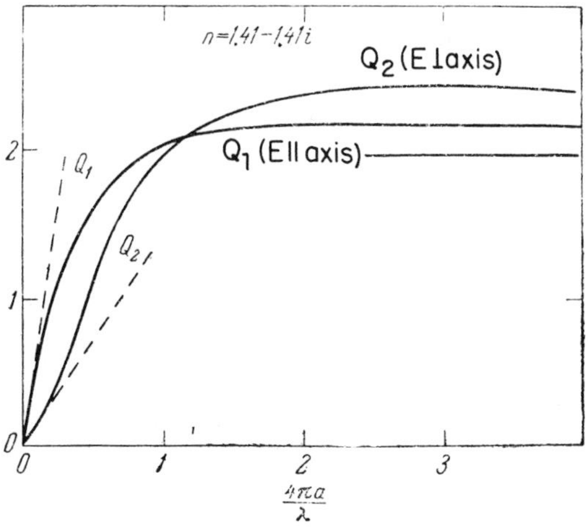

FIG. 35. Extinction factor for absorbing cylinders [196].

If the extinction and absorption coefficients are known, such an important parameter as the albedo of the particles may be determined,

$$\Lambda = \frac{\sigma_s}{\sigma_e} = \frac{\sigma_e - \sigma_a}{\sigma_e}, \tag{11.16}$$

which gives the fraction of radiation scattered during extinction. For dielectric particles, the albedo is equal to unity, but if the imaginary part of the refractive index is finite, although small, the albedo of the particles is appreciably decreased. It follows from Eq. (11.14) that when $\rho \ll 1$, the albedo of such particles is reduced to zero. With an increase in ρ, the albedo converges to $\frac{1}{2}$, as is easily seen from the graph of Fig. 32. When $\rho \tan \beta \gg 1$, for spherical particles we have the asymptotic equation

$$\Lambda = \frac{1}{2}\left[1 + \frac{1}{\rho^2}\left(\frac{1}{2}\cot^2 \beta + 4\cos^2 \beta \cos 2\beta\right) + \cdots\right]. \tag{11.17}$$

For intermediate values of ρ, the albedo is subject to large variations. For example, when $\beta = 15°$ and $\rho = 3$, we find from the graph of Fig. 32, $\Lambda = 0.8$. Figure 32 allows the ready calculation of the albedo of particles when $1 < \rho < 10$ for those cases when $\beta = 15°, 30°, 45°$.

The distribution in direction of scattered light is characterized by the indicatrix $x(\theta)$, specifying the fraction of radiation scattered through the angle θ

relative to the incident beam. The indicatrix may be calculated both by exact methods and by an approximate solution of the problem of light scattering by particles of refractive index close to unity.

The indicatrix of scattering depends on the refractive index and on the value $2\pi a/\lambda$, the ratio of particle dimension to the wavelength. Detailed results, even within the foregoing approximation, are not quoted here, because at the present time these calculations cannot be used completely for the study of interstellar absorption. The main features of the indicatrix of scattering consist of the following. For particles with $\rho \approx 2\pi a\xi/\lambda \gtrsim 1$, a considerable portion of the scattering is forward scattering. The indicatrix is not monotonic, with the first minimum (close to zero) being reached at the angle

$$\theta_1 \approx \frac{3.5\lambda}{2\pi a} \text{ rad}, \tag{11.18}$$

with the condition that $2\pi a/\lambda \gg 1$. The secondary and all subsequent maxima are much less than the original maximum when $\theta = 0$. The prolateness of the indicatrix of radiation scattering is conveniently described by means of a factor of asymmetry, defined as the ratio of the radiation scattered by the particles to its mean intensity. In other words, the asymmetry factor q is

$$q = \overline{\cos\theta}, \tag{11.19}$$

where the scattering index is used as a weighting function. For isotropic scattering, $q = 0$. From the definition of the asymmetry factor, it follows that the force of the light pressure on a cosmic dust particle is equal to

$$f = \sigma_e(1 - q)\frac{\pi F}{c}, \tag{11.20}$$

where πF is the radiation flux incident on the particle. From what has been said previously, the overwhelming portion of scattered radiation is concentrated in a solid angle of aperture θ_1, determined by Eq. (11.18); as an approximate expression for the asymmetry factor, the simple equation

$$q \approx \cos\left(\tfrac{1}{2}\theta_1\right) \approx \cos\left(\tfrac{\lambda}{a}16^\circ\right) \tag{11.21}$$

may be used, which gives good accuracy when $a > \lambda/4$.

The above extinction and absorption coefficients must still be averaged, using the particle-size distribution function. Let $dN_g(a) = N_g f(a)da$ be the number of particles of radius a to $a + da$; N_g is the total number of dust particles per unit volume. It is most convenient to determine the mean effective scattering factor from the equation

$$\bar{Q}_e = \frac{\int_0^\infty \sigma_e f(a)da}{\int_0^\infty s f(a)da} = \frac{\bar{\sigma}_e}{\bar{s}}, \tag{11.22}$$

where $\bar{s}$ is the average cross-sectional area of the particle. Similarly, the mean effective absorption factor $\bar{Q}_a = \bar{\sigma}_a/\bar{s}$ may be determined.

The actual distribution of particles in interstellar space according to size is unknown. From certain considerations that will be given in Sec. 15, Oort and van de Hulst [200] derived theoretically a distribution function that is a very rough approximation to the exponential function

$$f_1(a) = \frac{1}{a_1} e^{-a/a_1}. \tag{11.23}$$

According to Eq. (11.23), small particles more often occur; a_1 is the distribution parameter. If, however, the particle distribution function has a maximum at a definite size a_2, it can be approximated by the equation

$$f_2(a) = \frac{a}{a_2^2} e^{-a/a_2}. \tag{11.24}$$

Both distribution functions have the advantage that they can readily be applied to the calculation of the integrals in Eq. (11.22) for spherical particles whose extinction coefficients are determined by Eq. (11.11). Also, $\bar{Q}_a$ may readily be calculated in these cases. The results are as follows:

I. For the distribution function (11.23) [195]:

$$\bar{Q}_e(\rho_1) = 2\left\{1 - \frac{(1 + \rho_1 \tan \beta)^2 - \rho_1^2}{[(1 + \rho_1 \tan \beta)^2 + \rho_1^2]^2}\right\},$$
$$\bar{Q}_a(\rho_1) = 1 - \frac{1}{(1 + 2\rho_1 \tan \beta)^2}. \tag{11.25}$$

II. For the distribution function (11.24) [201]

$$\bar{Q}_e(\rho_2) = 2\left\{1 - \frac{1}{3}\frac{(1 + \rho_2 \tan \beta)^2 - \rho_2^2}{[(1 + \rho_2 \tan \beta)^2 + \rho_2^2]^2} - \frac{2}{3}\frac{(1 + \rho_2 \tan \beta)^3 - 2\rho_2^2(1 + \rho_2 \tan \beta)}{[(1 + \rho_2 \tan \beta)^2 + \rho_2^2]^3}\right\}, \tag{11.26}$$
$$\bar{Q}_a(\rho_2) = 1 - \frac{1}{3}\frac{1}{(1 + 2\rho_2 \tan \beta)^2} - \frac{2}{3}\frac{1}{(1 + 2\rho_2 \tan \beta)^3}.$$

Here ρ_1 and ρ_2 are equal to $4\pi\xi a_1/\lambda$ and $4\pi\xi a_2/\lambda$, respectively. From a comparison of Eqs. (11.25) and (11.26), it is easy to see that the choice of the distribution function has no marked effect on the mean values of the extinction and absorption coefficients, especially in the most interesting case when $\rho_{1,2} \tan \beta \ll 1$. The dependence of the mean coefficient $\bar{Q}$ on the choice of refractive index, especially its imaginary part, is more marked, but even here the mean value of the extinction coefficient at the extreme limits of refractive index and for $\rho > 1.5$ does not differ by more than a factor of 1.5.

Comparison with Observations

We shall compare theoretical calculations of the optics of dust particles with observational data on interstellar absorption. In principle, knowing the albedo Λ, their asymmetry factor q (from scattered light from reflection nebulae; see Sec. 13), and the selectivity of interstellar extinction, we can determine the characteristic particle sizes and their optical properties, as well as at least a qualitative estimate of the particle-size distribution function. Unfortunately, these possibilities have not yet been realized. There are no reliable data on the albedo, and the particle asymmetry factor and the mean extinction coefficients, as already noted, are not very sensitive to the choice of the distribution function. Nevertheless, certain conclusions can be drawn.

The most reliable determination from observations is the dependence of the extinction coefficient on the wavelength. Therefore, it is useful to proceed in the following fashion: we assume some value of the refractive index, choose a certain distribution function, and then compare the calculated mean extinction coefficient with Fig. 29 at various wavelengths, selecting a value of the parameter ρ_1 or ρ_2. If neither value of this parameter gives good agreement, the refractive index ($\tan \beta$) or the distribution function must be changed. Such a method was used by van de Hulst [195]. In this investigation, he tried out various distribution functions, among them (11.23) and a theoretical function that will be considered in Sec. 15. They both gave satisfactory results. Assuming that the dielectric particles have a refractive index $n = 1.33$, van de Hulst found that the value $a_1 \approx 0.4\ \mu$ corresponded best to the observations. If $n \neq 1.33$, then $a_1 \sim (n - 1)^{-1}$. For example, when $n = 1.25$, we find $a_1 \approx 0.53\ \mu$. In this assumption, we require that $\Lambda \approx 1$, $q \approx 1$.

There is the other possibility, investigated in detail by Schalen [197, 198] and also by Schoenberg and Jung [202] and Guttler [203], that interstellar absorption is caused by metallic particles. If the imaginary part of the refractive index does not depend on the wavelength, then the law $\kappa \sim 1/\lambda$ yields particles of small size ($\rho_1 \ll 1$). However, in reality, the refractive index of metallic particles changes with wavelength [which is equivalent to the presence of a dependence $\beta(\lambda)$], and therefore it is impossible here to draw a single graph $\bar{Q}(\rho_1)$. An analysis of the various possibilities leads to the conclusion that if the particles are iron, they have a mean size of $\sim 0.04\ \mu$ and an albedo $\Lambda \approx 0.10$–0.15. Extinction here is produced mainly by absorption, the asymmetry factor being also small.

The choice between these two possibilities can be made only from an estimate of Λ. The presence of scattered diffuse emission in the Galaxy and the relatively large brightness of reflection nebulae rather suggest that the albedo of the scattering particles cannot be very large (see Sec. 13).

Thus, the first possibility, that cosmic dust particles consist of dielectric crystals, but with a slight admixture of metallic atoms, is the most probable, since the albedo of all of them is less than unity. Moreover, the hypothesis of metallic impurities is required to explain interstellar polarization. Therefore, the refractive index must be complex. The most probable estimate of its value is

$$n \approx 1.25 - 0.05i. \tag{11.27}$$

It must be stressed again that this is only a preliminary estimate. It is particularly difficult to determine the imaginary part of the refractive index. The problem of the reason for the break in the extinction coefficient at $\lambda = 4500$ Å has so far not been solved.

12. Distribution of Interstellar Dust

The study of the distribution of cosmic dust in interstellar space occupies an important place in astrophysics and stellar astronomy. On the one hand, this distribution must be known for the accurate calculation of interstellar absorption, and on the other hand, the dynamics and evolution of gas-dust clouds are closely connected with the peculiarities of their distribution.

The Flat Dust Layer

As a first approximation, the dependence of density of dust on distance from the galactic plane may be described by the barometric equation

$$\rho_g(z) = \rho_g(0)e^{-z/\beta_g}. \tag{12.1}$$

Here, according to the barometric formula, the parameter β_g depends on the dispersion of the components of the velocity of dust clouds $\overline{v_z^2}$ perpendicular to the galactic plane and on the potential of the gravitational field. A study of stellar velocities readily gives the relation

$$\beta\ (\mathrm{pc}) = 1.2\overline{v_z^2}\ (\mathrm{km/sec})^2 \tag{12.2}$$

when v_z is in kilometers per second. Observations show that as a first approximation Eq. (12.1) satisfactorily describes the distribution of dust over the galactic disk if we assume that $\beta_g \approx$ 100–120 pc. From Eq. (12.2), it follows that $(\overline{v_z^2})^{1/2} \approx 9.1$ km/sec.

The density of cosmic dust near the galactic plane is determined from the total absorption. It is true that this value is liable to large fluctuations, but, on the average, absorption in the visual in the solar neighborhood near the galactic plane is close to one stellar magnitude per kiloparsec. For the determination of the mean dust concentration $\overline{N}_g$, we have the obvious equation

$$\sigma_e\overline{N}_g \times 1\ \mathrm{kpc} = \frac{A_V}{1.086} \approx 1. \tag{12.3}$$

From the data of the preceding section, we obtain for the visible spectral region

$\sigma_e \approx \pi a^2 \approx 8 \times 10^{-9}\ \mathrm{cm}^2$ when $a \approx 0.5\ \mu$. Then from Eq. (12.3) we obtain

$$\bar{N}_g \approx 4 \times 10^{-14}\ \mathrm{cm}^{-3}. \tag{12.4}$$

The mass of a single particle is $m_g \approx 4\pi a^3/3 \approx 5 \times 10^{-13}$ g, if we assume that the specific gravity of dust particles is about 1 g/cm³. Thus, the density of dust near the galactic plane is

$$\rho_g(0) \approx \bar{N}_g m_g \approx 2 \times 10^{-26}\ \mathrm{g/cm}^3, \tag{12.5}$$

which is two orders of magnitude less than the gas density. Thus, to a first approximation, interstellar dust forms a flat layer whose thickness increases with distance from the center and reaches about 200 pc at the position of the Sun. Its mean density is about $\rho \approx 10^{-26}\ \mathrm{g/cm}^3$. This layer is clearly visible in the photograph of a galaxy resembling ours, but seen edgewise (Fig. 36).

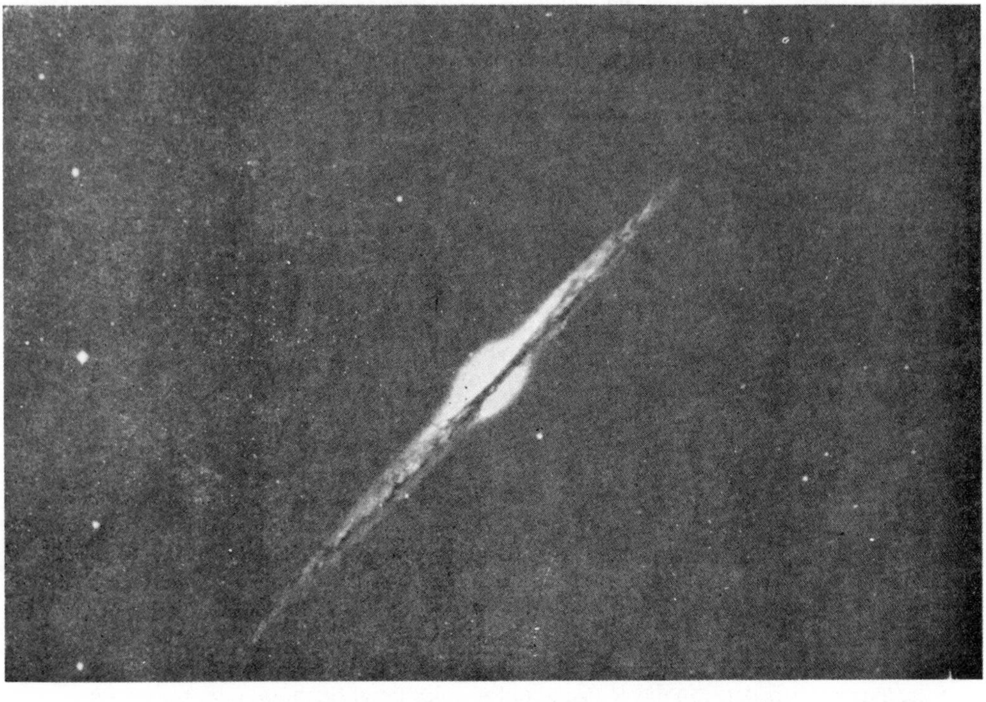

FIG. 36. Galaxy *Sb* NGC 4665. (Palomar Observatory.)

However, the uniform-layer model is an extremely rough approximation. Direct observations of interstellar absorption indicate a heterogeneous cloud-like structure for the distribution of cosmic dust. This structure, prominent in Fig. 37, for example, is considered below.

Statistical Analysis of a Cloudlike Structure

The following scheme is studied as a second approximation. Let the cosmic dust be concentrated in separate dust clouds distributed at random in interstellar space within the limits of the dust layer. The size and density of the clouds may be different, but we shall consider them to be independent of the position of a cloud in space. Statistical methods may be employed. These investigations were started by V. A. Ambartsumyan in 1939 [204, 205, 2]. The methods were elaborated by Chandrasekhar and Münch [206–211], T. A. Agekyan [212, 213], B. E. Markaryan [214], G. I. Rusakov [215], and others.

Statistical methods for the study of cosmic dust clouds belong more to the field of stellar astronomy than to astrophysics. Therefore, we shall limit ourselves here to the main characteristics of these methods and to a description of the results obtained.

In the first group of methods, the statistical relations are determined between the characteristics of dust clouds and the count of stars or galaxies per unit area on the celestial sphere. For example, a typical case of such a relation is the formula [213]

$$\overline{(\log N_\Gamma)^2} - (\overline{\log N_\Gamma})^2 = 0.36\bar{n}\mu^2. \tag{12.6}$$

Here N_Γ is the number of galaxies of a given stellar magnitude visible in a certain area of the celestial sphere, $\bar{n}$ is the mean number of absorbing clouds in this direction up to the limits of our Galaxy, and μ is the absorption in a single cloud expressed in stellar magnitudes. In the derivation of this relation it is assumed, in particular, that galactic distribution is uniform in space and that all clouds have identical optical thicknesses, while their distribution in the line of sight follows the laws of probability. None of these assumptions, strictly speaking, corresponds with the facts. Within the framework of statistical theory, galaxies appear to have a tendency to cluster [213] and the absorption of clouds to vary. The assumption of the random distribution of clouds along the line of sight forms the basis for statistical theory and therefore cannot be omitted.

In actual calculations with formulas of type (12.6), it is also assumed that the mean number of clouds is the same in all directions at a particular galactic latitude. Therefore, by estimating the fluctuations in the number of galaxies in areas with identical galactic latitudes, we can determine both μ and $\bar{n}$.

For the determination of cloud dimensions, correlation functions [2, 215] are established. In fact, if the lines of sight in two directions, separated by an angle ϕ, pass through a single absorbing cloud, the number of observed galaxies or stars must be correlated. This correlation disappears at an angle approximately equal to the ratio of the cloud diameter to the mean distance between them.

The application of all these methods shows that the observed cloud system is not well described by the statistical picture. For example, T. A. Agekyan showed that the mean size of a cloud and the absorption in it depend markedly on the galactic latitude (Table 12.1), although in the derivation of the formula

Table 12.1

b (deg)	Mean cloud diameter (pc)	μ (m)
15	110 ± 16	0.60 ± 0.07
32.5	17 ± 7	$.23 \pm .08$
52.5	4.5 ± 3.1	$.12 \pm .08$

for calculation it was assumed that all clouds were identical. However, it must be noted that, in spite of this, Table 12.1 reflects qualitatively the true characteristic peculiarities, a decrease in cloud size and density with increased distance from the galactic plane.

Different calculations lead to different values for the mean absorption in a cloud. Besides the values quoted in Table 12.1, T. A. Agekyan also obtained from other calculations, $0^{m}.24$, $0^{m}.39$, $0^{m}.46$. V. A. Ambartsumyan obtained $\mu = 0^{m}.18$, and so on. We must remember that all these different mean values of absorption for clouds in the northern sky were obtained for the same part of the Galaxy near the solar neighborhood.

Instead of the average number of absorbing clouds as a function of direction in the Galaxy, it is more convenient to introduce the quantity $\Delta m_{\perp} = \mu \bar{n}_{\perp}$, which gives the average absorption halfwidth of the dust in a direction perpendicular to the galactic plane. Various determinations of this quantity also lead to different results. The majority of them range from $0^{m}.3$ to $0^{m}.45$, although values of $0^{m}.26$ and $0^{m}.85$ [213] have also been obtained.

Another group of methods for determining the transparency coefficient of dust clouds employs measurements of brightness fluctuations in the Milky Way. These methods were proposed by V. A. Ambartsumyan [205] and were developed subsequently by Chandrasekhar and Münch [206] and other authors. A derivation of the basic equation of this theory, characterized by greater simplicity and clarity than in the papers quoted, is given below. We introduce the function $f(I, R)$, describing the probability that, in a given direction of the Milky Way where its extent is R, the observed brightness is greater than or equal to I. Then we can establish for this function the equation

$$f(I, R) = \int_0^\infty f\left[\left(I - \frac{\varepsilon}{\nu}\right) e^{\tau}, \quad \left(R - \frac{1}{\nu}\right)\right]\Psi(\tau)d\tau. \qquad (12.7)$$

Here ε is the energy emitted by stars per unit volume, ν is the mean number of clouds occurring per unit length, and $\Psi(\tau)d\tau$ is the normalized probability that the optical thickness of a cloud lies between τ and $\tau + d\tau$. The meaning of Eq. (12.7) is as follows. The probability of observing the brightness I at a given point must be comparable with the probability that at a distance $1/\nu$ from the observer (that is, beyond the first cloud of optical thickness τ), the brightness on the one hand is less than ε/ν, that is, less than the value of the energy emitted by stars in this region, and on the other hand, is e^{τ} times greater, that is, greater by the amount absorbed in the cloud. Equation (12.7) also includes the fact that, at a distance $1/\nu$ from the observer, the distance to the boundary is less, equal to $R - 1/\nu$, and takes into account the possibility of the dispersion of the optical thicknesses of the cloud. For the application of statistical analysis, we must satisfy the condition $\varepsilon/\nu \ll I$, $1/\nu \ll R$. Therefore, making the substitution $I - \varepsilon/\nu = I'$ and $R - 1/\nu' = R'$, expanding the left-hand side, and confining ourselves to the first terms of the expansion, we obtain

$$f(I, R) + \frac{\varepsilon}{\nu}\frac{\partial f}{\partial I} + \frac{1}{\nu}\frac{\partial f}{\partial R} = \int_0^\infty f(Ie^{\tau}, R)\Psi(\tau)d\tau. \tag{12.8}$$

Here the primes are omitted from I and R. A detailed study of Eq. (12.8) and its consequences was given in papers [206–209]. In particular, a solution can be obtained in explicit form if $R \to \infty$ and $\Psi(\tau) = ke^{-k\tau}$, where k is any number. In the general case, an equation may be derived from Eq. (12.8) for connecting moments of the function $f(I, R)$ with respect to I with the moments of the distribution function of clouds with respect to their optical thicknesses. In particular, a relation exists between the root-mean-square fluctuation of the observed brightness of the Milky Way, on the one hand, and the mean value of the transparency coefficient of a cloud $q = e^{-\tau}$ and the dispersion of this quantity, on the other. If $R \gg 1/\nu$, this relation has the form

$$\frac{\overline{(\Delta I)^2}}{I^2} = \frac{2(1 - \bar{q})}{1 - \overline{q^2}} - 1. \tag{12.9}$$

The brightness of the Milky Way can be directly measured or determined from star counts. Using the star counts given by Markaryan [214], Chandrasekhar and Münch found that

$$\bar{q} = 0.85, \qquad \overline{q^2} = 0.73. \tag{12.10}$$

Consequently, within the limitations of the method, the mean absorption in a cloud in stellar magnitudes is approximately equal to $0^{\mathrm{m}}.18$. Over a distance of 1 kpc, 8–10 dust clouds occur.

This method has the same defects as the first. In particular, it is assumed here that the distribution of stars in space is uniform and that clouds are distributed at random along the line of sight.

Moreover, it was implicitly assumed here that absorbing clouds cover on the average the total area over which the stellar counts were taken. Therefore, to make the theory adaptable, it is necessary to have stellar counts over areas measuring less than 0.01 deg^2 [209a]. In research up to now, stellar counts have always been taken over much larger areas. This also might explain the spread in the results.

We should note one statistical method of studying the fluctuations developed by Chandrasekhar and Münch [210]. In this method, the absorbing interstellar medium is considered to be continuous, but with random density fluctuations, that is,

$$\rho_g = \rho_g^{(0)}[1 + \delta(r)], \tag{12.11}$$

where $\rho_g^{(0)}$ is a constant, and $\delta(r)$ is the random function of the coordinates. Brightness fluctuations in the Milky Way can be connected with the characteristics of this function, in particular, with the quantity $\overline{\delta^2(r)}$. An analysis of direct measurements of brightness in the Milky Way by this method leads to a value of $\overline{\delta^2(r)} = 12$. A similar method was used by Serkowski [216] for a statistical study of the reddening of stars in the clusters h and χ Persei. He obtained $\delta^2(r) \approx 15$. The main drawback of this method rests in the assumption that $\delta(r)$ is a random variable, that is, that the probability of positive and negative values of $\delta(r)$ is the same. In fact, in the case where the fluctuations are large, the density of the medium, in accordance with (12.11), can turn out to be negative. This is explained by the fact that the variation in the density of the medium (3–4 times) obtained in this way is much less than the actual variation (15–20 times).

Thus, the statistical methods of analysis of light absorption by dust clouds give uncertain results, often differing among themselves. The main reason for this is that clouds of cosmic dust and gas are by no means uniformly distributed, but are concentrated in the arms and are stretched out along them. This is particularly noticeable in other galaxies. For example, galaxies and globular clusters are observed between the arms of the spiral structure of the Andromeda Nebula, while at the same time absorption in the arms themselves is considerable, this absorption being observed by the reddening of emission objects and associations. Close to the nucleus the arms are crossed by dark bands of dust (a remark by W. Baade at a conference on cosmic aerodynamics [217]). In our Galaxy and in other spiral dark nebulae, they are drawn out along the arms and have sizes of from 5 to 800 pc [218].

The elongation of nebulae along the arms gives errors in the calculations averaged statistically over galactic longitude, since the dimensions of clouds and their absorption are different in different directions. It is difficult, in a statistical study, to allow for the difference in distribution of stars of different groups, their concentration toward the spirals, and so on. The clouds themselves are different in various parts of the spirals. All these reasons can obviously contribute to the large scatter in the average values of the optical thickness of clouds obtained by statistical methods.

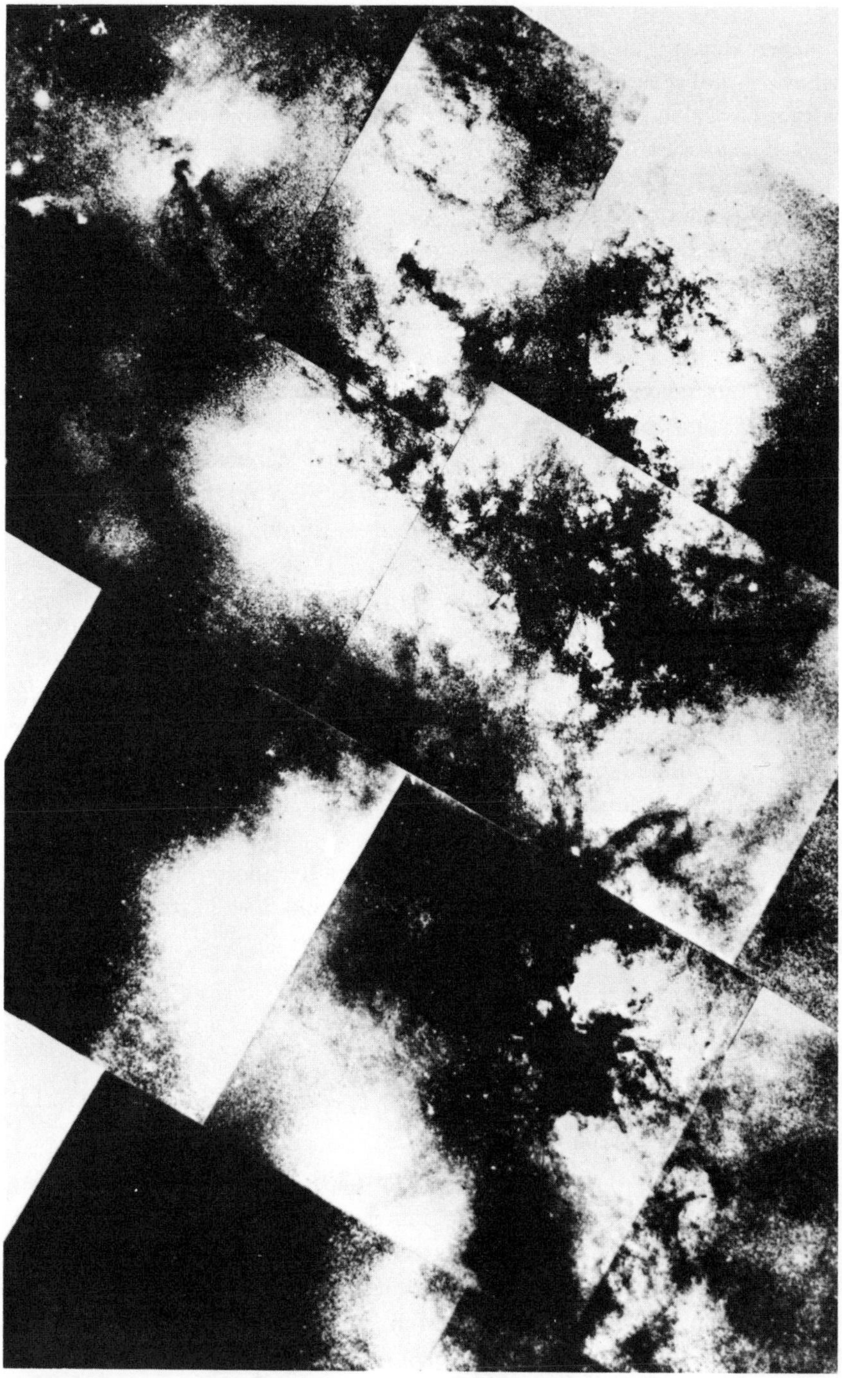

FIG. 37. Part of the Milky Way: *bottom to top*, the constellations Ophiuchus, Sagittarius, Scutum, and Aquila; bright clouds and dark nebulae are visible. (Mount Wilson Observatory.)

Results of a Detailed Study of the Distribution of Cosmic Dust

The inadequacy of statistical methods suggests that at the present time the distribution of dust should be studied from one cloud to another. These studies show that dust clouds are not uniformly distributed within the boundaries of arms. A considerable portion of dust forms large dense complexes, stretched out for 10 to 100 pc along the arms, whose masses range up to $10^5 M_\odot$, while ordinary clouds have masses of the order of $10^2 M_\odot$. These complexes provide strong absorption in large areas of the constellations Orion, Sagittarius, Cepheus, and Ophiuchus. In Fig. 37 a mosaic photograph is given of the region in the constellations Aquila, Scutum, Sagittarius, and Ophiuchus. In the lower right-hand portion a large nebula is seen in Ophiuchus with an area of about 10^3 deg^2, projected on a region with few stars. Conspicuous bands extend from it into low-latitude regions.

A classical method for studying individual dark nebulae is that of stellar counts proposed for the first time by Wolf. Let us call $N(m)$ the number of stars of visual stellar magnitude from $m - \frac{1}{2}$ to $m + \frac{1}{2}$ found in the solid angle Ω. We then have the equation

$$N(m) = \Omega \int_0^\infty r^2 D(r)\phi[m + 5 - 5 \log r - A(r)]dr. \tag{12.12}$$

Here $D(r)$ is the number of stars per unit volume at a distance r from the observer, $\phi(M)$ is a standard luminosity function, that is, the relative number of stars with an absolute luminosity over the range $M - \frac{1}{2}$ to $M + \frac{1}{2}$, and $A(r)$ is the absorption at a distance r, expressed in stellar magnitudes. Equation (12.12) is the fundamental equation of stellar statistics.

Let us suppose that a dark nebula, situated at a distance r_1, is characterized by complete absorption A_1, while absorption is small both in front of and beyond the nebula. Then Eq. (12.12) has the form

$$N(m) = \Omega\left[\int_0^{r_1} r^2 D(r)\phi(m + 5 - 5 \log r)\, dr + \int_{r_1}^\infty r^2 D(r)\phi(m + 5 - 5 \log r - A_1)\, dr\right]. \tag{12.13}$$

Having constructed from stellar counts the dependence of $N(m)$ on stellar magnitude (Fig. 38), we can select the parameters r_1 and A_1 to force agreement between the observed curve and calculated one for known values of $D(r)$ and $\phi(M)$. Thus, the total absorption in the nebula and its distance are determined, plus its size from its angular diameter. Many dark nebulae in the Milky Way, with absorptions of from one to four stellar magnitudes and greater, located at distances of several hundred parsecs from the Sun (see, for example, Bok [219, 220]), have been studied by this method.

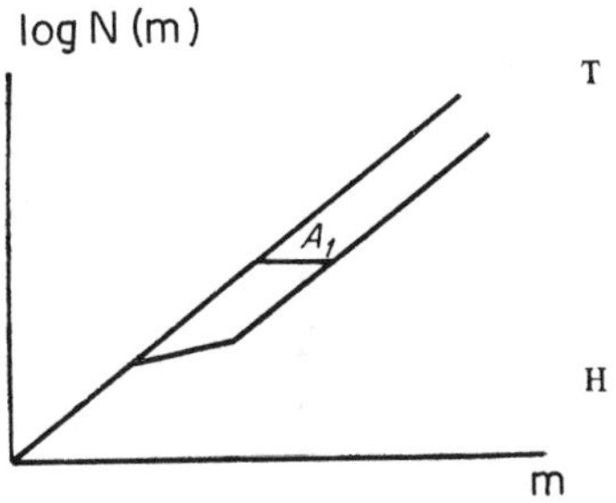

FIG. 38. A schematic representation of Wolf's method; A_1 is the absorption in the cloud.

Recently, with the introduction of the U, B, V photometric systems and with photoelectric standards that permit fairly high accuracy in determining the colors of stars, and with the increased precision in the absolute magnitude of stars of various classes, the measurement of color excess has been used more frequently in place of Wolf's method. If the absolute magnitude corresponding to a given spectral class, the observed magnitude, and the absorption are known, the stellar distance may be determined. Then a graph is drawn indicating change of absorption with distance. A series of such investigations was carried out, for example, at the Crimean Observatory. We give some illustrative examples of this work.

The strongly absorbing complex situated in the direction of the constellations Sagittarius and Scutum was studied by I. I. Pronik [221, 222]. This portion of the Milky Way is extremely heterogeneous, with many dark and luminous nebulae (Fig. 37). A region of 8 × 8.5° was studied [222], divided into 43 areas dictated by homogeneities in the complex. For each area, an absorption-distance relation was constructed. Some of the graphs of these relations, with a logarithmic scale along the x-axis, are given in Fig. 39.

It may be seen from these graphs that the change of absorption with distance has a steplike character. The first step, generally, for all graphs indicates that at a distance of 110–150 pc there is a cloud with absorption greater than one stellar magnitude, after which space remains relatively transparent, while at a distance of 200–600 pc absorbing clouds occur again in various regions. A mutual comparison of all parts of the cloud shows that the near-by cloud is extremely heterogeneous. Even at a distance of 3–5 pc (in the plane of the picture) and occasionally at a distance of 1 pc, absorption changes abruptly and the thickness of the cloud varies. Moreover, in separate areas absorption over a given distance does not remain constant, which explains in part the scatter of the points on the graphs. Apparently the near-by cloud contains much fine detail and many "cloudlets," and consists of a complex with an absorption of from five to seven times that of an average cloud. This complex is part of a still larger complex, connected with a dark nebula in Ophiuchus. The whole system of dark clouds borders the Orion arm, with the interior side at a distance

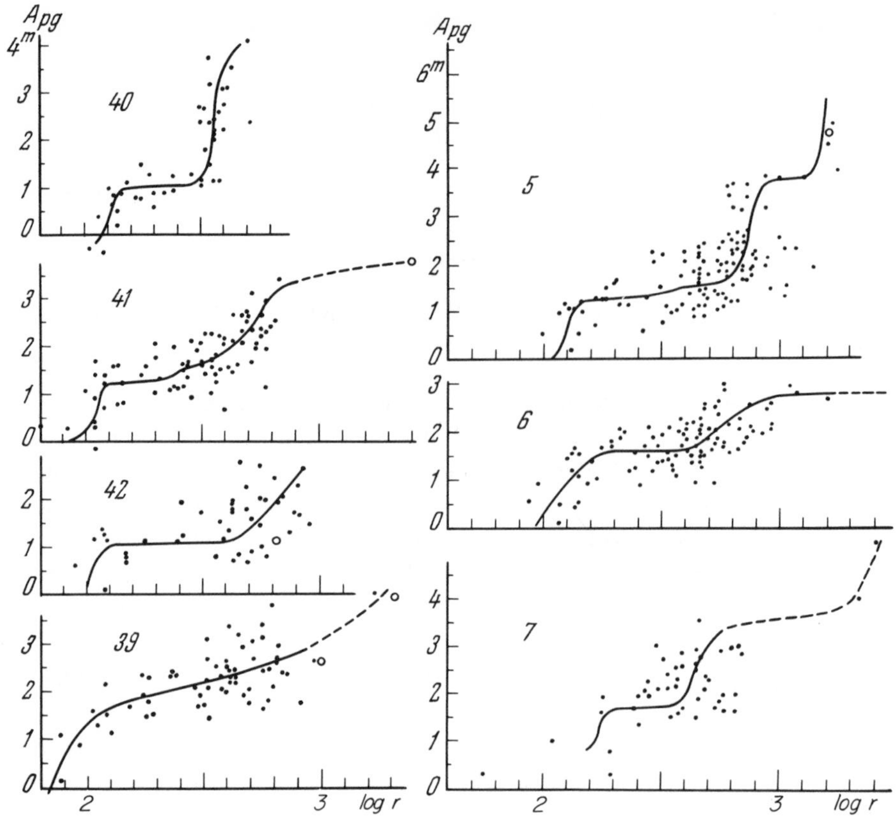

FIG. 39. The dependence of absorption on distance in certain regions of an area of the Milky Way with center at $l = 15°$, $b = 0°$ [221].

100–300 pc from the Sun [223]. Absorption in this belt amounts on the average to 1–3^m, but in particular places can reach up to 6–7^m.

Beyond the absorbing belt, as already remarked, there is a comparatively transparent zone, the interval between the arms, and then the Sagittarius arm is encountered, in which absorption begins to increase sharply, reaching 4–5^m at a distance of ~ 1 kpc, where the density of hot stars belonging to this arm is only beginning to increase [221, 222]. Absorption at greater distances in this direction is as yet uninvestigated; it may be greater farther on. Because of the strong absorption in this direction, experiments have failed to observe the galactic center in the infrared region and in the H_α line.

Hot stars are found in the region of Sagittarius–Scutum ($l \approx 15$–$30°$), beginning at a distance of ~ 1 kpc, that is, they belong to the Sagittarius arm. At the same time, gas, observed at a wavelength $\lambda = 21$ cm, and dust extend in this direction, beginning at the Sun itself. This means that there is a kind of gas-dust bridge across the arms, in the direction $l = 15°$ to $l = 35°$.

Dust distribution along the Orion arm in the direction of Cygnus was studied in papers by A. B. Numerova [224], R. N. Ikhsanov [225, 226], and L. P. Metik [227]. In these papers it is shown that dust is distributed very irregularly in the arm. In addition to relatively clear areas, there are areas where absorption exceeds 5^m at distances of 1300–1500 pc. Such areas occur in each of the regions studied. The complex structure of bright and dark matter in Cygnus is clearly shown in Fig. 9. Absorption in this region is closely connected with the distribution of hot stars and emission nebulae [226]. The relation between stars and polarization in this region was examined by V. A. Dombrovskii [228].

In the region of the anticenter (constellations Perseus and Cassiopeia), several areas were studied by E. S. Brodskaya [229, 230]. The mean absorption in these regions is weaker and more uniform than in Cygnus and in the direction of the galactic center. According to the distribution of hot stars and of neutral hydrogen, the Orion arm ends in this direction to a distance of about 1.5 kpc, but absorbing material continues farther. Absorption in the Perseus arm, located at a distance of 2–3 kpc, does not reach $3^m.5$, so that hot stars are observed up to a distance of 4–5 kpc.

In conclusion, we note that methods of statistical analysis may be refined by taking into account the presence of two kinds of clouds, dense complexes distributed along the arms, and more rarified formations that are randomly distributed.

The Connection Between Dust and Gas

The physical properties of gas in dust complexes are distinct from their properties in the average small gas cloud. A study of the $\lambda = 21$ cm line in absorption [165] showed that deep lines are formed, principally in massive clouds, where $\tau > 0.5$, whose temperatures are lower than the mean background temperature measured at $\lambda = 21$ cm, and, apparently, lower than 50°. These massive clouds are irregularly distributed, but on the average one cloud is encountered every 1.2 kpc. Their average size is 13 pc and they occupy about 1 percent of the volume of the flat galactic subsystem, that is, the same volume that is occupied by the dust. The gas concentration in a dense cloud is ~ 20 cm^{-3}, the mass about 10^3 solar masses. If these dense clouds could be "smeared" over the whole flat galactic subsystem, they would provide a mean density of hydrogen atoms $N_H \approx 0.2$ cm^{-3}, which constitutes an appreciable, but not a principal, part of all the gas.

Observations and theoretical considerations agree that gas and dust must be mutually associated. In fact, dust particles are formed, apparently, by the condensation of molecules, where the condensation rate must increase rapidly with a rise in gas density. The motion of gas clouds must drag along dust particles, since such factors as, for example, light pressure lead only to small velocity differences.

The connection between gas and dust may be estimated in principle by the comparison of the intensities of interstellar absorption lines of ionized calcium and neutral sodium with the reddening of those stars for which these lines are observed. Some correlation is in fact found, but it is not strong. For example, Binnendijk [231] found for separate areas of the Milky Way the dependence

$$W_\lambda \, (\text{Ca II}) = 0.18r + 0.7E, \tag{12.14}$$

where W_λ (Å) is the equivalent width, r (kpc) is the distance, and E is the color excess in the six-color photometer system. The weak correlation is explained in part by the fact that reddening is directly proportional to the number of dust particles, while the line intensities depend on the number of atoms in a more complex manner, mainly because the equivalent width of the more intense lines increases only logarithmically with an increase in the number of atoms.

More reliable results are given by a comparison of absorption from observations of radio emission at $\lambda = 21$ cm. Even the first measurements showed good correlation between hydrogen emission and total absorption, determined, for example, by the decrease in the number of galaxies in a given direction. From Fig. 40, taken from a paper by Lilley [232], it is clear that an increase in absorption (measurements made at a fixed galactic latitude $b = -15°$) corresponds to a triple increase of intensity in the spiral arms. The analysis of this association allows us to obtain for the ratio of gas and dust densities a value of about 100. A detailed study of the relation between optical and radio observations in a nebula near ρ Ophichus was made by Bok [171]. In the densest part of the nebula, with dimensions of about 10 pc, absorption changes from 8^m in the center to $1^m\!.5$ near the edge. The mean density of dust in this part of the nebula is $\sim 0.05 M_\odot/\text{pc}^3 \approx 3 \times 10^{-24}$ g/cm^3, which is 30 times the mean value for the whole Ophiuchus complex and 300 times the mean density of dust near the

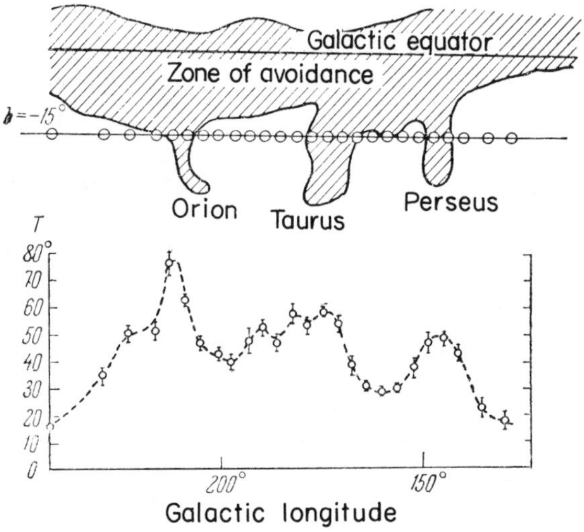

FIG. 40. The relation between absorption (zone of avoidance of galaxies) and intensity of the line $\lambda = 21$ cm [232].

galactic plane. In the center of the nebula, the density is four times greater still. Observations at $\lambda = 21$ cm also showed an increased density of neutral hydrogen, although not so clearly defined. The mean ratio of gas and dust densities in the Orion arm is equal to 400, in the observed complex about 100, and in the densest condensations, the line intensity at $\lambda = 21$ cm is not increased. This means that the ratio of the density of neutral hydrogen to the density of dust in these formations is not greater than 10. As Bok pointed out, it is quite probable that here most of the hydrogen atoms are transformed into H_2 molecules, and therefore he did not investigate the line $\lambda = 21$ cm. The density of the gas itself can be considerably greater in this case than the density of neutral hydrogen (see Sec. 9).

The density of dust is still greater in the globules, the small ($10^{16} - 10^{18}$ cm), dense, dark details, often spherical in shape (Figs. 20, 21), visible in nebulae and against the background of scattered galactic light. In the latter case, the globules themselves are larger. Evidently, the globules play a large part in the evolution of interstellar media (see Sec. 25).

A study of the correlation of emission at $\lambda = 21$ cm with absorption, carried out for a large number of regions, provided the following general conclusions [223]. The connection between gas and dust is of a highly statistical character and not a local one. The presence of a large amount of cold gas in a specific area suggests the probable presence of dust there, but the form and the separate details of the distribution of gas and dust do not agree. Sometimes even the reverse effect is seen, where a dense dust cloud promotes the formation of molecules which increase the size of the dust particles so that the relative gas concentration diminishes.

Dust not only is associated with clouds and neutral hydrogen complexes, but is also found in H II zones. Almost as a rule, a region with many hot stars contains many gas-and-dust clouds, although, of course, there are exceptions. At the same time, regions poor in dust seldom contain groups of hot stars. This correlation is not a localized one, but is valid for regions as a whole. Above all, it refers to such coarse formations as the spiral arms, in which gas, dust, and stars are grouped together. But the correlation also appears on a finer scale, as for example in complexes, which often contain hot stars.

The problem of the connection between gas and dust is examined in the following section.

13. Reflection Nebulae

A cloud of absorbing cosmic dust appearing in the vicinity of a bright star will reflect (scatter) the latter's radiation, and may therefore appear as a nebula with a continuous spectrum, similar to the spectrum of the star.

The continuous spectra of reflection nebulae differ sharply from the spectra of emission nebulae, which consist of distinct bright lines and a very weak continuous background whose origin has already been discussed in Sec. 2. We should note, however, that there are nebulae with composite spectra.

Several hundred reflection nebulae are known. Their size is small, in many cases less than 1′ and always less than 10′ (that is, their linear measurements are less than 1 pc). The majority of these nebulae are illuminated by stars of spectral classes B2–B8, and some by supergiants of later spectral classes. Thus, a nebula may be illuminated by any star of high luminosity. If the spectral class of the illuminating star is B0–B1, the nebula usually has a composite spectrum. Class O stars are associated with emitting nebulae.

The Statistical Connection with Stars

It is well known that if approximately one-third of all class O stars excite luminosity in the visible emission nebulae, then reflection nebulae are found with only one-tenth of all class B2–B8 stars. Therefore, we may assume that the connection between reflection nebulae and illuminating stars is of a rather random nature.

To check this assumption, the following calculation may be employed, first suggested by V. A. Ambartsumyan and Sh. G. Gordeladze [234] and then elaborated by Ambartsumyan [235]. Let us suppose that in a given region all reflection nebulae are observed whose surface brightness is greater than a given limit. The volume illuminated by stars of luminosity L such that nebulae inside this volume have a surface brightness greater than the given limit is proportional to $r^3 \sim L^{3/2} \sim 10^{-0.6M}$, since the surface brightness goes as $\sim L/r^2$. Now let n_i be the number of stars of a given spectral class in the region under study, $\phi_i(M)$ their luminosity function, and V_0 the volume illuminated by stars of absolute magnitude zero. Then for the total volume illuminated by stars in the ith spectral class we obtain, assuming that stars are uniformly distributed,

$$V_i = n_i V_0 \int_{-\infty}^{+\infty} \phi_i(M) \times 10^{-0.6M} dM. \tag{13.1}$$

Light absorption is not considered here. Writing Eq. (12.12) for stars of the ith spectral class, we change it to the form

$$N_i(m) = \Omega \int_0^{\infty} D_i(r)\phi_i(M)r^2 dr = \Omega D_i \int_0^{\infty} \phi_i(M)r^2 dr$$

$$= \frac{\Omega D_i r_0^3}{5 \log e} \int_{-\infty}^{+\infty} \phi_i(M) \times 10^{0.6(m-M)} dM. \tag{13.2}$$

Here also uniform distribution of stars in space and the absence of absorption are assumed. In the change from integration over r to integration over M, it was assumed that $5 \log r = 5 \log r_0 + m - M$, where $r_0 = 10$ pc. Dividing (13.1) by (13.2), we obtain

$$\frac{D_i V_i}{n_i} = \frac{2.16 V_0}{10^3 \Omega} N_i(m) \times 10^{-0.6m}. \tag{13.3}$$

The left-hand side is that part of the illuminated volume of the galactic subsystem where it may be assumed, although only to a first approximation, that stars are evenly distributed. Under the same conditions in the absence of absorption, the product $N_i(m) \times 10^{-0.6m}$ must remain constant under changes in m. For bright stars, this condition is fulfilled, although, to be sure, only very approximately.

Using star counts, it is possible to calculate from Eq. (13.3) that part of the volume of the flat portion of the Galaxy illuminated by stars of various spectral classes [235] (Table 13.1).

Table 13.1

Spectral class	B0	B1–B9	A	F	G	K	M
$(D_i V_i / n_i) \times 10^4$	0.2	2.2	0.6	0.5	0.1	0.45	0.05
Number of nebulae illuminated by stars	23	122	28	6	1	5	2

The conditions for observation of reflection nebulae (brightness threshold) determine only the parameter V_0. In Table 13.1, it is assumed that $V_0 = 2$ pc^3, which corresponds to a limiting observable surface brightness of $23^{\mathrm{m}}.25$ per square second of arc.

In the last line of Table 13.1, the numbers of nebulae illuminated by stars of specific spectral classes are given. It is easy to see that there is marked correlation between the second and third lines of the table (except for class B0 stars which are associated with emission nebulae). Thus it may be deduced that reflection nebulae are simply the result of the random proximity of a dust cloud and a bright star. The total fraction of the plane part of the Galaxy illuminated by bright stars is equal to approximately 5×10^{-4}. Hence, it follows that approximately one part in 2000 of all dust clouds are observed as reflection nebulae, so that the mean density of all clouds is $\sim 4 \times 10^{-4}$ pc^{-3}.

However, here, as well as in absorption studies (Sec. 12), statistical analysis turns out to be only a first approximation to the true situation.

A detailed study of the distribution of reflection nebulae was carried out by D. A. Rozhkovskii [236], who showed that they often occur in regions of considerable absorption, that is, in dense gas-and-dust clouds. This is especially noticeable for groups of such nebulae. At the same time, the dimensions of reflection nebulae are usually less than 1 pc, or considerably less than the dimensions of the total complex. Therefore the nebulae cannot be considered as small clouds distributed at random at various distances from a star. A better approximation is to assume that bright stars also occur inside large complexes and illuminate a relatively small region. The irregular structure obvious in reflection nebulae is explained by the fluctuations of density in the dust of the complex itself. Moreover, these fluctuations are large even in small volumes. The general shape of a nebula is often round. This may be explained by stars' being distributed in an expanded medium and illuminating the nearby surroundings. This picture agrees qualitatively with the data of Table 13.1, since the brighter the star, the larger is the size of the illuminated part of the complex and the greater the distance at which this nebula may be observed.

A more accurate estimate of the number of invisible nebulae must also be attempted. First, there is a close correlation between the class B stars and complexes concentrated in, and even between, the spiral arms. Therefore, where there are no hot stars, there are in general fewer interstellar clouds, so that the previous estimate of the total number of dark nebulae is probably much too high. Second, the idea of nebulae as small clouds does not correspond with reality. It is more correct to consider them as small parts of complexes. Therefore, we must speak of invisible complexes not illuminated by stars, of which naturally there are many more than complexes with reflection nebulae. Of course, there are also small clouds illuminated by a chance star, which in this case is not found near the center, but on the edge or even completely beyond the nebula.

Since only 10 percent of all class B stars are associated with reflection nebulae, we can assume that in the spiral arms of the Galaxy, where there are many of these stars, the complexes occupy a tenth of the volume, and approximately 5 percent of the volume of the entire plane part of the Galaxy. However, we must emphasize that these statistical characteristics were obtained from studies in the solar neighborhood, and therefore should not be extrapolated to the central and peripheral parts of the Galaxy.

Luminosity of Reflection Nebulae

If a reflection nebula is situated inside a complex, the optical thickness of the whole system may be estimated. Usually it is of the order of 2, rarely 3. Since the dimensions of the luminous portion are ten times less than the whole complex, the optical thickness of reflection nebulae rarely exceeds 0.2. These nebulae are relatively transparent. The light of the illuminating star penetrates beyond the apparent boundary of the nebula, but because of the low surface brightness, this portion is not observed. In this way, the visible nebula, as a rule, reflects a small part of the star light and its integrated luminosity is considerably less than the luminosity of the star.

A study of the emission from reflection nebulae can reveal both the properties of the dust and the structure of the nebula itself. However, the results obtained are rather unreliable. This is explained primarily by their usually extremely irregular structure. Moreover, the majority of diffuse nebulae possess a low surface brightness (often 10–20 percent of the brightness of the radiation from the night sky), which makes their photometry difficult. The location of a nebula relative to the illuminating star cannot always be determined. This means that detailed theoretical calculations of light reflection by dust in nebulae are scarcely justified and at the present time it is sufficient to restrict ourselves to the examination of simple models and qualitative estimates. A series of papers (they are reviewed, for example, by Dufay [237]), allows the following conclusions on typical observed properties of nebulae to be drawn.

1. The light from reflection nebulae, generally speaking, differs little from the light of the illuminating stars. As a rule, nebulae are somewhat bluer. From measurements of more than 50 nebulae, Greenstein concluded that the mean value for the difference between the color excess of the nebula and that of the star was $0\overset{m}{.}19 \pm 0\overset{m}{.}06$. It is true that there are exceptions. Certain nebulae associated with T Tauri stars are considerably bluer than the illuminating stars or, as it is better expressed, the star is much redder than the nebula. For example, in the nebula NGC 199, the difference between the color excess of the nebula and that of the star is $1\overset{m}{.}4$ [238].

2. In certain cases, it has been possible to detect circularly and sometimes noncircularly polarized light from reflection nebulae. In particular, in the nebula NGC 7023, the average circular polarization is ~12 percent [239], while in various places polarization reaches 20 and even 30 percent [240].

From these data we may reach the qualitative conclusion that the reflecting part of a nebula consists of dust particles whose dimensions are also of the order of a wavelength [241, 242]. For studying the quantitative characteristics of dust from reflection nebulae, we must first calculate the intensity of the star light scattered by the nebulae. For nebulae of regular shapes this is easy, but the actual shape of a nebula is often very irregular. It is true that for comparison with theoretical calculations a more regularly shaped nebula, which occurs in those cases when the star is situated in a comparatively homogeneous portion of the complex and illuminates a small portion of it [236], may be selected.

The emission intensity, taking into account first-order scattering, is readily calculated. Since the optical thickness of many reflection nebulae is about 0.2 to a first approximation, we can in fact restrict ourselves to single-stage scattering only. Let us consider a spherical volume, illuminated by a star at its center with a luminosity L (Fig. 41). Let us call the extinction coefficient per unit volume of the dust κ, and $x(\theta)$ the scattering index. Then the amount of energy per unit solid angle scattered by dust per unit volume situated at a distance R from the star will be equal to

$$\varepsilon(R, \theta) = \frac{\Lambda L \kappa}{16\pi^2 R^2} \chi(\theta) e^{-\kappa R}. \tag{13.4}$$

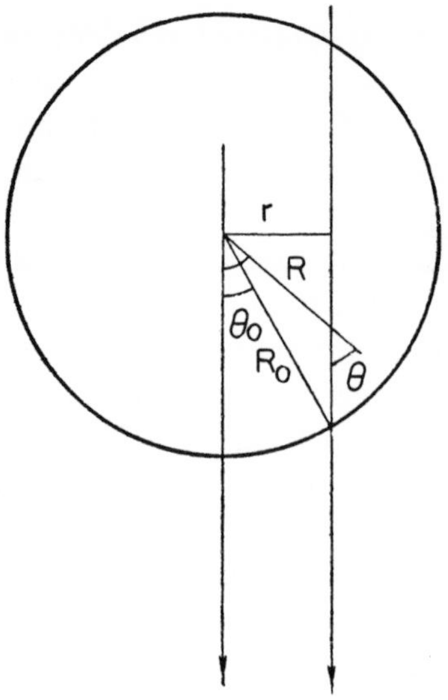

FIG. 41. Scattering of light by dust in a nebula.

Integrating the emission coefficient (13.4) along the line of sight intersecting the nebula at a perpendicular distance from the star r, and allowing for light absorption along this path, we obtain for the observed intensity the expression

$$I(r) = \int_{\theta_0}^{\pi-\theta_0} \varepsilon\left(\frac{r}{\sin\theta}, \theta\right) \exp\left[-\kappa\left(\sqrt{R_0^2 - r^2} - r\cot\theta\right)\right] d(r\cot\theta). \tag{13.5}$$

Here R_0 is the radius of the nebula, and $\sin\theta_0 = r/R_0$. Substituting Eq. (13.4) in Eq. (13.5), we find

$$I(r) = \frac{\Lambda L\kappa}{16\pi^2 r} \int_{\theta_0}^{\pi-\theta_0} x(\theta) \exp\left[\kappa\left(\sqrt{R_0^2 - r^2} - r\cot\theta - \frac{r}{\sin\theta}\right)\right] d\theta. \tag{13.6}$$

If the optical thickness of the nebula $\tau_0 = \kappa R_0$ is much less than unity, the exponent in Eq. (13.6) can be neglected. Then we obtain

$$I(r) = \frac{\Lambda L\tau_0}{16\pi^2 r R_0} \int_{\theta_0}^{\pi-\theta_0} x(\theta) d\theta. \tag{13.7}$$

Differentiating the product $rI(r)$, and remembering that $dr = R_0 \cos\theta_0 d\theta_0$, we obtain from Eq. (13.7)

$$\frac{d}{dr}[rI(r)] = -\frac{\Lambda L\tau_0}{16\pi^2 R_0^2} \frac{\kappa(\theta_0) + \kappa(\pi - \theta_0)}{\cos\theta_0}. \tag{13.8}$$

This formula, obtained by V. V. Sobolev [243], allows us to obtain, at least in principle, the scattering index from the two-dimensional intensity distribution of a nebula. It is true that Eq. (13.8) contains the sum of the scattering in two opposite directions. However, since for interstellar dust particles the scattering index is strongly biased in the forward direction, the second term in this sum is usually small.

By this method I. N. Minin [243a] obtained the scattering index of dust particles from the brightness distribution in nebulae IC 431 and IC 435 [236]. Mean values of the quantity $\kappa(\theta_0) + \kappa(\pi - \theta_0)$ are given in Table 13.2.

Table 13.2

θ_0 (deg)	10	20	30	40	50	60	70	80
$\kappa(\theta_0) + \kappa(\pi - \theta_0)$	11	5.0	3.1	2.3	1.4	1.0	0.8	0.7

The value $\kappa(0)$ (forward scattering) can be obtained only by extrapolation and it is therefore unreliable.

In calculations on the theory of light scattering, an approximate expression is often used for the scattering index having the form

$$x(\theta) = 1 + x_1 \cos \theta, \tag{13.9}$$

where the parameter x_1 is associated with the asymmetry factor q defined in Sec. 11, Eqs. (11.19)–(11.21), by the relation

$$q = \overline{\cos \theta} = \frac{1}{4\pi} \int x(\theta) \cos \theta d\omega = \tfrac{1}{3} x. \tag{13.10}$$

The approximation (13.9) is unsuitable when $q > \frac{1}{3}$, that is, for strongly biased indices. In papers [249] and [243a], from the brightness distribution of reflection nebulae, $x_1 \approx 1.5 - 1.9$ ($q \approx 0.6$) was obtained. Therefore, the value q corresponds to particles of size 10^{-5} cm. The reasons for the discrepancy from the universally adopted value $a \approx 5 \times 10^{-5}$ are not clear.

We note that the relation (13.8) will not give the asymmetry factor if the scattering index is used in the form (13.9), which may be seen by direct substitution of this equation in Eq. (13.8).

In the case of an isotropic scattering index, as well as in the case of (13.9), the integral in Eq. (13.7) is readily calculated. We obtain

$$I(r) = \frac{\Lambda L \tau_0}{8\pi^2 R_0^2} \frac{\frac{1}{2}\pi - \theta_0}{\sin \theta_0}. \tag{13.11}$$

This equation was employed by D. A. Rozhkovskii [236] for the analysis of the brightness distribution of reflection nebulae. The distribution of brightness and polarization in the central regions of nebulae, taking into account the total scattering index given by Mee's theory for spherical particles of various dimensions with a refractive index of $m = 1.33$, was calculated by I. N. Minin [243b]. It was shown that $rI(r)$ and, particularly, $rP(r)$ ($P(r)$ is the observed distribution in two dimensions of the degree of circular polarization, depends weakly on the distance from the illuminating star, but varies strongly with the dimensions of the particles. The circular polarization of 10–15 percent usually observed corresponds to particles with radii of 7×10^{-6} cm, which is several times less than that obtained from analysis of absorption by interstellar dust. The reason for this discrepancy has not been explained.

If the optical radius of a nebula is such that higher-order scattering may be neglected, but absorption in the nebula must be considered ($0.15 < \tau_0 < 0.7$), the exponent in Eq. (13.6) must be retained. The integral may then be calculated only by numerical integration. Values of the expression $U(\tau_0, r)$, defined by the equation

$$I(r) = \frac{\Lambda L}{8\pi^2 r R_0} U(\tau_0, r), \tag{13.12}$$

were calculated by G. A. Shain, V. F. Gaze, and S. B. Pikelner [244] for the index of (13.9). The calculated results show that the expression $U(\tau_0, r)$ is proportional to τ_0 when $\tau_0 < 0.25$, which follows from Eq. (13.11), and hardly depends on τ_0 when $0.5 < \tau_0 < 1$. Its dependence on r is approximately $r^{-1/2}$. Thus, for several dense nebulae we have

$$I(r) \sim \frac{U(\tau_0, r)}{r} \sim \frac{1}{r^{3/2}}. \tag{13.13}$$

For comparison of these calculations with observations, intensities in these equations are expressed in stellar magnitudes per square minute of arc. If Eq. (13.11) is used, that is, if it is considered that $\tau_0 \ll 1$, then from comparison of the observed and calculated intensities the product $\Lambda\tau_0$ may be calculated. If the optical thickness of the cloud is not small and Eq. (13.12) is used, τ_0 must first be estimated. In reflection nebulae that appear as part of a large dust complex, the limits are determined not by the absence of dust but by weakening of the radiation flux with distance from the star. In this case the quantity τ_0 depends on the dust concentration, where an increase in N also means an increase in the scattering coefficient, and regions remote from the star become visible, that is, the size of the luminous region is increased. Absorption per unit length is also increased.

As an illustration, we consider the results of D. A. Rozhkovskii [236]. Five reflecting nebulae were chosen, having a more rounded shape than the others, and from photographs in the two spectral regions 4400 Å $< \lambda <$ 5100 Å and 3700 Å $< \lambda <$ 4400 Å, isophots (Fig. 42) were constructed. Here brightness distribution is mapped along the radii, averaged in all directions (dots). Crosses mark the intensity values calculated from Eq. (13.11) when $\Lambda\tau_0$ is equal to 0.158 for the blue spectral region and 0.119 for the yellow region. Good agreement between the observed and calculated curves may be seen. With other nebulae, this agreement, with one exception, which may be explained by peculiarities

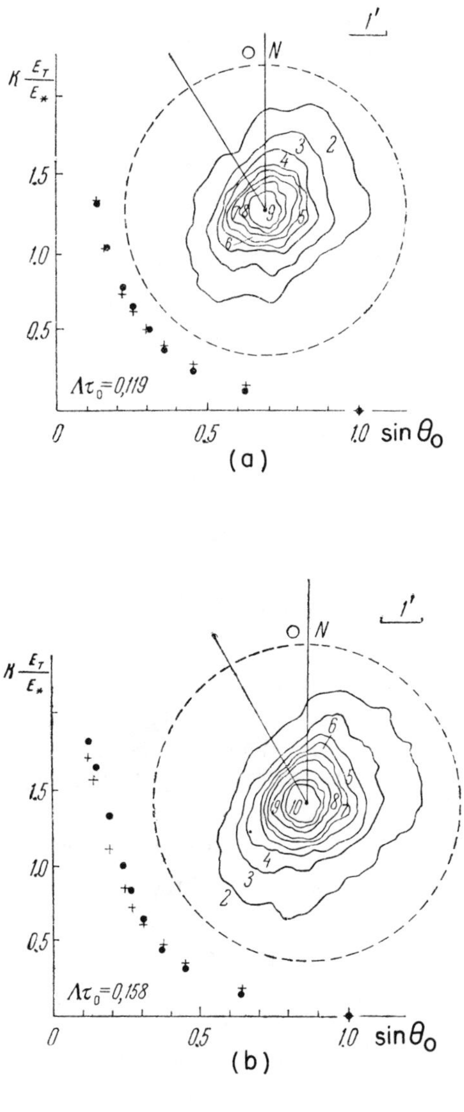

FIG. 42. Isophots of the reflection nebula Ced 167 in (*a*) yellow and (*b*) violet light. Dots and crosses are the observed and theoretical (at a specified value of $\Lambda\tau_0$) dependence of the average brightness on the distance from the star [236].

in its structure, is also quite satisfactory. From the results of [236] it follows that the calculated values of $\Lambda\tau_0$ occur over the range 0.08–0.4 for the blue spectral region and 0.06–0.25 for the yellow. Since the value of τ_0 must be small for these nebulae, which follows from the good agreement of the observed intensity distribution with Eq. (13.11), we may conclude that the albedo of the particles cannot be very small.

The Role of Higher-Order Scattering

If the optical radius is not small ($\tau_0 \gtrsim 1$) and the albedo of the particles is close to unity, higher-order scattering must be considered. This may be done when the scattering material is symmetric.

A similar problem was solved during a study of the intensity of diffuse galactic emission. From old measurements of Elwey and Roach [245] and Heney and Greenstein [246], the intensity of diffuse scattered radiation amounts approximately to 40 percent of the intensity of the total stellar radiation.

For calculating the intensity of galactic diffuse radiation, we consider the following arrangement. A flat layer of dust of optical thickness $2\tau_0$ scatters the radiation from stars arranged in space according to the barometric law (6.1). This problem and certain others were solved by Heney and Greenstein [246] and also Wang Shih-ku [247, 248]. Here we give a simpler solution, obtained by S. A. Kaplan.

Before writing down the transfer equation of the scattered radiation in this problem, we shall make some simplifications. First, we shall consider the radiation to be averaged over galactic longitude, that is, both the radiation intensity and the source functions depend only on the galactic latitude b and the optical distance τ from the galactic plane. Second, the scattering index is given by expression (13.9) approximately. Third, since the parameter β for the barometric distribution of stars is larger than the same parameter for dust, we shall assume that the total emission intensity of the stars depends only on the galactic latitude. For the region close to the galactic plane, which is of the greatest interest, these assumptions are sufficiently correct. Thus the transfer equation and the radiative-equilibrium equation take the form

$$\begin{aligned} \sin b \frac{dI(\tau, b)}{d\tau} &= -I(\tau, b) + B(\tau, b), \\ B(\tau, b) &= \frac{\Lambda}{2} \int_{-\pi/2}^{+\pi/2} [I(\tau, b') + I_*(b')]x(\gamma) \cos b' db', \end{aligned} \tag{13.14}$$

where $\cos \gamma = \sin b \sin b'$, that is, where γ is the angle between rays falling in the direction of galactic latitude b' and those reflected in the direction of galactic latitude b. Averaging over galactic longitude has already been taken into account here. We shall measure optical thickness from both sides of the galactic plane, considering the northern direction positive and the southern, negative. Substituting Eq. (13.9) in the second of Eqs. (13.14) and putting $\eta = \sin b$, we obtain

$$B(\tau, \eta) = \Lambda \bar{I}(\tau) + \Lambda x_1 \bar{H}(\tau)\eta + \Lambda \bar{I}_*. \tag{13.15}$$

Here the mean diffuse radiation intensity $\bar{I}(\tau)$, the flux of this radiation $\bar{H}(\tau)$, and the mean intensity of the total stellar radiation, $\bar{I}_*$, are expressed by the equations

$$\bar{I}(\tau) = \frac{1}{2}\int_{-1}^{+1} I(\tau, \eta')d\eta',$$
$$\bar{H}(\tau) = \frac{1}{2}\int_{-1}^{+1} I(\tau, \eta')\eta' d\eta', \tag{13.16}$$
$$\bar{I}_* = \frac{1}{2}\int_{-1}^{+1} I_*(\eta')d\eta'.$$

The condition $\bar{I}_*(\eta) = I_*(-\eta)$ is used here; it arises from the symmetry of the total stellar radiation relative to the galactic plane. Substituting Eq. (13.15) in Eq. (13.14), integrating the expression obtained, first with respect to $d\eta$ and then with respect to $\eta\ d\eta$, and using Eddington's approximation

$$\frac{1}{2}\int_{-1}^{+1} I(\tau, \eta)\eta^2 d\eta = \frac{1}{3}\bar{I}(\tau), \tag{13.17}$$

we obtain for $\bar{I}(\tau)$ and $\bar{H}(\tau)$ the system of equations

$$\frac{d\bar{H}}{d\tau} = -(1 - \Lambda)\bar{I}(\tau) + \Lambda\bar{I}_*,$$
$$\frac{1}{3}\frac{d\bar{I}}{d\tau} = -(1 - \tfrac{1}{3}x_1\Lambda)\bar{H}(\tau). \tag{13.18}$$

The solutions of these equations, meeting the demand of symmetry relative to the galactic plane and the limiting condition $2\bar{H}(\tau_0) = \bar{I}(\tau_0)$, which indicates that at the boundary of the medium there is only scattered radiation continuing outside, have the form

$$\bar{I}(\tau) = \frac{\Lambda\bar{I}_*}{1 - \Lambda}\left[1 - \frac{\cosh(k\tau)}{\cosh(k\tau_0) + 2k\sinh(k\tau_0)/(3 - x_1\Lambda)}\right], \tag{13.19}$$

$$\bar{H}(\tau) = \frac{\Lambda\bar{I}_*}{\sqrt{(1 - \Lambda)(3 - x_1\Lambda)}}\frac{\sinh(k\tau)}{\cosh(k\tau_0) + 2k\sinh(k\tau_0)/(3 - x_1\Lambda)}, \tag{13.20}$$

where we introduce the parameter

$$k = [(1 - \Lambda)(3 - x_1\Lambda)]^{1/2}. \tag{13.21}$$

For the case when $k\tau_0 \ll 1$, we have the simplifications

$$\frac{\bar{I}(\tau)}{\Lambda\bar{I}_*} = 2\tau_0\left[1 + \frac{(8\Lambda - 5 - x_1\Lambda)\tau_0^2 - (3 - x_1\Lambda)\tau^2}{4\tau_0} + \cdots\right],$$
$$\frac{\bar{H}(\tau)}{\Lambda\bar{I}_*} = \tau + \cdots. \tag{13.22}$$

In order to find the intensity of the scattered radiation near the galactic plane, we must calculate the integral

$$I(b) = \int_0^{\tau_0} B(\tau', \sin b) e^{-\tau'/|\sin b|} d\tau'/|\sin b|. \qquad (13.23)$$

This integral may be calculated in an explicit form from Eqs. (13.19) and (13.20), but very cumbersome expressions are obtained. Therefore, employing the approximate equation (13.22) and assuming for simplicity that $|\sin b| \ll 1$, we find

$$\frac{I(b)}{\Lambda \bar{I}_*} = \left[1 + 2\Lambda\tau_0 + \frac{\Lambda\tau_0^2}{2}(8\Lambda - 5 - x_1\Lambda)\right](1 - e^{-\tau_0/|\sin b|})$$
$$-\Lambda x_1 \sin^2 b \left\{(3 - x_1\Lambda)\left[1 - e^{-\tau_0/|\sin b|}\left(1 + \frac{\tau_0}{|\sin b|} + \frac{\tau_0^2}{2\sin^2 b}\right)\right]\right.$$
$$\left. -\left[1 - e^{-\tau_0/\sin b}\left(1 + \frac{\tau_0}{|\sin b|}\right)\right]\right\}. \qquad (13.24)$$

In accordance with the measurements of Heney and Greenstein [246], the intensity of the scattered radiation amounts to approximately 40 percent of the stellar radiation. Substituting $\tau_0 \approx 0.3$ in (13.24), we find that Λ must be about 0.3–0.4. Equation (13.24) gives a comparatively rapid fall in intensity with an increase in galactic latitude when $\sin b \gg \tau_0$, which agrees with the observations of Heney and Greenstein. However, it must be noted that subsequent measurements by Elsasser do not corroborate Heney and Greenstein's results. He did not detect appreciable traces of diffuse emission. A decision on this problem is most essential for an explanation of the properties of dust particles.

As seen from Eqs. (13.19) and (13.20), the solution of even simple problems taking into account higher-order scattering turns out to be extremely unwieldy. The complicated geometry of scattering media makes a solution difficult.

By the method employed above, on the problem of calculating the intensity of light scattered by a flat medium, V. V. Sobolev solved the problem of light scattering in a spherical nebula [243]. In general, when $\tau \gg 1$, an asymptotic expression for the source function has the form

$$B(\tau) = \frac{\Lambda^2 L\kappa}{16\pi^2 R}\left[\frac{3 + (1 - \Lambda)x_1}{2k} \ln\frac{1 + k}{1 - k} - x_1\right] e^{-\kappa k R} \sim \frac{e^{-k\tau}}{\tau}, \qquad (13.25)$$

where τ is measured from the center and k is determined by Eq. (13.21). When $\tau \lesssim 1$, the equation for $B(\tau)$ proves to be very cumbersome.

D. A. Rozhkovskii [236] obtained an answer to the problem of light scattering in a spherical nebula by expressing the source function and the intensity of scattered radiation in a series expansion about the optical radius of the nebula τ_0. The first term of the expansion agrees with Eq. (13.11).

Certain problems on multiple light scattering in flat and in spherical nebulae were solved by S. A. Kaplan and I. A. Klimishin [249, 250]. In particular, if we suppose that a nebula has the shape of a spherical envelope surrounding a star, we can use the solution of the problem of light passing through a thin layer to find the intensity of scattered emission and its distribution at the visible surface of the nebula [250]. For a study of reflected light from very dense nebulae, as, for example, globules, we can use determinations of the albedo of these objects [249], similarly to the introduction of these values for planetary atmospheres. For example, the spherical albedo of a globule, illuminated by the isotropic light of surrounding stars and emission nebulae, is changed from 0.07 to 0.28 during a change in Λ from 0.4 to 0.8 and equals unity when $\Lambda = 1$ (for the case of the scattering function with the form of (13.9) when $x_1 = 1$). Methods from the theory of radiation transfer also allow us to estimate the value of the scattered energy emerging from the elongated plane of the nebula at any angle, with the illuminating star inside [249] as well as outside [61] the nebula. We must state, however, that the observed data are still insufficient for comparison with the theoretical calculations. We shall therefore confine ourselves to a qualitative description.

Consider, for example, a nebula of finite dimensions and $\tau_0 \gg 1$. The radiation from a star and the nearby brightly illuminated dust does not emerge and we cannot see it. On the other hand, if the albedo of the particles is sufficiently high, a considerable portion of the radiation is not changed into another form of energy and emerges after a number of scatterings. The nebula will appear as a more or less uniformly shining sphere. If its dimensions are large, the surface brightness will be small and the nebula is not seen. This occurs in large dense dust clouds. More often, dense dust concentrations, the globules, are small in size. In this case the brightness of the nebula will be appreciable because of higher-order scattering and will be only a few times less than the luminosity of the star. Radiation from the star itself will be more strongly attenuated and the star may appear even weaker than the nebula if the albedo is sufficiently high.

As subjects for study, the so-called "comet-shaped" nebulae are of interest. These small objects, of about 0.03–0.1 pc, usually have the shape of a fan, at whose apex is situated a variable star of type T Tauri. The spectra of comet-shaped nebulae are continuous and their light is sometimes bluer than that of the star. Their radiation is circularly polarized, on the whole, relative to the star, but sometimes with certain variations. In one or two cases, it appears that the nebula is somewhat brighter than might be expected, considering the brightness of the star. Because of this, the assumption was made that the radiation from comet-shaped nebulae was nonthermal in nature, connected with the liberation from T Tauri stars of prestellar material possessing special properties [2]. However, this effect is explained more readily by the higher-order scattering referred to above; stellar radiation is strongly reduced by the small dense core and, on the whole, absorption has almost no effect on the brightness of the nebula. The asymmetric distribution of density produces deviations from circular polarization. The variability of the star in conjunction with the retardation of the reflected radiation produces certain Doppler effects in their emission, because of the finite velocity of light.

An interesting feature of many reflecting nebulae is their fine filamentary structure. In the Pleiades and other objects, the filaments are quite visible, their thickness being a few hundred or thousand parsecs. The filaments are approximately parallel to one another and slightly curved, while their concavity is turned toward the illuminating star. In this respect filaments differ from rings (hoops) at the boundary between dark and luminous material in emission nebulae, which are turned convexly to the exciting star. The concavity of filaments can possibly be explained by radiation pressure [187], but the actual formation of filaments cannot be explained by this effect [251]. A study of filaments shows a characteristic polarization (see Sec. 14).

The filamentary structure of reflective nebulae may be a consequence of some kind of magnetohydrodynamic instability when a field-free gas cloud intrudes into the interstellar magnetic field. Instability relative to interchange of lines of magnetic force induces the decay of the nebula into filaments. After this decay the fields penetrate into the filaments and prevent ions and charged dust particles from moving away [251a].

The Connection Between Emission and Reflection Nebulae

The difference between the spectra of these objects is explained by a temperature difference in the exciting or illuminating star, as well as by a different concentration of dust and gas. It may be shown that the continuous spectrum in the majority of emission nebulae is considerably more intense than might be expected from the dust occurring in the nebula [244]. This is partly explained because the brightness of reflection nebulae rapidly falls off with distance, while the intensity of the continuous emission spectrum hardly depends on the distance from the star (in H II zone boundaries). The principle reason is that dust shines on account of the reflection of stellar radiation in the visible spectral region and gas converts the ultraviolet radiation of exciting stars. With class O stars, radiation is much more intense in the ultraviolet region than in the visible. Thus the presence of dust in emission nebulae cannot noticeably increase the intensity of their continuous spectra. However, if the gas density is low, so that emission is weak, the reflected light can become appreciable. For example, Fig. 43 shows the relation between the brightness of nebulae in H_α and in the yellow region of the continuous spectrum. In the majority of nebulae, these values are proportional, which also follows from Sec. 2, but for weak nebulae, a shift to the right is visible. In this case, radiation in the continuous spectrum includes scattering by dust particles. The three small squares in Fig. 43 indicate nebulae associated with class B1 stars with mixed spectra.

Thus dustlike material also occurs in emission nebulae. This is borne out by

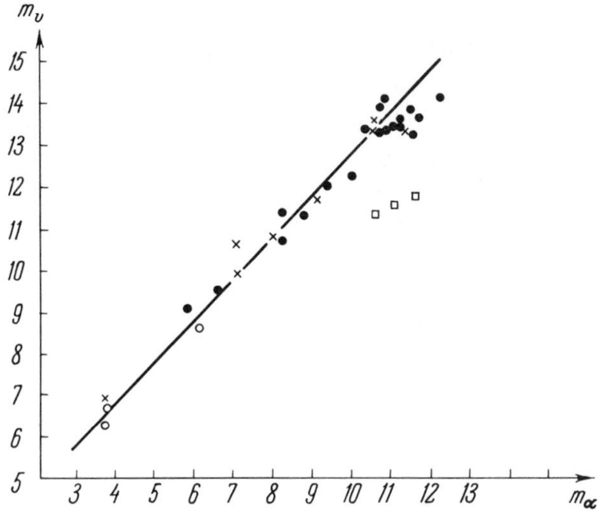

FIG. 43. Relation between the brightness of nebulae in H_α and in the continuous spectrum (stellar magnitude per square minute of arc) [244].

the fact that many emission nebulae are extremely opaque. In the Orion nebula, for example, absorption amounts to 10^m per parsec [406], which is appreciably greater than absorption in ordinary clouds, especially if we consider the difference in density. It is true that a large part of emission nebulae are almost free from absorption, or in any case they do not contain more dust than ordinary clouds. As an example, we can take nebula NGC 6523, the Lagoon Nebula, which causes hardly any reddening of the stars behind it. Only in the central portion of the nebula does absorption increase somewhat. On the other hand, gas found in reflection nebulae is not seen, since stars later than B1 have an H II zone that is in general small, and emission near B0–B1 stars is appreciable only at high gas density.

Generally, the relation between the amounts of dust and of gas differs in different regions. Many nebulae—NGC 6618, the Omega Nebula, for example—are contiguous with dense masses of cold gas, rich in dust. In a number of cases, dark formations are included in nebulae, for example in the Trifid (Fig. 26). Elephant trunks and globules also consist of rich dustlike material embedded in the nebula. As we have already remarked, dense nebulae rich in dust have a considerable amount of hydrogen, which can change into the molecular state and not be observed. Therefore, in such nebulae we cannot generally assert that the relative gas content is small.

Sometimes one part of a nebula is seen in emission and another part by reflected light. For example, the nebula IC 405 associated with the star AE Aurigae of the Orion complex, possesses a characteristic arch, visible in blue light but not noticeable in H_α [252]. The spectrum of this arch is continuous and is formed by scattering from dust [253]. The Trifid Nebula, which possesses a satellite almost equal to it in size and with a continuous spectrum, is another example. In Fig. 26 (photograph in H_α radiation) this satellite is not seen. A supergiant class A7 star illuminates a dust nebula [254]. It is interesting that this nebula is somewhat brighter than the star; probably the radiation from the star undergoes additional absorption. The polarization of the radiation from this nebula has distinctive features that we shall consider in the following section. The examples cited demonstrate clearly the variety and complexity of nebulae.

14. Interstellar Polarization of Light

In 1949 Hiltner [255, 256], Hall [257], and V. A. Dombrovskii [258] observed independently that light from certain stars is polarized, and that the maximum degree of polarization reaches 8–10 percent. Polarization is appreciable only for sufficiently distant stars.

Observational Data

Interstellar polarization has been studied from many observations (see Loden [259, 260], where a detailed bibliography is given). The results of these studies are basically as follows. In the visible region (up to $\lambda \approx 8000$ Å), within the limits of accuracy of the measurement, the degree of polarization does not depend on wavelength, although from certain data in the red region the degree of polarization is somewhat reduced (by approximately 20 percent). From the data of Gehrels [261] and Behr [262], the maximum degree of polarization lies in the region around 6500 Å. In the infrared spectrum around $\lambda = 2\mu$, the degree of polarization is several times less than in the visible region.

The plane of oscillation of the electric vector is close to the galactic plane, but this effect depends on the galactic longitude. It is seen most strongly in those directions in which the line of sight intersects a spiral arm at right angles and least in those directions where the line of sight skirts the arm longitudinally. The distribution with respect to galactic longitude of the angle θ between the galactic plane and the electric vector is shown in Fig. 44 (constructed by G. A. Shain [263]). The dots here represent all the stars with polarizations measured up to the present time with galactic longitude $|b| \leqslant 8^\circ$. It is readily seen that in the direction $l \approx 70$–80°, where the line of sight passes along the Orion arm, the scatter of θ is large, that is, polarization is random. The scatter of these values is greater also in the direction $l \approx 15$–25° (constellations Scutum and Sagittarius). Evidently, here the line of sight passes along the gas-and-dust bridge mentioned in Sec. 12, joining the spiral arms. In other directions, polarization is comparatively well organized, since here the angles θ are grouped around 0°. However, a careful examination of this diagram will show that the distribution curve of the average values of θ is not straight but rather sinusoidal, intersecting

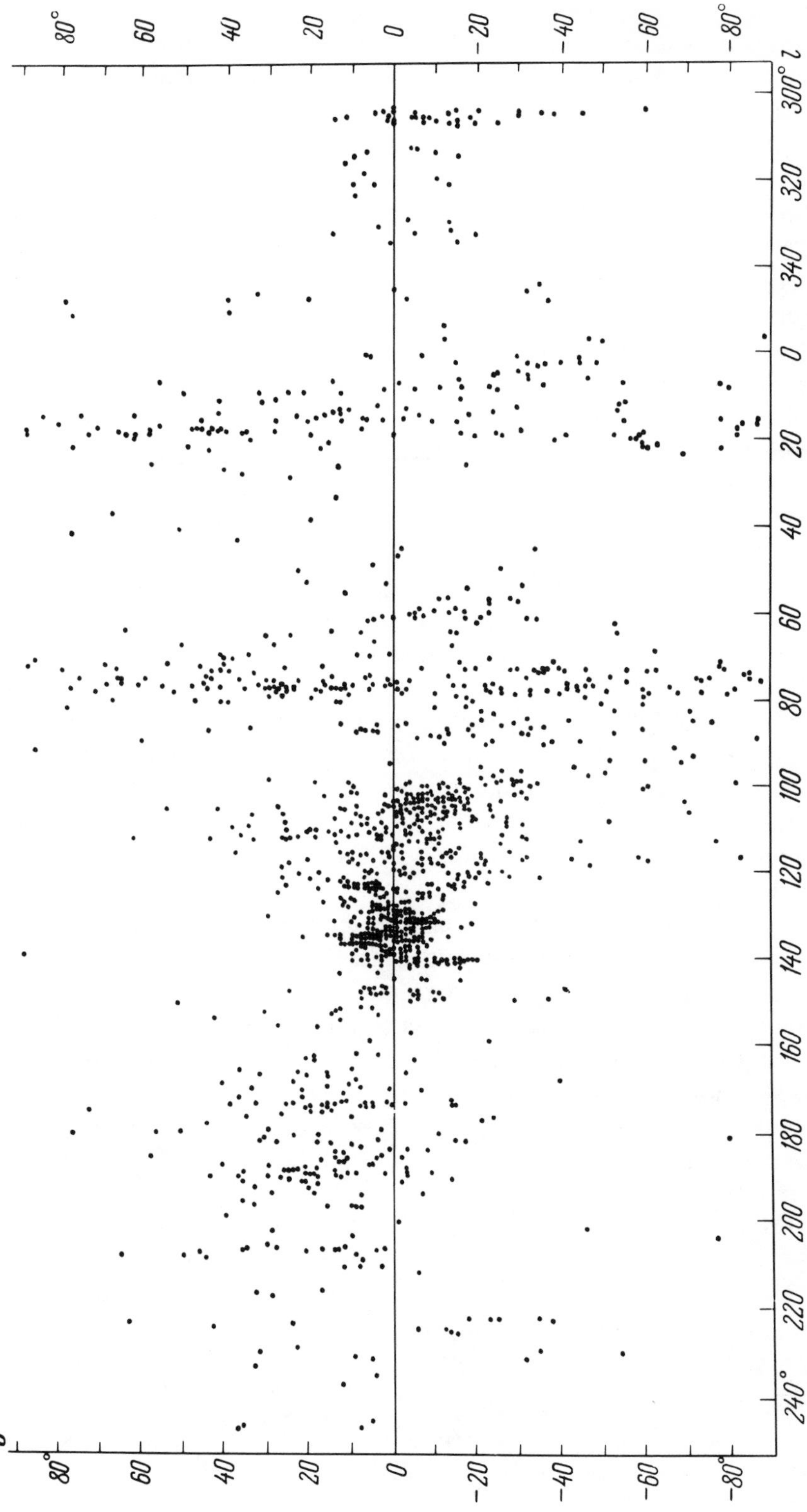

FIG. 44. Angle of the electric vector of polarization with respect to the galactic plane [263].

the axis of abscissas at $l \approx 135$–$140°$. This means that the polarization vectors are parallel not to the galactic plane but to a great circle inclined to it at 17–18°. The galactic coordinates of the pole of this circle are $l \approx 138°$, $b = 73°$, and, judging from the distances to the stars for which the polarization was observed, its radius is about 1000 pc. The mean plane of orientation of the particles that cause the polarized light does not coincide with the fundamental plane of the local class B star belt (Gould's belt). However, as radio observations in the line $\lambda = 21$ cm for the solar vicinity show, the direction where N_H is minimum agrees approximately with the pole of the great circle of polarization [118]. If we subtract from the quantities characterizing the interstellar gas distribution the average data related to the galactic plane, certain density "excesses" vanish, in particular, for the complexes of gas and dust distributed in the constellations Scorpius–Ophiuchus and Gemini–Taurus–Orion (Sec. 12). These complexes lie in the plane of a great circle, as found by G. A. Shain from the distribution of polarization, and apparently form a local system of gas, dust, and magnetic fields. It may be that such systems appear as fluctuations of the spiral arm.

A detailed analysis of light polarization in selected areas dispersed over the whole sky was made by Loden [259, 260]. In approximately 65 percent of all stars with polarized light, the plane of polarization is inclined to the galactic plane by less than 20°. For 22 percent of all stars, $21° \leqslant |\theta| \leqslant 40°$, and only in 4 percent of stars is the inclination of the plane of polarization to the galactic plane greater than 61°. It may be noted from Fig. 44 that the dispersion of the direction of the plane of polarization is not large ($\approx 10°$). In a number of cases a statistical dependence was observed of the degree of polarization on the angle of position; as a rule, the degree of polarization is larger in the direction of the principal direction.

In order to study the properties of particles that cause polarization of light, we must determine the relation between the characteristics of interstellar polarization and other parameters of the interstellar medium. In particular, it is important to find the degree of correlation of polarization with distance from the star, or with color excess. Such correlations have actually been discovered, but they are not always sufficiently marked.

According to Loden's data [260], correlation exists between the degree of polarization p, the distance modulus $m - M$, and the color excess E, given in Table 14.1.

Table 14.1

p (percent)	$\overline{m - M}$	$\bar{E}$
0.0–0.4	6.53	0.048
0.5–0.9	8.41	.082
1.0–1.4	8.56	.158
1.5–1.9	9.45	.298
2.0–2.4	10.50	.394

The mean dispersion of the values of $m - M$ is approximately 0.35 and the mean dispersion of color excess is 0.56.

Although Table 14.1 indicates marked systematic correlation, it must be emphasized that there is no direct functional relation. In many cases, no correlation was in general detected between, for example, color excess and degree of polarization. The correlation between degree of polarization and the distance modulus has a certain multistage character, as, for example, in the case of the selected area SA 19 (center coordinates $l \approx 114°$, $b = -1°$), where a rich accumulation of gas and dust is observed. Evidently the break is explained by the presence along the line of sight of regions with varying degrees of particle orientation. Some other results from observations on polarized light will be given below.

Scattering by Elongated Oriented Particles

Data on polarization may be explained qualitatively on the hypothesis that interstellar dust particles have an elongated shape. In this case, light with the electric vector directed along the major axis is scattered or absorbed more strongly than that perpendicularly polarized. The major axes of the dust particles are oriented perpendicular to the interstellar magnetic field. The vectors of these fields in different clouds can have, as will be shown below, different directions, but in the spiral arms the magnetic field is directed principally along their axes. The depolarizing effect of clouds with different alignments of their magnetic fields can bring the polarization to zero, although total absorption will be large. This explains the lack of correlation in several cases between p and E. The directional scatter of polarization in the constellation Cygnus (along an arm) is explained because here the major axis of the dust particles lies almost in the plane of the sky and small fluctuations sharply alter the direction of maximum angle to this plane.

The existence of interstellar magnetic fields was confirmed in 1949, both in connection with the discovery of interstellar polarization and on the basis of cosmic-ray studies. These fields will be examined in more detail in the following chapter; here we simply note that magnetic fields which orient particles are found in interstellar space.

As we remarked in Sec. 11, in order to calculate the polarization of light, an accurate solution to the problem of scattering by oblong particles is required. It is impossible here to restrict the investigation to the simple equation (11.5). The results of such a calculation for cylindrical particles whose diameter is comparable with the wavelength are given in Fig. 34 [264]. From these curves we see at once that polarization induced by scattering by such particles when $\rho > 1$ is small (~ 10 percent or less), especially if we consider the change in sign of polarization at large values of ρ. In the case of small particles, the degree of polarization can be high. For example, for thin cylindrical particles the effective absorption cross section in cases when the electric vector is parallel (σ_1) and perpendicular (σ_2) to the axis of the particle has the form

$$\sigma_1 = \frac{2\pi^2(1 + \xi)\xi \tan \beta}{\lambda} aS, \qquad \sigma_2 = \frac{8\pi^2(1 + \xi)\xi \tan \beta}{[(1 + \xi)^2 + 1]^2\lambda} aS. \tag{14.1}$$

Here α is the radius of the cylinder ($\alpha \ll \lambda$) and S is the area of the linear (along the axis) cross section of the cylinder. The quantity $n = 1 + \xi - i\xi \tan \beta$ is the refractive index of the particle. The polarization of the light absorbed and oriented by the particles is

$$p = \frac{\sigma_1 - \sigma_2}{\sigma_1 + \sigma_2} = \frac{[(1 + \xi)^2 + 1]^2 - 4}{[(1 + \xi)^2 + 1]^2 + 4} \approx \xi. \tag{14.2}$$

When $\xi \approx 0.5$, the degree of polarization of the absorbed light can reach 45 percent. However, such small particles cannot explain general light absorption for the reasons given in Sec. 11 and consequently do not explain the polarization of the light passing through.

We must therefore assume that both extinction and polarization of light in interstellar space are induced by oblong particles the size of a wavelength of visible light, except in those cases when polarization reaches 10 percent, when we must consider that all or nearly all of the particles are oriented by an interstellar magnetic field. The observed weak dependence of the degree of polarization on wavelength agrees qualitatively with the polarization curves in Fig. 34.

A calculation of the degree of polarization $p = (\tau_1 - \tau_2)/(\tau_1 + \tau_2)$ for oblate or prolate spheroids with a ratio of axes of 1:2, using a refractive index $m = 1.33 - 0.05i$ and assuming the particle-size distribution function given by Oort and van de Hulst (see Sec. 15), shows that in the case of fully oriented particles the degree of polarization of oblate ellipsoids may attain 20 percent and for prolate particles 9 percent [264a]. The dependence of the degree of polarization on wavelength is in good agreement with the observations [261].

The orientation of dust particles by a magnetic field not only causes polarization of light but affects the reddening characteristics. Since extinction by ellipsoidal particles depends on the angle between the particle axis and the direction of the light, absorption along an arm, where the oriented particles are turned toward the observer with maximum cross section, must depend on wavelength λ differently from that across the arm, where the particles are turned toward the observer with a smaller cross section. This effect was predicted theoretically by Ginzberg and Meltzer [265]. To calculate the effect we must average the effective cross-section functions (11.13) using the value of ρ from (11.15). The values obtained are then averaged with respect to the orientation angle $\chi = 90°$, for observations along an arm. For example, if the scattering function of particles is described by Eq. (11.23) and if we assume that particles scatter radiation only according to Eq. (11.25), we have for the extinction coefficient

$$\bar{Q}_e(\rho_1) = 2\rho_1^2 \frac{3 + \rho_1^2}{(1 + \rho_1)^2}. \tag{14.3}$$

For ρ_1, we must take the value $4\pi(n - 1)b_1/\lambda$ for an observation along an arm (b_1 is the mean value of the principle semiaxis) and a value approximately equal to $4\pi(n - 1)a_1b_1/\lambda[(a_1^2 + b_1^2)/2]^{1/2}$ for an observation across an arm (in this case $\cos^2 \chi = \sin^2 \chi = \frac{1}{2}$). It is obvious that the extinction coefficient depends differently on wavelength in the two cases.

A numerical calculation of this dependence was given by Greenberg and Meltzer for particles with the parameters $b = 2a$, $n = 1.33$. The distribution function was taken from the model obtained by Oort and van de Hulst [200]. The results are given in Fig. 45. The difference in dependence on wavelength of the average effective cross section for observations along and across the arms is extremely marked.

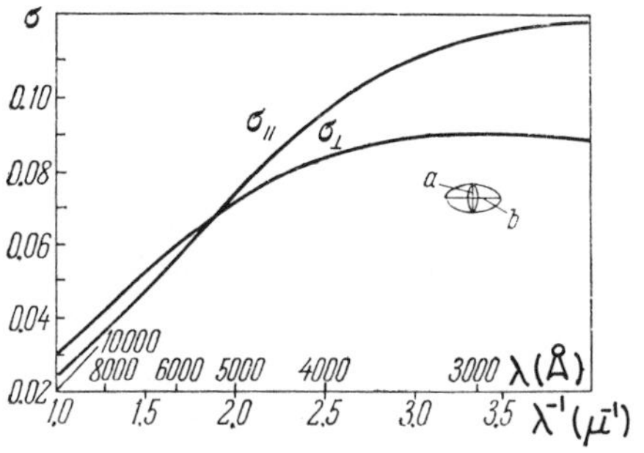

FIG. 45. The dependence of extinction on wavelength for differently oriented elongated spheroids rotating about their minor axes, oriented with respect to **H**, for observations along (∥) and across (⊥) the arms [265].

An analysis of the observational data [266, 267] does in fact disclose this effect in qualitative agreement with theory, but quantitative agreement is still difficult because of the smallness of the effect itself. The dependence of the quantities R and E_{U-B}/E_{B-V} on galactic longitude are given, for example, in Tables 11.1 and 11.2. The ratio of color excesses E_{U-B}/E_{R-I} in Perseus, taken across the arm, is equal approximately to 1.02, and in Cygnus, taken along the arm, the same ratio is approximately 1.27. According to theory, when the particles are completely oriented $E_{U-B}/E_{R-I} \approx 1.5$. The discrepancy may be due to the arbitrary choice of the ratio b/a, the refractive index, neglect of pure absorption, and the like. A qualitative dependence of absorption on longitude was detected [266, 266a, 267]. Along the arm, absorption in the ultraviolet is relatively greater than in the visible. This effect is difficult to measure quantitatively. We can investigate the dependence of color excess, not on the longitude, but on the quantity p/E (where p is the degree of polarization), which is directly connected with particle orientation. A positive correlation is also detected here. However, the maximum deviation from the absorption law does not coincide in the sky with the position of maximum polarization [266a, 266b]. This effect may be caused by differences in particle dimensions. For example, in Perseus, where the line of sight crosses an arm by a relatively short path, absorption occurs in clouds situated close to the stars. Here particle sizes may be less than the mean, because of evaporation due to radiation from the nearby stars and because of the disruptive action of the ionized gas. In addition, as was shown by Serkovskii [266b], this difference may be due to the dependence of color excess on stellar spectral class. In Cygnus, stars are considerably hotter, and therefore E_{U-B}/E_{B-V}, according to [266b], is considerably greater than 0.05 there than in other directions.

Nevertheless, it is possible that the hypothesis of the orientation of dust particles by an interstellar magnetic field to explain the many peculiarities of light absorption and polarization corresponds with reality. As we shall see below, it also explains qualitatively noncircularly polarized light in certain filamentary reflection nebulae.

Orientation Mechanisms

Let us consider possible mechanisms for the orientation of dust particles. We might assume that the particles have a natural magnetic moment (ferromagnetic particles). The interaction of such particles with an interstellar magnetic field causes them to precess. Because of thermal losses due to hysteresis, rotation must be retarded and the particles are gradually oriented along the magnetic field [268]. However, the existence of ferromagnetic properties in dust particles is unlikely, while damping of rotation by hysteresis is not entirely effective. Most important of all, according to observations, cosmic dust particles must be oriented perpendicular to the magnetic field and not along it, as would happen by this mechanism.

At the present time, the most probable theory is that of Davies and Greenstein [269], which assumes that interstellar dust contains ferromagnetic atoms in such small amounts that the particles have paramagnetic and not ferromagnetic properties. The rotation of a particle in a magnetic field causes continuous magnetic reversal, so that a fraction of its kinetic energy is changed to heat, resulting in a weak retarding torque on the particle. The action of this braking torque causes a periodic precession, whereby the particle is gradually oriented in a definite way.

The axis of rotation of the oriented particle must coincide with the magnetic field. In fact, change of torque **P** and of rotational energy K of a solid body is determined by the equations

$$d\mathbf{P}/dt = \mathbf{L}, \qquad dK/dt = \boldsymbol{\omega}\cdot\mathbf{L}, \tag{14.4}$$

where **L** is the angular momentum. In our case the angular momentum is determined by the magnetic field vector **H** and the particle magnetization vector **M**, that is,

$$\mathbf{L} = V\mathbf{M} \times \mathbf{H}. \tag{14.5}$$

Here V is the particle volume (the magnetization vector is defined per unit volume). The relation between magnetization and magnetic field is determined by the magnetic susceptibility χ_0. For a static field in an isotropic medium,

$$\mathbf{M} = \chi_0 \mathbf{H}. \tag{14.6}$$

The frequency of rotation of a particle, and consequently the change in its magnetic field, is ω. For a variable field, the magnetic susceptibility becomes complex (because of the presence of a dissipative field) and depends on frequency, that is, $\chi = \chi'(\omega) - i\chi''(\omega)$. This effect is completely analogous to the complex nature of the dependence of the dielectric constant on frequency. In a rotating particle only the component of the magnetic field perpendicular to the axis of rotation is changed. Therefore, the susceptibility remains equal to χ_0 along the axis of rotation. The magnetization along that part of the field perpendicular to the axis of rotation is determined by the term $\chi'(\omega)$, and, finally, the component of magnetization perpendicular to both ω and **H** is proportional to $\chi''(\omega)$, since the imaginary part of the magnetic susceptibility is equivalent to a phase lag

$$\mathbf{M} = \chi_0 \frac{\boldsymbol{\omega}}{\omega}(\boldsymbol{\omega}\cdot\mathbf{H}) + \chi'\left[\mathbf{H} - \frac{\boldsymbol{\omega}}{\omega}(\boldsymbol{\omega}\cdot\mathbf{H})\right] + \frac{\chi''}{\omega^2}(\boldsymbol{\omega}\times\mathbf{H}). \tag{14.7}$$

Substituting Eq. (14.7) in Eqs. (14.5) and (14.4), we have

$$\begin{aligned} \frac{d\mathbf{P}}{dt} &= \frac{V(\chi_0 - \chi')}{\omega^2}(\boldsymbol{\omega}\cdot\mathbf{H})\left[(\boldsymbol{\omega}\times\mathbf{H}) + V\frac{\chi''}{\omega}\{(\boldsymbol{\omega}\times\mathbf{H})\times\mathbf{H}\}\right], \\ \frac{dK}{dt} &= -\frac{V\chi''}{\omega}[\omega^2 H^2 - (\boldsymbol{\omega}\cdot\mathbf{H})^2]. \end{aligned} \tag{14.8}$$

Hence, it follows that a slowing down of rotation occurs as long as the axis of rotation does not coincide with the magnetic field, that is, as long as the equation $\omega^2 H^2 = (\boldsymbol{\omega}\cdot\mathbf{H})^2$ is not satisfied. We can show that the minor axis of a particle will be oriented along the field.

In order to solve Eqs. (14.8), that is, to determine the change with time of the angles that define the position of the axes of symmetry of a particle in space, we would have to rewrite these equations in their component forms. We shall not do this here, since the geometry is rather complex and the equations obtained are not solved in a straightforward manner. A study of these equations was undertaken by Davies and Greenstein [269, 270, 271; see also 10]. A simple calculation of the time of orientation of particles t_0 is obtained directly from the second of Eqs. (14.8). Since $K \approx \frac{1}{2}I\omega^2$, where I is the moment of inertia of a particle, we have

$$t_0 \approx \frac{I\omega}{2V\chi''H^2}. \tag{14.9}$$

The value of the complex part of the magnetic permeability at low concentrations of iron atoms does not depend on this concentration and at low frequencies is approximately equal to

$$\chi'' \approx 2.5 \times 10^{-12} \frac{\omega}{T_g}, \tag{14.10}$$

where T_g is the temperature of the particles. In order to retain readily obtainable quantities in the equation, we make the substitutions $I \approx a^5$, $V \approx a^3$, where a is the particle size. Then we obtain

$$t_0 \approx 2 \times 10^{11} T_g a^2/H^2, \tag{14.11}$$

which, when $a \approx 5 \times 10^{-5}$, $T_g = 10°$, $H \approx 10^{-5}$Oe, gives $t_0 \approx 5 \times 10^{13}$ sec $\approx 1.6 \times 10^6$ yr. Although this period is long, it is still less than the lifetime of typical interstellar clouds. In rapidly evolving complexes, the condition of complete orientation is not met.

Orientation is disturbed by collisions between the particles and atoms or molecules of hydrogen. The relaxation time for establishing the equilibrium energy distribution of particles is

$$t_{\text{rel}} \approx \frac{1}{\pi a^2 v_{\text{H}} N_{\text{H}}} \frac{m_g}{m_{\text{H}}} \approx \frac{3 \times 10^{19} a}{N_{\text{H}} T^{1/2}}. \tag{14.12}$$

Here m_g is the mass of the particles ($m_g = 10^{24} m_{\text{H}} a^3$), and T is the temperature of the interstellar gas. In particular, when $a \approx 5 \times 10^{-5}$, $N_{\text{H}} \approx 10^{-3}$, $T \approx 100°$, we have $t_{\text{rel}} \approx 1.5 \times 10^{13}$ sec, or the same order as the orientation time. The ratio t_{rel}/t_0 (when $t_{\text{rel}} < t_0$) must define the degree of orientation of the particles

$$\zeta = \frac{3 \times 10^{19} a}{N_{\text{H}} T^{1/2}} \frac{H^2}{2 \times 10^{11} T_g a^2} \approx \frac{1.5 \times 10^8 H^2}{N_{\text{H}} T^{1/2} T_g a}. \tag{14.13}$$

If $t_0 \ll t_{\text{rel}}$, the degree of orientation is practically complete. In the particular case of the values of the parameters given above, $t_{\text{rel}}/t_0 \approx \frac{1}{3}$, or approximately one-third of the particles are oriented. From Eq. (14.13) we see that the bigger the particle, the less the degree of orientation.

Degree of Polarization

Polarization may be quantitatively defined in different ways. The degree of polarization of the observed light (unabsorbed), in contrast to Eq. (14.2), is

$$p = \frac{I_\perp - I_\parallel}{I_\perp + I_\parallel}. \tag{14.14}$$

Here $I_\parallel$ and $I_\perp$ are the radiation intensities with the electric vector parallel and perpendicular, respectively, to the major axis of the particle. Each of these quantities may be expressed in terms of the corresponding mean effective extinction cross section:

$$I_\parallel = I_0 e^{-\bar{\sigma}_1 N_g r}, \qquad I_\perp = I_0 e^{-\bar{\sigma}_2 N_g r}. \tag{14.15}$$

Here r is the distance from the star. Substituting Eqs. (14.15) in Eq. (14.14), and remembering that always $(\bar{\sigma}_1 - \bar{\sigma}_2)N_g r \ll 1$, we find

$$p \approx \tfrac{1}{2}(\bar{\sigma}_1 - \bar{\sigma}_2)N_g r. \tag{14.16}$$

We denote by ϕ the angle between the direction of the major axis of an oriented particle and the line of sight. Then the angle between the magnetic field and the line of sight is $\frac{1}{2}\pi - \phi$. It is easy to see that the quantity $\sigma_1 - \sigma_2$ is proportional to $\cos^2 \phi$. One factor $\cos \phi$ takes into account the projection of the particle cross section on the plane perpendicular to the line of sight, which reduces equally σ_1 and σ_2, and the second factor $\cos \phi$ accounts for the change in polarizing properties of the particles themselves in relation to their orientation and reduces σ_1 with respect to σ_2. It is convenient to calculate the effective polarization cross section as the average over the particle-size distribution function:

$$\bar{\sigma}_p = \tfrac{1}{2}\int_0^\infty (\sigma_1 - \sigma_2)\zeta(a) f(a) da. \tag{14.17}$$

The factor $\zeta(a)$, determined from Eq. (14.13), introduces the different degrees of orientation of particles of different sizes. The factor $\frac{1}{2}$ accounts for the fact that, even for complete orientation, the particles rotate about their minor axes and therefore we observe only the average projection of this axis on the plane perpendicular to the line of sight.

Using Eq. (14.17), we shall rewrite Eq. (14.16) in the form

$$p = \tfrac{1}{2}\bar{\sigma}_p \cos^2 \phi N_g r. \tag{14.18}$$

The ratio of $\bar{\sigma}_p$ to the mean value of the effective extinction cross section σ_e, from what was previously said, is probably not greater than 0.1. Therefore, appreciable polarization (10 percent) can be observed only if the optical thickness of the medium for extinction is greater than unity. On the other hand, the quantity $\sigma N_g r$ is also proportional to the reddening. Therefore, we may expect to satisfy a relation of the form

$$\frac{p}{E} \approx \text{const.} \times \cos^2 \phi. \tag{14.19}$$

We note in connection with this that from observational data Hoag [272] arrived at the statistical dependence

$$\frac{p}{E} \approx 10.8[1 - |\cos(l - 68^\circ)|] \text{ percent,} \tag{14.20}$$

reminiscent of Eq. (14.19). Longitude $l = 68^\circ$ corresponds roughly to a direction along the Orion arm, where $\phi \approx 90^\circ$.

The calculation of $\bar{\sigma}_p$ or of the constant in Eq. (14.19) is difficult to carry out at the present time, since the necessary data on the properties of cosmic dust particles—their elongation, their refractive index, and their distribution function—are lacking. Therefore, up to now, we must be satisfied from present information with only qualitative correlation between polarization on the one hand and the distance and color-excess coefficients on the other, and also with the nature of the dependence of polarization on orientation of the magnetic field.

The Stokes Parameters

The characteristics of light polarization obtained from p and the slope of the plane of oscillation θ are not always convenient, since, together with the mean intensity, three quantities with different dimensions are determined. Therefore, during work on polarization, use is often made of the so-called Stokes parameters I, Q, U, V, which have the dimension of radiation intensity. These parameters may be introduced most readily in the following way. Let us consider plane-polarized radiation with an electric-field amplitude E_0. Let us now consider a certain direction forming an angle θ with the direction of the plane of polarization of this radiation. Project E_0 in the directions perpendicular and parallel to the x and y axes. Then we have $E_x = E_0 \cos\theta$, $E_y = E_0 \sin\theta$. We can now write the expressions

$$\begin{aligned} I &= \text{const.} \times (E_x^2 + E_y^2) = \text{const.} \times E_0^2, \\ Q &= \text{const.} \times (E_x^2 - E_y^2) = \text{const.} \times E_0^2 \cos 2\theta, \\ U &= \text{const.} \times (2E_xE_y) = \text{const.} \times E_0^2 \sin 2\theta. \end{aligned} \tag{14.21}$$

The constants here are identical. Since I is the total radiation intensity, we have $Q = I\cos 2\theta$ and $U = I \sin 2\theta$. In the case when there are two mutually perpendicular plane-polarized beams with intensities I_1 and I_2, the Stokes parameters are determined by the equations

$$\begin{aligned} I &= I_1 + I_2, \\ Q &= (I_1 - I_2)\cos 2\theta, \\ U &= (I_1 - I_2)\sin 2\theta. \end{aligned} \tag{14.22}$$

The minus sign in front of I_2 arises because the angle θ for this beam differs by $\pi/2$ from θ for I_1. Here also the additive property of the Stokes parameters is employed. The fourth parameter V is necessary for studying elliptically polarized radiation, which is practically never encountered in the physics of interstellar media. From Eqs. (14.22) we find as connections between the degree of polarization p, the angle θ, and the Stokes parameters the relations

$$p = \frac{(U^2 + Q^2)^{1/2}}{I}, \qquad \tan 2\theta = \frac{U}{Q}. \tag{14.24}$$

These equations may be used for the independent determination of the Stokes parameters. For a general study of polarization near the galactic plane, the angle θ must be measured from the direction of this plane. If polarization is to be considered from separate regions, it is more convenient to measure this angle from the direction of maximum polarization. Since the Stokes parameters have the dimension of intensity, they can be expressed in stellar magnitudes.

Besides the studies already noted at the beginning of this section, on the correlation of interstellar polarization in the Galaxy with distance and reddening, we must also note certain statistical investigations. In particular, V. A. Dombrovskii [273] and Serkovskii [216] studied the region near the double cluster h and χ Persei. For the mean values of the Stokes parameters in this region they found

$$\bar{Q} = 0^{\mathrm{m}}.079 \pm 0^{\mathrm{m}}.001, \qquad \bar{U} = 0^{\mathrm{m}}.000 \pm 0.001, \tag{14.23}$$

and an average reddening of $\bar{E} \approx 0^{\mathrm{m}}.57 \pm 0^{\mathrm{m}}.01$. Hence $p/E \approx 0.14$. Setting $\bar{U}$ equal to zero means that the planes of polarization are symmetrically arranged relative to the direction of the magnetic field in this cluster. A statistical investigation of polarization in the Milky Way was also carried out by Pacholczyk [274, 275].

In a series of papers [216, 276, 260], correlations were examined between the polarization characteristics for neighboring stars and the angular distances between them. Such a correlation exists, although with considerable scatter. Serkovskii discovered a correlation between the angular distances between stars in the cluster h and χ Persei and the difference between the corresponding values of the Stokes parameters. S. A. Kaplan and I. A. Klimishin found a significant correlation between the degree of polarization of stars in the general galactic field and angular distance in longitude. For longitude differences greater than 30–40°, the correlation disappeared [260]. The question of correlation will be discussed in more detail in Sec. 23.

We must emphasize that in the study of interstellar polarization there is a greater need, as we stated in Sec. 12, for the study of interstellar absorption. Very irregular structure and the presence of the spiral arms and complexes prevent us from obtaining by statistical studies an accurate idea of magnetic-field structure and the distribution of polarizing clouds. Here also we must carry out a detailed study of each region separately. We note that interstellar polarization is detected in extragalactic nebulae [277].

Polarization of Reflection Nebulae

Generally speaking, reflected light is polarized even by scattering from spherical particles, although in this case the principal direction of the oscillation of the electric vector is perpendicular to the direction of the illuminating star. This circular polarization is actually seen in several reflection nebulae [278, 279]. However, in a number of cases, especially in filamentary formations, noncircular polarization was observed. It may be explained by the same mechanism as for interstellar polarization during absorption [251]. In such a nebula, the particles are oriented in a magnetic field and scatter light with some polarization. It is essential that the polarization of reflected light complement the polarization of the absorbed radiation. In other words, the planes of polarization for reflection and extinction are perpendicular. This effect was actually observed in nebula IC 1287 [278], where the polarization of the star light was 1.8 percent and the polarization of the nebula reached 18 percent, which, generally speaking, is greater than the permissible theoretical limit. However, a not too accurate photographic method was employed here. Recently Elvius and Hall discovered in the filaments of the Pleiades polarization up to 10 percent [280]. In the Trifid Nebula, discussed at the end of the preceding section, light from the main nebula is circularly polarized, while light from the satellite is plane polarized in a direction parallel to the galactic plane, which is evidently explained by interstellar polarization [281]. Circular polarization is detected in the central region of the Orion nebula [279]; consequently, part of the continuous spectrum in this nebula is seen because of the reflection of the light by the cosmic dust particles in the brightest central parts.

15. Physical Properties of Cosmic Dust Particles

From what has been said in the preceding sections, it is clear that observations on interstellar absorption, scattering, and polarization of light have so far not provided any very specific data on the physical properties of cosmic dust particles. We can only say that they must in all probability be elongated dielectric particles with dimensions of about the wavelength of visible light and with small admixtures of metals.

Nevertheless, from theoretical considerations we can obtain a few (although unreliable) ideas about temperature, chemical composition, and size distribution of these particles. Similar investigations were carried out by Oort and van de Hulst [200], ter Haar [282], Kahn [283], and others. The results of such a theoretical analysis are given here.

Particle Temperature

To determine the particle temperature, we must construct a heat-balance equation that takes into account absorption and emission of energy and also heating due to collisions with atoms and cooling due to evaporation (sublimation) of the particles. An approximate contribution of these heating and cooling mechanisms to the energy balance is given in Table 15.1 (from the estimates by van de Hulst [195]).

Table 15.1

	percent
1. Heating due to absorption of interstellar radiation	95
2. Heating due to collisions with atoms and ions	0.5
3. Heating due to chemical reactions on the particle surface between adjacent atoms	4.5
4. Cooling due to thermal emission	100
5. Cooling due to evaporation of molecules from the particle surface	0.001

Therefore, for the determination of the temperature of dust particles, we need consider only light absorption and emission. Since we may assume that particle temperatures are low, we can take the following case. A particle absorbs radiation in the visible and ultraviolet regions and radiates in the infrared region. We shall estimate the infrared radiation of dust particles, assuming that it follows Planck's rule; in other words, we shall assume that a particle is found in thermodynamic equilibrium with the emission at a temperature T_g. Then for the total energy emitted by the particles per second we can write

$$W_e = 4\pi^2 a^2 \int_0^\infty Q_a \frac{2hc^2}{\lambda^5} \frac{1}{e^{hc/\lambda k T_g} - 1} d\lambda. \tag{15.1}$$

Here $Q_a = \sigma_a/\pi a^2$, where σ_a is the effective cross section of the pure absorption defined in Sec. 11. It is clear that the integral in Eq. (15.1) is, corresponding to Kirchhoff's law, the emission coefficient per square centimeter of particle surface, while $4\pi a^2$ is the area of this surface. For the sake of simplicity we shall confine ourselves to the case of spherical particles. Since a particle emits in the far-infrared region at $\lambda \gg a$, we can use Eqs. (11.14) and (11.7) for σ_a. We obtain

$$\begin{aligned} W_e &= \frac{2^7\pi^2 a^3}{3} \kappa' hc^2 \int_0^\infty \frac{d\lambda}{\lambda^6(e^{hc/\lambda k T_g} - 1)} \\ &= \frac{2^7\pi^2 a^3 \kappa'}{3} \frac{c(kT_g)^5}{(hc)^4} \int_0^\infty \frac{x^4 dx}{e^x - 1} = \frac{2^7}{3} \pi^2 a^3 \frac{c(kT_g)^5}{(hc)^4} 24.9\kappa', \end{aligned} \tag{15.2}$$

where 24.9 is the value of the integral, and $\kappa' = \xi \tan \beta$ is the imaginary part of the refractive index. We note that emission by dust particles is proportional to the fifth and not the fourth power of the temperature. Taking the logarithms of both sides of Eq. (15.2), we find

$$\log W_e = -0.98 + 5 \log T_g + 3 \log a + \log \kappa'. \tag{15.3}$$

The quantity of energy absorbed by particles is determined by

$$W_a = 4\pi \int_0^\infty \bar{I}_*(\lambda)\sigma_a(\lambda)d\lambda, \tag{15.4}$$

where $\bar{I}_*(\lambda)$ is the mean intensity of interstellar radiation. Here σ_a must be taken for the visible and ultraviolet regions where $\bar{I}_*$ is a maximum. The dependence of σ_a on wavelength, generally speaking, is unknown. However, as follows from the equations in Sec. 11, σ_a converges to πa^2 with decrease in wavelength when κ' is small. Using this limiting value, we obtain

$$W_a \approx \pi a^2 \times 4\pi \int_0^\infty \bar{I}_*(\lambda)d(\lambda) \approx \pi a^2 c\rho_0 \approx 3.3 \times 10^{-2}a^2. \tag{15.5}$$

Here ρ_0 is the mean density of interstellar radiation in the galactic plane (Sec. 11). In accordance with Table 15.1, the energy-balance equation may be written $W_e = W_a/0.95$. Substituting this value in Eq. (15.3), we find

$$5 \log T_g \approx 0.48 - \log a - \log \kappa'. \tag{15.6}$$

Taking $a = 5 \times 10^{-5}$ cm and $\kappa' = 0.05$, we obtain for the temperature of a dust particle

$$T_g \approx 17^\circ. \tag{15.7}$$

The temperature of a particle depends only slightly on the properties of the particle and on the intensity of interstellar radiation, since it enters into the equilibrium equation as the fifth power. Nevertheless, an investigation by Lichtenstein and Greenberg [284] showed that effects due to the dependence of refractive index and effective absorption cross section on wavelength markedly reduced the temperature of a particle. Depending on the parameters chosen, the latter may range from 2° to 12°.

It would be interesting to try to observe the specific infrared radiation from the dust. The maximum in this spectral curve is determined by the maximum of the expression under the integral (15.2), that is, at the wavelength

$$\lambda_m = \frac{hc}{5.98kT_g} = \frac{0.24}{T_g}. \tag{15.8}$$

Using Eq. (15.7), this gives $\lambda_m \approx 140\ \mu$. The total infrared radiation from dust must be very intense, since a large part of the starlight absorbed by dust is re-emitted in this spectral region. In fact, we find according to Eq. (15.2) that (for $a \approx 5 \times 10^{-5}$ cm, $T_g \approx 20^\circ$, and $\kappa' \approx 0.05$) the radiation per particle is approximately $W_e \approx 10^{-8}$ erg/sec. The emission coefficient per cubic centimeter for $N_g \approx 10^{-14}$ cm^{-3} is approximately 10^{-23} erg/cm^3 sec sterad. In a column of unit cross section along the galactic plane there is radiated approximately 1 erg/cm^2 sec sterad. An energy approximately 30 times less is radiated in a direction transverse to the plane of the Galaxy.

Adsorption of Gas and Evaporation from Particles

The adsorption coefficient is defined as the probability of the adhesion of an atom or molecule that strikes the surface of a particle. From the considerations enumerated below, we can consider that for most atoms and molecules in interstellar space the coefficient is almost unity. First, the kinetic temperature of atoms and molecules in H I zones is close to the Debye temperature of a crystal lattice. Therefore a transfer of energy from the striking atom to the lattice phonons is extremely probable. We must recall that during collisions between atoms and a crystal lattice vibrations are set up in the latter, usually described in terms of phonons. The Debye temperature Θ is the average energy of the phonons on a temperature scale. The adhesion of an atom is connected with transfer of kinetic energy to the particle lattice, that is, with excitation of the phonons. This process possesses maximum probability, if a single phonon is excited with energy $\sim k\Theta$. Second, interstellar molecules always occur in the lowest rotational and vibrational energy states, and therefore they must impart to the particles only the energy of their translational motion. This process also usually has a high probability.

It is a fact that theoretical analysis shows that, in pure (without inclusions) and regular lattices, the adsorption coefficient diminishes with a fall of temperature in the crystal. However, this result cannot be applied to cosmic dust particles, which are known to have an irregular shape and to be "contaminated" with various atoms and molecules. As the experimental results indicate, adhesion to such a surface is more probable.

Thus we see that every atom or molecule striking the particle adheres to it, giving up energy, which is subsequently emitted in the infrared region. Possible exceptions to this rule are helium, neon, and argon, which likewise give up their kinetic energy to the particle, but do not adhere to it because of their inertness. Calling N_μ the number of atoms or molecules of atomic weight μ per unit volume, we have for the number of particles adhering to 1 cm^2 of surface per second [195]

$$Z_\mu = \tfrac{1}{4} N_\mu v_T = \tfrac{1}{4}\left(\frac{8kT}{\pi \mu m_{\mathrm{H}}}\right)^{1/2} N_\mu . \tag{15.9}$$

We must remember that here T is the gas temperature, as distinct from T_g, the temperature of the particle. Taking the logarithm of both sides of Eq. (15.9) we obtain

$$\log Z_\mu = 3.53 + \tfrac{1}{2}\log T - \tfrac{1}{2}\log \mu + \log N_\mu . \tag{15.9$'$}$$

This refers only to H I regions; in H II zones, the kinetic temperature of the ions is much greater than the Debye temperature of the lattice. Therefore, to be stopped completely and subsequently to adhere, an ion must excite several tens of phonons, which is most improbable. In H II zones, the adsorption coefficient is small, probably about 0.01. Moreover, we may consider that collisions of ions, especially protons, with particles will lead to their more rapid destruction. However, it is reasonable to think that the probability of knocking off another atom from the lattice during the collision of an ion with a particle in a H II zone is also small. This can be seen from the following simple calculation. According to Eq. (15.9), the number of collisions of protons with 1 cm^2 of particle surface in dense gas-dust complexes is $\frac{1}{4} \times 10^3 \times 1.4 \times 10^6 \approx 3.5 \times 10^8\ cm^{-2}\ sec^{-1}$. Let us now assume that every proton ejects an atom from the particle surface. Then the particle will be destroyed after $10^{24} \times \frac{4}{3}\pi a^3/4\pi a^2 \times 3.5 \times 10^8 = 10^{15}a$ sec $\approx$ 1600 yr (here 10^{24} is the number of atoms per cubic centimeter of dust). At the same time, we know that the Orion nebula, for example, is about 10^6 yr old and that cosmic dust is present. If we assume that dust particles have existed here since the formation of H II zones, then the upper limit of probability for ejecting an atom from a particle by a proton is $\sim 10^{-4}$. We may assume that in H II zones both the probability of an atom's adhering and the probability of destroying the lattice of the particle by proton collisions are about 0.01, while both can be understood as the probability of exciting many phonons in a lattice [284, 285]. At the present time, it is not possible to answer the question whether dust particles are formed or destroyed in H II zones. It is only known that they are often present there.

We shall now consider the evaporation of atoms and molecules from the surface of particles. According to experimental data on the pressure of saturated steam at low temperatures, we have the relation

$$\log p_\mu = -\frac{A_\mu}{T_g} + B_\mu \log T_g + C_\mu, \tag{15.10}$$

where the constants A_μ, B_μ, and C_μ can either be calculated or determined experimentally. In a state of thermodynamic equilibrium, the number of particles evaporated is equal to the number adhering. Therefore, substituting (15.10) in (15.9′), where for T we must now substitute the particle temperature T_g, $p_\mu = N_\mu k T_g$, and, for simplicity, change $\log T$ to $1.6 - 6/T$, which over the temperature range from 10° to 40° does not give a large error, we obtain for the number of evaporating atoms and molecules

$$\log Z'_\mu = -\frac{A'_\mu}{T} + C'_\mu. \tag{15.11}$$

Here $A'_\mu = A_\mu + 6B_\mu - 3$, $C'_\mu = C_\mu - \frac{1}{2}\log\mu + 1.6B + 18.62$. Equation (15.11) determines the number of evaporation particles even in the absence of thermodynamic equilibrium. Estimates for A'_μ and C'_μ for certain atoms and molecules are given in Table 15.2 [195].

Table 15.2

Atom or molecule	Constant		
	μ	A'_μ	C'_μ
He	4	15	27.6
Ne	20	110	28.3
Ar	40	412	28.6
H_2	2	52	27.4
N_2	28	378	29.4
O_2	32	493	30.0
NO	30	829	30.0
CO	28	470	29.5
H_2O	18	2500	28.5
NH_3	17	1570	29.0
CH_4	16	496	28.2
Metals	20–200	5000–20,000	28

According to Eq. (15.9′), the logarithm of the number of adhering atoms or molecules changes from 4.5 for hydrogen to -3.5 for molecules that are rarely encountered. Here we take $T = 100°$, $N_H = 10$, $N_{mol} \approx 10^{-8}$; the range of $\log Z_\mu$ depends little on the choice of these parameters. From Eq. (15.11) we find that the adhesion of atoms or molecules to particles will predominate over evaporation if

$$\frac{A'_\mu}{T_g} > C'_\mu - \log Z_\mu \approx 25\text{–}35 \tag{15.12}$$

(for most atoms and molecules $C_\mu \approx 30$). From the data of Table 15.2, we may readily deduce that when $T_g = 17$ adhesion predominates over evaporation for all atoms and molecules except He, Ne, and H_2. It is true that when the grain temperature is raised, A, N_2, O_2, NO, and CH_4 may evaporate; however, we must remember that the data given in the table are for pure substances, while the actual particles of cosmic dust are very "impure." In this case parameter A_μ may be larger.

Thus, with the possible exception of the inert gases, all atoms adhere to the particles. Hydrogen is a special case. After striking a particle, H atoms react with other adsorbed hydrogen atoms to form molecules of H_2, which possess a high probability of evaporation. Therefore, there must be relatively few H_2 molecules in the dust particles. Hydrogen may, however, enter into other compounds.

With an increase in temperature the dust particles begin to disintegrate. At

$T_g \gtrsim 150^\circ$ the icy components evaporate, but the heavy parts remain, still causing considerable absorption of light. The complete grain is evaporated at temperatures $T_g \approx 1300^\circ$ [195a]. A study of the gradual destruction of particles with increase in temperature is of interest from the point of view of cosmogony. In a protostar, composed of clouds of gas and dust, the temperature rises as a consequence of the star's contraction, so that if evaporation takes place the opacity is diminished, which in turn eases the contraction.

The discussion above relates to H I regions. Conditions for the formation, development, and destruction of particles in H II zones remain obscure.

The Chemical Composition and Charge of Dust Particles

In agreement with the foregoing data on the probability of adhesion of atoms and molecules, and data on the mean chemical composition of interstellar gas, van de Hulst proposed the following chemical composition for cosmic dust particles: 100 molecules of H_2O, 30 of H_2, 20 of CH_4, 10 of NH_3, 5 of MgH. The refractive index of this mixture must be about 1.25, with a small imaginary part, which is difficult to estimate since it depends on metallic impurities. Interstellar dust particles resemble impure ice crystals. However, we must emphasize that the suggested chemical composition of interstellar dust is no more than a preliminary estimate, which may be considerably changed in the future.

The assumption of the existence of metallic particles can be valid only if the evaporation of the lighter molecules (H_2O, CH_4, NH_3, . . .) is high, which is improbable. We must remember that observational data on light scattering in interstellar space also favor nonmetallic particles.

In order to explain polarization data, the major axes of the particles must be at least twice the minor axes. Evidently the particles consist of extremely loose conglomerates of small crystals. The formation of regularly shaped crystals is difficult, since, because of its low temperature, the atoms cannot migrate over the particle surface.

Cosmic dust particles probably have a small negative charge, kept constant by collisions with electrons and ions. To determine the value of the charge we proceed in the following way. If the kinetic energy of an electron $K_e = \frac{1}{2}mv^2$ is less than its potential energy in the field eV of a charged particle, the electron, in general, will not collide with the particle. When $K_e > |eV|$, the electron can collide with a particle, but because of its hyperbolic motion the effective cross section of the particle is reduced $1 - eV/K_e$ times. The effective cross section for ionic collisions is increased $1 + e_iV/K_i$ times, where e_i and K_i are the ionic charge and kinetic energy. The particle potential V is found from the equilibrium condition between the number of incident electrons and ions

$$S\int_{\sqrt{2eV/m}}^{\infty}\left(1 - \frac{2eV}{mv^2}\right)vf_e(v)dv = S\int_0^{\infty}\left(1 + \frac{2e_iV}{m_iv^2}\right)vf_i(v)dv. \quad (15.13)$$

Here $f_e(v)$ and $f_i(v)$ are the velocity distribution functions for electrons and for ions, and S is the geometric cross section of the particle. Assuming $f(v)$ to be a Maxwellian function and integrating Eq. (15.13), we obtain

$$1 + \frac{eV}{kT} = \left(\frac{m_i}{m}\right)^{1/2} e^{-eV/kT}. \tag{15.14}$$

Positive carbon ions predominate in H I regions. Therefore, putting $m_i = \mu m_H = 12 m_H = 12 \times 1837 m$, we find from Eq. (15.14)

$$eV = 3.5kT, \tag{15.15}$$

which at a temperature $T = 100°$ gives a negative potential of about 0.03 V. The effective collision cross section for ions and dust particles is here $1 + eV/kT = 4.5$ times greater than the geometric cross section. If we consider H II zones, then for m_i we must substitute the mass of a proton. In this case it appears that $eV = 2.5kT$, which at the high electron temperature of these zones leads to a potential $V = 2$ V. However, these values may be reduced if photoelectric effects occur on the particle surface—a high negative potential reduces the work required for discharge. The photoelectric effect is very small on dielectric surfaces, but it may be greater in the case of loose cosmic particles. It has been difficult up to now to estimate the role of this effect. It is important in the future, then, that the charge of dust particles increase their effective cross section by a factor of 2 on the average.

The Formation and Disintegration of Interstellar Dust Particles

It is difficult not to agree that the particles are formed from the condensation of interstellar gas. The condensation nuclei are evidently interstellar molecules. As we noted in Sec. 9, since we are not able to calculate the rate of formation of even the simplest diatomic molecules, this is even more true of the more complex formation of dust condensation nuclei consisting of several atoms or molecules. The question remains open, but we shall assume that condensation nuclei are formed in interstellar space in such quantity that the mean number of particles they produce is about 10^{-13}–10^{-14} cm^{-3}.

If the primary dust nucleus is rapidly formed, its evolution is assisted further by the high probability of adhesion of atoms and molecules. Calling the particle mass m_g and designating by ρ the density of the interstellar gas, or, more accurately, the density of the components that can be concentrated on the particles, we have

$$\frac{dm_g}{dt} = 4\pi a^2 \delta \frac{da}{dt} = \left(1 + \frac{eV}{kT}\right) \pi a^2 \rho_1 v_T. \tag{15.16}$$

Here δ is the specific gravity of the particle, v_T is the thermal velocity of the atoms, and the quantity $1 + eV/kT$ takes into account the role of the particle's charge. For simplicity, we shall consider the particles to be spherical.

Substituting in Eq. (15.16) the numerical values

$$\rho_1 \approx 10^{-25} \text{ g/cm}^3,\ \delta = 0.5 \text{ g/cm}^3,\ v_T \approx 4 \times 10^4 \text{ cm/sec},$$

we obtain

$$\frac{da}{dt} = \frac{\rho_1 v_T}{2\delta} \approx 4 \times 10^{-21} \text{ cm/sec}. \tag{15.17}$$

The radius $a = 5 \times 10^{-5}$ will be reached after 1.25×10^{16} sec $\approx 4 \times 10^8$ yr. In typical gas-and-dust nebulae, particle growth requires a period of time that exceeds the period of their existence. In more dense nebulae, particle growth is more effective.

The fact that the dependence of absorption on wavelength is practically identical in different parts of the Galaxy indicates that particle dimensions must be almost identical. This implies that, just as there are factors that determine the growth of particles, other factors must exist that destroy them, while the mechanism for the destruction of particles must be extremely sensitive to their size. It is difficult to find such a mechanism acting within the limits of a single cloud. We emphasize that, according to observations, particle dimensions are approximately the same in both dense and more rarefied clouds. Oort and van de Hulst [200] proposed that particles are destroyed during a collision between clouds. During such collisions, dust particles from each cloud will permeate the other, preserving at first a high velocity v_0. The linear path of a particle up to the time when the velocity of the particle has not exceeded a certain limit, say v_m, may be obtained from the equation

$$l(v_m) = \int_0^{t(v_m)} v\, dt = \int_{v_m}^{v_0} \frac{v\, dv}{dv/dt} = \int_{v_m}^{v_0} \frac{m_g dv}{\pi a^2 \rho v} = \frac{4a\delta}{3\rho} \ln \frac{v_0}{v_m}. \tag{15.18}$$

Here we consider that the loss of momentum of a particle during its motion in a resisting medium is

$$m_g dv = -\pi a^2 \rho v^2\, dt; \tag{15.19}$$

$\pi a^2 \rho v$ is the number of atoms which a particle encounters per unit time. We note that, in contrast with Eq. (15.16), ρ is the total gas density. Substituting in Eq. (15.18) $a \approx 5 \times 10^{-5}$ cm, $\rho \approx 3 \times 10^{-23}$ g/cm^3, $\delta \approx 0.5$ g/cm^3, and $\ln (v_0/v_m) \approx 1$, we find that the path length of particles is $l \approx 0.3$ pc. Thus, during collisions between clouds that are not too dense and small, particles can penetrate them for comparatively great distances.

A collision between two particles moving at high relative velocities generally leads to their destruction. During a head-on inelastic collision of two particles of masses m_g amd m'_g, their kinetic energy of relative motion with reduced mass $m_g m'_g/(m_g + m'_g)$ is transformed into thermal energy. From an estimate by Oort and van de Hulst the energy required for the heating and complete vaporization of particles is about 1.1×10^{10} erg/g. Hence, both colliding particles will be destroyed if their relative velocity is greater than $v_{\min}$, determined from the relation

$$\frac{1}{2}\frac{m_g m'_g}{m_g + m'_g} v_{\min}^2 = 1.1 \times 10^{10}(m_g + m'_g). \qquad (15.20)$$

Putting $m_g = \frac{4}{3}\pi a^3 \delta$ and $m'_g = \frac{4}{3}\pi a'^3 \delta$ and using the designation $a'/a = x$, we find from (15.20)

$$v_{\min} = 1.5 \frac{1 + x^3}{x^{3/2}} \text{ km/sec}. \qquad (15.21)$$

We see that when their radii are comparable ($x = 1$) the particles are destroyed at velocities that are considerably less than the velocities common in interstellar media.

The probability of collision of clouds with velocities between v_0 and $v_0 + dv_0$ km/sec is written in the form

$$\frac{1}{t_r}\frac{v_0}{7.5} e^{-v_0/7.5}\frac{dv_0}{7.5}, \qquad (15.22)$$

where $t_r \approx 10^7$ yr is the mean time interval between two collisions. Multiplying (15.22) by the number of particles in the second cloud with which particles of the first cloud have collided, after a time $l(v_{\min})$, that is, up to the time when their velocity becomes greater than $v_{\min}$ given by Eq. (15.21), and integrating over all possible velocities v_0 and particle dimensions a of the second cloud, we obtain for the probability of the destruction of particles of a given radius, after collision with the first cloud, the expression

$$P(a) = \int_0^\infty N_g(a')\pi(a' + a)^2 da' \int_{v_{\min}(a')}^\infty \frac{4a\delta}{3\rho t_r} \ln \frac{v_0}{v_{\min}}$$
$$\times \frac{v_0}{7.5} e^{-v_0/7.5} \frac{dv_0}{7.5} = \frac{4\pi\delta a^4}{3\rho t_r}\int_0^\infty (1 + x)^2 N_g(xa)$$
$$\times \left[e^{-(1+x^3)/5x^{3/2}} + \mathrm{Ei}\left(\frac{1 + x^3}{5x^{3/2}}\right)\right] dx. \qquad (15.23)$$

We note that the dependence of the probability of particle destruction on particle radius is very strong ($\sim a^4$).

We obtain an equation for determining the particle-size distribution function $N_g(a)$ by equating the number of particles that fall in the range of radii a to $a + da$ due to their growth $[dN_g(a)/dt = dN_g(a)/da \times da/dt]$ to the number of particles destroyed by collisions $[2P(a)N_g(a)]$. The factor 2 takes into account the destruction of both particles after a collision. Using Eqs. (15.17) and (15.23), we obtain

$$\frac{\rho_g v_T}{2\delta}\frac{dN_g(a)}{da} = -\frac{8\pi a^4\delta}{3\rho t_r} N_g(a)\int_0^\infty (1 + x)^2 N_g(xa) \times \left[e^{-(1+x^3)/5x^{3/2}} + \mathrm{Ei}\left(\frac{1 + x^3}{5x^{3/2}}\right)\right] dx. \quad (15.24)$$

The solution of this integral-differential equation determines the particle-size distribution function. This equation was derived by Oort and van de Hulst [200] in a somewhat different form.

We obtain a very simple solution by setting x in Eq. (15.24) equal to unity, that is, by assuming that destruction occurs only after collisions between identical particles. Then we have

$$\frac{dN_g(a)}{da} = -\frac{92a^4\delta^2}{\rho\rho_g t_r v_T}[N_g(a)]^2. \quad (15.25)$$

Integrating, we obtain

$$N_g(a)da \approx \frac{N_g(0)da}{1 + \dfrac{18a^5\delta^2}{\rho\rho_g t_r v_T}N_g(0)} \approx \frac{\rho\rho_g t_r v_T}{18\delta^2} \approx 8 \times 10^{-9}\left(\frac{10^{-5}}{a}\right)^5 \frac{da}{10^{-5}} \quad (15.26)$$

if we use the numerical values employed before. Thus, the particle-size distribution function diminishes rapidly, as a^{-5}, with an increase in radius. Equation (15.26) gives a value $N_g \approx 5 \times 10^{-12}$ cm^{-3} for the number of particles with a radius in the range 0.4 to 0.6 μ. In Eq. (12.4) we obtained a value $N_g \approx 4 \times 10^{-14}$ cm^{-3}. Bearing in mind that Eq. (15.26) was evaluated for dense clouds, while Eq. (12.4) was valid for the mean density of the total galactic disk, the agreement between the calculated values may be considered qualitatively satisfactory.

The foregoing picture of particle formation and destruction is much simplified. After collision, not only can particles evaporate, but at lower velocities or by an off-center collision they can coalesce, or, on the contrary, can be split into smaller secondary particles. From the study by Oort and van de Hulst, the total number of secondary particles amounts approximately to 6 percent of the total number of particles, while particles measuring 10^{-4} cm or more are probably almost entirely formed by adhesion.

The hypothesis of Oort and van de Hulst assumes that the dimensions of the particles are determined by statistical growth and disintegration. There is an important fact, observed by Vesselinkov [286a], that speaks against this and all other similar hypotheses. He has shown that reddening of stars in the Magellanic Clouds follows the same law as in our own Galaxy. At the same time, conditions in the Magellanic Clouds, and consequently the rate of growth and disintegration, are different, so that from this point of view the average dimensions of particles must be different. If the data of Vesselinkov are confirmed; it follows that the dimensions of particles must depend on other factors.

The preceding discussion exhausts what can be said so far about the physical properties of cosmic dust particles. Much remains obscure, while a number of difficulties, such as the high density of dust material compared to the typical density of heavy elements, the presence of magnetic properties, polarization during scattering, and so forth, sometimes render doubtful even the correctness of interpreting interstellar absorption as light extinction by classical particles.

Other Ideas on the Nature of Particles

In particular, Platt [286] proposed that interstellar absorption and polarization are caused by light scattering by molecular complexes measuring about $3 \times 10^{-3}\,\lambda$, or considerably less than a wavelength. He assumed that these complexes have unfilled electron shells, that is, that they are analogous to unsaturated molecules. From experimental data we know, in fact, that radicals absorb strongly in visible light. The absorption coefficient per mole of certain substances reaches a value of 200,000. Hence, the continuous absorption coefficient is $\sim 3 \times 10^{-19}$ cm^2 per molecule. With this absorption coefficient, we need 10^{-3} such molecular complexes per cubic centimeter to give the mean observed interstellar absorption. If these molecular complexes absorb in such a way that the absorption coefficient is equal to the geometric cross section, the total mass of the particles may be two orders of magnitude smaller than the mass of dust; it is inversely proportional to the particle thickness. The presence of molecular complexes of different kinds and sizes would explain absorption over a wide spectral range.

The albedo of these particles is high, since for pure absorption, which requires structural dissociation or reorganization, high energy is needed and therefore a larger number of visible and infrared light quanta are commonly re-emitted. Finally, a breakdown of fluorescence-type quanta is possible, but such transitions are less probable in comparison with pure scattering.

Experiments show that odd electron systems and some even electron systems of unsaturated molecules are paramagnetic. The elongated form can be explained by random fluctuations in the number of atoms in the complex. If the number of atoms is about 500, the relative fluctuation is $1/500^{1/2} \approx 5$ percent. For total ordering of particles such elongation is nevertheless small. For smaller particle sizes, the fluctuations can give further elongation.

The size-distribution function of these particles can be estimated in the following way [287]. It follows from the observed absorption law that $N_k \sigma_k \sim \lambda^{-1}$, where N_k is the number of particles (complexes) per unit volume and σ_k is their effective scattering cross section. In resonance scattering, $\sigma_k \sim a^2 \sim \lambda^2$ and since for transitions in unfilled electron bands $\lambda \approx 400a$, where a is the size of the complex, we find

$$N_k \sim \frac{1}{\lambda \sigma_k} \sim \frac{1}{a^3}. \tag{15.27}$$

The uncertainty of the properties of these particles—complexes—gives us no basis for comparison with observations. Certain determinations obtained by Greenberg [287] indicate that the hypothesis based on complexes does not agree as well with observational data as the hypothesis of absorption by large particles. In particular, the dependence of the absorption law on galactic longitude, already discovered from observation, is much more weakly expressed here. Such particles also polarize radiation less effectively than elongated classical particles do. These particles require an elongation of ~ 12 percent, which is difficult to explain by random fluctuations during particle growth.

Another hypothesis on the nature of particles of interstellar dust [424c] is based on the fact that graphite particles of diameter 5×10^{-6} cm can give the observed dependence of extinction on wavelength. However, particles with a diameter of 8×10^{-6} cm already have another dependence [424a]. The albedo of graphite particles is about 0.3 to 0.4 at $\lambda = 0.4\ \mu$. These same graphite particles, according to Hoyle and Wickramasinghe [424], can be formed at the surface of cold variable carbon stars of class N. These particles in interstellar space are encased in a layer of ice which increases their diameter as much as three times and the albedo as much as twice. This icy coating has a weak influence on the form of the absorption law. The interstellar polarization is explained not so much by the anisotropy of the particles themselves as by that of the conductivity of the graphite. The ratio of the absorption coefficient for radiation with different directions of the electric vector may reach 10 for particles with $a \approx \lambda/2\pi$ and 100 for smaller particles, so that in this case the explanation of interstellar absorption encounters fewer difficulties. The difference in the absorption curve at various longitudes may be explained by the change of thickness of the ice layer surrounding the particle. Finally, assuming a change in the index of refraction of graphite with wavelength, Hoyle and Wickramasinghe [424b] give a qualitative explanation of the ultraviolet depression in stellar spectra observed from rockets. This hypothesis is also attractive because it easily explains the difficulty of the high abundance of heavy elements in cosmic dust, which is approximately equal to the abundance of these elements in interstellar space in general. However, there are, of course, also many difficulties with this hypothesis, primarily those connected with the formation, growth, and distribution of graphite particles in space.

The importance of graphite flakes is supported by rocket observations of interstellar extinction in the ultraviolet [287b]. These observations are shown in Fig. 46 together with the optical data [287c]. Points in the ultraviolet are high above the theoretical curve for classical dielectric grains. The theoretical curve for spherical graphite particles was calculated by Stecher and Donn [287d] from

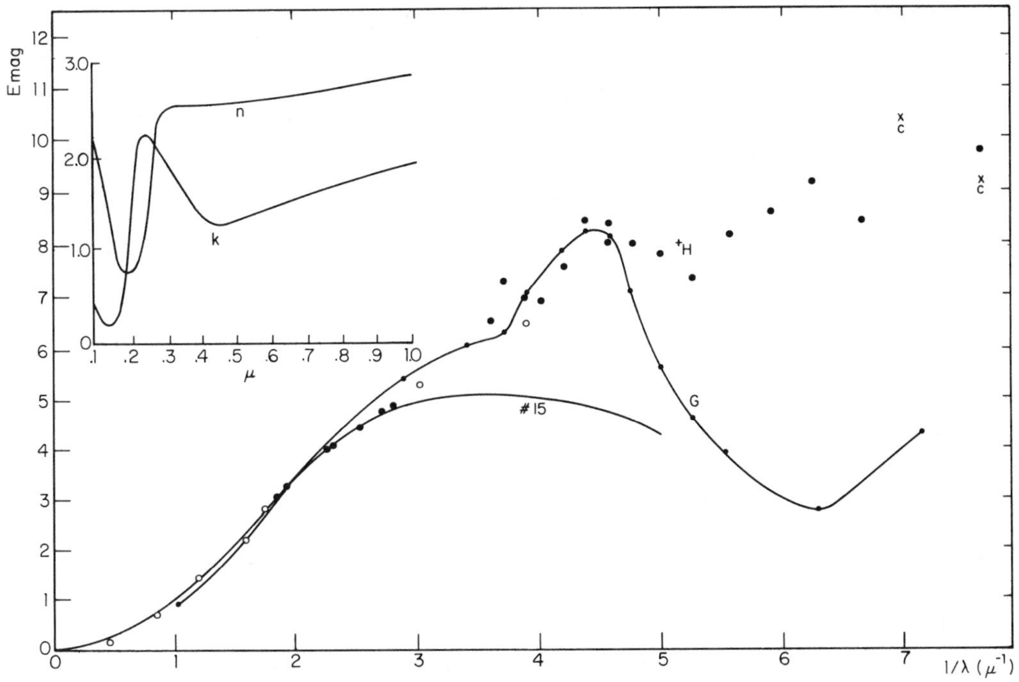

FIG. 46. Observed and theoretical interstellar extinction. *Dots and open circles*, mean extinction found by various observers. *Curve G*, theoretical interstellar extinction for graphite grains. *Curve 15*, theoretical extinction according to van de Hulst. *Inset*, optical constants for graphite as a function of wavelength. (From T. P. Stecker and Bertram Donn, *Ap. J. 142* (1965), 1682, copyright 1965 by the University of Chicago; by permission of the University of Chicago Press.)

the Mee theory. The size distribution was taken from calculations by Oort and van de Hulst, though this distribution may not be applicable to graphite grains. To fit the calibration to the optical observations a 50-percent increase in the relative radius of the particles was sufficient. The optical constants used are also given in Fig. 46.

It is very important that both observations and theory give the peak of absorption near $\lambda^{-1} = 4.4\ \mu^{-1}$. This coincidence can hardly be accidental. The albedo for the particles may be sufficiently high to account for reflection nebulae.

For shorter wavelengths the calculated curve is inconsistent with the observations. Dielectric particles can increase the absorption here. But the dielectric may be graphite itself; its conductivity for the electric vector perpendicular to the basal planes is 100 times smaller than in the longitudinal direction. Stecher and Donn [287d] show that a reversal in the sign of polarization would be expected in this spectral region.

Graphite flakes may grow in interstellar space. Their structure and mass are intermediate between classical dielectric grains and plates; their size is 1 to 5×10^{-6} cm.

Summarizing the foregoing discussion, we must emphasize that the question of the nature of the particles that make up the interstellar dust is still far from a satisfactory solution.

4. Interstellar Magnetic Fields and Nonthermal Radio Emission

16. Magnetic Fields in Interstellar Space

The existence of a magnetic field in interstellar space was first suggested by Alfvén in 1937 and then substantiated by Richtmyer and Teller [288], as well as by Fermi [289], in connection with the study of the motion of cosmic ray particles, their isotropy, and possibilities for their acceleration. A series of arguments connected with the explanation of nonthermal radio emission, rotation of the plane of polarization from distant radio sources, and polarization of star light, as well as with the interpretation of elongated types of nebulae, immediately arose in support of the presence of interstellar magnetic fields. Finally, success has recently been achieved in measuring the interstellar magnetic-field intensity directly from the Zeeman effect in the line $\lambda = 21$ cm.

Electric Conductivity of the Interstellar Medium

Interstellar gas is characterized, even in H I regions, by a reasonably high conductivity. As is known, magnetic fields can exist in this case for a long time, since they dissipate slowly, and in the presence of a moving medium they can be strengthened by the so-called crossing of the lines of magnetic force. Methods for calculating the coefficient of electric conductivity for different conditions are presented in [12]. Only a simplified, approximate development is given here, directed at the conditions of the interstellar medium.

First of all, we note that the interstellar medium cannot be considered to be a completely ionized gas, even in H II regions. Outside the He II zones, neutral helium occurs, which by number is only 5–6 times less concentrated than hydrogen. Even inside He II zones, a certain number, though small, of neutral hydrogen and helium atoms occur. We must also note that the parameter $\omega\tau$, where $\omega = eH/mc$ is the gyrofrequency and τ the time of free motion, is very large for all possible values of density and intensity of the magnetic field H in interstellar space, both for electrons and for ions.

For a simplified derivation of the coefficient of electric conductivity, taking these conditions into account, we shall consider the following problem. Suppose that a weak electric field of intensity $\mathbf{E}$ is superposed on a homogeneous incompletely ionized plasma in a strong magnetic field $\mathbf{H}$. Electrons and ions move under the action of this field with mean velocities $\mathbf{V}_e$ and $\mathbf{V}_i$. The movement of ions drags along some neutral atoms whose mean velocity is $\mathbf{V}_a$. We shall consider the field $\mathbf{E}$ to be so weak that acceleration may be neglected and assume that the state of relative motion of all the particles is stationary. The medium itself as a whole has an acceleration determined by the interaction between current and magnetic field, $d\mathbf{V}/dt = (1/\rho c)\mathbf{j} \times \mathbf{H}$. Because of the assumption that the relative motion is stationary, this quantity is equal to the acceleration of both atoms and ions. Neglecting the inertia of electrons, we can write down equations of motion for electrons and ions in a system of coordinates connected with the gas as a whole, in the form

$$-e\left(\mathbf{E} + \frac{1}{c}\mathbf{V}_e \times \mathbf{H}\right) + \frac{m}{\tau}(\mathbf{V}_i - \mathbf{V}_e) + \frac{m}{\tau_e}(\mathbf{V}_a - \mathbf{V}_e) = 0, \tag{16.1}$$

$$e\left(\mathbf{E} + \frac{1}{c}\mathbf{V}_i \times \mathbf{H}\right) + \frac{m_i}{\tau_i}(\mathbf{V}_a - \mathbf{V}_i) - m_i\frac{d\mathbf{V}}{dt} = 0. \tag{16.2}$$

Here τ and τ_e are the times of flight of an electron until collision with an ion or an atom, and τ_i is the time of free flight of an ion until collision with an atom, taking into account the factor due to the incomplete transfer of momentum from the ion to the atom. In other words, τ_i is $(m_i + m_a)/m_a$ times as large as the actual time of free flight of an ion. The mean velocity of the atoms may be determined from the condition $m_i N_i \mathbf{V}_i + m_a N_a \mathbf{V}_a = 0$. Calling the mass of neutral atoms F, we have $F = N_a m_a/(N_i m_i + N_a m_a)$. Therefore,

$$\mathbf{V}_a = -\frac{1 - F}{F}\mathbf{V}_i. \tag{16.3}$$

In H I regions $1 - F \approx 10^{-3}$, and in H II zones $F \approx 0.3$–0.4. Assuming the number of electrons and ions to be identical and equal to N_e, we have for the current density $\mathbf{j}$,

$$\mathbf{j} = N_e e(\mathbf{V}_i - \mathbf{V}_e). \tag{16.4}$$

From Eqs. (16.1)–(16.4) we can eliminate $\mathbf{V}_e$, $\mathbf{V}_i$, and $\mathbf{V}_a$. Then from simple calculation, remembering also that $eH\tau/mc \gg 1$, we obtain a generalized Ohm's law:

$$\mathbf{E} = \frac{m}{N_e e^2 \tau}\mathbf{j} + \frac{1}{N_e e c}(\mathbf{j} \times \mathbf{H}) - \frac{F^2 \tau_i}{N_e c^2 m_i}(\mathbf{j} \times \mathbf{H}) \times \mathbf{H}. \tag{16.5}$$

The physical significance of this formula is as follows. If the current is parallel to the field, that is, if $\mathbf{j} \times \mathbf{H} = 0$, the electric conductivity λ, given in this case by

$$\lambda = \frac{N_e e^2 \tau}{m}, \tag{16.6}$$

is determined only by collisions of an electron with ions or with atoms, which also may be taken into account in determining τ_e. In the conditions of interstellar space, this value is very large. The calculation of λ by the kinetic-equation method for a plasma, where the charge on an ion is unity, leads to the more accurate equation

$$\lambda = 2.3\left(\frac{2kT}{\pi m}\right)^{3/2} \frac{m}{e^2 L} \approx 6.5 \times 10^6 T^{3/2}. \tag{16.7}$$

Here $L \approx 40$–44 is the Coulomb logarithm. The conductivity λ does not depend on electron concentration, since with a reduction in N_e, τ in Eq. (16.6) is increased correspondingly. In this case, conductivity is high even in H I regions.

Instead of τ, it is more convenient to use the length of the mean free path $l \approx v_T \tau$, where v_T is the thermal velocity. In the general case, for collisions between identical particles with charge $Z_i e$, this quantity is

$$l_i = \frac{(kT)^2}{N_i e^4 L} \frac{1}{Z_i^4}. \tag{16.8}$$

In particular, it is approximately the same, 10^{11} cm, for both electrons and protons in H II zones and H I regions. Thus, a Maxwellian distribution is established in all cases (see Sec. 4).

The second term in Eq. (16.5) defines the Hall current in the magnetized plasma, or, in other words, the current due to particle drift. It is much larger in absolute value than the first term in this equation, but, as will be seen later, it does not lead to dissipation of the magnetic field, and therefore, in many cases, it can be discounted.

The third term in Eq. (16.5) is the most essential in the conditions of the interstellar medium. It also appears because of particle drift, but the presence of the factor τ_i shows that this term is connected with collisions and therefore leads to dissipation of the magnetic field. In fact, an electric field in a strongly magnetic incompletely ionized plasma causes not only relative motion of electrons and ions—that is covered by the first term in Eq. (16.5)—but also the general drift of electrons and ions relative to the neutral gas. Collisions of drifting ions with neutral atoms are described by Eq. (16.2); they are far more effective in scattering electromagnetic energy than collisions of light electrons with heavy particles. The last term in Eq. (16.5) disappears only in the case of currents parallel to the magnetic field (this is the so-called Bessel field, to be discussed below). In every remaining case, it appears for the interstellar medium by definition. The presence of this term reduces conductivity in H II zones 10^{10} times and in H I regions 10^{13} times (see below).

The magnetic energy of a current dissipated per unit volume is equal to the scalar product of current density and field intensity, that is,

$$q = \mathbf{j}\cdot\mathbf{E} = \frac{1}{\lambda} j^2 - \frac{1}{\lambda_3}\frac{j^2H^2 - (\mathbf{j}\cdot\mathbf{H})^2}{H^2}, \tag{16.9}$$

where

$$\lambda_3 = \lambda \frac{mm_i c^2}{F^2 \tau_i \tau e^2 H^2} = \frac{\lambda}{F^2 \dfrac{eH}{mc}\tau \dfrac{eH}{m_i c}\tau_i}. \tag{16.10}$$

The quantity proportional to $1/\lambda$ or $1/\lambda_3$ is usually called the magnetic viscosity ν_m or ν_m':

$$\nu_m = \frac{c^2}{4\pi\lambda}, \qquad \nu_m' = \frac{c^2}{4\pi\lambda_3}. \tag{16.11}$$

Using Eqs. (16.7) and (16.10), we find the following "standard" values for magnetic viscosities in the interstellar medium:

H II zones:
$\nu_m = 10^7$ cm^2/sec, $\quad \nu_m' = 10^{17}$ cm^2/sec, $\quad \nu_k = 10^{19}$ cm^2/sec;

H I zones:
$\nu_m = 10^{10}$ cm^2/sec, $\quad \nu_m' = 10^{23}$ cm^2/sec, $\quad \nu_k = 3 \times 10^{18}$ cm^2/sec.

We have included here data on kinematic viscosity coefficients ν_k. We note that, in the absence of magnetic viscosity, kinematic velocities of heavy particles (for example, in the case of the interstellar gas, by neutral atoms). We must also bear in mind that ν_m does not depend on density, whereas ν_k is inversely proportional to density and ν_m' is inversely proportional to the square of the density. The numerical values are for $N_i = 10$, $N_a = 1$ (H II zone) and $N_i = 10^{-2}$, $N_a = 10$ (H I cloud).

These data will be used later during the analysis of the motion and dissipation of magnetic fields in interstellar space.

Freezing of the Magnetic Field

Returning to Eq. (16.5), we recall that during its derivation the field **E** is determined in a system of coordinates in the center of inertia of the plasma. Movement of the plasma relative to the field, because of the laws governing the coordinate transformation from one inertial system to another, leads to the addition in Eq. (16.5) of the term $(1/c)\mathbf{V} \times \mathbf{H}$, where the velocity **V** is that of the total medium, that is, of its center of inertia. A pressure gradient and other nonelectric forces can lead to a charge separation, because of the difference in mass of ions and electrons, which is equivalent to the appearance of some field $\mathbf{E}''$ of nonelectric origin. For example, for a pressure gradient,

$$\mathbf{E}'' = \frac{1}{N_e e} \nabla p_e. \tag{16.12}$$

Bearing in mind these additions, the expression for the total field is written in the form

$$\mathbf{E} + \frac{1}{c}(\mathbf{V} \times \mathbf{H}) + \mathbf{E}'' = \frac{1}{\lambda}\mathbf{j} + \frac{1}{N_e ec}(\mathbf{j} \times \mathbf{H}) - \frac{1}{\lambda_3}\frac{(\mathbf{j} \times \mathbf{H}) \times \mathbf{H}}{H^2}. \tag{16.13}$$

One of the fundamental equations of magnetohydrodynamics, the induction equation, which measures the change of magnetic field with time, is derived from the Maxwell equations

$$\text{rot}\,\mathbf{H} = \frac{4\pi}{c}\mathbf{j}; \qquad \text{rot}\,\mathbf{E} = -\frac{1}{c}\frac{\partial \mathbf{H}}{\partial t}; \qquad \text{div}\,\mathbf{H} = 0. \tag{16.14}$$

The Gaussian system of units is employed here; the dielectric constant and the magnetic permeability are equal to unity. Substituting the expression (16.13) in (16.14), and neglecting the current **j**, we obtain

$$\frac{\partial \mathbf{H}}{\partial t} = \text{rot}\,(\mathbf{V} \times \mathbf{H}) + c\,\text{rot}\,\mathbf{E}'' + \nu_m \Delta \mathbf{H} + \frac{c}{4\pi e N_e}\text{rot}\,(\mathbf{H} \times \text{rot}\,\mathbf{H}) + \nu_m'\,\text{rot}\left[\frac{(\mathbf{H} \times \text{rot}\,\mathbf{H}) \times \mathbf{H}}{H^2}\right]. \tag{16.15}$$

Let us consider the separate terms of this equation. If in the right-hand part of Eq. (16.15) we can neglect all terms except the first, we obtain the induction equation for an ideal plasma, from which follows the well-known principle of freezing. From this "abbreviated" induction equation, and the continuity equation

$$\frac{\partial N}{\partial t} + \text{div}\,(N\mathbf{V}) = 0, \tag{16.16}$$

we easily obtain the relation

$$\frac{d}{dt}\left(\frac{\mathbf{H}}{N}\right) = \left(\frac{\partial}{\partial t} + \mathbf{V}\nabla\right)\frac{\mathbf{H}}{N} = \left(\frac{\mathbf{H}}{N}\nabla\right)\mathbf{V}. \tag{16.17}$$

We shall assume that the gas moves across the magnetic field. Then $[(\mathbf{H}/N)\nabla]\mathbf{V} = 0$, and from Eq. (16.17) it follows that $\mathbf{H}/N = \text{const}$. This means that every particle appears as if it were attached to its "own" magnetic line of force and carried along it. The density of the lines of force (designated by H) is proportional to the gas density. We must note that the lines of force will carry only the perpendicular component of the particle velocity.

The movement of conducting interstellar gas in a magnetic field stretches the lines of magnetic force, entangles them, and increases their density and, consequently, the magnetic-field intensity. This entanglement continues under favorable conditions until equilibrium is set up between the density of the kinetic energy $\frac{1}{2}\rho V^2$ and the density of the magnetic energy $H^2/8\pi$:

$$\frac{\rho V^2}{2} \approx \frac{H^2}{8\pi}. \tag{16.18}$$

At maximum velocities "entanglement" and "disentanglement" balance each other.

The increase of magnetic-field intensity during the motion of the interstellar medium may be explained by the formation of sufficiently strong magnetic fields, provided only that a very weak "seed" field is originally present. We must of course remember that because of the interlacing of random motions the field also will appear random. Nevertheless, an interstellar magnetic field has a comparatively ordered character. Stretching of the field by differential galactic rotation may possibly play a role here.

The second term in the right-hand member of Eq. (16.15) is usually small, but it helps to explain the formation of the original field. In fact,

$$\text{rot } \mathbf{E}'' = \frac{1}{e} \text{ rot} \left(\frac{1}{N_e} \nabla p_e \right). \tag{16.19}$$

If the isobaric and isothermal surfaces of an electron cloud coincide, then $(1/N_e)\nabla p_e$ may be represented in the form of the gradient of some function rot $\mathbf{E}'' = 0$. The potential part of $\mathbf{E}''$ is compensated by the formation charge and current is not generated. If these surfaces are not coincident, rot $\mathbf{E}'' \neq 0$. A rising eddy current slowly appears (because of self-induction), creating a magnetic field. A weak interstellar magnetic field can arise in other ways, for example, from matter with an inherent magnetic field ejected from stars, or from the motion of a stream of relativistic particles with unequal currents of ions and electrons [289a].

The following terms in Eq. (16.15) take into account the dissipation of magnetic fields (excluding the Hall-current term). Comparing the order of the left-hand side and the third or final terms (rot $\approx R^{-1}$), we can write for t_R, the dissipation time of the magnetic field on the scale R, the expression

$$t_R \approx R^2/\nu'_m. \tag{16.20}$$

For an H II zone of size $R \approx 10$ pc, the time $t_R \approx 10^{22}$ sec $\approx 3 \times 10^{14}$ yr is very large compared to the lifetime of the Galaxy. In H I regions of the same dimensions, $t_R \approx 10^{15}$ sec $\approx 3 \times 10^7$ yr, which is comparable with the lifetime of the cloud. Therefore, in this case, and especially on still smaller scales, consideration of dissipation appears to be essential. The explicit dependence of Eq. (16.20) on the parameters that characterize the medium is given by

$$t_R \approx \frac{8\sigma_{ia}(2\pi m_i kT)^{1/2}}{F^2 H^2} N_e N_a R^2. \tag{16.21}$$

On larger scales (> 100/pc) and in H II regions, dissipation of the magnetic field is unimportant.

This means that in the analysis of interstellar magnetic fields, their geometry and change with time, intuitive considerations can be made graphically, using the principle of freezing. This is especially valuable since the mathematical complexity of solving a system of nonlinear equations in magnetohydrodynamics prevents their use even in relatively simple cases.

A change in field due to magnetic viscosity may be regarded as the sliding or diffusion of lines of magnetic force relative to the medium. According to Eq. (16.20), the rate of this diffusion on the scale R is:

$$V_s \approx \frac{R}{t_R} \approx \frac{v'm}{R}. \tag{16.22}$$

Usually V_s is much less than gas velocities, but in some cases the diffusion of lines of force must be considered, for example, when crossed lines of force approach one another closely. Here the lines can lock, the topology of the field can change, closed loops can become isolated, or, at the other extreme, closed loops can blend together.

Bessel Fields

A general classification of interstellar magnetic fields is difficult to give at present. Their structure is probably very diverse. Of particular interest are the so-called Bessel fields [290, 291, 292]

$$\text{rot}\,\mathbf{H} = \frac{4\pi}{c}\mathbf{j} = \alpha\mathbf{H}, \tag{16.23}$$

where the current is parallel to the field. Here α is some scalar function of the coordinates and the Ampere force $(1/c)\mathbf{j} \times \mathbf{H}$ is equal to zero. It is obvious that a Bessel magnetic field does not exert pressure, and does not destroy hydrostatic equilibrium even if its energy is considerably greater than the thermal energy of the gas. Moreover, when α = const., a Bessel field has still more peculiarities [294]: (*a*) in a closed system, a Bessel field exists in a state of minimum energy compared to all other configurations that may be attained during any motion of the medium in the case of complete freezing; (*b*) the dissipation of energy from a Bessel field is minimum compared to that from all the other configurations mentioned; (*c*) a Bessel field is stable with respect to small disturbances, which follows from point (*a*). Therefore, we may assume that, if we isolate some system of gas and magnetic field from external influences and if in this system motion is damped owing to viscosity or dissipation of energy in shock waves more rapidly than the field is damped, then the latter must become a Bessel field.

Such a possibility may be found in nebulae, the remainders of envelopes discarded by supernovae. The original rapid motion at first intensifies the fields and then becomes dissipated in shock waves. The field that remains becomes a Bessel field. This point of view was developed by Woltjer [294]. Different solutions of Eq. (16.23) have been discussed in a number of papers. The general character of these solutions can be seen by the solution in the plane [293],

$$H_x = H_0 \sin(\alpha z), \qquad H_y = H_0 \cos(\alpha z), \qquad H_z = 0, \tag{16.24}$$

and for the cylindrical configuration [292],

$$H_z = 0, \qquad H_\phi = H_0 J_1(\alpha r), \qquad H_z = H_0 J_0(\alpha r). \tag{16.25}$$

Here J_0 and J_1 are Bessel functions and H_0 is the magnetic-field intensity. The solution of Eq. (16.25) has the character of cylindrical layers in each of which the magnetic lines of force are arranged in spirals whose pitch is decreased in proportion to the distance from the axis of the first zero of the Bessel function J_0. A spherical structure for the Bessel field may be constructed, consisting of concentric spherical layers with curved lines of magnetic force. A real Bessel field may have a more complex character. At the present time, it is difficult to ascertain whether fields in interstellar space possess a Bessel character or even whether the energy of the field is about the same as the energy of motion of the gas.

To avoid confusion we should like to emphasize that if the field is a Bessel field and is constructed according to the same law over all space, then it cannot induce motions. However, real magnetic fields always occupy finite volumes (with the exception, perhaps, of the metagalaxy). In this case the transition from the internal Bessel field to the space external to it or to a field of another type requires a transition layer between them in which there are currents. The interactions of these currents with the field give rise to forces, for example, the pressure of the internal field on the transition layer. One can show that this force depends only on the internal energy at the boundary and is the same for both Bessel and non-Bessel fields of a given strength. Therefore, a strong Bessel magnetic field also always tends toward expansion.

This can also be seen from the following reasoning. If the scale of the field is increased n times while its configuration is conserved, then H goes as n^{-2} and the total magnetic energy, $W_m \sim n^3 H^2 \sim n^{-1}$, being transformed into the acceleration of the restraining envelope, also decreases owing to the expansion.

The Zeeman Effect in the Line $\lambda = 21$ cm

The easiest method for the direct determination of magnetic-field intensity in interstellar space is by measurement of the Zeeman splitting of the line $\lambda = 21.11$ cm [295].

In the presence of a weak magnetic field, the upper hyperfine-structure sublevel is split into $2F + 1 = 3$ sublevels, with magnetic quantum numbers $m_F = -1, 0, +1$. Transitions between these sublevels and the lower unsplit hyperfine-structure sublevel give lines with frequencies

$$\pi \text{ component: } \nu = \nu_0 = 1420.41 \text{ MHz (transition } 0 \to 0); \tag{16.26}$$

$$\sigma \text{ component: } \nu = \nu_0 \pm \tfrac{1}{2}\nu_H = 1420.41 \pm \frac{eH}{4\pi mc} \text{ (transitions } 0 \to \pm 1). \tag{16.27}$$

Here $\nu_H = \omega_H/2\pi = eH/2\pi mc$ is the gyrofrequency. The difference between the two extreme σ components is

$$\Delta\nu = \frac{eH}{2\pi mc} = 2.8H \text{ MHz} \approx 30 \text{ Hz}, \tag{16.28}$$

if $H \approx 10^{-5}$ Oe. Such a negligible split will be completely lost, of course, in the dispersion of cloud velocities or the thermal motion of the atoms (Doppler width $\gtrsim 10^4$ Hz). However, it may be used to advantage, since both σ components are circularly polarized, although in opposite senses. For the determination of the Zeeman splitting in this case, narrow absorption lines, formed in a single cloud, are chosen, and the degree of polarization is measured near the edge of the lines at a point where the line profile has maximum steepness.

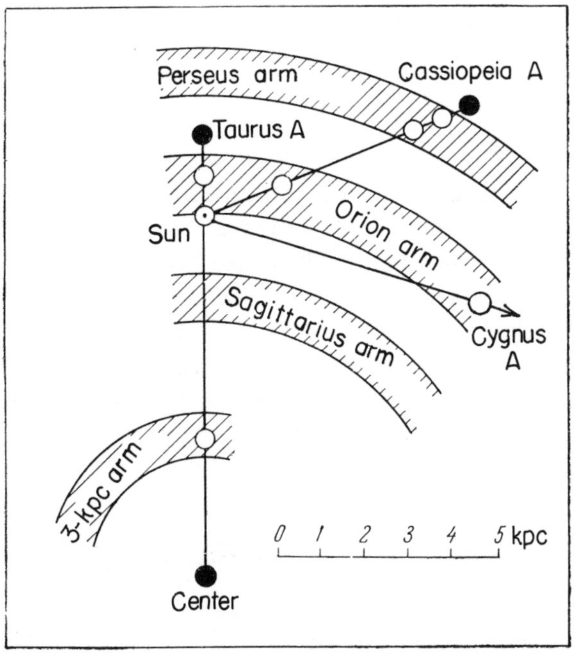

FIG. 47. Distribution of radio-emission sources and absorbing clouds in which the Zeeman effect in the line $\lambda = 21$ cm was observed [296].

By this method, only the linear component (along the line of sight) of the magnetic field may be determined.

Absorption lines instead of emission ones are chosen because a sharp line may be observed on a bright source background, which decreases the effect of noise and possible errors.

The Zeeman splitting of absorption lines was first studied in several clouds distributed as in Fig. 47 [296, 296b, c, d].

Some results of the measurements are given in Table 16.1. The plus sign indicates that the field is directed away from the observer. Observational values reduced under the assumption that magnetic fields in spiral arms are directed along the arms are presented in the last column of the table. It follows from the data in this table that the magnetic field of the spiral arms, if they are considered circular, is directed counterclockwise (looking from the northern galactic hemisphere) and has an average intensity of approximately -7×10^{-6} Oe.

Table 16.1

Spiral arm and source	Longitudinal component of field (10^{-6} Oe)	Angle between axis of arm and line of sight (deg)	Field strength along arm (10^{-6} Oe)
Perseus: Cas A	$+4.3 \pm 9.2$	52	$+7.2 \pm 15$
	-6.7 ± 3.0	52	-11.2 ± 5.0
Orion: Cas A	-4.3 ± 3.8	25	-4.7 ± 4.2
Tau A	$+6.4 \pm 4.5$	94	
Cyg A	-7 ± 54	45	-10 ± 80
3-kpc: Sag A	$+28 \pm 110$	270	

The relatively large value of the field perpendicular to the arm visible against the background in the source Taurus A can then be explained as a local fluctuation. However, an average field intensity of 10^{-5} Oe in a direction at about 20° from the arm corresponds even better to the observational data. Thus, a magnetic field with an intensity of approximately (7–10) $\times 10^{-6}$ Oe in interstellar space is very probable. However, it may have a complex structure, as exemplified by the observations of two clouds in the Perseus arm with oppositely directed magnetic fields. More detailed configurations of the magnetic field in the Galaxy will be discussed in Sec. 19.

17. The Motion of Charged Particles in Interstellar Space (Cosmic Rays)

The study of cosmic rays was not connected with astronomy for a long time, and only with the growth of ideas about interstellar magnetic fields and with the explanation of nonthermal galactic radio emission (1950–1953) did this problem take a place of its own in astrophysics.

Primary Cosmic Rays

We shall first give a short summary of the fundamental experimental data on primary cosmic rays observed near the Earth. More detailed information and a bibliography are given in [5].

One of the peculiarities of cosmic rays is their isotropy. For particles with energies $\mathscr{E} < 10^{17}$ eV, anisotropy is known to be less than 1 percent; when $\mathscr{E} > 10^{17}$ eV, this limit is increased to 3 percent; and only for very hard particles, when $\mathscr{E} \approx 10^{19}$ eV, can anisotropy be considerable. Such isotropy may come about in only two ways: either the cosmic-ray sources are distributed isotropically and uniformly in space, or there is a random magnetic field, distorting and entangling the motions of the particles so that their initial anisotropy (connected with the distribution of the sources) is completely obliterated after a relatively short time. The first assumption leads to isotropy of cosmic rays in extragalactic space. In this event the energy density of cosmic rays in extragalactic space must be the same as in the Galaxy, that is, of the order of 10^{-12} erg/cm^3. In the Galaxy this energy density is comparable with the radiation, kinetic, and magnetic energy densities, but in extragalactic space it may exceed all other forms of energy by several orders of magnitude. Sources known to us are not able to produce such quantities of energy in the form of relativistic particles in cosmogonic time spans. Of course, this is not a conclusive objection, but, all the same, it makes the extragalactic hypothesis rather improbable. There are a number of arguments against the "big-bang" origin of cosmic rays (that is, as residuals from the initial phase of evolution of the Universe).

The explanation of isotropy from the effect of magnetic fields means that the radius of curvature of the trajectory of a particle with charge Z,

$$r = \frac{\mathscr{E} \sin \theta}{300HZ} \tag{17.1}$$

(where $\mathscr{E}$ is expressed in electron volts), must be many times less than the dimension of the Galaxy (10^{22} cm), even for the most energetic particles ($\mathscr{E} \lesssim 19^{19}$ eV). Hence we obtain $H \gtrsim 3 \times 10^{-6}$ Oe when $Z = 1$. However, it is possible that the hardest particles enter the Galaxy from without and are bent not by galactic but by metagalactic magnetic fields. The condition of isotropy in cosmic rays places certain limitations on the nature of the magnetic field apart from its absolute value. For example, if the particles move along bundles of lines of force from the source toward the observer, their distribution will by no means be isotropic. Therefore, it follows that Eq. (17.1) is a necessary, but not a sufficient, condition. We shall return later to the question of isotropy of cosmic rays.

The chemical composition of cosmic rays differs from the relative abundance of heavy elements. The distribution of particles with energies greater than 2.5 BeV/nucleon in a flux of primary cosmic radiation outside the Earth's atmosphere is given according to their charge in Table 17.1.

Table 17.1

Group of particles	Charge	Number of particles relative to hydrogen	Mean chemical composition of Galaxy
P	1	1000	1000
α	2	68	152
L	3–5	1.46	1.5×10^{-6}
M	6–9	4.4	1.5
H	$\geqslant 10$	1.46	0.15
VH	$\geqslant 20$	0.41	.007

In group H are included all particles with charges $\geqslant 10$ (among them those with $Z \geqslant 20$). The total energy density of all particles with energies $\mathscr{E} > 2.5$ BeV/nucleon is 0.2 eV/cm^3. If we include the softer particles, the energy density is increased to ~ 1 eV/cm^3 or somewhat less. This value was obtained by measurements in a year of minimum solar activity. During a maximum period, the value of the interstellar magnetic field is greater, and the flux of primary cosmic radiation that reaches the Earth decreases by approximately 30 percent.

The energy spectrum of cosmic rays may be represented as a power function. In other words, the expression for $N(\mathscr{E})\, d\mathscr{E}$, the number of particles per unit volume with energies between $\mathscr{E}$ and $\mathscr{E} + d\mathscr{E}$, has the form

$$N(\mathscr{E})\, d\mathscr{E} = \frac{K}{\mathscr{E}^{\gamma}}\, d\mathscr{E}\,, \tag{17.2}$$

where $\gamma = 2.5 \pm 0.2$ for all groups of nuclei. The similarity of spectra for different groups is extremely significant. In regions of low kinetic energy (< 1 BeV) the spectrum becomes less steep, and when $\mathscr{E} \lesssim 400$ MeV the number of particles decreases with decrease in energy. The reason for this is not clear so far. Possibly, low-energy particles either are not produced in general or cannot pass through the magnetic fields of the solar system. Data on high-energy particles ($\mathscr{E} > 10^{15}$–10^{19} eV) are uncertain, since these particles are not observed directly. Evidently, here γ is increased (to 3). Relativistic electrons in primary cosmic rays, when $\mathscr{E} \geqslant 0.5$ BeV, were observed by means of rockets [297, 298]. After the effect of the magnetic field of the solar system was taken into account, it was found that the electron flux intensity constituted approximately 1 percent of the flux intensities of protons and heavy particles [5], while the number of positrons arriving from cosmic space is relatively small.

A study of the composition of cosmic rays allows the nature of their diffusion in the Galaxy to be determined. The heavy nuclei of cosmic rays are destroyed after collision with the nuclei of the atoms of interstellar gas, which explains the comparatively large number of nuclei of group L, the fragments of originally heavier nuclei. Since for heavy nuclei the effective destruction cross section is much greater than for light nuclei, the Fe nucleus, for example, cannot be among the long-lived cosmic-ray particles, even if iron nuclei alone were accelerated in the source. The upper limit of the path length may be determined by the presence of heavy atoms and the lower limit by the presence of group L nuclei, since all the atoms of this group certainly are of secondary origin. Instead of the length of the mean free path, expressed in centimeters, which depends on the density of the interstellar medium, it is more convenient to determine the mass in a column of cross section 1 cm^2 through which particles pass until they are destroyed. The corresponding data, taken from [5], are given in Table 17.2.

Table 17.2

	p	α	L	M	H	Fe
Path length (g/cm^2)	72	34	10	7.8	6.1	2.8
Mean lifetime of nucleus when $N_H \approx 10^{-2}$ (10^8 yr)	36	17	5	3.9	3.0	1.4

The average gas density in the Galaxy if the clouds and spiral arms are smeared over the volume of the corona is approximately (1–3) $\times$ 10^{-26} g/cm^3 (the density of the corona itself is negligible, being probably less than 10^{-26} g/cm^3). Assuming that there is free exchange between cosmic-ray particles of the galactic plane and its corona, we find that after the time t a particle has traversed the mass $ct\rho$. In particular, after 10^9 yr, it will have encountered an amount 10–30 g/cm^2. This would greatly reduce the number of iron nuclei. The large amount of Fe in cosmic rays shows that their mean life time in the Galaxy is less than 10^9 yr. In general, this means that cosmic rays cannot have persisted from an early stage of galactic evolution. On the other hand, the time cannot be much less than 10^8 yr, because it would then be difficult to explain the abundance of group L nuclei. It is true that group L may be formed, not in interstellar space, but in sources of cosmic rays. Although we do not know the conditions in sources and the composition of the particles produced there, we cannot rule out the possibility that the time that cosmic rays move in the Galaxy is less than 10^8 yr. Neither do we know the volume over which cosmic rays are distributed after their escape from the spiral arms. Therefore, the assumed values of ρ, and consequently of t, are only preliminary.

The Motion of Charged Particles in Interstellar Magnetic Fields

We shall now return to a theoretical analysis of the conditions of motion of charged particles in interstellar magnetic fields, without restricting ourselves to the case of cosmic relativistic particles. It is well known that in a magnetic field a particle with charge Ze and mass m_i moves in a spiral as if climbing the lines of magnetic force. The angular velocity of rotation and the radius of the projection of the trajectory on a plane perpendicular to **H** are

$$\omega_i = \frac{ZeH}{m_i c}, \qquad r_i = \frac{v_\perp}{\omega_i} = \frac{m_i c v_\perp}{ZeH} = \frac{m_i c v}{ZeH} \sin \theta \tag{17.3}$$

for nonrelativistic particles and

$$\omega_i = \frac{ZeH}{m_i c}\left(1 - \frac{v^2}{c^2}\right)^{1/2} = \frac{ZeH}{m_i c} \cdot \frac{m_i c^2}{\mathscr{E}}, \qquad r_i = \frac{c \sin \theta}{\omega_i} = \frac{\mathscr{E} \sin \theta}{300ZH} \tag{17.4}$$

for relativistic particles. Here θ is the angle between the velocity vector and the direction of the magnetic field **H**; $\mathscr{E}$ includes the rest energy expressed in electron volts. The factor $m_i c^2/\mathscr{E} = (1 - v^2/c^2)^{1/2}$ takes into account the relativistic change of the particle's mass. Such a representation of the helical trajectory of a charged particle is not, however, completely accurate. In the presence of an electric field or of external forces, and also in an inhomogeneous magnetic field, particles drift, as if crossing from one line of force to another, in a direction perpendicular to the force and to the field gradient. The rate of drift is, however, small. Calling l the characteristic dimension of the inhomogeneity of the magnetic field, or the distance over which the parameters of the medium are appreciably changed, we have for the rate of drift in order of magnitude

$$U \approx \frac{r_i}{l} v_\perp = \frac{m_i c v^2}{ZeHl} \sin^2 \theta, \qquad U = \frac{\mathscr{E} \sin^2 \theta}{ZHl} \cdot 10^8 \tag{17.5}$$

for nonrelativistic and relativistic particles, respectively. For example, in 10^{17} sec a proton with energy $\mathscr{E} = 10^{10}$ eV is displaced, owing to drift, by 30 pc, although during this time it will traverse 10^9 pc (here $l \approx 30$ pc, $H = 10^{-5}$ Oe). Thus, in many cases, displacement due to drift can be neglected and the convenient and graphic representation of the spiral motion of particles along a line of magnetic force can be employed with sufficient accuracy.

For the free motion of particles, that is, without considering collisions, we can find a series of simple integrals of the motion, using the concept of an adiabatic invariant. It is well known that, if the period of rotation of a particle is much less than the period of oscillation of the field, the value of the adiabatic invariant

$$\frac{1}{2\pi} \oint p\,dq = p_\perp r_i = \text{const.} \tag{17.6}$$

is conserved, where the integral is taken over the Larmor contour, q is the coordinate of the particle, and $p_\perp$ is its momentum. The integral in Eq. (17.6) is the mechanical moment of a particle. Consequently, in a slightly variable inhomogeneous magnetic field, the mechanical moment and, consequently, the magnetic moment, of a charged particle is conserved. This may also be shown directly.

If the magnetic field does not change with time, then the energy of the particle and the absolute value of its momentum are constant. Consequently, $p_\perp = p \sin\theta$, where θ is the angle between the direction of the velocity of the particle and the magnetic field. Since $r_i \sim \sin\theta$ [Eqs. (17.3) and (17.4)], it follows in this case from Eq. (17.6) that

$$\frac{\sin^2\theta}{H} = \text{const.} \tag{17.7}$$

both for relativistic and for nonrelativistic particles.

From Eq. (17.7) we can draw certain essential conclusions. During the motion of a particle from the region of a strong to a weak magnetic field, the value of θ must decrease, that is, the velocity vector of the particle approaches the direction of the magnetic field. The velocity component parallel to the magnetic field is then increased, and consequently the density of the particles is reduced; it is as if they pass more rapidly through a unit volume. We assume that an isotropic distribution of particle velocities occurs at a certain field intensity H_0. Then during the transition into a region with a field $H < H_0$ all the particle velocities must fall within the solid angle arc sin $(H/H_0)^{1/2}$ with its axis along the magnetic field. Assuming that the density of the particles is proportional to the size of the solid angle, we have

$$\frac{N(H)}{N(H_0)} = \frac{\displaystyle\int_0^{\arcsin(H/H_0)^{1/2}} \sin\theta\, d\theta}{\displaystyle\int_0^{\pi/2} \sin\theta\, d\theta}$$

$$= 1 - \cos\left[\arcsin\left(\frac{H}{H_0}\right)^{1/2}\right] = 1 - \left(1 - \frac{H}{H_0}\right)^{1/2} \tag{17.8}$$

or, if $H \ll H_0$, we obtain $N(H)/N(H_0) \approx H/2H_0$. We must remember that the particle flux intensity per unit solid angle remains constant according to Liouville's theorem. Equation (17.8) is correct when the total energy of a particle is unchanged, as, for example, in the absence of collisions. Therefore, it can apply only to relativistic particles.

In the opposite case, when a particle travels from a weak to a strong magnetic field, the particle velocity deviates from the field direction until θ reaches 90°, after which the particle is reflected from the strong field and proceeds in the reverse direction. The smaller the initial value of the angle θ_0, the stronger the field the particle can penetrate. In the limit, a particle with $\theta_0 = 0$ is not reflected by a strong field. If the distribution of particle velocities in a weak field is isotropic, during transition to a strong field a cone is formed from this distribution which then widens and the distribution becomes isotropic again. The number of particles in the flux is not reduced, because the intensity is conserved, and the solid angle in both cases is equal to 4π. Graphically, this apparent contradiction is explained by the fact that during an increase in H the cross section of the bundle of lines of force is decreased, which compensates for the role of reflecting particles with large angles θ.

The particle energy is changed as the field changes with time. For example, when the field contracts, the particle is accelerated—this is the so-called betatron acceleration. From Eq. (17.6) we can now obtain the momentum-transformation law for particles. In fact, it follows from Eq. (17.3) and (17.4) that $r_i \sim p_\perp/H$. Therefore, during a change of field, the quantity

$$\frac{p_\perp^2}{H} = \text{const.} \tag{17.9}$$

is conserved. In particular, in the case of a nonrelativistic particle, the "perpendicular" portion of the kinetic energy, $p_\perp^2/2m$, increases with intensity. For relativistic particles, growth is slower ($\sim H^{1/2}$). The "longitudinal" portion of particle energy is conserved if the field remains homogeneous in space.

Magnetic fields in the interstellar medium, generally speaking, are inhomogeneous and variable. Together with the betatron effect, another form of interaction of particles with a field takes place, namely, collisions with moving fluctuations of the field frozen in a cloud. Actually, if the field in the cloud is intensified, particles with large angles θ impinging on the cloud will be reflected by the more intense field. Upon reflection from the moving boundary, the particles acquire extra linear velocity. The value of the energy change of a particle, $\Delta\mathscr{E}$, is easily determined from the law of conservation of momentum upon reflection from a surface,

$$\Delta\mathscr{E} = -\frac{2\mathscr{E}}{c^2}\,\mathbf{U}\cdot\mathbf{v}. \tag{17.10}$$

Here $\mathbf{U}$ is the velocity of the inhomogeneity in the field and $\mathbf{v}$ is the velocity of the particle itself. We must remember that $\mathscr{E}$ includes the rest energy, and consequently in the nonrelativistic case $\Delta\mathscr{E} = -2m\mathbf{U}\cdot\mathbf{v}$, from the well-known equation for elastic reflection from a moving surface. Particle energy is increased by countercollisions (when the signs of $\mathbf{U}$ and $\mathbf{v}$ are different), and is decreased during "overtaking" collisions. If the velocities $\mathbf{U}$ of the inhomogeneities have a random distribution, then for $v \gg U$ the countercollisions are $(v + U)/(v - U)$ times greater than the overtaking collisions. Therefore, from a single collision, particle energy is increased on the average by $\sim U\Delta\mathscr{E}/v$, from which

$$\overline{\Delta\mathscr{E}} = \frac{U^2}{C^2}\,\mathscr{E}. \tag{17.11}$$

The angles between the vectors $\mathbf{U}$ and $\mathbf{v}$ are averaged here. This equation describes the well-known statistical acceleration of a particle, the so-called Fermi mechanism [289]. It must be stressed that Eq. (17.11) is correct only when the velocities of the inhomogeneities in the magnetic field are random. Moreover, the idea itself of a collision between particles and a field applies to particles with a Larmor radius much less than the dimensions of the cloud. Thus, charged particles can be accelerated in interstellar magnetic fields.

The Acceleration and Deceleration of Charged Particles in Interstellar Space

We shall present a summary of various mechanisms for accelerating or decelerating particles by interaction with a magnetic field.

(1) *Betatron acceleration in a time-dependent magnetic field.* Here the change of energy with time for relativistic particles is

$$\frac{d \ln \mathscr{E}}{dt} = \frac{1}{2}\frac{d \ln H}{dt}, \qquad \frac{d\mathscr{E}}{dt} = \alpha_3 \mathscr{E}. \tag{17.12}$$

Under usual interstellar conditions or in nebulae, this mechanism is probably of no great importance, since here the fluctuation of magnetic fields is not great and in the main is not monotonic. Betatron acceleration possibly occurs if a more or less homogeneous mass of gas with a frozen magnetic field is compressed in the process of condensation. During expansion of a gas with a frozen field, the particle energy is decreased by being expended on the acceleration of the gas with the field. This effect is essentially the adiabatic expansion of a relativistic gas with two degrees of freedom.

(2) *The statistical mechanism of Fermi*, described by Eq. (17.11), can operate in the interstellar medium or in nebulae, when intense random movements are found, while the system itself as a whole neither expands nor contracts. For the energy change of a relativistic particle with time we have in this case

$$\frac{d\mathscr{E}}{dt} = \frac{U^2}{lc}\mathscr{E} = \alpha_1 \mathscr{E}, \tag{17.13}$$

where l is the dimension of the moving inhomogeneities, since the number of collisions per unit time is approximately c/l.

Tsitovich [289b] has reported an interesting variant of the Fermi acceleration mechanism of particles by plasma waves. In plasmas not in equilibrium, for example, particle fluxes give rise to plasma waves presenting a more or less random ensemble. The random electric fields associated with them change the energies of the particles, just as the ordinary Fermi mechanism accelerates particles in a medium. In a concrete application of the mechanism to astrophysical objects it is necessary to know the energy density of the plasma waves. Preliminary estimates place this energy at approximately 0.001 the kinetic-energy density.

(3) *The deceleration of particles during expansion of the system due to their collision with inhomogeneities.* If a system containing inhomogeneities is expanded, the collision of particles with the inhomogeneities of the field will be predominantly of the overtaking type. Calling V the rate of expansion of a system of size R, we can take for the velocity of the inhomogeneities, assuming

linear expansion, $U = lV/R$. Then the energy change of the relativistic particles during the expansion is

$$\frac{d\mathscr{E}}{dt} = -\frac{V}{R}\mathscr{E} = -\alpha_2\mathscr{E}. \tag{17.14}$$

If a system with inhomogeneities in the magnetic field is compressed, the particle energies are increased. Adiabatic heating and cooling also has this same effect. A decrease in energy of type (17.14) probably occurs in the expanding envelopes of supernovae.

(4) *Acceleration in a magnetic trap.* In principle, a charged particle can be found between two strong convergent fluctuations of a magnetic field. The particle will then increase its energy considerably, because of repeated reflections from the convergent surfaces. Since only the longitudinal portion of the energy is increased, the angle θ is reduced and after a certain time the particle will be able to pass through the "walls" of the trap. This restricts the magnitude of the acceleration. If collisions between the particles themselves in this trap are sufficiently frequent, and invariants for each of them are not conserved, a more effective acceleration system of particles becomes possible. In this case, adiabatic heating of the system of particles under compression again takes place.

Several concrete models of cosmic magnetic traps have been proposed. Suppose, for example, that a magnetic loop is intersected by a shock wave so that lines of magnetic force emerge at one point of this front and enter at another [299a]. Then parts of the lines of magnetic force located in the path of the front will continually be decreased, in proportion to the motion of the front, until the whole magnetic loop is "absorbed" by the shock wave. Particles moving along lines in front of the shock are thus in a so-called trap with shifting magnetic surface. If the condition for reflection dictated by the requirement of the conservation of the adiabatic invariant is fulfilled for a sufficiently long time, acceleration of the particles is possible. Calculations show that the change in energy in the simplest case is determined by the equation

$$\frac{\mathscr{E}_1}{\mathscr{E}_2} \approx \left[1 + \frac{2}{3}\frac{H_2}{H_1}\left(1 - \frac{H_1}{H_2}\right)^{3/2}\right]\left[\frac{2}{3} + \frac{1}{3}\frac{H_1}{H_2}\right], \tag{17.15}$$

where H_2/H_1 is the discontinuity of the magnitude of the magnetic field in the wave. The second factor takes into account the decrease of energy in the expanding parts of the loop that remain in front of the shock. If the loop is not closed, this factor disappears. We note that in the case of a very strong discontinuity in the magnetic field, $H_2 \gg H_1$, and the change in energy is $\mathscr{E}_2/\mathscr{E}_1 \approx \frac{4}{9}H_2/H_1$, which is considerably more than for a single operation of the adiabatic invariant. A second example of a cosmic magnetic trap is the propagation of a shock wave alongside an increasing magnetic field.

(5) *Acceleration at the front of a magnetohydrodynamic shock wave.* We have been considering the case where the particles are reflected from the front of a

shock wave. However, acceleration is also possible when particles penetrate the front [299, 300, 301]. In fact, if the radius of the trajectory of a charged particle is much greater than the width of the front of the shock wave, and its velocity is appreciably greater than the wave velocity, the particle makes several revolutions, each of which intersects the wave front, before being carried away completely into the region behind the front. Since the magnetic-field intensities are different before and behind the front, the radii of curvature differ. As a result, a particle will intersect the front at a different angle during each revolution, which leads to a systematic change in its momentum. According to calculations by V. P. Shabanskii [300], the momentum of a particle that has passed through a shock-wave front is

$$\frac{p_2}{p_1} = \exp\left\{\int_{\psi_1}^{\psi_2} \frac{\sin^2 \psi \, d\psi}{[\pi H_1/(H_2 - H_1)] + \psi - \sin \psi \cos \psi}\right\}. \tag{17.16}$$

Here p_1 and p_2 are the initial and final particle momenta, H_1 and H_2 are the magnitudes of the magnetic field at opposite sides of the wavefront, and ψ_1 and ψ_2 are the angles relative to the front at which the particles enter the first time and leave the last time, respectively. If the velocity of the shock wave is small compared to the particle velocity, so that the particle can make many revolutions, penetrating the wave front each time, then in Eq. (17.16) we can set $\psi_1 = 0$ and $\psi_2 = \pi$. The integral is then obtained in explicit form and we get $p_2/p_1 = (H_2/H_1)^{1/2}$, that is, simply the condition for the conservation of the adiabatic invariant, which is to be expected, since the relatively slowly moving shock wave produces adiabatic acceleration. However, Eq. (17.16) can also be used in the study of charged particles when the velocity of the shock wave is comparable with that of the particle. Since in this case the integral in Eq. (17.16) is taken over a narrower range, the acceleration is as a rule less, which follows from the condition of adiabatic invariance.

We recall that for acceleration the radius of the path of a particle must be greater than the thickness of the wave front. This limitation is obviously not very critical for interstellar wave fronts, since according to present ideas the thickness of the front of a collisionless shock wave in a magnetized plasma is small and of the same order of magnitude as the Larmor radius of thermal particles. Shock waves are often met in interstellar space, and therefore substantial acceleration can result from them.

Besides the mechanisms given above for the interaction of charged particles with interstellar magnetic fields, there are a number of processes by which the particle energy is decreased. In one case a particle loses energy by magnetobremsstrahlung, in other cases, because of interaction with atoms and with interstellar radiation. For completeness, we shall present equations that describe these losses, with the goal in mind of using them in the analysis of the motions of cosmic rays.

(1) *Magnetobremsstrahlung, in the relativistic case the so-called synchrotron radiation.* It is well known that every charged particle undergoing acceleration is also radiating. The spiral motion of a charged particle in a magnetic field is no exception. The emission per unit time, from the well-known equation of electrodynamics, is

$$\frac{d\mathscr{E}'}{dt'} = -\frac{2e^2}{3c^3}\left(\frac{dv'}{dt'}\right)^2 \tag{17.17}$$

in an inertial system of coordinates having the instantaneous velocity of the electron. In this system of coordinates the acceleration of the electron is

$$\frac{dp}{dt} = \omega p_\perp = \frac{eH}{mc}p_\perp = \frac{eH_\perp}{mc^2}\mathscr{E}, \tag{17.18}$$

where $H_\perp$ is the component of the magnetic field perpendicular to the velocity of the particle. Substituting (17.18) in (17.17) we obtain

$$\frac{d\mathscr{E}}{dt} = -\frac{2e^4H_\perp^2}{3m^2c^3}\left(\frac{\mathscr{E}}{mc^2}\right)^2 = -\beta_1 H_\perp^2\mathscr{E}^2 = -3.8 \times 10^{-9}H_\perp^2\mathscr{E}^2 \text{ MeV/sec.} \tag{17.19}$$

Since the transformations of $\mathscr{E}'$ and t' in the observer's coordinate system follow one and the same law, $d\mathscr{E}/dt = d\mathscr{E}'/dt'$. Substituting numerical values and expressing energy in electron megavolts (MeV) we obtain

$$\frac{d\mathscr{E}}{dt} = \frac{2e^4H_\perp^2}{3m^2c^3}\frac{v^2}{c^2} = 10^{-9}\frac{H_\perp^2 v^2}{c^2} \text{ MeV/sec.} \tag{17.20}$$

In the case of relativistic particles, $v = c$. In the nonrelativistic case, $\mathscr{E} = mc^2 = 0.51$ MeV and, since $v^2 \ll c^2$, the magnetobremsstrahlung is small.

(2) *Energy losses by electrons due to photon emission (in the relativistic case by γ quanta) during collisions between electrons and other particles.* The effective cross section of the bremsstrahlung process after a collision with a hydrogen atom is

$$\sigma = \frac{4 \ln 137}{137}\left(\frac{e^2}{mc^2}\right)^2 = 1.1 \times 10^{-26} \text{ cm}^2. \tag{17.21}$$

Hence we obtain for the loss of energy due to bremsstrahlung in interstellar space the expression

$$\frac{d\mathscr{E}}{dt} = -2.3\sigma c N_{\mathrm{H}}\mathscr{E} = -\beta_2 N_{\mathrm{H}}\mathscr{E} = -8.0 \times 10^{-16}N_{\mathrm{H}}\mathscr{E} \text{ MeV/sec,} \tag{17.22}$$

where the factor 2.3 takes into account collisions with electrons ($N_e > N_{\mathrm{H}}$). This equation is valid for relativistic particles. In the nonrelativistic case, this energy loss is equivalent to the usual emission due to free-free transitions. The mass of hydrogen in a column of unit cross section which an electron encounters before an appreciable fraction of its energy is lost because of bremsstrahlung is $1.64 \times 10^{-24} \times 3 \times 10^{10}/8 \times 10^{-16} \approx 62$ g/cm^2.

(3) *Ionization losses.* During collisions with free or atomic electrons of the interstellar medium, fast particles (both electrons and protons) impart to them a portion of their own energy—on the average, about 32 eV for each collision. The rate of energy loss of a particle with charge Z, moving with a velocity v, is given by the equation

$$\frac{d\mathscr{E}}{dt} = -\frac{2\pi N Z^2 e^4}{mv} L, \tag{17.23}$$

where N is the total number of electrons, free and bound, per unit volume. The Coulomb logarithm L takes on different values, depending on whether or not the particle is relativistic, and on the degree of ionization of the gas. Moreover, L varies logarithmically with energy and density. In all the conditions encountered in the interstellar medium and in nebulae, this value is about 40–75. Taking $L \approx 60$ and substituting for v in Eq. (17.23) the speed of light, we obtain

$$\frac{d\mathscr{E}}{dt} = -\frac{2\pi N Z^2 e^4}{mc} L = -\beta_3 N_{\mathrm{H}} Z^2 = -4.5 \times 10^{-13} N_{\mathrm{H}} Z^2 \text{ MeV/sec}. \tag{17.24}$$

The role of ionization losses decreases with increase in particle energy.

(4) *Energy losses from the inverse Compton effect.* Under this term are grouped the energy losses of electrons during "collisions" with the photons of the interstellar radiation. The effective scattering cross section of photons, with certain limitations, is equal to the Thompson cross section

$$\sigma_0 = \frac{8\pi}{3}\left(\frac{e^2}{mc^2}\right)^2 = 6.6 \times 10^{-25} \text{ cm}^2. \tag{17.25}$$

The mean energy transferred by an electron to a photon per collision is approximately

$$\overline{\Delta\mathscr{E}} = \frac{\mathscr{E}}{mc^2}\bar{\varepsilon}' = \left(\frac{\mathscr{E}}{mc^2}\right)^2 \bar{\varepsilon}, \tag{17.26}$$

where $\bar{\varepsilon}'$ is the mean energy of a photon in a system of coordinates connected with the electron and $\bar{\varepsilon}$ is the mean energy of a photon in a system of galactic coordinates. The loss of energy by an electron per unit time due to scattering by photons is

$$\frac{d\mathscr{E}}{dt} = -\sigma_0 N_f c \overline{\Delta\mathscr{E}} \approx -\sigma_0 \rho_r c \left(\frac{\mathscr{E}}{mc^2}\right)^2, \tag{17.27}$$

where N_f is the number of photons per unit volume and $\rho_r = N_f \bar{\varepsilon}$ is the interstellar radiation density. We point out, incidentally, that the inverse Compton effect is simply the magnetobremsstrahlung and electrobremsstrahlung of relativistic electrons in the electromagnetic field of the thermal radiation from stars. In fact, substituting expression (17.25) in Eq. (17.27) and setting $\rho_r = H^2/4\pi$, we can easily reduce this formula to Eq. (17.19). Inserting the values of the numerical coefficients, we have

$$\frac{d\mathscr{E}}{dt} = -\beta_4 \rho_z \mathscr{E}^2 = -4.8 \times 10^{-8} \rho_z \mathscr{E}^2 \text{ MeV/sec}, \tag{17.28}$$

since $\beta_4 = 4\pi\beta_1$. Obviously, losses due to the inverse Compton effect predominate where the density of normal thermal radiation is larger than the density of magnetic energy, as in the case of nearby hot stars (or supergiants), or in the galactic halo.

Equation (17.26) also has an analogue in the theory of magnetic damping of synchrotron radiation. If we omit in Eq. (18.6) the factor 3/2 and multiply both sides by Planck's constant $\hbar$, it can then be written in the form

$$\hbar\omega_c \approx \hbar\omega_H \left(\frac{\mathscr{E}}{mc^2}\right)^2. \tag{17.29}$$

Here $\hbar\omega_c$ is the energy of the magnetically damped quantum, and $\hbar\omega_H$ is the energy of the quantum "rotated in the magnetic field." The analogy between Eqs. (17.26) and (17.29) is obvious. The difference between synchrotron radiation of radio quanta in a constant magnetic field and the inverse Compton effect in the rapidly oscillating magnetic field of the total radiation from stars is only that the higher the frequency of the oscillating magnetic field above the frequency of rotation in the constant magnetic field, the higher will be the frequency of the radiated quanta from the inverse Compton effect. The total energy radiated remains constant for a given average field strength. Only for one x-ray or gamma quantum from the inverse Compton effect, there are $\omega_s/\omega_H \approx 2 \times 10^8/H = 10^{14}$ radio quanta (ω_s is the frequency of the visible starlight).

(5) *Energy losses by protons and heavy particles during nuclear collisions.* In contrast to the preceding mechanisms, a heavy particle or a proton in the process of a nuclear reaction immediately dissipates its energy and can disintegrate. Therefore, it is easier in this case to determine, not the energy change of a particle with time, but the change in the number of particles per unit volume.

The mean lifetime of particles from each group can be defined as $T = 1/\sigma N c$. The effective cross section for nuclear collisions with hydrogen for all particles except hydrogen and helium is $\sigma \approx \pi(1.26 \times 10^{-13} A^{1/3})^2$ cm^2, where A is the atomic weight. For hydrogen, $\sigma \approx 2.3 \times 10^{-26}$ cm^2, and for helium, $\sigma \approx 10^{-25}$ cm^2. Values for the mean lifetimes of different groups for $N_{\rm H} \approx 0.01$ cm^{-3} are listed in Table 17.2. During the course of nuclear reactions, particles can not only disappear, but be modified. For example, during nuclear reactions with particles of group H, there is a certain probability that particles from all the lighter groups (p, α, L, M) can be formed. The number of particles from group i formed during the nuclear reaction with a particle from group k is called the fragmentation coefficient P_{ki}. An estimate of the values of these coefficients, obtained by Lohrman and Teucher [302], is given in Table 17.3 (see also [5]).

Table 17.3

P_{ki}	L	M	H
α	1.5	1.5	2.0
L		0.40	0.1
M		0.16	0.7
H			0.5

A fragmentation coefficient with equal indices indicates the possibility during nuclear reactions of the formation of another lighter nucleus, but belonging to the same group.

It is now easy to write as an equation that determines the change in the number of particles of each group the expression

$$\frac{dN_i}{dt} = -\frac{N_i}{T_i} + \sum_{i<k} P_{ki}\frac{N_k}{T_k}. \tag{17.30}$$

Particle lifetimes are given in Table 17.2 (if $N_{\mathrm{H}} \neq 0.01$, they must be correspondingly increased or diminished), while fragmentation coefficients P_{ki} are given in Table 17.3.

In addition to the processes of disintegration of particles already mentioned, there is still one more which is important for nuclei of ultrahigh energies ($\mathscr{E} > 10^{16}$ eV). A collision of such a nucleus with a light quantum may disrupt it, since the energy of the quantum in the system of the relativistic nucleus is very high. This is a possible explanation of the decrease in the fraction of nuclei among extremely hard particles.

X-rays and γ-rays in the interstellar medium

The interaction of relativistic particles with other components of the interstellar medium produces electromagnetic waves of various wavelengths. Observation of this radiation can lead on the one hand to a knowledge of the energy of relativistic particles in interstellar space, and on the other hand to certain information on the distribution and physical properties of the various components of the interstellar medium. A particularly large amount of data has been obtained during the analysis of radio magnetobremsstrahlung (synchrotron radiation) from relativistic electrons in interstellar magnetic fields. These questions will be discussed in the following section.

X- and γ-radiation arising in interstellar space, and possibly in extragalactic space, has recently been observed. So far these data are preliminary and do not contain much information. Therefore, we limit ourselves here to a short description of the results obtained up to the present.

It has been shown by methods of rocket astronomy that the flux of x-ray quanta with energies greater than 1.7 keV from the area of the galactic center is approximately equal to 5 quanta/cm^2 sec ster, and from the direction of the anticenter approximately one-third as much [302a, b]. The flux of γ-ray quanta with energies higher than 5×10^7 eV (averaged over direction) is probably about 3.5×10^{-4} quanta/cm^2 sec ster, although it is not clear at present whether this is actually the value of the flux or an upper limit [303c].

A number of works connected with this subject have been produced in which an analysis of the possible x- and γ-radiation of the interstellar medium is given [303d, e]. It is clear that the energy losses of particles by radiation discussed above—Eqs. (17.19), (17.22), and (17.28)—also determine for a given energy and density of relativistic particles in interstellar space the observed x- and γ-radiation. There are, however, additional mechanisms, such as the emission of γ quanta from the decay of π^0 mesons formed from the interaction of cosmic-ray particles with atoms of the interstellar medium, from the annihilation of positrons, and from other processes [303e]. X-rays may arise from transitions to inner electron shells of heavy atoms in highly ionized states [303d].

The observed flux of x-rays can be explained, for example, by synchrotron radiation from electrons with energies of about 10^{15} eV. This explanation requires a flux of such particles with an intensity of about 5×10^{-14} electrons/cm^2 sec ster. This is entirely possible, since the flux of all cosmic particles with this energy is a few thousand times greater [303e]. Gamma-rays are most probably generated by the inverse Compton effect for relativistic electrons, which is also responsible for the synchrotron radio emission. In fact, since the frequency of the observed synchrotron radio emission is of the order of 10^8 Hz, then, according to the previous discussion, the frequency of the x-ray quanta must be in the neighborhood of 10^{22} Hz, which corresponds to quanta with an energy of 4×10^7 eV. The intensity of γ-rays is determined by the explicit expression

$$I_\gamma(\mathscr{E}_\gamma) = \frac{N_f c}{4\pi} \int_{\mathscr{E}_\gamma}^{\infty} (\mathscr{E}_\gamma, \mathscr{E}) N(\mathscr{E}) dE, \tag{17.31}$$

where, according to Eqs. (17.25) and (17.26), the effective cross section for emission of quanta with energy $\mathscr{E}_\gamma$ by electrons of energy $\mathscr{E}$ can be expressed in the form $\sigma_0 \delta[\mathscr{E}_\gamma - \bar{\varepsilon}(\mathscr{E}/mc^2)^2]$ (that is, by the delta function). A more accurate expression containing a numerical multiplicative factor of order unity has also been given [303e]. With the help of Eq. (17.2), further calculations are elementary. It has been shown as a result that the flux of γ quanta with energies greater than 5×10^7 eV must be of the order of 10^{-4} quanta/cm^2 sec ster (the electron distribution function was determined from the observed synchrotron radio emission). The inverse Compton effect gives an approximately isotropic distribution of γ radiation. The damping mechanism and the decay of π^0 mesons produces a strong anisotropy. The intensity of radiation in the direction of the center of the Galaxy must be some 30 times that in the directions of the poles. The total intensity of γ radiation due to these mechanisms is approximately one order of magnitude less than that due to the inverse Compton effect. The extragalactic x- and γ-radiation has also been estimated [303e]. Here the flux density of relativistic electrons remains undetermined. Also, neither the density of matter nor the intensity of the magnetic field is known in extragalactic space.

Motion of Cosmic Rays in Interstellar Space

It is customary to consider that the motion of cosmic rays in interstellar space is of the nature of diffusion, with a certain coefficient D whose value has so far not been determined. However, the definition of diffusion as random motion is, strictly speaking, not very well founded itself, since the particles move along lines of magnetic force. Certain elements introduce statistically particle reflection from inhomogeneities in the magnetic field or from disruptions due to shock waves [303], displacement of particles by drift, which shifts them from one magnetic line of force to another, random cloud motion, which drags along lines of magnetic force together with cosmic-ray particles, changes in magnetic-field topology due to the finite conductivity of the gas, and, finally, simple instability of the plasma, which leads to particle scattering. Therefore, here the invariance of (17.7) cannot be conserved [5]. Because of this, at least as a first approximation, the motion of these particles may be thought of as diffuse, giving to D the meaning of a semiempirical parameter that characterizes the diffusion.

An appreciable fraction of cosmic rays are evidently formed during outbreaks of type II supernovae in the spiral arms, after which they diffuse into the corona and to the galactic boundary. We may assume two models, closed and open, for the galactic magnetic field. In the first, the lines of magnetic force are joined inside the corona, while outside, beyond the galactic limits, only a small portion emerges, approximately equal to the ratio of the field intensity outside and inside the corona. In this case, the galactic-field boundary is comparatively sharp. In the open model, there is no sharp boundary and the magnetic-field intensity decreases gradually because of the increase of the surface and the divergence of the lines of magnetic force and the cosmic rays slowly diffuse to the boundaries of the Galaxy.

Special attention should be paid in this respect to the spiral arms, whose fields probably emerge into extragalactic space after several tens of revolutions, and therefore are often entangled with their outer parts. In this case the escape time is generally less than 10^7 yr, and supernovae cannot furnish the observed quantities of cosmic rays. In addition, it is difficult in this case to explain isotropy. The structure of the galactic magnetic field will be discussed in Sec. 19.

From the point of view of the energy of cosmic rays, the two models are equally correct (except in the case of the arms opening into the metagalaxy); we can always select coefficients for diffusion and reflection of particles from the galactic boundary such that for sources of given output we obtain cosmic-ray energies approximately equal to those observed. However, these models are not equivalent from the point of view of the chemical composition. In the open model, other conditions being the same, the escape time of the particles is less than in the closed model. Therefore, a smaller number of nuclei are able to disintegrate.

The chemical composition of cosmic rays can be studied with the help of Eq. (17.30). Three-dimensional diffusion of particles is accounted for by adding to this equation the term div (DN_i). The solution of this equation was studied by V. L. Ginzburg and S. I. Syrovatskii [5].

A particle path must be limited, so that nuclei of groups H and VH are not completely destroyed; on the other hand, its length must be sufficient for the formation of group L nuclei. Since the original composition of cosmic rays is unknown (there is a basis, for example, for the assumption that, for the most part, nuclei of the heavy elements are accelerated), it is more convenient to proceed in the following way. Given the observed composition of cosmic rays near the Earth, we can "retrace" their paths by diffusion in the reverse direction, determine in this way the original distribution in composition, and compare this distribution with the cosmic abundance of the elements. The results of such estimates by Levis [304] and by V. L. Ginzburg and S. I. Syrovatskii [5] definitely showed that the closed model was inconsistent with the observed composition. For a retraced path of sufficient length, the number of protons, alpha particles, L nuclei, and so forth formed in this way is too large. The open model with slow diffusion agrees with the observed composition. The diffusion coefficient D must then be about 10^{30} N, where N is the concentration of atoms in the interstellar medium. Regular (nondiffuse) particle motion along lines of magnetic force also agrees with the observed composition on the assumption that the route taken by the particles from the source to the Earth is approximately 6.5 g/cm^2. However, it is difficult to explain isotropy on this model, and, in addition, the route taken by the particles will also depend on the value of the adiabatic invariant and on the angle θ. Therefore, the anisotropy must be different for different groups and different energies, which also is not observed.

This means that it has been impossible up to now to derive from cosmic-ray studies reliable data on their motion in interstellar magnetic fields, or on the acceleration of these particles in sources. It is possible that the magnetic field, at least in the corona, is entangled and open, gradually blending into the metagalactic magnetic field.

So far, it has not been possible to say much about the mechanism of particle acceleration in sources. It is described possibly either by Fermi's statistical mechanism or by acceleration at the fronts of shock waves. Since the energy during statistical acceleration increases with an increase in $\mathscr{E}$, while ionization losses decrease, only particles of sufficiently high energy can be accelerated. The preliminary acceleration is called injection. A. A. Korchak and S. I. Syrovatskii [306] showed that, if initial ionization is small, heavy ions have a low injection threshold, almost in the thermal range. However, the present state of the problem of the acceleration of cosmic-ray particles in sources is such that it is impossible to obtain from theoretical considerations even a rough idea of the efficiency of different mechanisms.

The spectrum (17.2) can be derived, after certain assumptions, from the condition of equilibrium between the kinetic- and magnetic-energy densities and cosmic-ray particle energies [307].

Relativistic Electrons

We turn now to the problem of the formation of relativistic electrons, the particles responsible for nonthermal cosmic radio emission. At the present time, direct observations on relativistic electrons give little information about their total energy, ~ 1 percent of the energy of heavy particles. Information about relativistic electrons comes mainly from observations in radio astronomy, which are better treated in the following section.

There are two possible mechanisms for the formation of relativistic electrons: acceleration of the source (primary electrons) and formation from the decay of $\pi^{\pm}$ mesons created during collisions between cosmic-ray protons and gas protons, both in sources and in the interstellar medium (secondary electrons). The effective cross section of this collision is $\sim 4 \times 10^{-26}$ cm^2. If a π^0 meson is formed during proton collisions, it decays into two γ-quanta. Hence, if secondary electrons are formed, their number must be proportional to the number of γ-quanta. Strictly speaking, secondary electrons are always being formed. The question is whether they constitute the major portion of relativistic electrons. As a rough estimate, secondary electrons acquire approximately 5–10 percent of the energy of the nuclei [5]. If we consider that cosmic rays in the Galaxy disappear as the result of nuclear reactions, then the energy of the secondary electrons formed each second must be approximately equal to $0.1\text{–}1 \times 10^{39}$ erg. This is approximately the quantity of energy emitted in the radio band. However, if an appreciable quantity of cosmic radiation leaves the Galaxy without undergoing collision, there will be fewer relativistic electrons.

The discovery in cosmic rays of positrons and γ-quanta, where the number of positrons is somewhat larger than the number of electrons, would be direct proof of the secondary formation of relativistic electrons. However, as a recent calculation by V. L. Ginzburg and S. I. Syrovatskii shows, in contrast to what is said in [5], the role of secondary electrons must be small. Recent measurements from rockets have actually shown that the number of positrons in cosmic rays is considerably less than the number of electrons, since they are formed primarily in the atmosphere. Electrons are probably accelerated directly in their sources.

The possibility of observing the relativistic-electron spectrum by the methods of radio astronomy poses the problem of estimating it theoretically. This may be done by solving the kinetic equation for the energy distribution function, $N(\mathscr{E}, t)$ for relativistic electrons. In this equation account is taken of three-dimensional diffusion [308] of particles and their "diffusion" along the energy axis [309]. The general kinetic equation is rather cumbersome. Although it may be studied with sufficient accuracy by known methods of diffusion theory [5], we shall limit ourselves here to a simplified formula. If we neglect the fluctuating energy changes of a particle due to the effect of Fermi's mechanism and consider only its systematic change, the equation determining the distribution function of electrons is written in the form

$$\frac{\partial N(\mathscr{E}, t)}{\partial t} = -\frac{\partial}{\partial \mathscr{E}}\left[N(\mathscr{E}, t)\sum_{\mathrm{I}}\frac{d\mathscr{E}}{dt}\right] + \frac{N(\mathscr{E}, t)}{\mathscr{E}}\sum_{\mathrm{II}}\frac{d\mathscr{E}}{dt} + Q(\mathscr{E}, t). \quad (17.32)$$

Here $N(\mathscr{E}, t)d\mathscr{E}$ is the number of cosmic-ray electrons with energies between $\mathscr{E}$ and $\mathscr{E} + d\mathscr{E}$ per unit volume and $Q(\mathscr{E}, t)d\mathscr{E}$ is the number of secondary electrons injected or being formed per unit time in the same energy range. The summation index I indicates the summation of expressions (17.12), (17.13), (17.14), (17.19), and (17.24), that is, all processes in which the energy changes continuously. Summation II includes (17.22) and (17.28), where, as the result of collisions with particles or photons, the electron energy is immediately reduced to a value comparable with the initial energy. Since the dependence of each of these expressions on the energy is known, we can write

$$\begin{aligned}\frac{\partial N(\mathscr{E}, t)}{\partial t} &= -\frac{\partial}{\partial \mathscr{E}}\{N(\mathscr{E}, t)[-\beta_1 H_\perp^2 \mathscr{E}^2 + (\alpha_1 - \alpha_2 + \alpha_3)\mathscr{E} - \beta_3 N_{\mathrm{H}}]\} \\ &\quad - [\beta_2 N_{\mathrm{H}} + \beta_4 \rho_r(z)\mathscr{E}]N(\mathscr{E}, t) + Q(\mathscr{E}, t) \\ &= -\frac{\partial}{\partial \mathscr{E}}\{N(\mathscr{E}, t)[a\mathscr{E}^2 + b\mathscr{E} + c]\} - (d + f\mathscr{E})N(\mathscr{E}, t) + Q(\mathscr{E}, t). \quad (17.33)\end{aligned}$$

Energy is expressed in this case in electron megavolts. We must emphasize that the parameters a, b, c, d, and f may depend on time. For the proper account of particle energy fluctuations due to collisions with the moving inhomogeneities in a magnetic field, we must add to (17.32) a term of the form

$$\frac{\partial}{\partial \mathscr{E}^2}[D_{\mathscr{E}} N(\mathscr{E}, t)], \quad (17.34)$$

where $D_{\mathscr{E}}$ is the "diffusion" coefficient along the energy axis. If the energy spectrum has the characteristic of a power law, that is, if $N(\mathscr{E}, t) \sim \mathscr{E}^{-\gamma}$, then, instead of adding the term (17.34) to (17.33), we can replace the numerical coefficient α_1 by a factor, depending on the value of the power γ of the spectrum, of approximately unity. Since the value of α_1 is determined accurately to an order of magnitude, this effect need not at first be considered.

The solution of Eq. (17.33) is found in simple cases without difficulty. For example, in a steady state, when $\partial N/\partial t = 0$ and all the parameters a, b, c, d, and f, together with the source function Q, are independent of time, we can write for the general solution of Eq. (17.33)

$$N(\mathscr{E}) = \frac{\exp\left[\int \frac{(d + f\mathscr{E})d\mathscr{E}}{a\mathscr{E}^2 + b\mathscr{E} + c}\right]}{a\mathscr{E}^2 + b\mathscr{E} + c} \int Q(\mathscr{E}) \times \exp\left[-\int \frac{(d + f\mathscr{E})d\mathscr{E}}{a\mathscr{E}^2 + b\mathscr{E} + c}\right] d\mathscr{E}. \tag{17.35}$$

The limits of integration depend on the boundary conditions imposed on the spectrum. In a rarefied medium, where there is no need to consider the collisions of electrons with atoms, ions, and photons (the inverse Compton effect), the solution is simplified, since in this case $d = f = c = 0$. Thus we have

$$N(\mathscr{E}) = \frac{\int Q(E)d\mathscr{E}}{a\mathscr{E}^2 + b\mathscr{E}} \sim \frac{\mathscr{E}^{-(\gamma-1)}}{|b\mathscr{E} - \beta_1 H_\perp^2 \mathscr{E}^2|}, \tag{17.36}$$

if the spectrum of injected electrons is described by a power law with the exponent γ. We recall that here $b = \alpha_1 - \alpha_2 + \alpha_3$, which introduces macroscopic acceleration and deceleration of particles.

Investigation of nonstationary spectra is more complex. Certain solutions to this problem were given by S. A. Kaplan [309], while a more complete study was provided by N. S. Kardashev [310]. In the latter work, various combinations of mechanisms for acceleration and deceleration of particles were considered, together with a series of hypotheses on the distribution function of injected electrons.

A characteristic property of these solutions is the presence of cusps in the spectra, which may be observed, for example, in the steady-state spectrum (17.36) at an energy of

$$\mathscr{E}_k \approx \frac{b}{\beta_1 H_\perp^2}. \tag{17.37}$$

In other words, the spectral index is equal to γ for $\mathscr{E} \ll \mathscr{E}_k$ and to $\gamma + 1$ for $\mathscr{E} \gg \mathscr{E}_k$. Another cusp occurs because of the inverse Compton effect. In nonstationary spectra, the cusp is displaced with time to the high- or low-energy side.

If we consider a spectrum change under fixed external conditions in a rarefied medium (a and b do not depend on time, $c = d = f = 0$), we can obtain the spectrum $N(\mathscr{E}, t)$, without solving Eq. (17.33). In fact, writing the expression

$$\frac{d\mathscr{E}}{dt} = b\mathscr{E} - \beta_1 H_\perp^2 \mathscr{E}^2 = b\left(\mathscr{E} - \frac{\mathscr{E}^2}{\mathscr{E}_k}\right), \tag{17.38}$$

and integrating, we find the dependence of $\mathscr{E}$ on time and on the original energy $\mathscr{E}_0$. We have

$$\frac{1}{\mathscr{E}} = \frac{1}{\mathscr{E}_k} - \left|\frac{1}{\mathscr{E}_k} - \frac{1}{\mathscr{E}_0}\right| e^{-bt}. \tag{17.39}$$

Thus, in the course of time, electrons with $\mathscr{E}_0 > \mathscr{E}_k$ decrease their energy to $\mathscr{E}_k$, while electrons with $\mathscr{E}_0 < \mathscr{E}_k$ increase theirs to the same value. The spectrum becomes monochromatic with time. If the original spectrum has the form $N(\mathscr{E}_0, 0) \sim \mathscr{E}^{-\gamma}$, then at a time t we have

$$N(\mathscr{E}, t) \sim \left[\frac{1}{\mathscr{E}_k} - \left|\frac{1}{\mathscr{E}_k} - \frac{1}{\mathscr{E}}\right| e^{bt}\right]^\gamma. \tag{17.40}$$

Since such spectra have so far not been observed, we may assume that constant acceleration is rarely encountered. If particle acceleration is unimportant, $b = 0$ and the solution of Eq. (17.38) has the form

$$\mathscr{E} = \frac{\mathscr{E}_0}{1 + \beta_1 H_\perp^2 \mathscr{E}_0 t}. \tag{17.41}$$

Substituting $\mathscr{E}_0$ found from Eq. (17.41) into the original spectrum, we find

$$\begin{aligned} N(\mathscr{E}, t) &\sim \left(\frac{1}{\mathscr{E}} - \beta_1 H_\perp^2 t\right)^\gamma, & \mathscr{E} &< \frac{1}{\beta_1 H_\perp^2 t}, \\ N(\mathscr{E}, t) &= 0, & \mathscr{E} &> \frac{1}{\beta_1 H_\perp^2 t}. \end{aligned} \tag{17.42}$$

In using Eqs. (17.42), the following circumstance must be kept in mind. It is obvious that in any system there are always electrons moving at different angles to the magnetic field, and consequently with various values of $H_\perp$. Therefore, the solution of Eq. (17.42) must again be averaged according to $\sin\theta \sim H_\perp$. This means that the total spectrum never returns to zero and will have a cusp close to the energy value [310]

$$\mathscr{E}_k' \approx \frac{1}{\beta_1 H^2 t} = \frac{8.2}{\beta_1 H^2 t\,(\text{yr})}\ \text{MeV} \tag{17.43}$$

while the spectrum will have the form

$$\begin{aligned} N(\mathscr{E}, t) &\sim \frac{1}{\mathscr{E}^\gamma}, & \mathscr{E} &\ll \mathscr{E}_k', \\ N(\mathscr{E}, t) &\sim \frac{1}{\mathscr{E}^{\gamma+1} H^2 t}, & \mathscr{E} &\gg \mathscr{E}_k'. \end{aligned} \tag{17.44}$$

In Eqs. (17.43) and (17.44) H is the total magnetic field intensity. We note that a change of exponent by unity occurs here, as in the case of Eq. (17.36). Thus, in summary, we can say that practically always, within certain limits, the energy spectrum of relativistic particles must approximate a power function. The location of the cusp in the spectrum defines the "initiation" or the "shutdown" of new mechanisms of acceleration or of loss of particles.

It has recently been recognized that the collective processes in a plasma may play a significant role in the motion of cosmic-ray particles. For instance, if in the sources there are excited plasma waves with energy density W_l, the relativistic particles double their energy in the time $(N_e m_e c^2/\omega_0 W_l)(c/V_p)^2(\mathscr{E}/mc^2)^2$, where N_e is the electron density, $\omega_0 = (4\pi e^2 N_e/m_e)^{1/2}$ is the plasma frequency, and V_p is the phase velocity of plasma waves [310a]. This time is very short, but the actual density of plasma energy is usually unknown and must be computed on the basis of some assumption. The relativistic particles may also be accelerated by electromagnetic radiation under special conditions [310a].

Another effect of plasma physics related to cosmic-ray particles is the isotropization of these rays by scattering on plasma waves and inducing the so-called beam instability. The time of isotropization is also very short [310a, b]. The study of plasma aspects of astrophysics is only beginning.

18. Nonthermal Radio Emission from the Interstellar Medium

As we have discussed in Sec. 4, a certain amount of continuous cosmic radio emission, whose intensity (in the case of an optically thin layer) depends only slightly on frequency, is formed in H II regions owing to free-free transitions of electrons in a proton field. The brightness temperature of this emission cannot exceed 10,000°. However, observations show that the intensity of a considerable portion of galactic radio emission, especially at meter wavelengths, depends strongly on frequency, the intensity increasing with increasing wavelength. The brightness temperature of this emission reaches a value of some hundreds of thousands and millions of degrees. Obviously, such temperatures cannot occur in the stationary or even quasi-stationary conditions of the interstellar medium, and therefore this emission is of a nonthermal nature.

The spectrum of nonthermal radio emission can be approximated by the power law

$$I_\nu \sim \nu^{-\alpha}, \qquad T_b \sim \nu^{-\alpha-2}, \tag{18.1}$$

where the parameter α is called the spectral index. Generally speaking, α is somewhat different for various frequency ranges. In the decimeter and meter range, apparently $\alpha \approx 0.8$. At longer wavelengths ($\lambda > 10$ m) the spectrum becomes flatter ($\alpha \approx 0.35$), in the frequency interval from 6 to 1.5 MHz the spectrum becomes almost constant ($\alpha \approx 0$) [310c], and at still lower frequencies the spectrum possibly drops off (the spectral index becomes negative). For an approximate value over the whole observable range of the spectrum we can assume $\alpha \approx 0.5$.

At the present time two mechanisms for nonthermal radio emission are known: excitation of plasma oscillations changing into electromagnetic radiation, and magnetobremsstrahlung (synchrotron radiation). The frequency of plasma oscillations in the interstellar plasma when $N_e \approx 1\ \text{cm}^{-3}$ is $\omega_0/2\pi = (Ne^2/\pi m)^{1/2} = 9.0 \times 10^3$ Hz ($\lambda = 33$ km), which is beyond the present limits of observation. It is well known that the second mechanism for nonthermal radio emission completely and satisfactorily explains its main features, the intensity and spectrum of observed cosmic radio emission. This explanation was proposed by Alfvén and Herlofson and by Kippenhauer in 1950, and was developed in the work of V. L. Ginzburg, I. S. Shklovskii, G. G. Getmantsev, and others.

Magnetobremsstrahlung

Let us consider the emission from a relativistic electron orbiting in a magnetic field. We shall first of all assume that the electron is moving in a circular orbit perpendicular to the magnetic field. The vector potential of a rapidly moving charge (the Lenard–Wiechert potential) is described by the equation

$$\mathbf{A} = \frac{e\mathbf{v}}{c\left(R - \dfrac{\mathbf{R}\cdot\mathbf{v}}{c}n\right)}, \tag{18.2}$$

where $\mathbf{v}$ is the electron velocity, $\mathbf{R}$ is the radius vector from the observer to the instantaneous position of the electron, and n is the refractive index. Denoting by ϕ the angle between the vectors $\mathbf{v}$ and $\mathbf{R}$, assuming that it is small, and using the well-known expression for the refractive index in a plasma,

$$n^2 = 1 - \frac{4\pi N_e e^2}{m\omega^2},$$

where ω is the frequency of the radiation (anisotropy in the refractive index may be completely neglected, since in the interstellar medium the gyrofrequency is not greater than 30 Hz), we can rewrite Eq. (18.2) in the form

$$\mathbf{A} = \frac{ev}{cR\left(1 - \dfrac{v}{c} + \dfrac{v}{c}\dfrac{\phi^2}{2} + \dfrac{v}{c}\dfrac{2\pi N_e e^2}{m\omega^2}\right)} \approx \frac{ev}{R\left[\dfrac{1}{2}\left(\dfrac{mc^2}{\mathscr{E}}\right)^2 + \dfrac{\phi^2}{2} + \dfrac{2\pi N_e e^2}{m\omega^2}\right]}. \tag{18.3}$$

Here the energy of the electron is $\mathscr{E} = mc^2/(1 - v^2/c^2)^{1/2} \gg mc^2$. From Eq. (18.3) we see first that almost all the radiation from a relativistic electron is concentrated in the narrow solid angle $mc^2/\mathscr{E}$, and second that the difference between the refractive index and unity need not be taken into account until the energy of the emitting electrons meets the condition

$$\frac{mc^2}{\mathscr{E}} \gg \left(\frac{4\pi N_e e^2}{m\omega^2}\right)^{1/2} \approx \frac{\omega_0}{\omega} \approx \frac{\lambda\sqrt{N_e}}{33\text{ km}}, \tag{18.4}$$

where λ is in kilometers. In the radio band so far observed, this condition is met, but at long wavelengths or in denser nebulae it may be violated. Since the electron emission is centered in a narrow cone $\phi = mc^2/\mathscr{E}$, an observer located in the plane of the orbit will record, not continuous emission, but short-lived bursts of duration Δt repeated at a frequency ω'_H, where

$$\Delta t = \left(1 - \frac{v}{c}\right)\frac{r\phi}{c} \approx \frac{r}{c}\left(\frac{mc^2}{\mathscr{E}}\right)^3 \approx \frac{mc}{eH}\left(\frac{mc^2}{\mathscr{E}}\right)^2$$
$$\omega'_H = \frac{eH}{mc}\frac{mc^2}{\mathscr{E}}. \tag{18.5}$$

Here the orbit radius r and the rate of revolution ω'_H are determined by Eq. (17.4) when $\theta = \pi/2$ and $Z = 1$. The first of Eqs. (18.5) incorporates a term to allow for the Doppler effect; because of the motion of the electron toward the observer with speed v, the signal from the end of the burst seems to overtake the signal from the beginning and the corresponding time displacement differs by a factor v/c from the duration $r\phi/c$ of this burst in the system of coordinates of the electron. In Eq. (18.5), no account has been taken of the divergence of the refractive index from unity. If necessary, it is readily done by changing $mc^2/\mathscr{E}$ to $[(mc^2/\mathscr{E})^2 + 4\pi Ne^2/m\omega^2]^{1/2}$. The decomposition of short bursts, of duration Δt and occurring with a frequency ω'_H, gives a large number of high overtones of the frequency ω'_H. Since $\Delta t \ll \omega_H'^{-1}$, these overtones combine to give a practically continuous spectrum, where the maximum density of the emission is concentrated in the frequency range close to $1/\Delta t$, or, more precisely, at the frequency 0.3 ω_c, where

$$\omega_c = 2\pi\nu_c = \frac{3}{2\Delta t} = \frac{3}{2}\frac{eH_\perp}{mc}\left(\frac{\mathscr{E}}{mc^2}\right)^2 = 2\pi \times 16.2 \times H_\perp\mathscr{E}^2\text{ MHz}. \tag{18.6}$$

Here, as before, $\mathscr{E}$ is measured in electron megavolts. The factor 3/2 is added for convenience in writing further formulas. Moreover, H is replaced here by $H_\perp$; in this form, Eq. (18.6) is correct not only when the orbit of the electron is circular but also when it is a spiral. Assuming that the frequency range of the emission from a relativistic electron is $\Delta\omega \approx \omega_c$ and dividing Eq. (17.19) by Eq. (18.6), we obtain for the emission coefficient of a single electron per unit frequency range the expression

$$P(\omega, \mathscr{E}) \approx \frac{e^3 H_\perp}{mc^2}. \tag{18.7}$$

It is essential that the emission intensity close to maximum intensity should not depend on energy, since the spectral width and the total energy emitted are proportional to $\mathscr{E}^2$.

An accurate solution to the problem of the emission from a relativistic electron moving in a homogeneous magnetic field was first obtained by Schott in 1912, and has been repeatedly studied in the years following. An account of the solution is given in [311 and 312]; polarization of the radiation is investigated in [313, 314, and 319a] and the effect of the refractive index in [315–318] and elsewhere. The solution is straightforward in principle, but computation is cumbersome. Therefore we shall present here only the final expressions.

The energy emitted by an electron of energy $\mathscr{E}$ over the frequency range from ν to $\nu + d\nu$ is

$$P_1(\nu, \mathscr{E})\, d\nu = \frac{\sqrt{3}e^3 H_\perp}{mc^2} \left[\frac{\nu}{\nu_c} \int_{\nu/\nu_c}^{\infty} K_{5/3}(\eta)\, d\eta \right] d\nu, \tag{18.8}$$

where $K_{5/3}(\eta)$ is a Bessel function of imaginary argument.

Magnetobremsstrahlung from a relativistic electron is generally elliptically polarized. The major axis of the polarization ellipse of the electric vector, perpendicular to the projection of the magnetic field in the pictorial plane, is small and parallel to it. Therefore, the ratio of the semiaxes of the polarization ellipse depends on the angle between the vectors $\mathbf{v}$ and $\mathbf{R}$. In particular, if this angle is zero, polarization is linear. By averaging over the duration of a burst, linear polarization is also obtained. As in the case of interstellar-polarization studies, it is more convenient here to introduce the Stokes parameters Q and U. For linear polarization, the parameter V is zero. Since the Stokes parameters have the dimensions of intensity, it is best to determine the emission coefficients P_Q and P_U similarly to P_I. As a result, we have [314]

$$\begin{aligned} P_Q(\nu, \mathscr{E})\, d\nu &= \frac{\sqrt{3}e^3 H_\perp}{mc^2} \left[\frac{\nu}{\nu_c} K_{2/3}\left(\frac{\nu}{\nu_c}\right) \right] \sin 2\chi\, d\nu, \\ P_U(\nu, \mathscr{E})\, d\nu &= \frac{\sqrt{3}e^3 H_\perp}{mc^2} \left[\frac{\nu}{\nu_c} K_{2/3}\left(\frac{\nu}{\nu_c}\right) \right] \cos 2\chi\, d\nu, \end{aligned} \tag{18.9}$$

where χ is the position angle between certain fixed directions and the semiaxis of the polarization ellipse perpendicular to the projection of the magnetic field in the pictorial plane.

Substituting numerical values, we obtain

$$\frac{\sqrt{3}e^3 H_\perp}{mc^2} = 2.34 \times 10^{-22} H_\perp \text{ erg/sec Hz}. \tag{18.10}$$

A graph of the function in brackets in Eq. (18.8) is given in Fig. 48. The maximum of the emission spectrum,

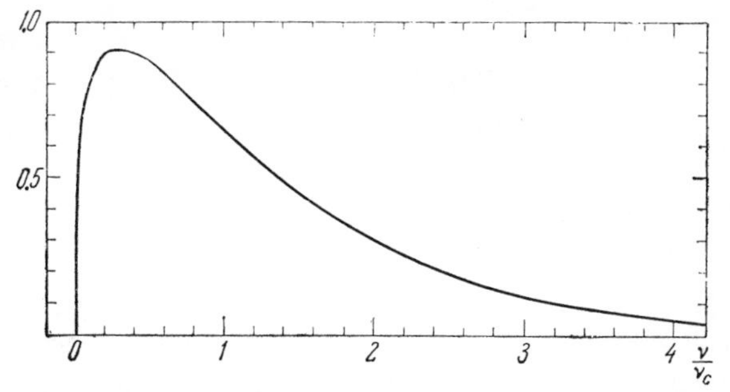

FIG. 48. Emission spectrum from a relativistic electron in a magnetic field, that is, the function $(\nu/\nu_c) \int_{\nu/\nu_c}^{\infty} K_{5/3}(\eta) d\eta$.

$$P_I(\nu_m, \mathscr{E}) = \frac{1.6 e^3 H_\perp}{mc^2} = 2.16 \times 10^{-22} H_\perp \text{ erg/sec Hz}, \qquad (18.11)$$

is reached when $\nu = 0.3\nu_c$. Asymptotic expressions for $P_I(\nu_m, \mathscr{E})$, for both high and low frequencies, are determined from the equations

$$\begin{aligned} P_I(\nu, \mathscr{E}) &= 5.04 \times 10^{-22} H_\perp (\nu/\nu_c)^{1/3}, & \nu < 0.01\nu_c, \\ P_I(\nu, \mathscr{E}) &= 2.94 \times 10^{-22} H_\perp (\nu/\nu_c)^{1/2} e^{-\nu/\nu_c}, & \nu > 10\nu_c. \end{aligned} \qquad (18.12)$$

For intermediate values of ν/ν_c, Fig. 48 may be used. Approximate expressions for the Stokes parameters may be similarly obtained; except for the numerical factors, they agree with Eqs. (18.12). It is interesting to note that the degree of polarization of the radiation from a single electron is

$$p = \frac{(P_U^2 + P_Q^2)^{1/2}}{P_I} = \frac{K_{2/3}(\nu/\nu_c)}{\int_{\nu/\nu_c}^{\infty} K_{5/3}(\eta)\, d\eta} = \begin{cases} \frac{1}{2}, & \nu \ll \nu_c, \\ 1, & \nu \gg \nu_c. \end{cases} \qquad (18.13)$$

The refractive index becomes essential if the last term in Eq. (18.3) is not very small compared with the first. Knowing that the maximum emission occurs at the frequency $\nu = 0.3\nu_c$, we can rewrite the condition under which we can neglect the difference of the refractive index from unity in the form

$$\left(\frac{mc^2}{\mathscr{E}}\right)^2 \gg \frac{10 N_e e^2}{\pi m \nu_c^2}, \qquad \left(\frac{\mathscr{E}}{mc^2}\right)^2 \gg \frac{N_e mc^2}{H_\perp^2/8\pi} = \frac{2 \times 10^{-5} N_e}{H_\perp^2}, \qquad (18.14)$$

changing ν to $0.3\nu_c$ and using expression (18.6). If inequalities (18.14) are not satisfied, the emission spectrum of a relativistic electron bunches up, so to speak, in regions of low frequencies. Indeed, in this case we can use Eqs. (18.8)–(18.13), but the argument of the function, ν/ν_c, must be replaced [315, 316] by the quantity

$$\frac{\nu}{\nu_c'} = \frac{\nu}{\nu_c}\left[1 + (1 - n^2)\left(\frac{\mathscr{E}}{mc^2}\right)^2\right]^{3/2}; \qquad (18.15)$$

and we must multiply the expression in brackets in Eq. (18.8) by $[1 + (1 - n^2)(\mathscr{E}/mc^2)^2]^{-1/2}$. If the second term in Eq. (18.15) is small, this equation leads to the determination of ν_c from Eq. (18.6). If, on the other hand, the second term is larger than the first,

$$\frac{\nu}{\nu_c'} \approx \frac{4}{3} \frac{ce^2}{(\pi m)^{1/2}} \frac{N_e^{3/2}}{H_\perp} \frac{\mathscr{E}}{mc^2} \frac{1}{\nu^2} \tag{18.16}$$

and with a decrease in the frequency ν, the argument of the Bessel function passes over into the region of the exponentially decreasing function $P_I(\nu, \mathscr{E})$. This steep fall in the spectrum can be explained graphically. Because the phase velocity of the radiation, $v_f = c/n$, is considerably greater than the electron velocity, $v \approx c$, the directivity of the radiation is weaker, which also reduces the number of high overtones.

The observed sharp fall in the spectrum will depend on the energy-distribution function of the electrons. However, its characteristic frequency may be directly determined by setting the second term in Eq. (18.15) equal to unity and eliminating $(\mathscr{E}/mc^2)^2$ with the help of Eq. (18.6). As a result we obtain

$$\nu \approx \frac{\nu_c}{\nu} \frac{4N_e ec}{3H_\perp} \approx 7 \frac{ecN_e}{H_\perp} \tag{18.17}$$

since from $\nu/\nu_c \approx 0.3$ at maximum emission it follows that $\nu/\nu_c \approx 2^{3/2}/0.3$. Different hypotheses regarding the distribution function can change the coefficient in (18.17) from 3 to 10.

So far we have considered emission from individual electrons. In order to change to emission from a system of electrons, we must average the expressions obtained over energy as well as over the direction of motion, that is, over the components $H_\perp$.

Averaging over the direction of motion requires a knowledge of the geometry of the magnetic field, together with the initial conditions, since the invariant $\sin^2 \theta/H = \mathrm{H}_\perp^2/H^3 =$ const. is conserved during the motion of an electron. Both are usually unknown. We shall therefore restrict ourselves to two limiting cases: (1) all the electrons have the same component $H_\perp$ in a homogeneous magnetic field; (2) the distribution of fields is isotropic.

Let us consider the first case. The electron emission coefficient per unit volume per unit solid angle per unit frequency interval is

$$\varepsilon_I(\nu) = \frac{1}{4\pi} \int_0^\infty P(\nu, \mathscr{E}) N(\mathscr{E})\, d\mathscr{E} \text{ erg/cm}^3 \text{ sec Hz}. \tag{18.18}$$

As we noted in Sec. 17, the distribution function of relativistic electrons can be approximated by the power law

$$N(\mathscr{E}) = \frac{K}{\mathscr{E}^\gamma}, \tag{18.19}$$

at least in a fixed range of energy whose limits are $\mathscr{E}_1$ and $\mathscr{E}_2$ ($\mathscr{E}_1 < \mathscr{E}_2$). If these limits are such that we can assume $\mathscr{E}_1 \to 0$ and $\mathscr{E}_2 \to \infty$, then if Eqs. (18.19) and (18.8) are substituted into Eq. (18.18) the integral assumes an explicit form. The result of this calculation is [319]

$$\varepsilon_I(\nu) = a(\gamma)\frac{e^3}{mc^2}\left(\frac{3e}{4\pi m^3 c^5}\right)^{\frac{1}{2}(\gamma-1)} H_\perp^{\frac{1}{2}(\gamma+1)} K \nu^{\frac{1}{2}(\gamma-1)}, \tag{18.20}$$

where

$$a(\gamma) = \frac{2^{\frac{1}{2}(\gamma-1)}\sqrt{3}}{4\pi(\gamma+1)}\Gamma\left(\frac{3\gamma-1}{12}\right)\Gamma\left(\frac{3\gamma+19}{12}\right). \tag{18.21}$$

Values of the function $a(\gamma)$ for various values of γ are given in Table 18.1 (something will be said later about the other quantities in this table).

Table 18.1

γ	0.28	1.5	2	2.5	3	4	5
$a(\gamma)$	0.325	0.195	0.145	0.122	0.111	0.116	0.158
$b(\gamma)$	0.283	0.147	0.103	0.0852	0.0742	0.0725	0.0922
$y_1(\gamma)$	0.80	1.3	1.8	2.2	2.7	3.4	4.0
$y_2(\gamma)$	0.00045	0.011	0.032	0.10	0.18	0.38	0.65

Substituting in Eq. (18.20) numerical values for the constants, and setting $\nu = c/\lambda$, we obtain

$$\varepsilon_I(\lambda)d\nu = a(\gamma)1.3 \times 10^{-22}(2.8 \times 10^8 \lambda)^{\frac{1}{2}(\gamma-1)} H_\perp^{\frac{1}{2}(\gamma+1)} K\, d\nu. \tag{18.22}$$

The nature of the frequency dependence of the coefficient ε_I is readily understood from the following considerations. The major contribution to the integral (18.18) is given by electrons for which ν_c from Eq. (18.6) is close to ν, while $P(\nu, \mathscr{E})$ is constant, and the interval range is $\Delta\nu \approx \nu_c$. Therefore, to a first approximation, $e(\nu) \sim N(\mathscr{E})(d\mathscr{E}/d\nu)\nu_c \sim N(\nu_c^{1/2})\nu_c^{1/2} \sim \nu_c^{(1-\gamma)/2} \sim \nu^{(1-\gamma)/2}$, which gives the same frequency dependence as Eq. (18.20). Thus, the power of the spectrum from magnetobremsstrahlung is completely determined by the power of the spectrum of the relativistic electrons, independently of the character and intensity of the magnetic field. In this connection, we emphasize that Eq. (18.6) determines at the same time the energy of the electrons that make the greatest contribution to the emission at a given frequency.

In a similar way we can calculate emission coefficients for the Stokes parameters, $\varepsilon_Q(\nu)$, $\varepsilon_U(\nu)$ [314]. They differ from Eq. (18.20) only by a numerical factor depending on γ and also, of course, by the factors sin 2χ and cos 2χ. A simple equation [319a] for the degree of polarization is

$$p = \frac{(\varepsilon_U^2 + \varepsilon_Q^2)^{1/2}}{\varepsilon_I} = \frac{\gamma+1}{\gamma+7/3}. \tag{18.23}$$

Equations (18.20)–(18.23) can also be used over narrower limits for the spectrum (18.19). In particular, the error in Eq. (18.20) for a definite frequency ν will not be greater than 10 per cent if $\mathscr{E}_1$ and $\mathscr{E}_2$ satisfy the conditions [320]

$$\begin{aligned} \mathscr{E}_1 &\leqslant mc^2\left[\frac{4\pi mc\nu}{3eH_\perp y_1(\gamma)}\right]^{1/2} = 2.5 \times 10^{-4}\left[\frac{\nu}{H_\perp y_1(\gamma)}\right]^{1/2} \text{MeV}, \\ \mathscr{E}_2 &\geqslant mc^2\left[\frac{4\pi mc\nu}{3eH_\perp y_2(\gamma)}\right]^{1/2} = 2.5 \times 10^{-4}\left[\frac{\nu}{H_\perp y_2(\gamma)}\right] \text{MeV}. \end{aligned} \tag{18.24}$$

The functions $y_1(\gamma)$ and $y_2(\gamma)$ are given in Table 18.1. As the data from this table show, an overwhelming part of the emission at a given frequency is in fact bound to a comparatively small energy range of the electrons.

We shall now consider the second case, where the electron-velocity distribution may be regarded as isotropic. Here, also, during the calculation of the integral in Eq. (18.18), we must still average out $H_\perp$. It is clear that a formula of type (18.20) will be obtained, where instead of $H_\perp^{(\gamma+1)/2}$ we have the factor

$$\overline{(H_\perp^{\frac{1}{2}(\gamma+1)})} = \frac{1}{2}\int_0^\pi (H \sin\theta)^{\frac{1}{2}(\gamma+1)} \sin\theta\, d\theta = \frac{\frac{1}{2}\pi^{1/2}\Gamma[\frac{1}{4}(\gamma+5)]}{\Gamma[\frac{1}{4}(\gamma+7)]} H^{\frac{1}{2}(\gamma+1)}. \tag{18.25}$$

Therefore, the final equation in this case becomes

$$\varepsilon_I(\nu) = b(\gamma)\frac{e^3}{mc^2}\left(\frac{3e}{4\pi m^3 c^5}\right)^{\frac{1}{2}(\gamma-1)} H^{\frac{1}{2}(\gamma+1)} K \nu^{\frac{1}{2}(\gamma-1)} \tag{18.26}$$

(values of $b(\gamma)$ are also given in Table 18.1). The Stokes parameters, after averaging over the angle χ, go to zero, and consequently the degree of polarization is zero. Certain particular cases of the geometry of the magnetic field and of the distributions of electron velocities, giving lower polarization than Eq. (18.23), are discussed by A. A. Korchak and S. I. Syrovatskii [314].

Nonthermal magnetobremsstrahlung, as well as the thermal component, can be absorbed by the interstellar medium. The absorption coefficient for free-free transitions was determined in Sec. 4. However, we can now determine absorption of radio emission by magnetic damping [5, 318], although this absorption (so-called reabsorption) has not much significance in interstellar space.

To determine the reabsorption coefficient, it is convenient to derive the Einstein coefficients for spontaneous magnetobremsstrahlung A_{21}, induced emission B_{21}, and absorption B_{12}. Using the known relation between these coefficients, we can write

$$A_{21} = \frac{8\pi h\nu^3}{c^3} B_{21} = \frac{8\pi h\nu^3}{c^3}\frac{g_1}{g_2} B_{12} = \frac{P_I(\nu, \mathscr{E})}{h}, \tag{18.27}$$

where $P_I(\nu, \mathscr{E})$ is the function (18.8) as before, and g is the statistical weight proportional to $\mathscr{E}^2$. Using these values, the expression for the absorption coefficient per unit volume is written in the form

$$\kappa(\nu) = \frac{1}{c}\int_{h\nu}^{\infty}\left[B_{12}N(\mathscr{E} - h\nu) - \frac{g_2}{g_1}B_{21}N(\mathscr{E})\right]h\nu\, d\mathscr{E}$$

$$= \frac{c^2}{8\pi\nu^2}\int_0^{\infty}\left(-\frac{dN}{d\mathscr{E}} + \frac{2N}{\mathscr{E}}\right)P_I(\nu, \mathscr{E})\, d\mathscr{E}, \qquad (18.28)$$

since $h\nu \ll \mathscr{E}$. Substituting expressions (18.8) and (18.19) into (18.28), we obtain

$$\kappa(\nu) = c(\gamma)\frac{e^3}{6\pi m}\left(\frac{3e}{2\pi m^3 c^5}\right)^{\gamma/2} K H_{\perp}^{\frac{1}{2}\gamma + 1}\nu^{\frac{1}{2}(\gamma+4)}, \qquad (18.29)$$

where

$$c(\gamma) = \frac{3\sqrt{3}}{4}\Gamma\left(\frac{3\gamma + 2}{12}\right)\Gamma\left(\frac{3\gamma + 22}{12}\right). \qquad (18.30)$$

This function depends only weakly on γ. For a change in γ of from 1 to 5, the function $c(\gamma)$ is changed from 2.88 to 2.17. Transforming Eq. (18.9), we have

$$\kappa(\nu) = c(\gamma) \times 6.5 \times 10^{-3}(3.5 \times 10^9)^{\gamma} K H_{\perp}^{\frac{1}{2}(\gamma+2)}\nu^{-\frac{1}{2}(\gamma+2)}. \qquad (18.31)$$

It is curious that the ratio $\varepsilon(\nu)/\kappa(\nu)$, which in a system in thermodynamic equilibrium is proportional to ν^2, here is given by

$$\frac{\varepsilon(\nu)}{\kappa(\nu)} = \frac{a(\gamma)}{c(\gamma)}\frac{6\pi}{2^{\gamma/2}}\left(\frac{4\pi m^3 c}{3eH}\right)^{1/2}\nu^{5/2}. \qquad (18.32)$$

In small, but intense, synchrotron sources, the role of reabsorption may be significant. It follows from Eq. (18.32) that this absorption is the principal effect at long wavelengths where the cutoff in the spectrum occurs. In other words, if reabsorption is present, there must be a maximum in the synchrotron spectrum. It is not difficult to obtain a formula that connects the frequency of the maximum (ν_m) with the angular dimension of the source θ. In fact, we have for the frequency ν_m the obvious condition $\kappa(\nu_m)R \approx 1$, where R is the linear dimension of the source, and for the flux at that same frequency the order-of-magnitude quantity $F_{1/2} \approx \varepsilon(\nu_m)R\theta^2$. From this we obtain $\theta \approx [F(\nu_m)\kappa(\nu_m)/\varepsilon(\nu_m)]^{1/2}$, or, from Eq. (18.32) after some numerical substitutions [314a],

$$\theta \approx 4.3 \times 10^{16}[F(\nu_m)\nu_m^{-5/2}H^{1/2}]^{1/2}\left(1 + \frac{\Delta\lambda}{\lambda}\right)^{1/4} \text{ rad.} \qquad (18.33)$$

The flux F is expressed in watts per square meter per cycle per second. The factor $(1 + \Delta\lambda/\lambda)^{1/4}$ is added to take into account the decrease in flux due to the red shift. The accuracy of the angular dimensions determined from this formula is usually sufficient, since the unknown parameter H enters as $H^{1/4}$. If the flux in the region of the maximum of the spectrum is undetermined, but the magnitude of the flux is known at other frequencies $\nu_1 > \nu_m$ where reabsorption is small, it is then possible to modify the formula (18.33) by replacing $F(\nu_m)$ by $F(\nu)(\nu/\nu_m)^\alpha$, where α as usual is the spectral index. Using these formulas, Slish has determined the angular dimensions of a number of sources. For example, the source 3C48 has an angular diameter of $0''.14$. A maximum is not observed in the spectra of the majority of radio sources, but there are a number of sources with "bulges" in their spectra. Applying Eq. (18.33) to such sources (for example, CTA 21 and CTA 102), we find angular diameters possible up to $0''.01$.

The question arises why, in a system consisting of a very "hot" gas, self-absorption occurs with an intensity that is considerably less than that given by the Planck curve. As is seen from Eq. (18.28), absorption depends not on energy, but on the slope of the $N(\mathscr{E})$ curve. Therefore, if the distribution is not Maxwellian, the equilibrium intensity is determined not by "kinetic temperature" but by "color temperature," as it were, which is not high for a power-law spectrum.

The Intensity of Nonthermal Radio Emission

If we exclude very long waves ($\lambda \gg 10$ m) and individual dense H II clouds, the Galaxy is transparent to radio emission. Therefore, by direct integration of the emission coefficient along the line of sight, we obtain for the observed intensity of nonthermal radio emission the expression

$$I_\nu = \int_0^R \varepsilon_I(\nu)dr = a(\gamma) \times 0.4(2.8 \times 10^8 \lambda)^{\frac{1}{2}(\gamma-1)} K H_\perp^{\frac{1}{2}(\gamma+1)} R, \quad (18.34)$$

or a corresponding expression with $b(\gamma)$ instead of $a(\gamma)$ and H instead of $H_\perp$ in the case of the isotropic distribution of field directions. The distance R to the galactic boundary is here expressed in kiloparsecs.

We suppose that the intensity of the nonthermal emission is $I(\nu) \sim \nu^{-0.5}$. Hence, $\frac{1}{2}(\gamma - 1) \approx 0.5$ and $\gamma \approx 2$. Thus, Eq. (18.34) takes the form

$$I(\nu) \approx 10^3 \lambda^{1/2} K H_\perp^{3/2} R \text{ erg/cm}^2 \text{ sec ster Hz}, \quad (18.34')$$

which corresponds to $T_b(\nu) = 3.5 \times 10^8 \lambda^{5/2} H_\perp^{3/2} KR$ deg. By means of Eq. (18.34′) we can determine from the observed intensity or brightness temperature either $H_\perp$ or the energy density of cosmic rays or, finally, the dimensions of the Galaxy. Naturally, in order to find one of these parameters, the other two must be given.

For the index $\gamma = 2$, the approximate formula for the spectrum of the relativistic electrons has the form

$$N(\mathscr{E})d\mathscr{E} \approx \frac{K d\mathscr{E}}{\mathscr{E}^2}. \quad (18.35)$$

The energy density of the relativistic electrons in the range $\mathscr{E}_1$ to $\mathscr{E}_2$

$$\int_{\mathscr{E}_1}^{\mathscr{E}_2} N(\mathscr{E})d\mathscr{E} = K \ln \frac{\mathscr{E}_2}{\mathscr{E}_1}. \tag{18.36}$$

The spectrum (18.35) does not qualitatively contradict the anticipated spectrum of relativistic secondary electrons, since the index of a proton spectrum is almost 2. Assuming that the energy density of relativistic electrons amounts to 1 percent of the energy density of relativistic protons (~ 1 eV/cm^3) and that their spectra are similar, we find for the value of the parameter K (assuming ln $\mathscr{E}_2/\mathscr{E}_1 \approx 3$),

$$K \approx 5 \times 10^{-15} \text{ erg/cm}^3 \approx \tfrac{1}{3} \times 10^{-2} \text{ eV/cm}^3. \tag{18.37}$$

If we assume that $H_\perp \approx 3 \times 10^{-6}$ Oe, it then follows from Eq. (18.34′) that

$$T_b \approx 10(\lambda \text{ m})^{5/2} R, \tag{18.38}$$

that is, of the order of 30,000° at $\lambda = 10$ m and $R = 10$ kpc, which agrees approximately with observations. We note that when $H_\perp \approx 3 \times 10^{-6}$ Oe in the frequency range 300–1000 MHz, the radiation is primarily from electrons with energies

$$1.4 \times 10^3 < \mathscr{E} < 2.5 \times 10^4 \text{ MeV}. \tag{18.39}$$

As we have already remarked, the spectral index is not a constant over the whole observable spectrum. Therefore, the true distribution of relativistic electrons in the Galaxy cannot be described by a single function (18.35) or, in general, in the form (18.19) for a sufficiently large energy interval. The simplest refinement rests on the assumption that the real electron distribution is described by two functions of the type (18.19) matched at a certain energy. For example, it was assumed in [314b] that for electrons radiating at frequencies greater than 150 Mc/sec the index is $\gamma = 2.8$, and for electrons with lower energies $\gamma = 1.3$ in the halo and 1.8 in the disk. In principle, one could construct from more precise observations the true electron-energy spectrum, which may differ in various parts of the Galaxy. However, such data are not at present available, and Eq. (18.38) should be considered only preliminary.

The Distribution of Nonthermal Radio Emission

By constructing isophots of galactic radio emission (lines of equal intensity) we can determine the dimensions of the system, as well as relative density changes of the relativistic particles and the distribution of magnetic fields. A very complete survey of a similar kind was carried out by Mills [16:422] by observations with his so-called cross at $\lambda = 3.5$ m. A qualitative picture of the distribution of continuous nonthermal cosmic radio emission, except for discrete sources, which will be discussed in Sec. 20, may be described in the following way.

The general radio emission from the Galaxy is divided into radiation from the disk near the galactic plane and radiation from the corona and extragalactic space. Such a separation was first made by Shklovskii in 1952. To determine the radiation from the corona, one must isolate the extragalactic background. To treat it as isotropic is not reliable, since the corona may be elongated and somewhat irregular. The extragalactic background may be obtained from the observations of the nebula 30 Doradus in the Large Magellanic Cloud. This large and dense nebula is opaque at long wavelengths, but its thermal radiation can be easily calculated. The difference in brightness between this nebula and neighboring parts of the sky gives, according to Shane [16], the value $T_b \approx 250^\circ$ at the wavelength $\lambda = 3.5$ m. However, this value seems unrealistically high and is not confirmed by other data, although measurements of 30 Doradus are unfortunately few in number.

Recently Baldwin [320a] has shown that at $\nu = 178$ MHz the smooth radio emission of the Galaxy may be represented by the expression $T_B = 7^\circ \sec b + 4^\circ$. Baldwin calculated first the composition due to the disk, and then that of the extragalactic background (at a wavelength of 3.5 m this quantity corresponds to 26°), and on the basis of this argument he concluded that the Galaxy has no corona. However, there is insufficient evidence for such a conclusion. The corona may be elongated or have a shell-like structure with an increase in brightness toward the periphery. In addition, it is not clear from the theoretical point of view whether there is in general strong radiation from extragalactic space in addition to the background caused by radio galaxies and other sources.

In spite of what has been said, we shall assume that the Galaxy has a corona with a brightness of about 25° at a wavelength of 3.5 m. This value is substantially less than that assumed earlier from the data of Mills [16]. This means that the intensity of the field and the concentration of relativistic electrons in the corona are considerably less than in the disk, that is, $H \approx 10^{-6}$ Oe.

In radio emission from the disk a certain concentration is observed toward the galactic plane and toward the center of the Galaxy. The total radiation from the disk at $\lambda = 3.5$ m is approximately 10^{28} erg/sec Hz. The main feature of the distribution of brightness with galactic longitude is its steplike character (Fig. 49). If we consider the emission to be isotropic, we can change from a brightness to a volume emission coefficient. Distinct maxima are obtained at distances of 0.53 R_0 and 0.78 R_0 and less distinct ones at 0.98 R_0, 0.26 R_0, and 0.38 R_0 (R_0 is the Sun's distance from the galactic center). The position of the maxima evidently agrees with tangents to the spiral arms. However, possibly because of errors in estimating the distance, the agreement with $\lambda = 21$ cm data is rather poor. Furthermore, the maxima at $l = 280^\circ$ (0.98 R_0), 337° (0.38 R_0), and 40° are of a thermal nature [317]. However, we can consider that the nonthermal emission of the disk is closely linked with the spiral arms. We should remember, incidentally, that supernovae of type II, which produce relativistic electrons, are also concentrated in the arms. The distances of the maxima from

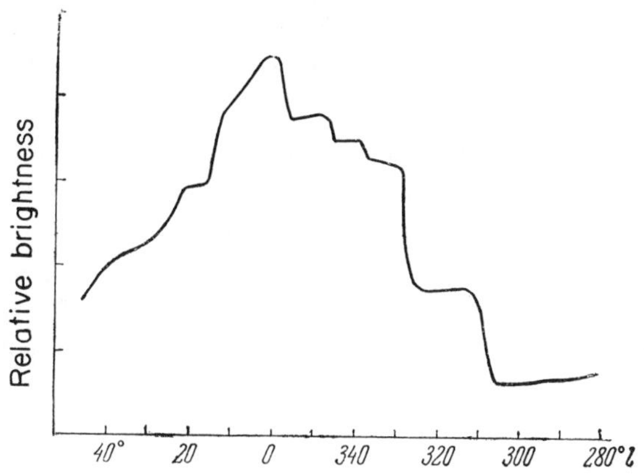

FIG. 49. Brightness distribution of nonthermal radio emission along the galactic equator [16:422].

the galactic center are multiples of 0.13 R_0 (since the Galaxy is a spiral of type $R \approx 0.13\ R_0\phi/2\pi$). The coefficient of nonthermal radio emission in the disk is ≈ 2 deg/pc. In the galactic center, a source of nonthermal emission is observed whose diameter is about 300 pc and whose thickness is 130 pc.

Observations with large dispersions have shown that the galaxy Andromeda has a corona. Here there is an appreciably complex structure. Four projections, resembling galactic spurs, issue from its central part in various directions. The radio center is displaced 10′ northwest of the optical center of this galaxy [321b]. The dimensions of the corona are 100 × 60 kpc at frequencies of 38 and 178 MHz [321], and 80 × 35 kpc at 1417 MHz [321a]. Coronas have been observed around a few other galaxies, but in general they are not present. In particular, many galaxies have been identified in which the region of radio emission is considerably smaller than the region of optical radiation.

The distribution of nonthermal radio emission characterizes the change from place to place of the magnetic field intensity and the concentration of relativistic electrons, or, more accurately, the change in the quantity $KH_\perp^{(\gamma+1)/2}$. Since $KH_\perp^{3/2}$ is 10–20 times less in the corona than in the disk, and since H is evidently less there also, the concentration of relativistic electrons in the corona must be several times less than close to the galactic plane.

Relativistic electrons are evidently formed in type II supernovae, concentrated in the spiral arms, from where they diffuse into the corona. If the electrons are secondary, then an equally large part of them are formed in the arms, where a considerable amount of the galactic gas is concentrated. Therefore, the increase of radio emission in the arms is due, not only to intensification of the field there, but also to the greater number of relativistic electrons. The thickness of the arms, according to radio-emission data, is 500 pc, which is twice the thickness of the gas layer. Therefore, it is possible that the bulk of the emission from the disk is emission from regions of a magnetic field both in and around the arms. We note that, since the nonthermal radio emission from a relativistic particle is proportional to $H_{\perp}^{(\gamma+1)/2} \approx (H \sin \theta)^{(\gamma+1)/2}$, when the particle moves along the field it does not emit. Since a particle emits in the direction of motion, we cannot "see" emission from a region where **H** is directed toward the observer. If the spiral arms represent the true lines of force of a magnetic field, the brightness distribution of the disk for $\gamma > 1$ must not be steplike, but more sloping. The observed steplike distribution could correspond to the case $\gamma = 1$ on the assumption of a homogeneous field. In fact, in this case the emission is proportional to $(\sin \theta)^{(\gamma+1)/2} = \sin \theta$, that is, inversely proportional to the length of the line of sight in the arm. The intensity here does not depend on θ, which might give a step during the transition from one arm to the other. Since γ is known to be greater than unity, we have to assume that the field in an arm is not homogeneous [323], and therefore that the intensity of the radio emission, even when the line of sight is tangential to the axis of the arm, does not fall to zero. In practice, even the observed dispersion of interstellar polarization directions, 5–7°, is sufficient to explain the steplike distribution.

If we assume the adiabatic invariant to be conserved, that is, if we assume that during the transition from a strong field in the arms to a relatively weak field in the corona, we fulfill the condition $\sin \theta \sim H^{1/2}$, then $H_{\perp} = H \sin \theta \sim H^{3/2}$, and the observed radio emission intensity is $I_{\nu} \sim H^{3(\gamma+1)/4} = H^{2.25}$, which can also explain why emission in the corona falls tenfold with a threefold reduction in the field.

The Polarization of Nonthermal Radio Emission

In addition to the spectrum and intensity distribution of nonthermal radio emission throughout the celestial sphere, its degree of polarization is also of interest. According to Eq. (18.23), the degree of polarization of magnetobremsstrahlung perpendicular to a homogeneous field must be considerable. When $\gamma = 2$, the value of p amounts to 69 percent, while the electric vector of the principal direction must be perpendicular to the projection of the magnetic field on the celestial sphere. If the magnetic field is only partially homogeneous, the degree of polarization is consequently reduced. In particular, if the field homogeneity is low, polarization of nonthermal radiation is proportional to the ratio of the squares of the homogeneous and random constituents of the field [314].

The Faraday effect is another reason for a decrease in the observed degree of polarization of nonthermal radio emission. The interstellar medium with a magnetic field is a magnetoactive plasma, in which, as we know, the plane of polarization rotates. The plane of polarization of a radio wave traversing a distance R turns through an angle

$$\psi = \frac{1}{2}\frac{4\pi N_e e^2}{m\omega^2}\frac{eH_{\perp}}{mc}\frac{R}{c} = 8 \times 10^5 N_e H R \lambda^2 \cos \theta. \qquad (18.40)$$

Here θ is the angle between the direction of wave propagation and the magnetic field; the angle of rotation ψ is measured in radians, the wavelength in meters, and the distance R in parsecs. It is convenient to introduce here the rotation measure $\mathrm{RM} = \psi/\lambda^2$ rad/m^2. The dependence of ψ on the path traversed means that, even in the case of a completely homogeneous field, the emission will be partially or completely depolarized because of the differences in rotation of the plane of polarization of the radiation at various points along the line of sight. In order to calculate the degree of polarization and to allow for this effect, we must integrate the Stokes emission coefficient $\varepsilon_U(\nu)$ over the line of sight, assuming the angle χ to be proportional to distance. Omitting the trivial calculations, we proceed immediately to the result. The degree of polarization p of magnetobremsstrahlung in an expanded, homogeneous magnetic field, uniformly filled with relativistic electrons, allowing for the Faraday effect, is given by the expression [324]

$$p = \frac{\gamma + 1}{\gamma + 7/3}\left|\frac{\sin 2\psi}{2\psi}\right|, \tag{18.41}$$

where the angle ψ is determined from Eq. (18.40) and R is the total extent of the emitting system. Finally, a further reduction in the degree of polarization is caused by the presence along the line of sight of several regions of the magnetic field that are differently aligned.

Depolarization due to the Faraday effect is readily separated from depolarization in a complicated field because of the strong dependence of the first on frequency. In particular, this effect is substantial only in the meter wave band. For example, when $N_e = 10^{-3}$, $H = 3 \times 10^{-6}$ Oe and $R = 1$ kpc, it follows from (18.40) that when $\nu = 10^3$ MHz, $\psi = 0.24$ rad, and when $\nu = 10^2$ MHz, the angle is already equal to 24 rad. In agreement with Eq. (18.41), when ψ is appreciably greater than unity, polarization decreases rapidly. We note that, if the observed degree of polarization does not change sharply with frequency, it means that the depolarizing Faraday effect is small ($\psi \ll 1$), so that, using Eq. (18.40), we can determine the upper limit of the electron concentration.

From what has been said, the polarization of interstellar nonthermal radio emission is probably small. In separate discrete sources the degree of polarization may be considerably greater, approaching the limit (18.23). Observations at $\lambda = 75$ cm [325] and also at $\lambda = 1.45$ m and 3.3 m [324] verify that, except for small portions, polarization of the galactic background is low. The polarization detected correlates on the whole with the interstellar polarization of star light. Regions of small dimensions with considerable polarization are evidence of the irregular distribution of N_e, which causes depolarization. Another possibility, that of a fine-scale structure of the field, seems less probable.

More than 90 percent of the polarized radiation comes from the region along approximately 50° of the great circle on the celestial sphere passing through the pole and intersecting the galactic plane at longitudes of about 340° and 160°, that is, in directions perpendicular to the axis of the arm (we recall that the latter runs from the constellation Cygnus, $l \approx 70°$, to the constellation Carina, $l \approx 250°$). These data were obtained from observations at 408 MHz [325c].

A comparison between observations at the frequencies 408 and 610 MHz allows an estimate from the Faraday effect of the value of the quantity $\overline{N_e HR}$. Close to the galactic plane we have $\overline{N_e HR} \approx -7 \times 10^{-7}$ Oe pc/cm^3 ($b = 8°$), while at high latitudes $\overline{N_e HR} \approx 53 \times 10^{-7}$ Oe pc/cm^3 ($b = 59\overset{\circ}{.}3$). Using this latter value (here there is a smaller probability of depolarization) we obtain for $R = 100$ pc and $H = 3 \times 10^{-6}$ Oe, the value $N_e = 2 \times 10^{-2}$ cm^{-3}.

19. The Spiral Arms, the Corona, and the Magnetic Field of the Galaxy

We have already encountered in previous sections certain properties of the spiral arms, primarily their radio emission at 21 cm and in the continuum. We have also noticed their connection with the magnetic field. We now discuss this question in more detail.

To begin with, the spiral arms are large-scale extensions of gaseous condensations. They are divided into finer details, clouds and complexes of clouds, but they present themselves somewhat as a whole, and not simply as an aggregate of clouds. Cold stars are also connected with the gas, about which more will be said in Chapter V.

The Magnetic Field and the Equilibrium of the Spiral Arms

Observations of the Zeeman effect in absorption at 21 cm show the presence of a magnetic field along the spiral arms. In addition to these direct observations, existence of the magnetic field along the arms follows from data on nonthermal radio emission and from interstellar polarization of stars. The magnetic field in the arms also manifests itself in the elongated form of diffuse emission nebulae, as well as in the form of dark nebulae. This problem was investigated by G. A. Shain [336, 337]. Nebulae that are expanding in strong magnetic fields must extend along the field. A comparison of the shape of nebulae with pictures of the magnetic field drawn from measurements of interstellar polarization (Fig. 50) in fact confirm this connection, so that we can conclude, first, that in the spirals there is a sufficiently strong magnetic field with a magnetic-energy density greater than the density of internal energy in the cloud, and second, that a regular field exists not only in clouds, but between them, so that a field external to the clouds defines their shape as if they were being stretched from within and leads to the orientation of dust particles. This means in particular that clouds do not rotate with respect to the arms.

From Fig. 50 and other data, we see that the magnetic field inside the arms is

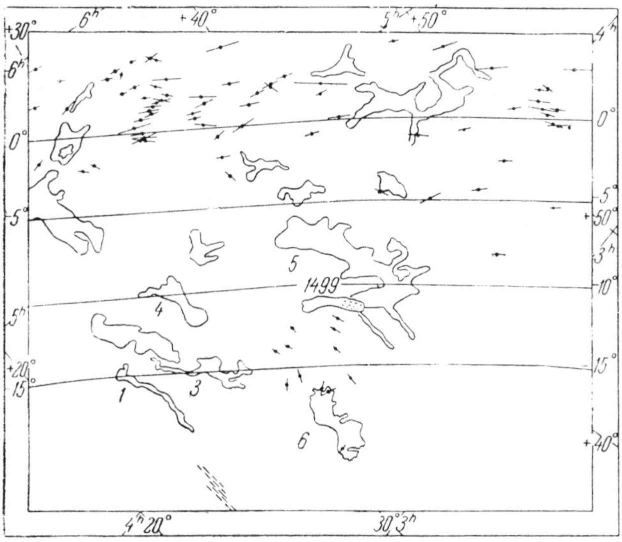

FIG. 50. A comparison of interstellar polarization and elongated nebulae in the Perseus–Taurus region [336].

not entirely homogeneous; its fluctuations measure from 10 to 100 pc. Essentially, the arm itself constitutes a very large fluctuation in the magnetic field. The need for a certain irregularity in the field in the spiral arms follows, as we noted before, from the dispersion of the angles of the plane of interstellar polarization and from the interpretation of the steplike distribution of nonthermal radio emission.

The magnetic field of the arms controls not only the expansion, but the motion of clouds. Analysis of the radial velocities of rapid clouds shows that velocities along an arm are, on the average, higher than across it [338].

The magnetic field is the basis for the existence of the arms. It guarantees their conservation as separate entities. Without the field the arms would first fragment, and then gather into denser condensations which would later transform into stars. The problem of the stability of the arms as gaseous cylinders was first considered by Chandrasekhar and Fermi [339] and then generalized in numerous works to include various effects. We give here only the basic results.

For stability to radial oscillations, while preserving the shape of the cylinder, the root-mean-square magnetic-field intensity must satisfy the condition

$$(\overline{H^2})^{1/2} \lesssim \pi D(\rho\rho_t G)^{1/2} = \tfrac{1}{2}H_s, \tag{19.1}$$

where D is the diameter of the cylinder, ρ the gas density, and ρ_t the total density of the gas and of the stars. When $\rho \approx 2 \times 10^{-24}$ g/cm^3, $\rho_t \approx 3\rho$ and $D \approx 500$ pc, we have from this equation $(\overline{H^2})^{1/2} \lesssim 5 \times 10^{-6}$ Oe. On the other hand, sinusoidal disturbances alter the shape and always render the cylinder unstable, while the wavelength of the disturbances causing instability increases with an increase in H. For example, when $H = \frac{1}{2}H_s$, disturbances with a wavelength of $\lambda > \pi D/0.480$ are unstable, while disturbances of $\lambda = \pi D/0.28$ have maximum instability. In the absence of a magnetic field, disturbances are unstable when $\lambda > \pi D/1.067$ and maximum instability is reached when $\lambda = \pi D/0.58$. The time required for a cylinder to break up into separate portions in the two cases is $(1/0.133)(4\pi G\rho)^{-1/2}$ and $(1/0.246)(4\pi G\rho)^{-1/2}$, respectively.

In this way, a gaseous cylinder in its own gravitational field as well as a magnetic field is always unstable, but the presence of the field increases the time of disintegration and the length of the pieces produced by disintegration. For example, when $H \approx 5 \times 10^{-6}$ Oe, the time of disintegration is $\sim 10^9$ yr and the length of the pieces formed by disintegration of an arm is $\sim 2.8 \times 10^4$ pc. Therefore, we may consider that the stability of the spiral arms and their preservation as linear connected filaments is due to the presence of the magnetic field.

Condition (19.1) describes essentially the equilibrium between gravitational and magnetic pressure. This means that the radius of the arm is not arbitrary, but is determined by the conditions of equilibrium. The inequality sign in (19.1) takes into account the possibility of gaseous and kinetic pressure (macroscopic motion of the clouds) in addition to the magnetic pressure. However, these effects are not significant in the spiral arms.

Observations show that the arm has a flattened form with its dimension in the z-direction less than those in the galactic plane. In addition, the thickness of the arm, and to a lesser degree its breadth, grows with distance from the galactic center (see Table 5.1). These peculiarities may be explained with the help of condition (19.1) [339a], while at the same time the influence of stars is taken into account. The predominant mass of stars in the Galaxy forms the intermediate and spherical subsystems with strong concentration toward the center. It is easy to show that, if the distribution of stars is symmetric about the axis of rotation, then they cannot influence the breadth of an arm in the galactic plane, but do determine its thickness in the z-coordinate. Therefore, the arms have a flattened form, particularly for $R < 10$ kpc, where the density of stars is greater than that of the gas. The thickness ΔZ of an arm is given to within a factor of the order of unity by the equation, similar to (19.1),

$$\Delta Z \approx \frac{H}{2\pi\rho G^{1/2}}\left(\frac{\rho}{\rho_t}\right)^{1/2}. \tag{19.2}$$

The quantity ρ/ρ_t strongly increases with approach toward the galactic center, going from 1 at $R = 15$ kpc to 100 at $R = 2$ kpc. If we assume that an arm consists of a bundle of lines of force of width d, then $H \sim \Delta Z^{-1}d^{-1}$, and consequently $\Delta Z \sim (\rho d)^{-1/2}(\rho/\rho_t)^{1/4}$. This formula explains qualitatively the data in Table 5.1 if we consider that ρ slowly increases toward the center, and that, because of the growth in magnetic pressure, the width d does not increase. In the outer regions the cross section of an arm becomes circular and expands because of the general decrease in the density of the gas.

The decrease in the cross section of an arm toward the galactic center is explained by the decrease there in the quantity of gas. In fact, the density of gas in the arms cannot be above a certain limit determined by the conditions of star formation.

One must keep in mind in an analysis of the equilibrium of an arm in a magnetic field that the density of the gas is not uniform, but is higher in clouds than the average value by an order of magnitude. The lines of force intersect many clouds consecutively, which primarily helps to stabilize the arms. Massive clouds deflect the lines of force, so that the magnetic pressure and intensity maintain them against the force of gravitation. We point out again here that the determination of H by the Zeeman effect and by polarization with respect to clouds, and its estimate from radio emission and from the distorted forms of nebulae, characterize only the average field in the arms.

The motion of clouds twists the lines of force, causing a restoring force. Therefore, we may suppose that the clouds oscillate within the limits of an arm. This explains the stability of the arms in spite of the large velocity dispersion. Stars formed in the arms, and, having the same dispersion as the clouds, are not restrained and leave the arms after approximately 50×10^6 yr. Deformation of the lines of force may be the reason for the observed dispersion of interstellar polarization of light (about 7 percent) and for the irregular fields which are postulated to account for the nonthermal emission along the arms.

We may suppose the following picture for the motion of clouds in the spiral arms. Magnetohydrodynamic waves are propagated along the arms, causing oscillations of clouds as if they were attached to lines of force. Because of these oscillations, clouds cannot rotate owing to the resistance of the magnetic field. The velocity of gaseous motion in these waves must be greater between clouds than inside them, because of the lower density. Probably this arrangement persists on a smaller scale for gaseous motion inside clouds. Thus, an arm is considered as a single dynamic system in which energy is transmitted along the field and is broken down (by dissipation) by smaller-scale motion. The properties of this "oscillatory" dissipation have so far not been altogether clear, but dissipation must occur, both as indicated by general observations of its character, if only on account of its partially irregular motion [340], and because the sources of energy for the motion of the interstellar medium, as we shall see below, are large-scale ones. We shall return to the problem of cloud motion in the following chapter.

The spiral arms and the magnetic field of the galactic disk must be closely associated with its rotation. It is quite clear that differential galactic rotation must lead to an elongation of the magnetic field of the disk. Possibly even the arms are formed in this manner. The nature of the time scale needed to elongate the field follows from the well-known Oort formula for radial velocities,

$$V_r = Ar \sin 2l \cos^2 b. \tag{19.3}$$

Differentiating Eq. (19.3) with respect to the radius vector r near the galactic plane ($b = 0$) and replacing $\sin 2l$ by the mean of its absolute value, we find for the characteristic time for winding up the field

$$t_3 \approx \frac{1}{dV_r/dr} = \frac{1}{A|\sin 2l|} \approx \frac{2}{A} \approx 10^8 \text{ yr}. \tag{19.4}$$

Here $A = 15$ km/sec kpc, Oort's constant for the solar neighborhood. In the central portion of the Galaxy, the time for winding of the field is much less.

Thus, elongation and coiling of the lines of magnetic force in the arms do indeed occur, but, on the other hand, it has so far not been possible to explain their shape more exactly, since the Galaxy has existed much longer than 10^8 yr and before this period the arms must have been in a process of transformation into more or less true circles. It is possible that either the age of the arms is much less than the age of the Galaxy, or the motion of the gas is not completely circular.

Up to the present time the spiral arms have been regarded as bundles of lines of force directed along the axis of the arm. Hoyle and Ireland [339b] have suggested that these bundles are twisted, that is, that the lines of force themselves are spirals tightly wound around the axis of the arm. However, such twisted arms have nevertheless a unidirectional magnetic field. Moreover, the situation is also possible where lines of magnetic force with opposing directions are found within the limits of a spiral arm. This possibility is closely connected with the origin of the galactic magnetic field and the spiral arms. For example, Heisenberg and Weizsäcker [339c] proposed that the arms form as condensations of gas with a random field. Differential rotation stretches these condensations, giving them their spiral form, but in this model the magnetic field remains random, although the magnetic lines of force form small angles with the axis of the arm. In another model, suggested in general form by Hoyle and developed in a series of papers [112, 339a, d, e], it is proposed that the galactic field is part of the extragalactic field that became contracted during the condensation of the Galaxy. In this case the arms are compressed bundles of the initial large-scale field in which the lines of force were unidirectional.

Structure of the Galactic Magnetic Field in the Solar Neighborhood

The Zeeman effect at 21 cm has already shown that the directions of the lines of magnetic force differ in different clouds. We note that no other method of investigating the magnetic field is able to measure its direction. However, the most complete information on the direction of the magnetic field will be obtained from an analysis of the Faraday effect. As will be shown in more detail in a later section, the radio emission from many galactic and extragalactic sources is linearly polarized. This radiation has passed through the plasma of the spiral arms, experiencing the rotation of the plane of polarization given by Eq. (18.40), where the sign of the rotation depends on the direction of the field.

A statistical analysis of the polarization of 37 sources, predominantly extragalactic sources located at high latitudes, has been carried out by Morris and Burge [339f]. The result of a determination of the "rotation measure" RM $= \psi/\lambda^2$ for these sources is presented in Fig. 51, where the quantity RM sin b is given separately for the northern ($b > 0$) and southern ($b < 0$) hemispheres.

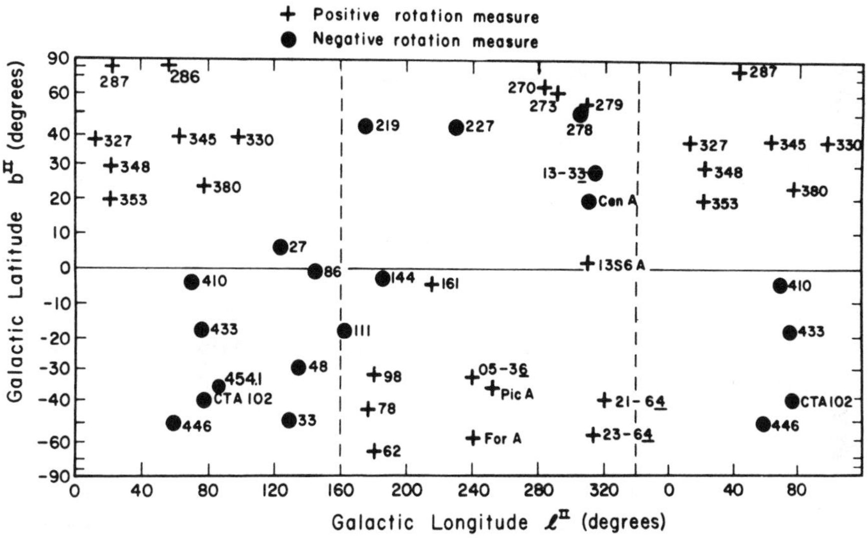

FIG. 51. The variation of the sign of the rotation measure with galactic coordinates: $+$RM > 0; ● RM < 0. The sources are identified by their numbers in the 3C catalog or the Mills, Hill, and Slee catalog, or, in the case of the most intense sources, by constellation.

Looking at this graph, we can draw the following conclusions. First, RM sin b changes but little in regions of different longitude and on the average is 10–20 rad/m^2. This means that RM $\sim$ cosec b, that is, that rotation actually takes place, although in a relatively thin layer. Second, the sign of RM sin b does not change on crossing the galactic plane, while, of course, sin b does change sign. This means that the rotation measure, and consequently the direction of the magnetic field, change signs. Thus, the magnetic fields on opposite sides of the galactic plane oppose one another. Third, the sign changes in the lower hemisphere at $l \approx 160^\circ$ and 340°. This means that the magnetic field is directed there toward longitudes $l \approx 160^\circ - 90^\circ = 70^\circ$, which corresponds with other determinations of the directions of the spiral arms (in the constellation Cygnus).

In the northern hemisphere the change takes place at $l \approx 250°$; here the direction of the magnetic field is the reverse.

Using Eq. (18.40) and knowing that RM $\sin b = 10$–20 rad/m^2, we can estimate the average value of the field strength. Setting $R \sin b \approx 100$ pc, we obtain $\overline{N_e H} \approx 2 \times 10^{-7}$ Oe/cm^3. If $N_e \approx 2 \times 10^{-2}$ cm^{-3}, then $H \approx 10^{-5}$ Oe.

In the paper [339g] the average value of the rotation measure was also determined. The larger value of RM ≈ 60 rad/m^2 was obtained.

Thus, the investigation of the rotation of the plane of polarization of radio emission leads to a very important result of great cosmogonic interest. The fact that the directions of the magnetic field are opposite on opposite sides of the galactic plane answers some of the questions connected with the formation of the Galaxy. These questions will be discussed in more detail in Sec. 25. Formation of interstellar gas clouds may also be connected with this peculiarity of the galactic magnetic field [339a].

Consider the condition about a neutral surface dividing two fields of different signs. The gradient of magnetic pressure cannot support a gas in this case, but, on the contrary, compresses it. Equilibrium is established when the gas pressure close to the neutral surface is of the same order of magnitude as the magnetic pressure in the arms. At $T \approx 100°$ this corresponds to a density $N \approx 300$ cm^{-3}. The thickness of the layer of compressed gas is so small that considerable diffusion of the field through the gas and annihilation of the field close to the boundaries of the division take place in a cosmogonic interval of time. The thickness of the compressed gas layer without the field grows with time, and after 10^{10} yr it may attain some tens of parsecs.

Disturbances of the compressed layer by hot stars found in H II zones, by supernova outbursts, and by other phenomena lead to the ejection of clouds of dense gas without a field into the layers of the arms. The external field penetrates into the cloud because of "ripple" instabilities. These are connected, apparently, with the finely rippled structure of interstellar clouds, particularly noticeable, for example, in reflection nebulae [339a]. According to this picture, the field in the clouds must be of the same order of magnitude as that between the clouds. This explains why synchrotron radio emission is formed at all layers of the arms.

We must again remind ourselves of the deviation of the plane of symmetry of the gas distribution from the plane of rotation in the direction of the Magellanic Clouds and in the opposite direction (Sec. 5). It is impossible to explain this feature by the tidal action of the Magellanic Clouds, since their mass is small. The tension of the lines of force connecting the galaxies may be acting here [112]. Moreover, we might suppose [331, 335] that the departure of gas from the galactic plane is connected with the action of the extragalactic medium by the motion of the galactic corona through this medium. In fact, during the motion of the Galaxy relative to the extragalactic medium, circulations are set up in the corona that mix gas from the plane of symmetry.

The Galactic Corona

The spherical component of nonthermal galactic radio emission is formed in the extended regions surrounding the disk and has a quasispherical form. The necessity of the presence of such a corona also follows from an analysis of the conditions for confining cosmic rays of high energy [326].

Knowing the approximate intensity of the radio emission from the corona and its radius, we can estimate from Eq. (18.34) the quantity $KH_{\perp}^{(\gamma+1)/2}$. This quantity is several hundred times less in the corona than in the disk. If we assume that the basic part of relativistic electrons issue from the disk while the adiabatic invariant is conserved, their concentration is by Eq. (17.8) $\sim 1 - (1 - H/H_0)^{1/2}$, and the value of $H_{\perp}$ is by Eq. (17.7) $\sim H^{3/2}$. Therefore, $KH_{\perp} \sim H^{(7+3\gamma)/4} \approx H^3$. From this it follows that the field intensity in the corona is in all 5–8 times less than in the disk.

An investigation of the intensity of the nonthermal radio emission does not yield information about the character of the magnetic field of the corona. In principle it may be random, as well as regular. An assumption of a regular field blending into the extragalactic field runs into several difficulties. Such a field cannot contain cosmic rays within the Galaxy. The magnetic field of the corona is probably twisted by the rotation of the disk. In this case it forms a spiral with some tens of loops over and under the disk. However, such a field can hold relativistic particles for several million years. The character of the field in the corona remains an open question at the present time.

The magnetic field of the corona must be connected with the gas. Various effects may be used to estimate the density of this gas, such as the condition for hydrostatic equilibrium [327], the depression of the spectrum at high frequencies (Eq. (18.17)), and depolarization. The condition for hydrostatic equilibrium must include, in addition to the gas, the dynamics, magnetic effects, and pressure of the cosmic rays. This condition must also be satisfied approximately for an expanding corona. All methods accurate to within half an order of magnitude lead to concentrations of 2×10^{-3} cm^{-3}. The temperature of the corona may be of the order of $10^{6\circ}$ [328].

There are some recent investigations of the structure of the galactic magnetic field. Gardner and Davies [328a] found a region of relatively strong magnetic field near the galactic plane. Bologna *et al.* [328b] showed that the depolarization of extragalactic sources increased toward the galactic plane and with increase in angular size of these sources. It may be explained as the existence of small fluctuations of the galactic magnetic field with dimensions ~ 0.3 pc and a value of the parameter of depolarization $N_e H \approx 3 \times 10^{-6}$ G/cm^3. Even smaller fluctuations were suggested by Maltby [328c].

The irregular structure of the galactic magnetic field was investigated by Pronik [328d]. She compared the polarization both with data on interstellar clouds and with the shape of dark nebulae. The magnetic field turned out to be irregular; often its direction is different in the two hemispheres of the Galaxy, sometimes being parallel neither to the galactic plane nor to the spiral arm.

From all this it follows that now it is difficult to form a definite picture of the galactic magnetic field. The simple assumption that the lines of magnetic force are only parallel to the spiral arms and are stretched by galactic rotation does not explain many facts [328e]. It may be assumed [328f] that there is a weak homogeneous field ($H \approx 5 \times 10^{-6}$ G) superposed on a strong fluctuating field (with dimensions $\sim$0.1 pc), but in this case there would be high-velocity motions and very dense fluctuations connected with magnetic forces. Moreover, the rate of dissipation of the field would be too high [328g].

The Boundaries of the Corona and the Intergalactic Medium

The solution of the following problem has important consequences. Is the corona confined by a closed magnetic field and sharp boundaries, or does it gradually change into the intergalactic gas? We may assume that the corona has no sharp boundary. In fact, in a closed, static corona the density of cosmic rays is only a little less than their density in the disk, and therefore the cosmic-ray pressure gradient close to a sharp boundary would be large and there is nothing to balance it, since there is relatively little gas in this region. With uniform density, the force of gravity is also uniform and its equalizing gradient must be small everywhere [330]. Hence we conclude that the galactic corona cannot have sharp boundaries and must gradually blend into intergalactic space.

In this connection, we may try to relate the corona with the intergalactic medium, or, more accurately, with the medium in the Local galactic system. It is well known that our Galaxy, the Andromeda nebula, the Magellanic Clouds, and a number of other galaxies (about 16 in all) form a tight group with a radius of about 600 kpc. The intergalactic medium in the Local system, as indeed in extragalactic space, is not observed directly, but its existence is conjectured from the following considerations put forward by Kahn and Woltjer [331]. (An obvious exception is the absorbing cloud in the constellation Microscopium [332]. It strongly reduces the number of visible galaxies but not the number of distant stars of the spherical subsystem. Absorption in the cloud is $A_v \approx 0^m\!.5$–$1^m\!.2$, while in size and shape it is similar to the Magellanic Clouds. It belongs to the Local system.) The two major members of the Local group, our Galaxy and the Andromeda nebula, together with their satellites, have a combined mass of $10^{11} + 4 \times 10^{11} \approx 5 \times 10^{11}$ solar masses. This amounts to approximately 80–90 percent of all the galactic mass of the Local group, which therefore is about $6 \times 10^{11}\ M_\odot$. At the present time, the two galaxies are converging. Assuming that they originated together, then separated, and are now drawing together again (they oscillate around a common center of mass with a period not greater than 15×10^9 yr), we can determine the mass of the whole system, which is approximately equal to $18 \times 10^{11}\ M_{\odot}$. The required excess mass is evidently dispersed in the form of the intergalactic medium. Another consideration leads to the same result; the kinetic energy of the Local system of galaxies is greater than 12×10^{57} erg, while the gravitational energy (if we ignore the intergalactic medium) is less than 6×10^{57} erg. According to the virial theorem, there must be a difference between these two energies of not less than 24×10^{57} erg. Thus, in the Local system of galaxies, there exists an intergalactic medium of total mass around $12 \times 10^{11}\ M_{\odot}$ and with a mean density of about 10^{-4} cm^{-3}.

From the conditions necessary for the formation of galaxies from the extragalactic medium during gravitational instability, and assuming that the intergalactic rarefied gas cools very slowly, we can estimate the temperature of the medium to be 500,000° [331].

From an analysis of the properties of cosmic rays, Sciama [333] suggested a model in which there is no galactic corona but a gradual transition from the interstellar to the intergalactic medium. Lines of magnetic force emerge from the galaxies, but join up in the Local system. There is free exchange of cosmic rays between galactic and surrounding space, but the escape of particles from the Local system into extragalactic space must be difficult, since otherwise the well-known difficulty of the high total energy of cosmic rays arises.

In this model, difficulties in explaining the composition of cosmic rays do not arise; most of the time the particles exist in the rarefied extragalactic medium, where changes in chemical composition occur slowly. According to this hypothesis, conditions in the "corona" (that is, close to the Galaxy) are approximately: $N \approx 10^{-3}$ cm^{-3}, $T \approx 1{,}000{,}000°$; the energy density of cosmic rays is $\sim 10^{-13}$ erg/cm^3, with the magnetic and kinetic energy $\sim 3 \times 10^{-14}$ erg/cm^3, and the thermal energy $\sim 3 \times 10^{-13}$ erg/cm^3.

This and similar hypotheses meet with the difficulty that the magnetic fields of clusters and groups of galaxies cannot be twisted or tangled within cosmogonic times; this will be discussed further at the end of Sec. 25. In addition to the stationary models of the corona discussed, nonstationary ones are possible. For example, the corona may consist of gas collected by the Galaxy from the intergalactic medium [334]. The idea was proposed that the corona was disintegrating as the result of an eruption of a mass of gas, that is, that it is a relatively transient phenomenon [343a]. In this connection, it is of interest that certain galaxies of class *Sc* do not have a corona, while the frequency of supernova outbursts in them is rather low. This means that the presence of cosmic rays alone is still insufficient to account for a corona. We shall speak about explosions originating in galactic centers in Sec. 25.

In postulating the existence of motion in the corona, we must indicate the sources of energy. Motion in the corona can be maintained by various mechanisms. For example, a disturbing effect is caused by waves from type II supernova explosions in the disk and from type I supernovae of intermediate distribution (between the spherical and plane galactic subsystems). The interaction of the rapidly rotating disk and the slowly rotating corona can also lead to disturbances in gas distribution, which are evidently observed as an intensification of radio emission in the vicinity of the disk. Motion in the corona can also be maintained by the interaction of gas from the corona with extragalactic media if their relative velocity is high [335]. Plasma instability, supported by the pressure of the field and cosmic rays, can also lead to vigorous motion [327a].

In addition to the thin and apparently ionized gas in the corona, there are separate denser clouds and unionized gas. In particular, Münch and Zirin [238a] have observed interstellar absorption lines in the spectra of distant stars at high galactic latitudes. Fast H I clouds were also observed in the corona, according to observations at $\lambda = 21$ cm [238b, c]. They were all observed in the small region of the sky $70^\circ < l < 165^\circ$, $+15^\circ < b < +50^\circ$. All these clouds have negative radial velocities in the interval from -20 to -174 km/sec. The average number of atoms in the line of sight in one cloud is 6×10^{19} cm^{-2}. In the work quoted various hypotheses connected with the formation of these clouds and their relation to the corona are discussed. However, definite conclusions about their nature, or even an estimate of their distance, could not be made.

In recent investigations of these clouds some new details have been found [340a, b, c]. They may be interpreted as a flow of neutral gas with mass $\sim 2\text{–}3 \times 10^3\ M_\odot$ from the region with galactic coordinates $l \approx 110\text{–}180^\circ$ and $l \approx +15^\circ$. The angular diameters of these clouds (or flows) are several degrees. There are such clouds or flows in the southern hemisphere of the Galaxy. Some possible explanations of this phenomenon are discussed by Oort [340d] and Shklovskii [340e] but up to now there is no satisfactory answer.

Coherent Synchrotron Radiation

In this whole section it has been assumed that the synchrotron radiation is not coherent, that is, that each relativistic electron emits radiation with an arbitrary phase. Therefore we add the intensity of radiation of each electron, not its amplitude.

But in some cases it is possible that synchrotron radiation may be coherent; then it is necessary to add amplitudes, not intensities, of the radiation of each electron. In other words, the intensity in this case is determined by induced, not spontaneous, radiation, as in the case of incoherent radiation.

Coherent synchrotron radiation is possible when there is a special distribution of energy of relativistic particles and in a plasma of a certain density.

Zeleznyakov [340f] noted that there is instability of a relativistic electronic gas to synchrotron radiation at the frequency $2\ ecN_e/H_\perp$ (compare Eq. (18.17)). Therefore, if there is the appropriate distribution of energy of relativistic particles and the optical depth to the radiation is great enough, this instability can lead to the generation of very strong coherent radiation. The necessary optical depth was calculated by Kaplan [340g]. For the operation of this effect it is necessary that the density of relativistic particles and electrons in the plasma be unusually high. The most efficient mechanism of coherent synchrotron radiation is the inverse Compton scattering of relativistic particles by excited plasma waves. This mechanism, first proposed by Kovriznych and Tsytovich [340h] was discussed in connection with cosmic radiation by Kaplan [340i]. This mechanism may be used for interpretation of very strong but small sources in quasars or the so called "point source" in the Crab Nebula.

20. Discrete Galactic Sources of Nonthermal Radio Emission

Besides the more or less continuous background of galactic nonthermal emission, separate, so-called discrete, sources of nonthermal radio emission are observed; they differ from discrete sources of thermal emission (emitting H II regions) in that their spectra resemble the spectra of galactic nonthermal radio emission. Most of them consist of galaxies with abnormally intense radio emission, while certain discrete sources are situated inside our Galaxy and are identified with gaseous nebulae, the remnants of supernova outbursts.

We shall first consider discrete galactic sources, the remnants of supernova outbursts, of which, up to the present time, more than ten have been positively identified [343b]. Their characteristic peculiarity is the presence of an envelope.

Table 20.1.

	Type I	Type II
Light	Red	Bluish
Spectrum	Wide, intense, unidentified bands	Emission lines of hydrogen and certain ions
Mass of star	Average (old stars)	Massive (young stars)
Mass of envelope (solar masses)	0.1	A few
Rate of dispersion of envelope (km/sec)	1000–2000	5000–7000
Interval between outbursts (yr)	400–600	100–200
Optical luminosity at maximum brightness (units of solar luminosity)	10^9–10^{10}	10^7–10^8
Emission energy during period of outbreak (erg)	10^{49}	10^{48}
Energy of envelope (erg)	10^{48}	10^{50}–10^{51}
Energy lost in the form of cosmic rays (erg)	10^{48}	10^{50}

All supernovae may be separated into two types according to their distribution in space. Type I supernovae constitute an intermediate system more concentrated toward the center, while type II supernovae belong to the plane component. Statistical data on these stars are given in Table 20.1.

Type II supernovae are observed mostly in class *Sb* spiral galaxies, less often in *Sc*; type I can occur both in spiral and in elliptical galaxies.

Of the nebulae remaining from type I outbursts, the one best studied is the well-known Crab Nebula (the discrete source is Taurus A); of the residues from type II supernovae, there is a young nebula (source, Cassiopeia A) formed about 250 yr ago, although the outburst itself was not observed. Another interesting nebula, the remains of a type II supernova outburst, is the so-called Loop Nebula in Cygnus. This nebula is older, the supernova outburst having occurred about 70,000 yr ago.

We shall restrict ourselves to a discussion of these three typical examples, together with the best-studied galactic sources.

The Crab Nebula

The supernova outburst that led to the Crab Nebula was observed in China and Japan in A.D. 1054. The structure of the nebula is well known; it consists of an external network of filaments with an amorphous mass inside (Fig. 52), which, however, is gradually becoming weaker as it escapes beyond the limits of the filamentary network.

The filamentary envelope has an oval form, with ratio of semiaxes equal to 1.5. The proper motions at the ends of the major and minor axes have been measured, and the radial velocity of the central parts of the filamentary structure shown on visual photographs of the nebula is 1150 km/sec. If we make some assumptions about the three-dimensional shape of the nebula, we can obtain from these data an estimate of its distance. For example, according to Woltjer, the Crab Nebula has a spindle-shaped form, and, therefore, the radial velocity should be associated with the proper motion at the ends of the major axis. The distance to the nebula is accordingly approximately 1500 pc, its major semiaxis is of the order of 1.3 pc, and the expansion velocity at the limits of this axis is approximately 1400 km/sec [353]. If, on the other hand, we assume the form of the nebula to be an oblate ellipsoid, then the measured radial velocity would correspond to the proper motion at the ends of the major axis, and the distance to the nebula would be 1.5 times less than in the former case.

The spectrum of the filaments is similar to the usual spectrum of gaseous nebulae; the forbidden lines [O II], [O III], [S II], and so forth are observed. Lines of neutral atoms, [O I] in particular, are also seen. Lines of the Balmer series are relatively weak. Because of the presence of lines of He I and He II in the spectrum, we can determine roughly the temperature of the filaments, which is probably about 15,000°. A temperature estimate based upon the

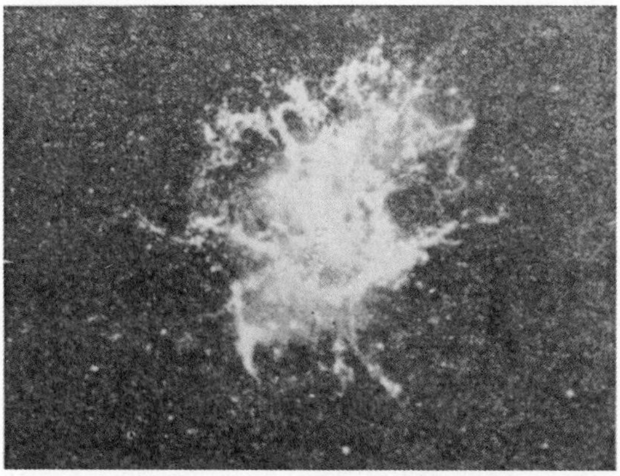

(*a*)

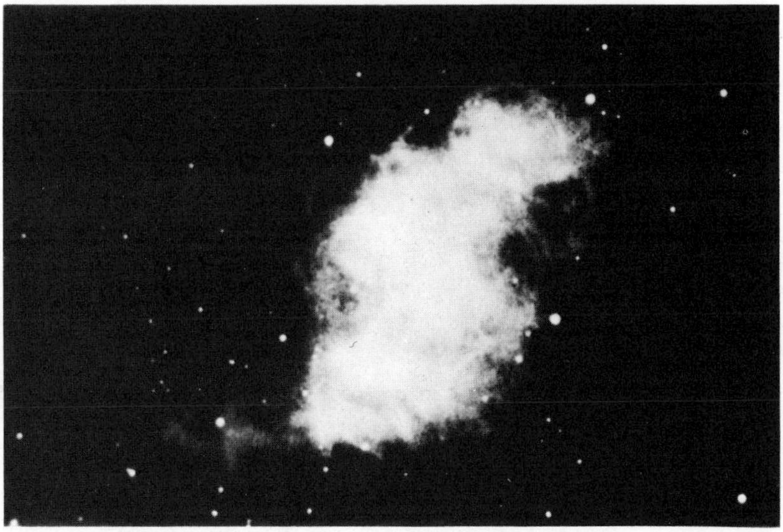

(*b*)

FIG. 52. The Crab Nebula in (*a*) H_α + (N II) lines and (*b*) the continuous spectrum [354].

intensity ratios of lines $\lambda = 4363$ Å and $N_1 + N_2$ also leads to a value of $\sim 17{,}000^\circ$ [353, 344]. The value of N_e in the filaments, determined from the intensity ratios of the components of line O II, $\lambda = 3727$ Å, varies between 550 and 3700 cm^{-3}; on the average it is $\sim 10^3$ cm^{-3} [345]. The chemical composition is apparently normal, although there is a slight deficiency of hydrogen. The mean ionization is single or double (hence there is one-third electron per unit atomic weight), and the gas density in the brightest filaments is approximately 5×10^{-21} g/cm^3. The mean length of filaments is about 0.7 pc, and their mean width is 3×10^{16} cm or 0.01 pc. The mass of a single filament is ~ 0.05 solar mass and the total mass of the filamentary envelope is of the order of 0.2 $M_\odot$. This value, obtained by summing the masses of the separate filaments found from their dimensions and density, agrees with an estimate of the mass from dynamic considerations (see below).

The spectrum of the amorphous portion is continuous, with no trace of spectral lines. The intensity increases to the red side, but it is difficult to investigate accurately the energy distribution in the spectrum because of interstellar reddening, which in this direction is poorly known. According to [345a], absorption in this direction is about $0^{\rm m}.7$/kpc, that is, from the Earth to the nebula $A_v \approx 1^{\rm m}.0$. If the spectrum of the amorphous region is to be described with the help of a spectral index, as was done for radio emission in the preceding sections, then, considering the effect of interstellar absorption, we must take $\alpha \approx 0.8$. The brightness of the amorphous portion falls off approximately twofold from the center to the periphery. On large-scale polaroid photographs (Fig. 53), we see that the amorphous mass also consists of comparatively regular filaments, although they are somewhat entangled [346]. The degree of polarization of an optical cross section near the center of the nebula is about 18 percent, while the polarization diminishes, on the average, toward the periphery. In certain small discrete areas, the degree of polarization rises to 70 percent, so that the picture of the amorphous portion of the nebula is very different accord-

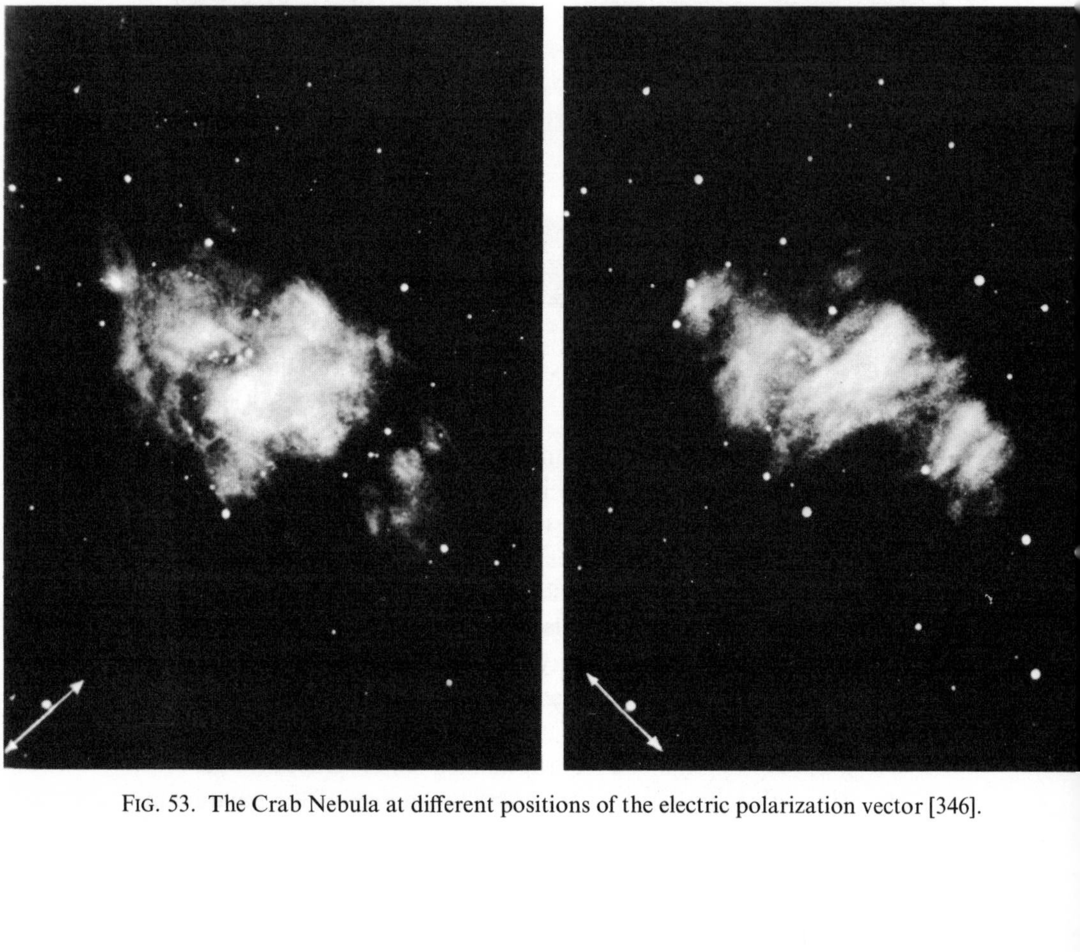

FIG. 53. The Crab Nebula at different positions of the electric polarization vector [346].

ing to the position of the polaroid. From the direction of the plane of polarization, we can estimate the direction of the magnetic field; the filament structure appears to coincide with the lines of magnetic force.

In the center of the nebula are two weak stars, the brighter of which is of sixteenth magnitude. The lines in the spectra of these stars are not unusual. One of these stars may possibly be the former supernova. Near the stars, and in approximately the same place, luminous plaits sometimes appear. They move through their own length, recede from the stars, and then disappear. The velocity of these plaits is about 0.1 the speed of light, and the interval between outbursts is about 3 months; no emission spectrum has been observed in the plaits. Smoother irreversible changes in brightness are observed in the amorphous portion, which alter its appearance in different localities [347]. No noticeable changes have been seen in the filaments.

The Crab Nebula is an intense source of radio emission. The dimensions of the nebula in radio emission at a wavelength of 9.1 cm are $4'.3 \times 2'.7$, with the position angle of the major axis at 149°, that is, approximately the same as in visible light. Although a certain tendency has been noticed toward an increase in dimensions with an increase in wavelength [122], it is not clearly defined. The radio-emission spectrum of the nebula follows a power law with spectral index $\alpha \approx 0.27$. The corresponding spectral index of relativistic electrons is $\gamma \approx 1.55$. The slope of the spectrum indicates that there are many high-energy electrons in the nebula. In 1953, I. S. Shklovskii showed that the optical emission also consisted of magnetobremsstrahlung from relativistic electrons with energies of 10^{11}–10^{12} erg. Both optical and radio emission can be combined in a single spectrum, given in Fig. 54.

Radio emission from the Crab Nebula is also polarized. In particular, at $\lambda = 3$ cm, polarization is about 10 percent and at $\lambda = 9.4$ cm, it is appreciably less, while in this case the plane of polarization is somewhat rotated [348]. At

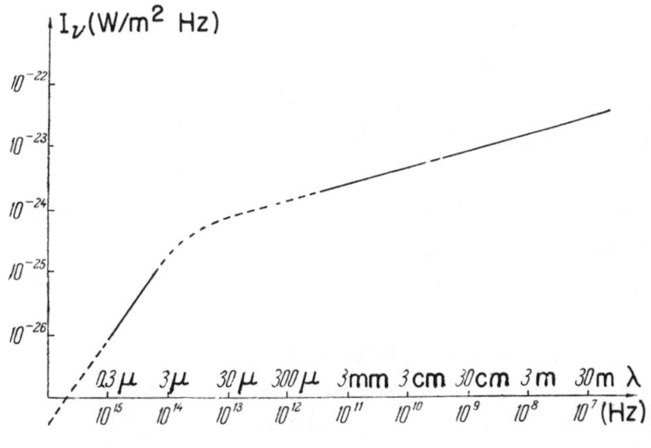

FIG. 54. Spectrum of nonthermal emission in the Crab Nebula [352].

$\lambda = 21.6$ cm the degree of polarization of the Crab Nebula is 1.5 ± 0.2 percent at a position angle of $92^\circ \pm 4^\circ$. The rotation of the plane of polarization may be explained by the Faraday effect in the nebula. Using Eqs. (18.40) and (18.41) and substituting $|\sin\psi/\psi| \approx 0.2$ at $\lambda = 10$ cm, we find $N_e H \approx 3 \times 10^{-4}$. If $H = 10^{-3}$ Oe, then in the amorphous part of the nebula $N_e \approx 0.3$ cm^{-3}. Consequently, its mass will be about $0.03\ M_\odot$.

We shall now proceed to the determination of the physical conditions, and in particular of the magnetic-field intensity, in the Crab Nebula. As we saw in Eq. (18.34′), magnetobremsstrahlung from a system of known dimensions allows us to determine the product $KH_\perp^{(\gamma+1)/2} \approx KH_\perp^{1.3}$ ($\gamma \approx 1.6$ for the Crab Nebula). To determine K and $H_\perp$ separately, we need one more relation. We usually invoke the condition of equality between the magnetic-energy density and the density of the total cosmic-ray energy [347]. Generally speaking, this assumption requires justification. We shall proceed from the following considerations.

In order to retain cosmic rays in a vacuum, a strong magnetic field must exist, but in the presence of gas the field may be weaker. In fact, because of the adherence principle, even a weak field couples the gas and cosmic-ray particles, and therefore cosmic-ray pressure may be compensated for by gravitational forces, as in the galactic corona, or by the inertia of the gas. However, in order that the system, including gas, cosmic rays, and magnetic field, may be more or less stable, the magnetic-field energy must not be less than the cosmic-ray energy. There is no absolute stability in this case, but the dimensions of unstable disturbances increase with increased field strength and the development of instability is retarded. We may draw directly from observations on the Crab Nebula conclusions about the approximate energy balance [330]. As observations on polarized optical emission show, the magnetic field seems to "protrude," although only slightly, from between the filaments. The lines of force bend the filaments and emerge in the form of arches. This also explains the somewhat larger dimensions of the luminous amorphous parts in comparison with the filamentary structure. The intensity of the field in the arches is less, and therefore the luminosity is also less there. Since the curvature of an arch is relatively low, it follows that on the one hand gas in the filaments would assume the cosmic-ray pressure and restrain the field, while on the other the magnetic-field energy is comparable with the cosmic-ray energy. If the magnetic field were weaker, the curvature of the lines of force between the filaments would be much more noticeable. Thus, the assumption of an energy balance between the different components of the Crab Nebula has received qualitative verification from observations.

However, a new unknown quantity now appears, the relative amount of cosmic-ray energy in the electrons, since K refers to the density of the relativistic electrons, while cosmic-ray energies are determined chiefly from protons and nuclei. We assume for a starting point that this energy is about 0.1 the energy of the relativistic protons and heavy nuclei, although we must bear in mind that it may yet be 10 times less.

The emission flux F_ν from the Crab Nebula at a frequency $\nu = 10^8$ Hz ($\lambda = 3$ m) is 1.9×10^{-20} erg/cm^2 sec Hz. The distance of the nebula is about 1500 pc. Hence, the total emission J_ν is $4\pi R^2 F_\nu \approx 5 \times 10^{24}$ erg/sec Hz. On the other hand, according to Eq. (18.22) for the total emission by the synchrotron mechanism, we have for $\gamma = 1.6$ and $a(1.6) \approx 0.16$,

$$J_\nu = 4\pi\varepsilon_\nu V \approx 5 \times 10^{-19} VKH_\perp^{1.3}, \tag{20.1}$$

where V is the volume of the emitting region. For the Crab Nebula $V \approx 10^{56}$ cm^3. Comparing the observed and the calculated values of J, and using the foregoing value for the volume, we find that $KH_\perp^{1.3} \approx 10^{-13}$. The energy density W_e of relativistic electrons having a power spectrum with $\gamma = 1.6$ and with energies in the range $\mathscr{E}_1$ to $\mathscr{E}_2$ is $2.5K(\mathscr{E}_2^{0.4} - \mathscr{E}_1^{0.4})$. Assuming somewhat arbitrarily that the spectrum of magnetobremsstrahlung extends with the given slope only as far as the near ultraviolet, we define $\mathscr{E}_2$ so that the maximum emission from electrons with these energies takes place at a frequency $\nu = 10^{15}$ Hz. Hence we find that, when $H_\perp \approx 10^{-3}$, $\mathscr{E}_2 \approx 4.5 \times 10^{11}$ eV ≈ 0.7 erg. An accurate figure for $H_\perp$ is not necessary for this estimate. Since $\mathscr{E}_1$ is known to be much less than $\mathscr{E}_2$, it may be disregarded. We note, however, that if $\gamma > 2$ we must also estimate the lower limit of the energy spectrum. From the foregoing value of $\mathscr{E}_2$, we have for the energy density of the electrons $W_e \approx 2K$ erg/cm^3. Assuming that the energy density of relativistic protons is 10 times greater, we obtain $W_p \approx 20K$. Finally, equating W_p to the energy density of the magnetic field ($H^2 \approx \frac{2}{3}\bar{H}^2$), we obtain the second relation,

$$\frac{H_\perp^2}{8\pi} \approx \frac{2}{3}\frac{H^2}{8\pi} \approx (20 + 2)\frac{2}{3}K = 15K. \tag{20.2}$$

From Eqs. (20.1) and (20.2), we find $H_\perp \approx 7 \times 10^{-4}$ Oe, $H \approx 9 \times 10^{-4}$ Oe, and $K \approx 1.5 \times 10^{-9}$. Strictly speaking, an average over angles of type (18.25) is included in the expression for $H_\perp$, but it is not required here for obtaining only the order of magnitude. The mean energy density of electrons is $W_e \approx 3 \times 10^{-9}$ erg/cm^3.

Similar calculations were made by Burbidge [16] and A. A. Korchak [350] for a number of sources. However, we must remember that so far they are of use only for a rough estimate, chiefly because of the arbitrariness in the choice of the ratio W_p/W_e.

Another procedure for estimating the intensity of the magnetic field may be indicated. During magnetobremsstrahlung, electron energy decreases with time according to Eq. (17.40). We assume here that $t \approx 800$ yr, that is, we assume that the present value of the magnetic field was established 100 yr after the explosion. When $\mathscr{E} \approx \mathscr{E}_0/(1 + 10^{-4}\mathscr{E}_0 H^2)$, the electrons in the Crab Nebula with $\mathscr{E}_0 \gtrsim 10^4 H^{-2}$ eV must have lost after 800 yr a large part of their energy. This means that the break in the spectrum [351] must be near frequency $\nu = 5.4 \times 10^{-6} H_{\perp} \mathscr{E}^2 \approx 540 H^{-3}$ Hz. The character of the break depends on whether new relativistic electrons are formed, whether the field changes, and so on, but, in any case, there must be some special feature in the spectrum. Here, of course, a constant value of H is the essential assumption, but even if the magnetic field slowly and regularly changes there still must be a peculiarity in the spectrum at those electron energies for which a loss in energy over the lifetime of the nebula is sufficiently large.

A comparison of the spectral indices in the radio and optical bands shows that in the infrared region at about 3–4 μ a certain cusp actually occurs [352]. In fact according to measurements by V. I. Moroz, the spectral index in the optical band (0.3–2 μ) is $\gamma \approx 4$. Approximately the same value is obtained from a theoretical estimate [353] of the spectrum in the far ultraviolet (see (20.3) below). The intersection of this spectrum with the radio spectrum, where $\gamma \approx 1.6$, occurs in the infrared wavelength range quoted above (Fig. 54). This corresponds to $H \approx 2 \times 10^{-4}$ Oe, which seems low in comparison with (20.2). However, if the estimate of interstellar absorption made above is true, then $\alpha \approx 0.8$ and $\gamma = 2.6$. Consequently, the optical spectrum is flatter and the cusp occurs at longer wavelengths. This also leads to an increase in the value of the magnetic field obtained from the position of the cusp. Measurements of the infrared radiation flux from the nebula are necessary for further improvement of the determination of H by this method.

We consider now the origin of the field in the Crab Nebula. It could have been merely ejected from the star in conjunction with the envelope, since in this case the field in the star up to the time of the explosion must have been very strong ($H \sim r^{-2}$). On the other hand, observations indicate its relative regularity and, therefore, it cannot have been formed from the entanglement of an original weak field. Finally, it cannot have been formed by a strengthening of the interstellar magnetic field, since the latter cannot penetrate the envelope because of its high conductivity. We might assume that the field is formed because of cosmic-ray energies and that it is associated with the anisotropy of the flux from relativistic particles. However, this hypothesis encounters considerable difficulties [289a].

A more probable hypothesis as to the origin of the field in the nebula has been suggested by Kardashev [339d]. During the collapse of a supernova, its radius decreases until it reaches a value determined by rotational instability. According to the calculations of Kardashev, the final radius of the collapsing star is of the order of 200 Schwartzschild gravitational radii, corresponding to a period of rotation of ≈ 0.2 sec. If we suppose that the magnetic field initially connected the core with the envelope, then during rotation of the core the field will become coiled many times and will be strengthened at the same time. The increase in the magnetic pressure may also be the reason for the ejection and acceleration of the envelope. In recent times the weakening of the field due to expansion is partially compensated for by coiling of the field. Assuming that the initial field at the surface of the star until the collapse was of the order of 1 Oe, we might expect that this mechanism can provide the present intensity of the magnetic field in the Crab Nebula. A toroidal coiled field does not contradict the observed regular structure of the field in the nebula. According to this hypothesis, the radius of the envelope increases in time as $t^{4/3}$.

Let us turn now to the interpretation of the filaments. If the filamentary network had always been expanding with the present velocity, the size of the Crab Nebula would be greater than observed. Hence Baade found a filament acceleration of about 0.0011 cm/sec^2 [354]. Evidently this acceleration arises from the pressure of the magnetic field and of cosmic rays. On this assumption, we can determine the mass of the expanding envelope, which is 0.1 solar mass [351] when $H = 3 \times 10^{-4}$ Oe. The cosmic-ray pressure on the envelope, together with the freezing of the field, probably creates the filamentary structure. This phenomenon may be described qualitatively in the following manner. Where the envelope is less massive, its acceleration is greater, "protruding magnetic mounds" are formed, and gas, under the action of the acceleration field equivalent to a gravitational force, "rolls down" the lines of force into the "troughs," forming the complicated filamentary structure [330].

It is necessary for their stability that the filaments become elongated along the field. However, the filaments can be accelerated by the pressure of the magnetic field only when their direction does not coincide with that of the magnetic field. It may be that the lines of magnetic force in the region of the filamentary structure form a complicated interwoven grid. Another proposal, the ionization of gas in the filaments by electron collisions, is less probable. In fact, for appreciable ionization, the electron temperature must be of the order of 45,000°, which does not agree with the temperature determined from the ratio of line intensities in oxygen [353e]. In addition, the determination of temperature in different filaments shows it to be relatively constant, which is highly improbable in the case of a kinetic origin.

Calculations of the ultraviolet radiation from the Crab Nebula made by Woltjer apparently confirm the observations of its radiation. As is well known, Friedman and his collaborators [353a] have observed two intense sources of radiation in the sky in the range 1.5–8 Å, one of which has been shown to be associated with the Crab Nebula, while the other is located in the constellation Scorpius. The flux F from the Crab Nebula in this wavelength interval is $\approx 2 \times 10^{-9}$ erg/cm^2 sec Å. The angular dimensions and the identification of the source were made more precise by the method of lunar occultation [353d]. The diameter of the region radiating the x-ray is of the order of 1′, that is, one-fifth the visual dimensions. The source is located close to the center of the nebula and about the central star. Thus, the source of the x-radiation is not the central star, possibly a neutron star, and is not the entire nebula. The x-radiation may be synchrotron radiation from very hard electrons emitting only close to their place of origin. In this case, the spectral index is $\alpha = 3$ and $\gamma = 7$. Recently Giaconni and his group [353b] measured x-rays in the range of $(2\text{–}4) \times 10^{-2}$ Å. The value of the flux at 0.04 Å corresponds approximately to the extrapolation of the synchrotron spectrum with the same index $\alpha \approx 3$, but at 0.02 Å the spectrum declines sharply. Apparently this corresponds to the upper limit of the electron energy. With $H = 0.7 \times 10^{-3}$ Oe, electrons with energies of the order of 10^{14} eV emit radiation at $\lambda = 0.04$ Å. The de-excitation time for such electrons is of the order of 1 yr, and therefore they must continually be reproduced. Incidentally, in a year's time relativistic electrons do not travel far from their place of origin, which is probably close to the central star. This also explains the small dimensions of the x-ray source.

It is interesting that southwest of the center of this nebula, at approximately one-half the radius, there is a small region with dimensions of 30″ which has a very steep radio spectrum ($\alpha \approx 1.2$) [353c], so that at low frequencies this region contributes a considerable part of the total radiation. Its nature is unknown.

Woltjer [353] assumes that the line emission of filaments is excited by ultraviolet magnetobremsstrahlung. Thus, from the line intensity, we can estimate approximately the distribution of the exciting emission. Very roughly, Woltjer's calculation for total emission from the nebula gives

$$J_\nu \approx 10^{22}\left(\frac{7 \times 10^{14}}{\nu}\right)^{1.5} \text{erg/sec Hz}. \tag{20.3}$$

So far, we can say nothing about the interaction of the Crab Nebula with the interstellar medium. There are no dense clouds in its vicinity, so that resistance to expansion of the envelope is negligible. On the contrary, because of internal pressure, expansion of the envelope is accelerated. The elliptical shape of the envelope can be explained by a regular component of H directed along the minor axis, and therefore creating an anisotropy in the pressure [314]. A region of hot gas, heated by shock waves, might be expected to surround the nebula. This region is not observed because of low density.

To sum up, we can state that, although in principle emission from the Crab Nebula is explained, many difficulties remain, even in the qualitative interpretation of many of the phenomena. The origin of relativistic particles is not clear, especially of optical and relativistic electrons, which are rapidly de-excited. Moreover, it is not understood why no reduction in total particle energy due to expansion is observed. We cannot estimate the amount of cosmic-ray energy included in the electrons, and so forth.

From the most recent results of investigations of the Crab Nebula we may notice the following. Shklovskii [354a] noted that the acceleration mechanism must operate chiefly on electrons. Actually, the observation of x-rays from this nebula shows that there are a significant number of high-energy relativistic electrons which emit their energy very quickly, within a few months. Protons, on the other hand, do not lose energy at all. It is impossible that in this nebula the total energy of protons is much greater than the total energy of relativistic electrons; otherwise, the pressure of protons must produce a very fast expansion of the nebula. From this it follows that the protons may not be accelerated at all and the mechanism of acceleration operates mainly with electrons.

Much attention has been paid to the above-mentioned small detail in the radio emission of this nebula at a wavelength $\lambda \approx 7.5$ m. This small object, with a dimension $\sim 10^{15}$ cm, radiates at this wavelength range approximately the same quantity of energy as the whole Crab Nebula. It is impossible to explain this radiation by the usual synchrotron mechanism; perhaps it is due to coherent synchrotron emission (see the end of Sec. 18).

Cassiopeia A

This is a typical example of a nebula resulting from a type II supernova explosion and it is the brightest discrete source of nonthermal radio emission. Its distance is 3400 pc, and its diameter about 4 pc. Its intense brightness is evidently explained by its youth, since it flared up only 250 yr ago. In contrast to the Crab Nebula, the brightness of radio emission from Cassiopeia A increases toward the periphery. This source resembles a half shell.

Cassiopeia A also possesses a filamentary structure and is expanding at a rate of 7500 km/sec. The rate of the front side (turned toward the observer) is less, about 5000 km/sec. This is probably due to the slowing down of the envelope by the interstellar medium. In contrast to the Crab Nebula, where all the filaments are more or less the same, in this case we can separate them into fast filaments, which form chiefly the envelope, and slow filaments, which hardly move and which are situated inside the envelope.

The spectrum of the fast filaments is in general similar to the spectrum of the filaments of the Crab Nebula. Lines of [O I] and [O III] are observed in addition to the strong lines of [S II] and [Ne III]. The [O II] line is weak and traces of H_α and [*N* II] have not been observed. The spectrum of the slow, dense, and somewhat extended filaments is of a different type. Strong [N II] and H_α lines are seen plus traces of [O I] [355]. A density estimate gives $N_e \approx 10^5$, assuming that collisions of the second kind quench the forbidden O II lines (they are weak in comparison with the Balmer lines), and the estimated temperature is $T \approx 10{,}000^\circ$, since traces of auroral lines are not observed [356]. However, we must emphasize that the interpretation of the spectra of these filaments offers serious difficulties. With a density $N_e \approx 10^5$ and a filament thickness of about 0.02 pc, the emission measure is 2×10^8, so that the nebula should be exceptionally bright in the H_α lines, which has not been observed, while the H_β line has generally not been detected. The steep Balmer decrement is possibly explained by the accumulation of L_α quanta, which render the nebula opaque in H_β. The comparatively low brightness may result because only a thin layer in each filament is emitting, for example, the front of a shock wave spreading inside the filament [357].

The slow filaments are evidently formed from the interstellar medium. During its motion, the envelope of the supernova may enclose the density fluctuations of the interstellar gas, compress them, and then pass on. Even inside the envelope, these formations must still be strongly compressed by the magnetic-field and cosmic-ray pressure.

The sources of luminosity, both for fast and for slow filaments, are so far unknown. The magnetobremsstrahlung spectrum in this nebula is steeper than in the Crab ($\gamma \approx 2.6$), and therefore there is no visible or ultraviolet magnetobremsstrahlung in this case. Heating by cosmic-ray particles is insufficient, and, what is more, they hardly penetrate the filaments because of the magnetic shield that compresses the filaments. Heating due to shock waves possibly occurs. We saw that the cooling period of filaments is about 1 yr, so that the heat sources of the gas in the filaments must be fairly intense.

From the gas pressure in the slow filaments we can estimate the combined pressure due to the magnetic field and cosmic rays, which evidently preserves these filaments [330]. This is the third relation, which, together with the relation between the energy of the relativistic electrons, the magnetic field, and the radio emission and the assumption of the equality of the energy density of relativistic protons to that of the magnetic field, allows us to estimate the portion of cosmic-ray energy included in the electrons. In fact, the gas pressure in slow filaments is $p \approx NkT \approx 10^{-7}$ dyne/cm^2. Assuming $W_p + H^2/8\pi \approx H^2/4\pi \approx 10^{-7}$ dyne/cm^2, we find for the magnetic field $H \approx 10^{-3}$ Oe. According to Burbidge, from the observed value of the radio-emission flux when $H \approx 10^{-3}$, W_e is approximately equal to 5×10^{-10} erg/cm^3. Thus, $W_p \approx 10^2 W_e$, which agrees qualitatively with the theoretical estimate.

The number of fast filaments in Cassiopeia A is about 200, and the mass of a single filament is on the average 2×10^{30} g. Hence the total mass of the filamentary network is ~ 0.1 solar mass. The mass of the entire envelope is probably about 1 $M_\odot$. Such an envelope can be slowed down by the interstellar medium to 7500 to 5000 km/sec in 250 yr if the density of the medium is about 0.8 cm^{-3}. The mass of one slow condensation is on the average 50 times the mass of a fast filament, that is, about 10^{32} g. Such condensations, produced by the magnetic field and by cosmic rays, may possibly create, after cooling, gravitationally unstable globules.

The expansion of the envelope reduces the magnetic-field intensity and the energy of the cosmic-ray particles. Therefore, the radio-emission flux must diminish. Because of conservation of magnetic flux, $H \sim 1/r^2$, where r is the radius of the nebula. The energy of relativistic particles is reduced by expansion of the nebula. If the adiabatic invariant is conserved, $\mathscr{E} \sim H^{1/2} \sim r^{-1}$. The change in the number of relativistic particles of a given energy as each particle loses in energy is equal to

$$\frac{\Delta N}{N} \approx \gamma \frac{\Delta \mathscr{E}}{\mathscr{E}} \approx -\gamma \frac{\Delta r}{r}, \qquad N(\mathscr{E}) \sim r^{-\gamma}. \tag{20.4}$$

Hence, for the radio-emission flux [358] we have

$$F_\nu \sim N(\mathscr{E}) H^{\frac{1}{2}(\gamma+1)} r \sim r^{-2\gamma}. \tag{20.5}$$

In particular, in the Cassiopeia A source $\gamma \approx 2.6$ and $F_\nu \sim r^{-5.2}$. If its expansion is uniform, the brightness of the nebula must fall on the average by $5.2/250 \approx 2$ percent per year, or, taking deceleration into account, by ~ 1.6 percent. Measurements have given values of 1.1 to 1.7 percent, in qualitative agreement with this prediction. For the Crab Nebula, the change of F_ν in the visible is much less, partly because of slow expansion and partly because of greater age.

It follows from Table 20.1 that nebulae of the same type as Cassiopeia A, the remnants of type II supernovae outbursts, must be considerably more powerful objects than nebulae of the Crab type, the remnants of type I supernovae explosions. In this case the kinetic energy of the envelope, as well as the total energy of the cosmic-ray particles, is larger by two orders of magnitude. Therefore, we may also expect a considerable difference in the fluxes of radio emission.

The Loop in Cygnus

There is a well-known expanding nebula in Cygnus, the so-called Loop, whose bright portions are designated NGC 6960 and 6992–6995 (Figs. 55 and 56). The Loop consists of a nebula measuring about 20 pc formed by a supernova outburst about 70,000 yr ago. It also has a filamentary structure. The filament thickness is about 0.01 pc while the length is a hundred times greater. The spectrum of the filaments consists of the forbidden lines of O II, N II, and O III and lines of the Balmer series H_α, H_β. The rate of expansion of the filaments is about 100 km/sec. The velocity gradient observed from the thickness of the envelope is from 40 km/sec at the inside boundary to 116 km/sec in the exterior, brighter portion [18:227]. Besides the filaments, a weakly diffuse zone with slightly brighter regions is visible. The density of the bright filaments determined from the doublet $\lambda = 3727$ Å is about 500 cm^{-3}, and the average for the entire filamentary network is about 200 cm^{-3} [359]. The average mass of a filament is 2×10^{-4} solar mass, and the total mass of all the filaments is about 0.1 solar mass. The total mass of the entire expanding gas system [360] is much greater and amounts to several hundred solar masses. Evidently nonionized hydrogen from the interstellar medium is picked up by the moving envelope.

The integrated radio emission from the Loop in Cygnus is approximately 1000 times weaker than the Cassiopeia A emission. This radio emission is discussed in [360a]. The spectral index of the principal maximum of the radiation is $\alpha = 0.47 \pm 0.1$. The value of the flux at the frequency 38 MHz is of the

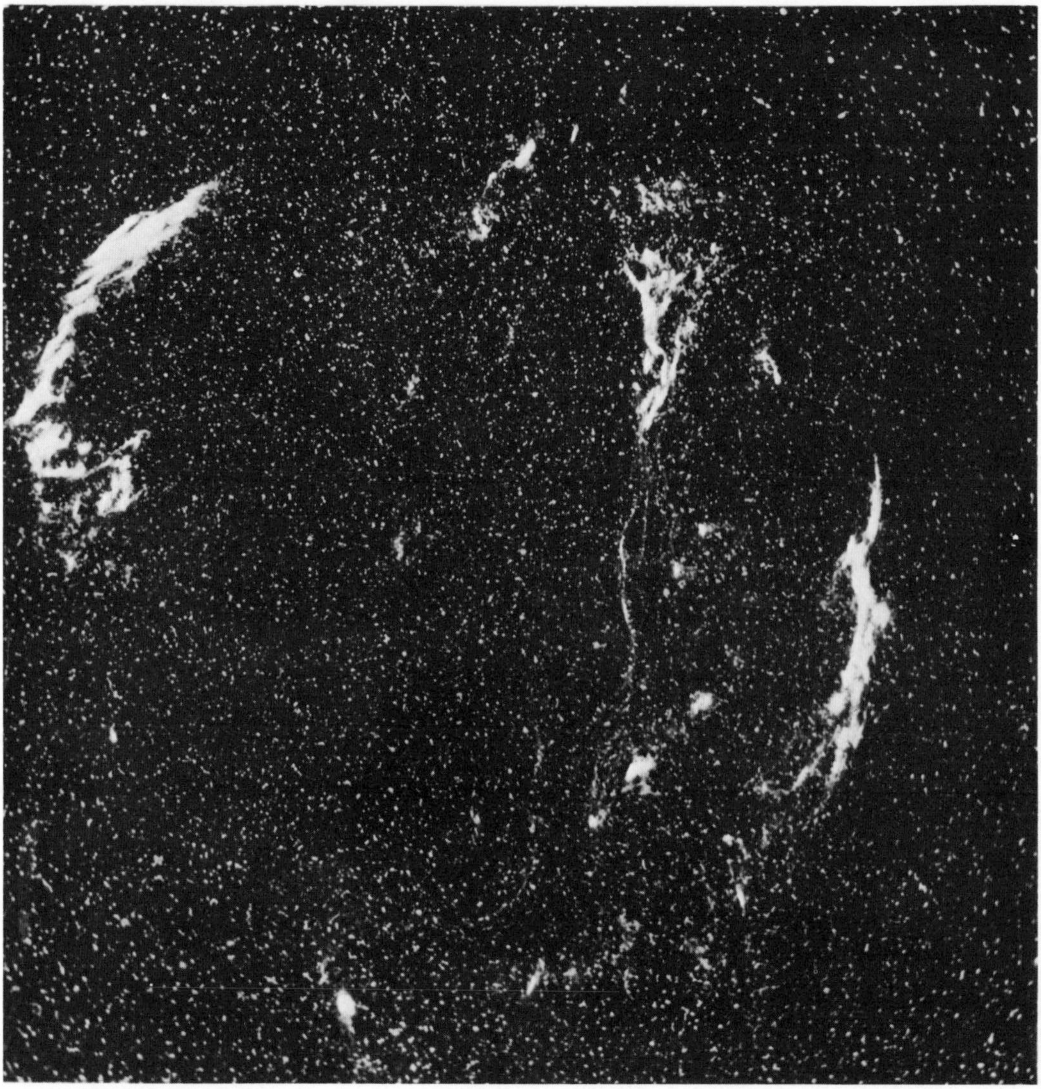

Fig. 55. The filamentary Loop Nebula in Cygnus, NGC 6960 and 6992–6995. (Crimean Observatory.)

order of 8.1×10^{-24} W/m^2 Hz. In addition to the principal maximum close to the center of the nebula, there are two other maxima with considerably lower intensities. The weakest of these has a distorted spectrum, possibly a source of thermal radiation. The determination of magnetic-field intensity by the method described previously in application to the Crab Nebula leads to a value of about 2–3 $\times 10^{-5}$ Oe, which, however, seems too low. The magnetic field frozen in by the compression of the interstellar gas by shock waves must alone be about 10^{-4} Oe. Therefore, the densities of magnetic energy and of cosmic-ray energy

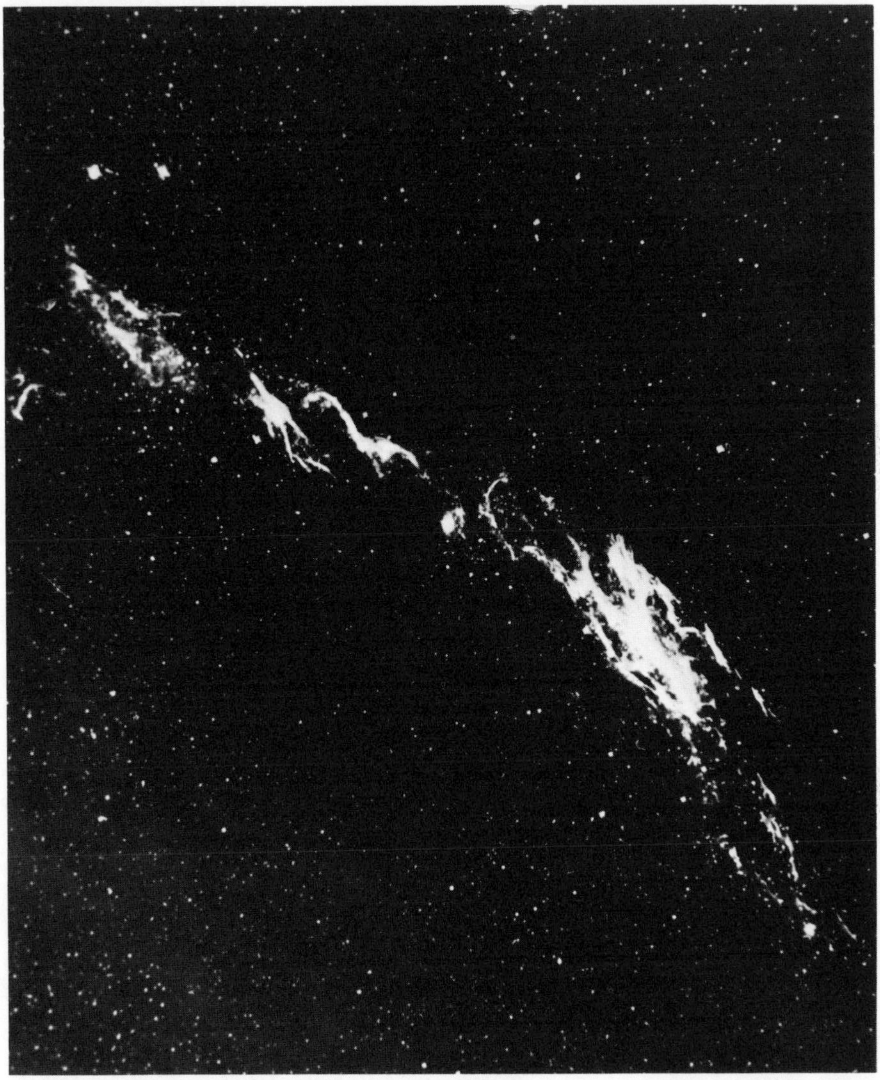

FIG. 56. Detail in the filamentary nebula NGC 6992 in Cygnus. (Palomar Observatory.)

are not identical here. Taking $H \approx 10^{-4}$ Oe, Burbidge [16] found the energy of cosmic-ray particles in the Loop to be 10^{48} erg, while the magnetic-field energy is $\sim 10^{50}$ erg. We recall that in Cassiopeia A the energy of cosmic-ray particles is $\sim 10^{50}$ erg. A decrease in cosmic-ray energy can be only partly associated with expansion of the nebula. The size of the Loop in Cygnus is 12 times that of the Cassiopeia A source, so that expansion here must reduce the particle energy by the same amount (since the energy of relativistic electrons changes under adiabatic cooling according to the law $\mathscr{E} \sim \rho^{1/3} \sim R^{-1}$). The observed reduction of yet an order of magnitude is possibly associated with the de-excitation of the electrons, which now provide a smaller portion of the energy of the particles.

The Galactic Spur

The remains of a supernova particularly close to the Sun, called the Spur, are still observed, appearing as two arches of intense radio emission covering a considerable part of the sky. Therefore, the Spur was considered earlier as part of the general galactic background, but now it is identified as a supernova remnant.

The Spur is ring-shaped, having a radius of approximately 56° and with its center at the coordinates $l = 330°$ and $b = 19.5°$. Strictly speaking, not the whole ring is observed, but only two sections, of lengths 88° and 21° [430b]. The outer boundary of the Spur is quite sharp, with its intensity halfwidth equal to only 0.5°. Its extended structure parallel to the outer boundary is to be noted, so that this source consists of a multitude of envelopes similar to the Loop in Cygnus. In general, the parameters of radio emission (brightness temperature, spectrum, structure) are consistent for this nebula. Assuming a radius of 20 pc, we place the distance from its center to the Sun at approximately 30 pc and the thickness of its envelope at 2 pc [430a]. The observed emission from the Spur, as from the Loop, is greater than that expected from an estimate of the radiation from cosmic-ray electrons of the general galactic background in the enhanced magnetic field of the nebula envelope. Obviously, generation of cosmic-ray particles in these objects has gone on continuously until the present time. A study of the polarization of the radio emission from the Spur also confirms its proximity to the Sun and the synchrotron character of this radio emission.

Former supernovae are one of the principal sources of cosmic rays in the Galaxy. Even the Sun, which is a completely quiescent star, ejects during periods of chromospheric outbursts clouds of soft cosmic rays that fill the solar system. In the Galaxy, many stars change their brightness, their spectrum, and their magnetic field, these changes being attended by pulsation, the ejection of matter, and so on. Undoubtedly, relativistic particles must be formed in these variable stars.

Radio emission from exploding stars of the type UV Ceti has actually been observed and forms the whole basis for the assumption that optical as well as radio outbursts are synchrotron in nature. However, all these stars furnish many times fewer cosmic rays than exploding supernovae.

If a supernova of type II explodes once in 100 to 200 yr, and yields after expansion about 10^{48} erg in the form of relativistic particles, then the total energy of cosmic rays produced in the Galaxy by these stars is approximately 3×10^{38} erg/sec. The total energy of cosmic rays in the Galaxy is of the order of $1\text{–}3 \times 10^{56}$ erg. If their production time is 10^8 yr, the input necessary for their replenishment is 3×10^{40}–10^{41} erg/sec, which is considerably greater than that furnished by supernovae. It is true that, because the estimates of the parameters are uncertain, the influx may be larger, since some particles are injected into the Galaxy before the distintegration of nebulae, while their generation continues for a long time. The time of escape from the Galaxy of relativistic particles may also be longer, since the nature of the field at the periphery is not clear at the present time. On the other hand, if the magnetic field of the spiral arms simply expands at the periphery and blends into the extragalactic field, the escape time may be less by up to 10^7 yr. Of course, this is the least probable case, since the asymmetry of the cosmic-ray flux would be marked.

The most probable hypothesis at present is that cosmic rays originate as the result of explosions or other similar violent processes in the nuclei of galaxies. Evidently these processes differ only in scale from the powerful explosions in radiogalaxies and the distant quasistellar sources. The nature of these explosions is not clear. Radiogalaxies will be considered in more detail in Sec. 25. We only emphasize here that, in view of the short escape time, these explosions must take place sufficiently frequently in the nuclei of ordinary galaxies. In this connection, we recall that violent motions are also observed in the central region of our own Galaxy.

5. Interstellar Gas Dynamics and the Evolution of the Interstellar Medium

Interstellar gas dynamics deals with the study of the physical features of gas motion in interstellar space: its formation, the properties and attenuation of ionizing and shock waves, the statistical laws of interstellar turbulence, the condensation of interstellar gas clouds, and so forth. Thus, it is distinguished, on the one hand, from the general dynamics (if we may use the term) of the interstellar medium, where the problem consists mainly of constructing on the basis of observational and theoretical premises a general idea of the distribution and motion of gas in galaxies and, on the other hand, from the theory of evolution of the interstellar medium, where the general development of such material with time is studied. Of course, results from interstellar gas dynamics are employed in this and in other problems, but we emphasize again that particular attention will be paid here to a study of the concrete, physical peculiarities of interstellar gas motion.

A description of the principles of interstellar gas dynamics is presented in two monographs by the authors of the present book. In one of them [8], the fundamental aspects of this area of the physics of the interstellar medium are described directly, and in the other [12], particular attention is paid to the effect of magnetic fields on gas motion under cosmic conditions in general, and in interstellar space in particular. Therefore, we limit ourselves here to a more restricted account of results, using the monographs quoted for any details.

21. Shock Waves in Interstellar Space

The motion of interstellar gas is determined by the major role of supersonic or near-sonic flow. In H I regions the velocity of sound, c_s, is 1 km/sec at $T \approx 100°$, while in H II zones, $c_s \approx 15$ km/sec ($T = 10{,}000°$). Since the mean velocity dispersion of interstellar gas clouds is 7–8 km/sec, their motion in H I regions is supersonic. In H II regions the velocity dispersion is usually higher, about 10–15 km/sec, so that here near-sonic flow predominates. However, as we know, motions occur in interstellar space at considerably higher velocities, reaching 6000–7000 km/sec, for example, in the dispersion of envelopes from novae and supernovae.

When gas is in supersonic motion, shock waves are almost inevitably formed. We must therefore assume that shock waves are often encountered in interstellar space. As is known [375], directional gas motion is converted in a shock wave into random thermal motion, so that dissipation of energy occurs. In a monatomic gas behind a strong shock front, in which the total gas energy is conserved, the temperature, density, and velocity are determined by the well-known equations

$$T_2 = \frac{3\mu_2 m_H v_1^2}{16k} = \frac{\mu_2 m_H (v_1 - v_2)^2}{3k},$$
$$N_2 = 4N_1, \qquad v_1 - v_2 = \frac{3}{4} v_1. \tag{21.1}$$

Both here and later, the index 1 indicates the values of the parameters before the passage of the wave front, and the index 2, those after the passage of the front. If the wave is moving through a stationary gas, v_1 is the velocity of the wave front and $v_1 - v_2$ is the gas velocity behind the front. If we consider the helium content of H I regions, the molecular weight there is $\mu \approx 13$, while in H II zones it is $\mu \approx 0.7$.

The Structure of a Shock Wave

The structure of a shock front, that is, the region where dissipation of kinetic energy occurs, may be calculated by kinetic-theory methods. The most suitable is the method of Mott-Smith, in which the particle-distribution function inside the shock front is described as the sum of two Maxwell functions. The reader will find details in the original paper [376], so that here we shall only quote the results for a shock wave in a monatomic gas. The density distribution inside the front of a strong shock wave in a neutral gas in a system of coordinates moving with the wave front is described by the equations

$$N(x) = \frac{N_1}{1 + e^{x/l}} + \frac{4N_1}{1 + e^{-x/l}}. \tag{21.2}$$

Here the origin of the coordinates is represented by a point where the concentration is $N(0) = 5N_1/2$. The width of the wave front is characterized by the parameter l, given by

$$l = l_a \approx \frac{1}{4\sqrt{2}\sigma N_1} = \frac{1}{\sqrt{2}\sigma N_2}, \tag{21.3}$$

where σ is the gas-kinetic effective cross section. In the derivation of these equations, atomic collisions are treated as those of elastic spheres. If we assume $\sigma = 10^{-15}$ cm^2, then $l_a \approx 6 \times 10^{-5}/N_1$ pc.

Equation (21.2) may be interpreted in the following way. Suppose a stream of hot gas of density N_2 flows into a cold quiescent gas of density N_1. At the front of this stream particles of the quiescent gas (first term) increase their energy and concentration because of atomic collisions with the inflowing stream (they are bunched up) and are incorporated into the second stream, described by the second term of (21.2). Thus l in this case is the mean free path of particles from the first stream relative to the second.

The structure of a shock wave in a fully ionized plasma was considered by Tidman [377] by the same method. We note here that the exchange of energy between electrons and protons is much less effective, because of their difference in mass, than the exchange of energy between identical particles. Therefore, inside the boundary of the shock front in a fully ionized plasma, thermal equilibrium between an electron gas and an ionized gas is upset, or, in other words, the electron temperature T_e and the ion temperature T_i are different.

An ion gas behaves approximately as a monatomic gas; at the front of a strong shock wave the density distribution is described by Eq. (21.2). However, the width of the front is now determined by the Coulomb interaction of the ions. Calculation of the so-called "collision term" in the Boltzman kinetic equation for this case leads to the expression

$$l \approx l_i = \frac{(m_H v_1^2/e^2)^2}{690 N_1 L} = \frac{0.06}{N_1}\left(\frac{v_1}{10^8 \text{ cm/sec}}\right)^4 \text{pc.} \tag{21.4}$$

We recall that the expression $(e^2/m_H v_1^2)^2$ is approximately equal to the effective cross section of the Coulomb interaction, and $L = 2 \ln \Lambda$ is the Coulomb logarithm which allows for interactions of distant encounters of the ions. In (21.4), we take $L \approx 40$ (see Sec. 16). While $v_1 \ll 1000$ km/sec, the size of the "ion discontinuity" at the front is much less than representative formations in the interstellar medium, but when $v > 1000$ km/sec the value of l becomes so much greater that, in these cases, the concept itself of a shock wave becomes inapplicable. It is true, as we shall see below, that there are reasons for considering that the actual width of the shock front must be less than that given by (21.4).

The width of the front of a strong plasma shock wave, in particular, may be considerably less if the gas is not completely ionized. In fact, in this case with very high velocities the interaction of ions with even a small number of neutral atoms becomes more effective than the interaction of ions among themselves, and therefore the dissipation of energy is determined by collisions between ions and atoms whose cross sections depend very little on velocity. The structure of the shock wave in this case is determined as before by Eq. (21.2), only now, instead of l, we must substitute the smallest of the values

$$\frac{l_a}{(1 - X)^2}, \quad \frac{l'_a}{(1 - X)X}, \quad \frac{l_i}{X^2}, \tag{21.5}$$

where X is the degree of ionization. Within the limits of the distances in (21.5), the degree of ionization hardly changes, since, as we shall see later, the electron temperature here remains low. The values l_a and l'_a are estimated from (21.3) using the effective cross sections for atom–atom and atom–ion collisions. If the masses of the atoms and ions are different, $l'_a/(1 - X)X$ must be multiplied by the square root of the ratio of the larger to the smaller mass. For example, if during the passage of a shock wave helium remains unionized, the thickness of the ion–atom front cannot exceed the value

$$2 \frac{6 \times 10^{-4}}{N_1 \times 0.15 \times 0.85} \approx \frac{10^{-2}}{N_1} \text{ pc.} \tag{21.6}$$

The behavior of an electron gas in plasma shock waves has certain peculiarities. The interchange of energy between the light electrons proceeds very rapidly, so that the electron velocity-distribution function in interstellar space can be represented nearly always by the single-valued Maxwell equation. This condition is often used, as, for example, in Chapter I. Also, because of the relative smallness of the particle mass, the electron stream is easily decelerated, but this changes only the linear momentum of the electrons. An ion front striking a stationary electron gas carries it along in its wake. We must emphasize that the temperature of the electron gas is almost unchanged during this motion. The distance through which the electron gas is carried in a fully ionized plasma is determined by the equation [377]

$$l_v \approx \frac{m_i^{3/2} m^{1/2} c_s^3 v_1}{10 e^4 N_1 L} \approx 69 \left(\frac{m}{m_i}\right)^{1/2} \frac{c_s^3}{v_1^3} l_i \approx 1.6 \frac{c_s^3}{v_1^3} l_i, \tag{21.7}$$

where c_s is the velocity of sound in the quiescent plasma. Since in strong shock waves $v_1 \gg c_s$, and consequently $l_v \ll l_i$, the velocity of the electron gas is practically everywhere the same as that of the ion gas. It follows from the condition of quasi neutrality that their densities also are identical. A similar circumstance arises when collisions occur between charged particles and neutral atoms.

As we have noted already, the exchange of energy between ions or atoms and the electron gas proceeds quite slowly. If the relative velocity between an ion and the electron gas is zero, and if the condition $T_e m_i \gg m T_i$ is met (as it always is under interstellar conditions), then the change in electron temperature due to collisions between electrons and ions is described by the equation [377, 80, 379]

$$\frac{d}{dt}\left(\frac{3}{2} k T_e\right) = \frac{4\pi e^4 N_i L}{(2\pi m k T_e)^{1/2}} \frac{m}{m_i} \frac{T_i - T_e}{T_e}. \tag{21.8}$$

The meaning of this equation is apparent if we write the expression on the right-hand side in the form (accurate to a numerical factor)

$$N_i\left(\frac{e^2}{mv_e^2}\right)^2 Lv_e \frac{mv_e^2}{2}\frac{m}{m_i}\frac{T_i - T_e}{T_e} \approx \frac{mv_e^2}{2\tau}\frac{m}{m_i}\frac{T_i - T_e}{T_e}, \tag{21.9}$$

where v_e and τ are the thermal velocity and the time of traversal of the mean free path of the electrons, and $m(T_i - T_e)/m_i T_e$ takes into account the probability of energy transfer from ions to electrons, and vice versa. From Eq. (21.8) we find

$$\frac{dT_e}{dt} = 0.16N_i \frac{T_i - T_e}{T_e^{3/2}} \text{ deg/sec}. \tag{21.10}$$

The distance at which electron and ion temperatures converge is approximately

$$l_e \approx \frac{(2\pi mkT_2)^{1/2}}{8\pi e^4 N_i L}\frac{m_i}{m}\frac{3kT_2}{2}v_2 \approx 13l_i. \tag{21.11}$$

Here Eqs. (21.1) are used with $\mu = 0.7$ and assuming that $N_i = N_2 = 4N_1$. Thus the thickness of the front of the electron shock wave considerably exceeds that of the region of sudden density change.

Let us now consider in more detail the behavior of an electron gas in a shock wave. We shall assume at first that the gas is completely ionized and we shall neglect electron thermal conductivity and emission. Here the structure of the front is as described in [8]. First, the front of the ion discontinuity, where the temperature of the ionized gas jumps to $T_2 = 3m_H v_1^2/16k$ while the electron temperature hardly changes, expands in the undisturbed plasma. Then follows a region whose thickness is of the order of l_e given by (21.11), where the electron and ion temperatures converge. At the end of this region, $T_e \approx T_i \approx T_2 \approx 0.16 m_H v_1^2/k$, if luminescence is neglected. Inside the temperature-equalization region, owing to conservation of energy, the equality

$$T_i(x) + T_e(x) = T_2 = \tfrac{1}{2}T_2, \tag{21.12}$$

where x is the distance from the front of the ion discontinuity, is always obeyed, while the detailed structure of the region can be calculated from Eq. (21.8). In particular, since $T_e \ll T_i$, the relation $T_e \sim x^{2/5}$ is found. When T_e is close to T_i, the temperature change can be described by an exponential function. However, we must note that luminescence or de-excitation, about which we shall speak later, upsets Eq. (21.12). Moreover, if the shock wave is not propagated in an ionized gas, some energy is dissipated by ionization.

We know that an electron gas has good thermal conductivity. This means that in a region where T_e is large the electron gas appears to be heated even before the shock wave itself passes. Therefore, the increase in electron temperature begins earlier, even before the passage of the ion discontinuity. Calculations by Jukes [378] showed that where the electron gas is heated it is changed into a noncompressed gas over a distance of a few l_i.

Energy transfer by radiation leads to the same effect, except that here the heating of the electron gas in front of the shock is caused by the natural radiation of the gas behind the wave front. The width of the heated region is about $\sim 1/a_1 N_1$, where a_1 is the continuous absorption coefficient and N_1 is the number of absorbing atoms. If a strong shock wave is propagated in unionized hydrogen, the width of the heated region is

$$l_r \approx \frac{0.05}{N_{\mathrm{H}}} \text{ pc.} \tag{21.13}$$

However, in order that heating by radiation should be substantial, the gas behind the wave front must emit in the continuous spectrum with sufficient intensity. Under the conditions of the interstellar medium, this requirement is obviously not met.

Luminescence and Ionization in Shock Waves

If gas ionization and energy losses occur, the structure of shock waves is considerably complicated because of excitation of atoms and ions. On one hand, ionization processes change the relative numbers of ions and electrons, and on the other, some of the wave energy is dissipated in this way. The kinetic energy of the shock wave is also dissipated in exciting atoms and ions. Emission arising from downward transitions, especially in forbidden lines, is not absorbed in a region where motions are taking place, and therefore a shock wave seems to dissipate energy irreversibly (as distinct from the heating phenomenon), that is, the region becomes luminous. The dissipation of wave energy by ionization and luminescence leads to a temperature drop behind the front and to an additional, but slower, increase in gas density.

The detailed analysis of the structure of shock waves with ionization and luminescence requires numerical integration of the appropriate equations [379, 12, 8]. We have considered so far only certain examples of relatively weak shock waves ($V_1 \approx$ 20–200 km/sec). The difficulty of a numerical study of shock waves is aggravated by the great diversity of conditions in interstellar space. We have to take into account the ionization state of interstellar gas before the wave front passes in H I or H II regions, while it is possible for emission from the wave itself or from a stellar outburst to change the ionization state. The luminous energy in the wave depends on the degree of ionization of a number of elements, primarily oxygen, with low-lying metastable levels. The wave structure in this case also depends on the origin of the wave. If a shock wave is caused by sustained pressure on the gas, then, as the temperature falls, the density correspondingly rises because of luminescence. If, however, a shock wave is due to a momentary impulse, the external pressure far behind the front may be diminished and the temperature drop does not then correspond to the additional increase in density. Such a shock wave is found to be nonstationary.

Mathematical difficulties are encountered in the solution of the equations for the structure of shock waves incorporating ionization and luminescence, mainly because of the complex and cumbersome expressions needed to describe the energy losses of the gas.

Before we consider certain results obtained in this direction, we shall introduce a number of general relations. The classical equations that describe approximately the excitation and ionization of atoms and ions by electron collisions are given in Appendix I, Eqs. (I.39)–(I.41). They are used in the estimation of energy losses by the electron gas in the shock wave. Let N_a be the number of atoms and ions of type a that have low-lying levels or have been ionized by electron collisions. Assuming that for every ionization and for every excitation of a level the energy χ_a is dissipated, we can write

$$\frac{d}{dt}\left(\frac{3}{2}kT_e\right) = -\sum_a N_a \chi_a q_a. \tag{21.14}$$

In this case emission by absorption and by collisions of the second kind, that is, the restoration of energy to the gas, is neglected. Combining Eqs. (21.8) and (21.14), we obtain

$$\frac{d}{dt}\left(\frac{3}{2}kT_e\right) = \frac{8\pi e^4 N_p}{(2\pi mkT)^{1/2}}\left\{\frac{Lm}{2m_p}\frac{T_i - T_e}{T_e} - \frac{3}{2}\sum_a \frac{N_a f_a}{N_p}\left[e^{-\chi_a/kT} - \frac{\chi_a}{kT}\,\mathrm{Ei}\left(\frac{\chi_a}{kT}\right)\right]\right\}. \tag{21.15}$$

Equation (21.15) is correct when ionization is not too low, or when T_e is much less than 1,000,000°. In the opposite case, we have to take into account in the first term the transfer of energy from atoms to electrons. The role of the various terms in the sum in Eq. (21.14) changes when the electron temperature throughout the whole shock-wave structure is altered.

Equation (21.15) allows us to examine qualitatively the structure of the fronts of interstellar shock waves. If we disregard the heating of the electron gas in front of the shock wave, the second term in the right-hand member of (21.15) is considerably less immediately behind the front of the ion discontinuity than the first term, that is, luminescence and losses due to ionization are still not substantial. Therefore, the electron temperature rapidly increases in this case. We must stress that the larger $T_i = T_2$, or the stronger the shock wave, the less the relative energy losses at the earlier stage.

As T_e rises and T_i simultaneously falls, the role of luminescence is increased, and when

$$\frac{Lm}{2m_p}\frac{T_i - T_e}{T_e} \approx \frac{3}{2}\sum \frac{N_a f_a}{N_p}\left[e^{-\chi_a/kT_e} - \frac{\chi_a}{kT_e}\,\mathrm{Ei}\left(\frac{\chi_a}{kT_e}\right)\right] \tag{21.16}$$

the electron temperature reaches a maximum. We can estimate the order of magnitude of the quantities in this equation. We may assume that $Lm/m_p \approx 1/50$

and $\frac{3}{2}\sum N_a f_a/N_p \approx 10^{-3}$. We suppose at first that the maximum of T_e is reached when $\chi_a \ll kT_e$ for the most abundant elements. Then from (21.16)

$$\frac{T_i - T_e}{T_e} \approx \frac{1}{10}, \tag{21.17}$$

that is, in this case the electron temperature is practically the same as the ion temperature and they are both equal to $T_2 \approx 0.16 m_H v_1^2/k$. This case arises only in very strong shock waves where the velocity v_1 satisfies the condition

$$0.16 m_H v_1^2 \gg \chi_a \approx 10\text{–}100 \text{ eV}; \qquad v_1 \gg 100\text{–}300 \text{ km/sec}. \tag{21.18}$$

Since the right-hand member of (21.16) can only diminish with an increase in the ratio χ_a/kT, the maximum electron temperature is always close to the ion temperature, but for a determination of the electron temperature itself and to take into account luminescence, a more precise calculation is necessary.

An investigation of the structure of a stationary shock wave with the consequent compression of gas and production of luminescence was made in [8] and later in more detail in [379a, b]. The equations for the structure of a stationary shock wave were solved on an electronic computer from relation (21.15), the changes in the degree of ionization in the wave itself being taken into account. To facilitate the calculation of the last sum in Eq. (21.15), the terms describing ionization were chosen in a form to approximate the experimental data, and the remaining members of the sum were replaced by a constant, γ, called the luminescence coefficient. Calculations were made for $3 \times 10^{-5} \leqslant \gamma \leqslant 3 \times 10^{-3}$, and for velocities of the shock wave $36 \leqslant v_1 \leqslant 360$ km/sec. Various values of the initial ionization of hydrogen were also considered.

The results of these calculations are as follows. For small γ, the electron temperature toward the maximum is almost independent of the initial luminescence and may be determined with sufficient precision from Eqs. (21.1) and (21.12) if the shock wave is strong, and from the corresponding equations from the ordinary theory of shock waves for small Mach numbers. For a hundredfold increase in the parameter γ, the maximum electron temperature in the wave is decreased by only 30 percent for strong waves and by 60 percent for the weakest waves considered ($v_1 = 36$ km/sec). The weak dependence of the electron temperature at maximum justifies, incidentally, the introduction of the "luminescence coefficient" in place of a very complicated calculation with the explicit form of the function in the sum (21.15). The thickness of the region of heating of the electron gas (from the atom–ion discontinuity to the maximum electron temperature) also depends weakly on the parameter γ and for $v_1 = 100$ km/sec is approximately equal to $3 \times 10^{15}/N_H$. However, for large velocities of the shock wave, this quantity increases rapidly, approximately as v_1^3. It is interesting that the optical thickness of the region of heating of the electron gas for thermal radiation depends weakly to a first approximation on the velocity, even for strong shock waves, and may be determined from the approximate equation

$$\tau(\lambda) \approx 3 \times 10^{-7}\lambda^2 N_e, \tag{21.19}$$

where λ (cm) is the wavelength and N_e is the electron concentration behind the wave front. At those wavelengths for which $\tau \approx 1$ it is possible, obviously, to observe quite intense radio emission from these shock waves, with brightness temperature close to the maximum electron temperature of the shock wave.

A region of luminescence and the corresponding compression of the gas follows the maximum electron temperature. The decrease of the electron temperature sets in relatively slowly. However, at the end of the region of luminescence the temperature drops sharply and, consequently, so does the density. The decrease in temperature continues until equilibrium is established with the interstellar radiation field at the same degree of ionization that was attained in the wave (for strong shock waves the ionization of hydrogen is almost complete).

The width of the region of luminescence depends on the velocity of the shock wave (as a rule, it increases as v_1), but it depends more strongly on the choice of the luminescence coefficient, rapidly decreasing as γ increases. For small v_1 and large γ (within the limits considered), the width of the region of luminescence is several times that of the zone of heating of the electron gas, but for large v_1 and small γ it may be several tens of times larger.

Detailed calculations for an impulsive shock wave being propagated at a velocity of 100–150 km/sec in an undisturbed H I region, or in the same region when the hydrogen was first partially ionized by another shock wave, were given by S. B. Pikelner as an explanation of the structure of filaments in nebulae, while considering the filaments as regions of interception of the shock waves [379]. It was assumed that the subsequent compression of the gas behind the front caused by the fall in temperature does not yield an appreciable amount of energy, since it occurs slowly. It was shown that the electron temperature reached a maximum value of about 100,000° and then fell rapidly because of luminescence. The ion temperature at first fell rapidly to the maximum electron temperature; after the maximum, $T_e \approx T_i$. The degree of hydrogen ionization in these waves reached 20–60 percent if the hydrogen had not been ionized before the passing of the shock wave, and approached 100 percent when some original ionization existed. In this calculation, the heating of the gas in front of the shock wave was not taken into account.

Heating of Gas by Shock Waves in H I *Regions*

The case of a relatively slow shock wave ($v_1 < 10$ km/sec) in H I regions is of particular interest. In this case, the temperature behind the front of the atom discontinuity is $T_2 \approx 0.3 m_H v_1^2/k \lesssim 4000°$ and neither hydrogen nor oxygen is ionized in such a wave. However, since there are also free electrons ($N_e \approx$

$10^{-3}N_H$) in H I regions, the structure of the shock wave is complex. Following the atom discontinuity we have a density change which is determined from Eqs. (21.2) and (21.3) with $\sigma_H \approx 3 \times 10^{-16}$–$10^{-15}$ cm^2. As before, we obtain after the atom discontinuity a region of equal atom and electron temperatures, which may be described with the help of the equation

$$\frac{d}{dt}\left(\frac{3}{2}kT_e\right) \approx 8\sigma_{ae}\left(\frac{2kT_e}{\pi m}\right)^{1/2} N_H k(T_H - T_e)\frac{m}{m_H}, \tag{21.20}$$

where $\sigma_{ae} \approx 6.3 \times 10^{-15}$ cm^2 is the effective collision cross section of electrons with hydrogen atoms [137, 18]. The relation (21.20) may be rewritten in the form

$$\frac{dT_e}{dt} \approx 5.8 \times 10^{-12} T_e^{1/2}(T_H - T_e)N_H \text{ deg/sec.} \tag{21.21}$$

Since the density of the electron gas is here much less than the density of the atomic gas, we may assume, as long as $T_e \neq T_H$, that $T_H \approx T_2 \approx$ const. However, an intense expenditure of shock-wave energy occurs here in exciting atoms and molecules and subsequently in luminescence. We shall assume, first of all, that there are no hydrogen molecules in the H I region being considered. Then the expenditure of energy in exciting the principal metastable levels of C^+ and Fe^+ is determined from Eq. (8.18). Therefore, the total temperature change of the electron gas with time is described by the equation

$$\frac{dT_e}{dt} \approx 5.8 \times 10^{-12} T_e^{1/2}(T_H - T_e)N_H - 3.4 \times 10^{-4}\frac{N_C}{T^{1/2}}$$
$$\times (e^{-92/T_e} + 10e^{-413/T_e} + 2.6e^{-554/T_e} + 3.4e^{-961/T_e}), \tag{21.22}$$

where N_C is the number of carbon ions per unit volume. If $T_H > T_e \gtrsim 1000°$, the first term in (21.22) is much larger than the second, and the temperature of the electron gas is raised. After equalization of the electron and atom temperatures, intense luminescence of the gas is set up in the H I region, which is accounted for in (21.22), where we can now assume that $T_e \approx T_H$. Such a calculation was carried out by Seaton [137]. As a result he obtained the dependence of T_H on time shown in Fig. 57.

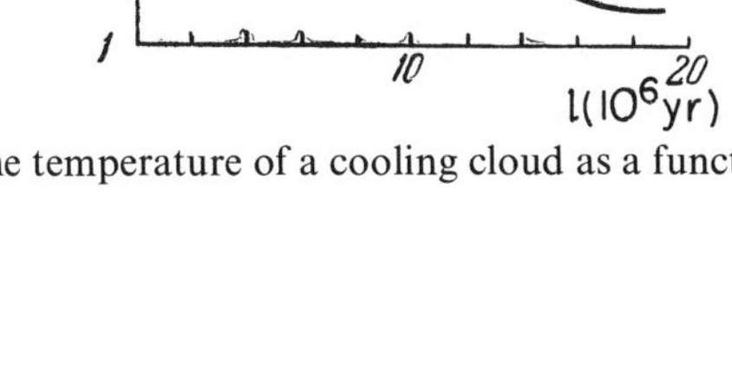

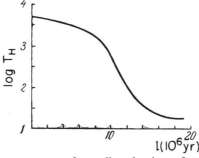

FIG. 57. The temperature of a cooling cloud as a function of time.

As we know, the gas temperature in H I regions is observed from radio emission in the line $\lambda = 21$ cm, while observations give the harmonic mean of T. In this connection, calling t_1 the period during which cooling proceeds, we find for the harmonic mean (over time) of the temperature T_H, determined from (21.22),

$$\overline{\left(\frac{1}{T}\right)} = \frac{1}{t_1}\int_0^{t_1} \frac{dt}{T_H(t)}. \tag{21.23}$$

According to observations, $\overline{1/T} = 1/125°$. This provides an equation that connects the original conditions with the cooling time (Table 21.1). Along with

Table 21.1.

T_2 (°)	t_1 (10^6 *yr*)	$\overline{T}$ (°)	T_k (°)
5000	15.8	1850	22
4000	12.6	1420	24
3000	9.6	1000	27
2000	6.9	600	33
1000	4.3	300	43

(21.23), we can determine the mean temperature,

$$\overline{T} = \frac{1}{t_1}\int_0^{t_1} T_H(t)\,dt. \tag{21.24}$$

The dependence of t_1, $\overline{T}$, and T_k (final temperature) on the initial temperature T_2 and $N_H = 10\ \text{cm}^{-3}$ is given in Table 21.1.

In connection with these results, Kahn [17:60] proposed the following interpretation of the fact that the observed gas temperature in H I regions ($\sim 125°$) is much higher than the equilibrium temperature of unionized hydrogen in interstellar space ($\sim 20°$). Collisions between H I clouds moving with velocities of ~ 7–8 km/sec lead to shock-wave formation that heats the gas to a temperature of $\sim 3000°$. The shock wave (an atomic discontinuity) rapidly passes through the cloud, leaving behind a region of temperature adjustment and of luminous gas, which from Table 21.1 cools over several million years. Since the period between the collision of two clouds is also about 10^7 yr, we might suppose that H I clouds hardly attain a state of thermal equilibrium, and that they are therefore observed in a nonsteady state. The mean gas temperature in an H I region may in such a case be higher than 1000°. Evidently this account agrees qualitatively with reality, although we must note that the idea of a simple collision between clouds is a very rough one, since we ought to consider here the effect of the magnetic field, whose energy between the clouds is comparable with the energy of their motion.

If there are H_2 molecules in H I regions, they also make a contribution to the cooling of the gas behind the front of the shock wave. The absence of data on the concentration of these molecules prevents us from estimating their role with any assurance, but obviously molecular cooling is no more effective than cooling from ionic excitation.

High-compression Shock Waves

We remarked before that, with the intense luminescence behind the front of a stationary shock wave moving under the action of sustained pressure on the medium, strong compression of the gas must take place. In fact, in the ordinary shock wave without luminescence, the increase in density is limited by [375]

$$\frac{\rho_2}{\rho_1} = \frac{\gamma + 1}{\gamma - 1}, \tag{21.25}$$

where γ is the adiabatic exponent. In a shock wave with luminescence, this condition is not fulfilled, and therefore, in principle, we can have any compression here. In fact, the temperature drop during luminescence reduces the internal gas pressure, which can no longer counteract further compression behind the wave front. The theory of such waves was discussed by S. A. Kaplan [8]. The density change in the shock wave with luminescence is readily obtained from the equations

$$\rho_1 v_1 = \rho_2 v_2; \qquad \frac{\rho_1 k T_1}{\mu_1 m_H} + \rho_1 v_1^2 = \frac{\rho_2 k T_2}{\mu_2 m_H} + \rho_2 v_2^2, \tag{21.26}$$

which describe the conservation laws for the flux of mass and momentum in the gas during the passage of the disturbance. Here v_1 and v_2 are the gas velocities relative to the wave front. In a strong shock wave with luminescence, $\rho_1 \ll \rho_2$, $T_1 \lesssim T_2$. Then from (21.26) it follows that

$$\frac{\rho_2}{\rho_1} \approx \frac{\mu_2 m_H v_1^2}{k T_2} \approx \gamma \frac{v_1^2}{c_2^2} = \gamma M^2, \tag{21.27}$$

where c_2 is the velocity of sound behind the front of the shock wave with luminescence and M is the Mach number. In an H II region, the final temperature behind the luminescence region is not greater than 20,000°, since at a higher temperature an ionized gas emits intensely in the forbidden lines. Therefore, here ($\mu \approx 0.7$) we have

$$\frac{\rho_2}{\rho_1} > \left(\frac{v_1}{15 \text{ km/sec}}\right)^2, \tag{21.28}$$

and, if $v_1 > 100$ km/sec, gas compression must be considerable. Strong compression waves due to luminescence can be propagated in H I regions, only here it is more difficult to determine the final temperature behind the luminescence region. The value of T_k in Table 21.1 is hardly reached because it requires a long time. It is improbable that a shock wave would compress the gas over a period of 10^7 yr. If we set $T_2 \approx 1000°$ in (21.27), then ($\mu \approx 1.5$) we obtain

$$\frac{\rho_2}{\rho^1} \approx \left(\frac{v_1}{2.4\,\text{km/sec}}\right)^2. \tag{21.29}$$

A shock wave with luminescence moving in an H I region with a velocity of 20 km/sec compresses the gas by approximately fifty times.

Radiation from Interstellar Shock Waves

The question of the luminescence of shock waves is closely related to the problem of their observation. We shall consider separately medium-intensity waves traveling with a velocity of 100–200 km/sec, more powerful waves with velocities exceeding 200 km/sec, and, finally, very weak waves with velocities less than 20 km/sec.

In the first case, appreciable ionization arises only at maximum electron temperature. Prior to this maximum, at relatively high temperatures, many neutral hydrogen atoms exist in the wave. It is therefore very probable that in this region lines of the Lyman series are excited, owing to electron collisions. These quanta are degraded as they scatter and are most noticeable in the Balmer series. A considerable amount of energy is used in ionizing hydrogen and helium, while luminescence of the gas near the maximum electron temperature in the observed lines is relatively low. Here luminescence may be increased in the resonance lines of the most abundant elements. After the fall in electron temperature below its maximum, hydrogen luminescence is reduced, although the hydrogen recombination spectrum appears. Its intensity cannot be great, since the recombination probability is low. Here the principal role is played by the excitation of forbidden lines, which become especially intense at temperatures of about 25,000–40,000°. This luminescence then falls off. Emission intensity in the continuous spectrum must be weak because of the low optical thickness of the wave front. The region of the Lyman continuum is an exception, since the optical thickness there is large. However, Lyman quanta are degraded during diffusion and therefore carry away hardly any energy. Owing to large Doppler shifts, the buildup of L_α quanta evidently does not occur.

Shock-wave spectra in the visible must be similar in general, since luminescence originates primarily at a later stage through which any moderate shock wave passes, regardless of the initial conditions. There is one special feature, the degree of ionization of elements, which depends on shock-wave intensity. For example, the relative intensity of the lines λ 3727 and [O III] can be greatly changed, depending on whether the maximum temperature was sufficient for substantial secondary ionization of oxygen. Calculations show [379] that when T_2 is increased from 200,000° to 400,000° the intensity of [O III] is raised by almost an order of magnitude, while at the same time the intensity of other lines is hardly altered.

In very intense shock waves propagating at velocities much greater than 200 km/sec, ionization of the principal elements is almost complete. Therefore, the luminescence from these waves is relatively low and they are practically adiabatic waves in a totally ionized gas [380]. Here, primarily lines from elements many times ionized are excited, all in the far-ultraviolet spectrum, many of them beyond the Lyman limit. Such shock waves are difficult to observe by light in the visible spectrum. Incidentally, calculations by Heiles of the excited levels of the hydrogenlike ion O VIII indicate that there may be considerable cooling of the gas in filaments whose distance from the supernova exceeds 30 pc [380a].

Weak waves, propagating at velocities of about 10 km/sec, do not alter the state of ionization and therefore do not lead to the appearance of new lines. However, an increase in temperature, and especially in density, may lead to intensification of luminescence in lines that were previously weak because of the low probability of excitation at low temperature and density. In H II regions, this results in an intensification of a fluctuating luminescence. In H I regions, weak shock waves intensify luminescence in the unobserved infrared spectral region (Sec. 8).

Luminescence due to shock waves may be observed, apparently, in filamentary nebulae that border the expanding envelopes of supernovae. One of these nebulae, the Loop in Cygnus, has already been discussed in Sec. 20. It is expanding with a velocity of 50 km/sec at the inner boundary and at 110 km/sec at the outer [18:227]. We may interpret the filamentary structure as the lines of intersection of shock waves, diversely refracted by the discontinuities of the interstellar medium (focusing of the front by discontinuities), forming, therefore, a network of lines of intersection, and not a continuous front. Here, luminescence originates not in the envelope itself, but in the compressed shock wave of the interstellar gas. The computation of spectra for such a model and comparison with observed data given by S. B. Pikelner lead to the results summarized in Table 21.2 [379].

Table 21.2

Lines	Theoretical spectrum, $v_1 = 105$ km/sec	Observed spectrum
[S II] 6716–6730	1	1.1
[N II] 6584–6548	1	1.1
[O III] $N_1 + N_2$	6	1.9
[O II] 3727–3729	10	10
H_β	0.8	0.4

Agreement is satisfactory in general, except for the lines $N_1 + N_2$, but the intensity of this doublet, as we noted before, depends very much on the original conditions at the wave front. A slight decrease in v_1 can bring about agreement with the observations of $N_1 + N_2$. In addition, according to data by Code [18:229], lines of [O II] in bright filaments are somewhat weaker than in the table. This also means that we must assume somewhat lower velocity.

So far, no other serious attempts to associate the spectra of any formation in interstellar space with shock waves have been made.

We can obtain by a simple procedure an equation that determines the total amount of energy radiated per square centimeter of wave surface per second [8]. We have

$$J = (\rho_1 v_1)^3\left(\frac{1}{4\rho_1} - \frac{1}{\rho_2}\right)\left[\frac{5}{2}\frac{p_1 + \rho_1 v_1}{(\rho_1 v_1)^2} - \frac{1}{2\rho_1} - \frac{2}{\rho_2}\right]$$

$$\approx 2(\rho_1 v_1)^3\left(\frac{1}{4\rho_1} - \frac{1}{\rho_2}\right)\left(\frac{1}{\rho_1} - \frac{1}{\rho_2}\right), \quad (21.30)$$

where p_1 is the gas pressure before the passing of a wave front with velocity v_1, and ρ_2 is the gas density behind the region of luminescence. In the second equation we use the condition $\rho_1 \ll \rho_1 v_1^2$. If $\rho_2 \gg 4\rho_1$, then

$$J \approx \frac{1}{2}\rho_1 v_1^3, \quad (21.31)$$

as we expect, since the expression on the left-hand side is simply the flux of kinetic energy across the front of the shock wave, which in this case is almost entirely transformed into radiation. When $\rho_1 \approx 2 \times 10^{-23}$ g/cm^3 and $v_1 \approx 100$ km/sec, Eq. (21.31) gives $J \approx 10^{-2}$ erg/cm^2 sec. However, it is difficult to estimate what portion of this radiation falls in the visible spectrum. The increase in radiation intensity with higher shock-wave velocities is apparently still less than would be given by the cubic relation (21.31), because in more powerful shock waves a relatively larger amount of energy is emitted in the nonvisible spectrum and, as we noted before, in such waves the role of luminescence is a small one, while in (21.30) it is impossible to satisfy the condition $\rho_2 \gg 4\rho_1$.

To avoid misunderstanding, we emphasize the substantial difference between the filamentary structure of young nebulae like the Crab or Cassiopeia A and old nebulae like the Loop in Cygnus. In these latter nebulae the envelope no longer radiates, but rather a relatively thin layer of cold interstellar gas carried along by the motion. Here radiation is excited only forward of the front. Nebulae from recent outbursts of supernovae emit from the envelope itself. Other excitation mechanisms act here as well (ultraviolet rays, as in the Crab Nebula, ionization by cosmic-ray particles, or other similar mechanisms). That the spectra in the two cases are similar is explained by the fact that the most effective emitting layer has a temperature of approximately 20,000°, and that there is such a layer in each envelope, independent of the conditions of its formation. Observations in the far ultraviolet would allow us to distinguish the different conditions of formation of the filamentary structure.

The formation of filaments in these objects also differs. If the filamentary structure of the envelope is due to instabilities in the envelope itself caused by penetration ("bulging") of the magnetic field across its boundary, then the formation of filaments in nebulae like the Loop is connected with intersections of shock waves.

Shock Waves in Interstellar Magnetic Fields

The theory of magnetohydrodynamic shock waves is similar to the theory of ordinary shock waves, but more complex. Therefore, we shall restrict ourselves to certain theoretical results necessary for the analysis of the magnetohydrodynamics of interstellar gas.

The effect of a field on a shock wave propagated along the lines of magnetic force is unimportant. The nature of this wave does not differ from the cases already discussed. If the velocity of the shock wave is perpendicular to the direction of the lines of magnetic force, the magnetic field impedes compression of gas in the wave. Because of the adherence principle, an increase in magnetic-field intensity in this case must be proportional to an increase in density.

In the general case, when the normal to the front of the shock wave makes an angle ϕ with the direction of the magnetic field, the change of field intensity in the wave satisfies the condition [8]

$$\frac{H_2}{H_1} \leqslant \frac{\rho_2}{\rho_1} \sin \phi. \tag{21.32}$$

This can be considered an equality if $H_2 \gg H_1$. Thus, in a magnetic shock wave, part of the kinetic energy is transformed into magnetic energy. This leads on the one hand to a pressure drop and consequently to dissipation of energy as heat, while on the other hand, after the passage of the wave front, the "compressed" magnetic field can reversibly regain its original condition. This gives a partially elastic character to a shock wave in a magnetic field. We note that the gas pressure behind the front of the shock wave in interstellar conditions cannot return the gas to its original state, not only because of the irreversibility of the compression process (entropy increase), but also because its energy is reduced owing to luminescence.

The greatest increase in the magnetic field, and consequently the smallest dissipation of energy and the greatest elasticity, occur in so-called perpendicular shock waves ($\phi = \pi/2$), where the field is parallel to the wave front. We shall consider this case in more detail. If we assume infinite conductivity of the gas, then $H_2/H_1 = \rho_2/\rho_1$. In a strong shock wave without luminescence, the magnetic energy is increased by $[(\gamma + 1)/(\gamma - 1)]^2 = 16$ times. If luminescence plays an important part, field intensification may be considerably greater, but we must stress that the compression of the gas itself is less here than in a wave without a field under the same initial conditions. For example, if $H_1^2 \gg 4\pi N_1 k T_2$, then instead of (21.27) we obtain

$$\frac{H_2}{H_1} = \frac{\rho_2}{\rho_1} = \frac{(8\pi\rho_1)^{1/2}}{H_1} = \frac{v_1\sqrt{2}}{a_1}, \tag{21.33}$$

where a_1 is the velocity of the Alfvén wave. We note that here compression is proportional to the first power of the wave velocity and not to the second power as it is without the field. A more general equation, suitable for any relation between H_1 and T_2, is given in [8].

If we ignore luminescence, the dissipation of kinetic energy into heat in strong perpendicular waves, generally speaking, may be large, since in this case the growth of the magnetic field is limited and the work required to compress the field is relatively small. The equation that describes the dissipation of energy in a strong perpendicular shock wave, unaccompanied by luminescence, is [8]

$$\Delta\mathscr{E} = \frac{5}{2}\frac{kT}{\mu m_{\mathrm{H}}} - \frac{17H_1^2}{64\pi\rho_1}. \tag{21.34}$$

If the velocity of the shock wave is close to the Alfvén wave velocity, $a_1 = H_1/(4\pi\rho)^{1/2}$, then dissipation is noticeably reduced. In this case [379a], the irreversible temperature increase is equal to

$$T_2 \approx 0.2\mu m_{\mathrm{H}} a_1^2/3k. \tag{21.35}$$

An estimate of the heating and ionization of the gas by magnetohydrodynamic shock waves under such conditions is given in [379a].

In oblique shock waves with a magnetic field ($\phi \neq \pi/2$), we can, by a suitable choice of the system of coordinates, represent the gas as moving along the lines of magnetic force, both before and after the passage of the wave front. At the front of the wave itself the lines of magnetic force and the direction of motion of the gas are refracted. The greater the discontinuity in density in an oblique shock wave with luminescence, the closer the lines of force in the compressed region are to the plane of the front. Compression of only the perpendicular component of H explains qualitatively the reduced growth of the field in the wave and consequently its reduced elasticity and reduced effect on dissipation. Quantitative calculations on oblique shock waves are cumbersome and therefore they are not introduced here. We need only to note from (21.32) that in a strong magnetohydrodynamic shock wave accompanied by luminescence the change in the magnetic field can be quite large.

The structure of magnetic shock waves has certain peculiarities that are still not completely clear. Our discussion has referred to shock waves in a plasma of high conductivity. If the magnetic viscosity is less than the kinematic viscosity or the coefficient of thermal conductivity, the magnetic field follows everywhere behind the gas and the foregoing conclusions remain correct. If, on the other hand, the magnetic viscosity is comparatively high, the magnetic field seems to detach itself from the gas particles and the discontinuity in the field is smoother than the discontinuity in density. In interstellar space, both cases can obviously arise, depending on the relation between v_m' and v_k (see Sec. 16), in view of the changed ionization state in a shock wave. No detailed analysis has so far been made.

There is another important feature of magnetic shock waves in plasmas. We know that the parameter $\omega_H \tau$ is very large in interstellar space. In other words, interstellar shock waves can be described as collisionless. Many papers have been devoted to the study of collisionless shock waves [381–384, 18].

If in a nonionized gas the transmission of an impulse can originate only through collisions between particles, then transmission in a plasma takes place through the magnetic field and through collective particle interactions caused by the effect of the electromagnetic fields on the motion of the particles themselves. Therefore, the thickness of the front may be considerably less than the mean free path. In practice, it is restricted by dispersion effects, in which the velocity of the disturbance begins to depend on wavelength and is equal to c/ω_0, where ω_0 is the natural frequency of the plasma. Behind the region of growth of a field of such a size or somewhat larger, a series of pulsations follow, rapidly changing into random plasma oscillations. Thus, the increase in entropy that must occur in a shock wave appears here not as a temperature rise, but as oscillations which only much later are converted into heat by collisions. Since it is not possible to observe the structure of a front in the interstellar medium, we shall not consider this in detail. We must note, however, that the mechanism for conversion of directed motion into random motion at the boundary of the front fixes the initial conditions for calculating shock-wave luminescence. If this mechanism is connected with ionic instability, we may assume, as before, that the ion temperature is high and the electron temperature is at first low. If the oscillations are connected with the bunching instability of electrons, the electron temperature may be high from the beginning.

The theoretical conclusions concerning the abruptness of the front of collisionless shock waves were verified both by the sudden appearance of disturbances in the Earth's magnetic field due to the solar flux moving with a velocity of 1000 km/sec (here the mean free path is about the distance from the Sun to the Earth) and by experimental data obtained under laboratory conditions [384].

We note that the magnetic field has no appreciable effect on the character of the region where electron and ion temperatures converge or on the luminescence zone. As already stated, the magnetic field can impart an interaction between particles in the absence of collisions, simply by particle adherence to the magnetic lines of force. If the gas pressure is relatively low, we may obtain an isolated symmetric impulse of finite amplitude with a sharpness of front of $\sim c/\omega_0$. Because of the adherence principle, the pressure is now $p \approx H^2/8\pi \sim \rho^2$; hence $\gamma = 2$ and the maximum increase in intensity and density in the perpendicular wave is given by the expression

$$\frac{\rho_2}{\rho_1} = \frac{H_2}{H_1} = \frac{\gamma + 1}{\gamma - 1} = 3. \tag{21.36}$$

With respect to the intensity of the wave, its velocity falls between the limits

$$\frac{H_1}{(4\pi\rho_1)^{1/2}} < v < 2\frac{H_1}{(4\pi\rho_1)^{1/2}}. \tag{21.37}$$

However, no dissipation of energy occurs, so that this wave is completely elastic; behind the wave front the medium reverts to its original state [385]. If the impulse is not caused by a brief shock but by a slow piston motion, we obtain a shock wave with the oscillatory structure already mentioned. When the piston moves rapidly, the velocity of the front is greater than $2a$, the condition (21.37) is not fulfilled, and the front does not consist of isolated impulses, but is a highly unstable region containing motions with many different velocities. One particular property of shock waves in a magnetic field, acceleration of fast particles at the front, was discussed in Sec. 17.

Plasma clouds and streams of relativistic particles ejected by radio galaxies move in extragalactic space with velocities close to the speed of light. If the pressure of such a flux is considerably greater than the energy density of the stationary extragalactic medium ρc^2, this flux creates a relativistic shock wave. A relativistic shock wave is also formed if the Alfvén velocity in intergalactic space is comparable with the velocity of light.

The properties of relativistic shock waves in extragalactic space were investigated in [384a], where it was shown that, if the Alfvén velocity is close to the velocity of light, such a wave can be characterized to a first approximation by the value of a single parameter, the amplification factor $(p_s/\rho c^2)^{1/2}$, where p_s is the dynamic pressure of the flux or the pressure of the magnetic field. The discontinuities in a relativistic shock wave of the concentration of particles, the magnetic-field intensity, and the average particle energy are determined to within factors of the order of unity by the relations

$$\frac{N_2}{N_1} \approx \frac{H_2}{H_1} \approx \frac{\bar{\mathscr{E}}_e}{m_e c^2} \approx \frac{\bar{\mathscr{E}}_p}{m_p c^2} \approx \left(\frac{p_s}{\rho c^2}\right)^{1/2}, \tag{21.38}$$

where m_e and m_p are the masses of the electron and proton, respectively. The structure of a relativistic shock wave with a magnetic field has not yet been studied. Since the mean free path of relativistic particles is long, we may assume that the structure of such shock waves is analogous to the collisionless shock waves in plasma physics. Estimates show that under certain conditions the amplification factor may reach values of the order of 1000. Fairly intense synchrotron radiation is excited behind such wave fronts.

The propagation of shock waves in the interstellar medium is influenced by the cosmic dust. This problem was investigated in a series of papers [384b, c, d]. When the shock wave moves through the gas-dust medium, the dust particles at first do not take part in the motion of the gas. This leads to the appearance of differences in the velocities of the gas and dust components. After some time the dust particles are accelerated to the velocity of the gas. Under different conditions there are some possibilities for explanation of the condensation of dust, driving dust particles away from the region, and so on.

In conclusion, let us enumerate the phenomena in interstellar space associated with the formation of shock waves.

1. Envelopes from novae or supernovae moving in interstellar space "rake up," as it were, the interstellar medium behind them. As a result, a very intense shock wave passes through the interstellar gas. These shock waves will be considered in Sec. 23.

2. During the ionization of the interstellar medium (conversion of H I regions into H II zones), the gas pressure is increased, which begins to compress the gas in the H I region. Shock waves that will be discussed in Sec. 22 also arise under these circumstances.

3. Interstellar clouds in random motion collide with one another. When this occurs, a region of hot compressed gas results, which propagates as shock waves through both the colliding clouds. As a result, the dissipation of energy in the interstellar medium is increased and the gas temperature in H I regions is raised. Here, the magnetic field plays an especially important part, since it renders collisions between clouds partially elastic and thus reduces dissipation and gas heating in H I regions. This problem is discussed in more detail in Sec. 23.

22. Ionization Fronts

An unusual kind of disturbance in the motion of interstellar gas not encountered under terrestrial conditions is the ionization front separating H I and H II regions. We noted in Sec. 1 that the transition region between H I and H II zones was comparatively narrow, or $\sim 0.05/N_H$ pc; in many cases, this allows us to treat it as a surface disturbance.

Another important peculiarity of an ionization front is the sharp temperature changes. It is well known that the gas temperature in H II regions is almost always about 10,000°, while in H I regions it is evidently less than 1000°. This inequality in temperature creates at the ionization front a discontinuity in pressure that can set the interstellar gas in motion. The study of ionization disturbances was started by Kahn [386].

The radius of the Strömgren zone, determined in Sec. 1, characterizes the quasi-stationary position of the ionization front. In fact, it is determined from conditions of ionization equilibrium, which are established after a recombination time $t \approx 1/N_e\alpha_t(T)$. Since this time ($\sim 10^4$ yr) is usually much less than the characteristic time scales of motion, an ionization-recombination equilibrium can become established. During this same time, the pressure of the hot gas causes expansion of the H II region and motion of the cold gas. As the density falls, absorption in the Lyman continuum decreases and the quanta can be propagated further, ionizing fresh layers of gas. In the future the motion of the ionization front will be referred to a coordinate system connected with the front.

Ionization fronts may be observed as bright circular hoops separating the H II emission zones and the dark H I regions (Sec. 10). A detailed statistical study of the hoops was made by Pottasch [18:235]. Their width is of the order of a few hundredths of a parsec and their mean density is 10^3 cm^{-3}.

A qualitative picture of hoop formation is the following. When the pressure of hot gas sets the cold unionized gas in motion, condensation is set up in the cold gas and consequently a shock wave, traveling before the ionization front, develops. Thus, the density of gas directly ahead of the ionization front is greater than in an H II region. The ionization of this condensation increases the pressure there, which in turn causes the gas to emerge at the side of the H II zone. The observed hoops, as a rule, are convex toward the star. They obviously constitute denser H I regions where shock waves and ionization fronts are retarded and then flow around the condensation. We shall show later that this qualitative representation agrees with quantitative calculations.

The Parameters of the Ionization Front

We shall derive certain relations that allow us to calculate the parameters of ionization fronts.

The number of hydrogen atoms intersected by an ionization front per second must be equal to the number of ionizing $\mathrm{L_c}$ quanta striking this front in the same time. Neglecting for the present absorption of $\mathrm{L_c}$ quanta in an H II zone and ignoring diffuse $\mathrm{L_c}$ quanta (these partially compensate one another), we can write these conditions in the form of the equality

$$N_{\mathrm{H}}v_{\mathrm{H}} = N_{\mathrm{p}}v_{\mathrm{i}} = \frac{R_*^2}{4r^2}\mathrm{N_L}, \tag{22.1}$$

where v_{H} is the velocity of hydrogen atoms relative to the ionization front, v_{i} is the velocity with which protons emerge from the front, r is the distance from the star of radius R, and $\mathrm{N_L}$ is the flux of $\mathrm{L_c}$ quanta per square centimeter of stellar surface.

During the flow across the ionization front, the conditions of conservation of momentum and energy flux must be satisfied while the interaction with the radiation is taken into account. These conditions may be written in the form of the equations

$$N_{\rm H}(kT_{\rm H} + m_{\rm H}v_{\rm H}^2) = N_{\rm p}(2kT_{\rm i} + m_{\rm H}v_{\rm i}^2) + \frac{(\bar{\varepsilon}_0 + \chi_1)R_*^2}{4cr^2}\,{\rm N_L}, \qquad (22.2)$$

$$N_{\rm H}v_{\rm H}(5kT_{\rm H} + m_{\rm H}v_{\rm H}^2) = N_{\rm p}v_{\rm i}(10kT_{\rm i} + m_{\rm H}v_{\rm i}^2) - \frac{\bar{\varepsilon}_0 R_*^2}{4r^2}{\rm N_L} + J. \qquad (22.3)$$

Here $T_{\rm H}$ and $T_{\rm i}$ are the temperatures in H I and H II zones, $\bar{\varepsilon}_0$ is the mean energy acquired by an electron during ionization, the factor 2 in both equations takes into account the pressure of the electron gas resulting from ionization, and J is the energy flux inside the ionization front converted to luminescence. Two simplifying assumptions were made in the derivation of Eqs. (22.2) and (22.3): first, the pressure and energy of all elements other than hydrogen were neglected, and second, it was assumed that the electron and ion temperatures are approximately equal to one another inside the boundary of the ionization front. The relations (22.1)–(22.3) connect the parameters on the two sides of the ionization front. There have been no detailed calculations made on the structure of ionization fronts similar to those on the structure of shock waves described in the preceding section.

A solution of the system (22.1)–(22.3) with given values of ${\rm N_L}$, R_*, $r\bar{\varepsilon}_0$, $N_{\rm H}$, $T_{\rm H}$ ($\sim 1000°$), and $T_{\rm i}$ ($\sim 10{,}000°$) allows us to determine $V_{\rm H}$, $V_{\rm i}$, and $N_{\rm p}$, the velocity of the gas, its rate of flow across the ionization front, and its density behind the front.

However, in addition to satisfying the system (22.1)–(22.3), the parameters of the ionization front must often satisfy yet another equation describing the so-called Jouguet-point condition. In fact, if the escape of gas from the ionization front is caused by only the gas pressure near the front, it must take place with the local speed of sound, that is, it must satisfy the condition

$$v_{\rm i} = c_s = \left(\gamma \frac{kT_{\rm i}}{\mu m_{\rm H}}\right)^{1/2}, \qquad (22.4)$$

where γ is the adiabatic index and μ is the molecular weight. A similar case arises in shock waves from detonations, where, because of the motion, energy appears inside the wave front [375]. Gas motion in the case of ionization fronts is associated with heating inside the front, in contrast to ordinary shock waves, whose motion is associated with the external pressure or energy of the incident gas stream. We note that the escape of gas, and consequently the Jouguet-point condition, occur only after the formation of the quasi-steady H II zone. Until this time the velocity of the ionization front is determined by the number of ionizing quanta and is not connected with energy released inside the front.

Addition of Eq. (22.4) to the system of Eqs. (22.1)–(22.3) changes the system so that solution of these equations is possible only for certain values of some of the given parameters. For example, gas can cross an ionization front of this type only when its density ahead of the front, N_H, has a completely definite value, which we call $N_H^{(0)}$. As we shall show below, this means that if the true density is $N_H < N_H^{(0)}$, a shock wave precedes the ionization front, compressing gas in the H I region to the required value N_H. On the contrary, when $N_H > N_H^{(0)}$, a wave diluting the gas may precede the ionization front.

However, we must emphasize that the Jouguet-point condition in the form (22.4) is correct only if the temperature inside the ionization front increases monotonically from T_H to T_i. If the temperature inside the front rises at first from T_H to some maximum value T_{max} and then falls to $T_i < T_{max}$ because of luminescence, we must substitute T_{max} for T_i in (22.4). In principle, the existence of such maximum temperatures is possible, even though luminescence is not very intense and is begun only after some higher temperature is attained, or if ε_0 is large.

The quantity T_{max} in turn depends more strongly on the structure and parameters of the ionization front than does T_i, which is determined by the conditions of thermal equilibrium in the H II region and therefore depends but little on the motion of the front. Hence, such a definition of the Jouguet point, essentially introducing a new parameter T_{max}, does not change the system (22.1)–(22.4), and allows us to obtain a solution for arbitrary values of the original parameters (for example, N_H). A qualitative analysis of the structure of such ionization waves was given by Axford [387], using a simplified expression for the function $L_{ei} = 0.97 \times 10^{-31} (T - 4000)N_e N_p$, which determines the luminescence of the gas and consequently the behavior of the temperature inside the ionization front.

The Jouguet-point condition cannot be satisfied if the gas behind the ionization front cannot flow freely, as, for example, when an H II zone is bounded on all sides by a dense H I region and a flux into the H II zone raises its pressure.

Thus many diverse conditions are possible at ionization fronts, especially if we consider the different geometries of these disturbances (spherical, cylindrical, or plane waves) and the change with time of the flux of L_c quanta as a result of evolution of the star or absorption in the outflowing gas. During successive analyses of the structure of the ionization discontinuity, we must allow for the change in $\bar{\varepsilon}_0$ with depth inside the front [62], the difference between the electron and ion temperatures, the change in the ionization state of other elements, and the losses due to luminescence and other effects. All this makes the general study very complex. Therefore we shall restrict ourselves to a simplified picture, which proves to be sufficiently accurate to solve many problems associated with a study of the motion of ionization fronts [8].

We shall assume that cooling inside the front is so effective that the gas temperature does not reach the maximum and only gradually rises from T_H to $T_i = 10{,}000^\circ$. The validity of this assumption is substantiated by the following considerations. The cooling time for the gas in both H II regions and H I zones is less than the time required for recombination or ionization, so that to the extent that the gas flows across the ionization front a quasi-equilibrium state whose temperature is fixed by the number of free electrons N_e will be set up continuously. Since this number increases monotonically inside the limits of the ionization front, we may assume the temperature rise to be monotonic. In particular, in the work by Axford cited above, too low a value for the rate of luminescence was used because the possibility of intense cooling in the H I region was neglected, which also led to the appearance of a pronounced temperature maximum. Finally, we note that observations of hoops have not shown any increase in their temperature above the ordinary temperature of an H II zone [188]. We shall assume further that the interaction of gas with radiation is so effective that the temperature T_i is determined only by the radiation and does not depend on the state of motion. Under these conditions we may take $T_i = 10{,}000^\circ$ and use the isothermal velocity of sound, that is, $c_s = (2kT_i)^{1/2}/vm_H = 11$ km/sec. We also disregard Eq. (22.3), since in this case the luminescence J just compensates the remaining terms. Finally, taking $T_H \ll T_i$ and neglecting the small last term in Eq. (22.2), we obtain from the system of Eqs. (22.1), (22.2), and (22.4) the two possible solutions [8]:

$$\text{I.} \qquad v_H = 2v_i \approx 2c_s \approx 22 \text{ km/sec},$$

$$N_H \approx \frac{1}{2} N_p \approx \frac{\mathrm{N_L} R_*^2}{8c_s r^2} \approx 0.028 \frac{[U(\text{sp. cl.})]^3}{r^2} \text{ cm}^{-3}; \tag{22.5}$$

$$\text{II.} \qquad v_H \approx c_s \frac{T_H}{4T_i} \approx 0.27 \text{ km/sec } (T_H = 1000^\circ), \tag{22.6}$$

$$N_H \approx \frac{4T_i}{T_H} N_p \approx 40 N_p \approx \frac{5\mathrm{N_L} R^2}{c_s r^2} \approx 1.1 \frac{[U(\text{sp. cl.})]^3}{r^2} \text{ cm}^{-3}.$$

Here U(sp. cl.) is the function of the stellar spectrum defined by Eq. (1.4), and r (pc) is the distance from the star. In agreement with considerations discussed in the preceding section, the temperature of the gas in the H I region to be used in Eq. (22.6) is taken as 1000°. A more general solution, correct when T_H is comparable with $2T_i$, is given by Kaplan [8] and Olinik [288].

The solution of Eqs. (22.5) corresponds to a compressive ionization front of type R, according to Kahn's terminology [386]. Such a front can be propagated through the tenuous gas in H I regions. An R-type front moves with supersonic speed in H I regions and therefore does not create sonic or shock-wave-type disturbances ahead of it. Such fronts appear either in the initial formation of H II zones or when an H II zone expands into the extended medium between clouds.

The solution of Eqs. (22.6) corresponds to a tenuous ionization front of type D in Kahn's terminology. In this case, the dense gas of H I regions intersects the ionization front at low velocity and is then greatly expanded. The structure and the motion of the gas are determined by its flow outward. Obviously, for the formation of a D-type front, either the ionization front must enter a dense H I cloud or a shock wave must precede the front, condensing the gas in H I. Bright hoops evidently occur with D-type ionization fronts.

A completely definite value of N_H was obtained from Eqs. (22.5) and (22.6), which, as we have already noted, results from the redefinition of the system of Eqs. (22.1)–(22.7) to include the Jouguet-point condition. We can obtain more "flexibility" in the value of N_H if this condition is somewhat modified. For example, if the gas in the H II region is sufficiently rarefied, then during the motion of a D-type front the velocity of outward flow can also be greater; for example, it can correspond to the adiabatic velocity of sound, $c_s = (10kT_i)^{1/2}/3m_H \approx 20$ km/sec. The value of N_H is correspondingly reduced. Finally, as we noted previously, if a temperature maximum occurs inside the front, the velocity c_s is also somewhat increased. However, we must stress that all these factors cannot change N_H by more than a factor of 2.

We need not, in general, allow for luminescence in the case of an ionization front in a highly rarefied gas ($\bar{J} = 0$). In this case, temperature inside the front increases monotonically from T_H to T_i, which is now determined from the common solution of the system (22.1)–(22.4) under the assumption that c_s is the adiabatic velocity of sound.

We give here only the results of Kaplan [8]. For an R-type ionization front,

$$T_2 = \frac{\gamma(\gamma - 1)}{\gamma + 1}\frac{\mu_2 \varepsilon_0}{Rm_H}; \qquad \frac{\rho_2}{\rho_1} = \frac{v_1}{v_2} = \frac{\gamma + 1}{\gamma}. \tag{22.7}$$

In the case of a D-type ionization front,

$$T_2 = \frac{\gamma - 1}{\gamma(\gamma + 1)}\frac{\mu_2 \varepsilon_0}{Rm_H}; \qquad \frac{\rho_2}{\rho_1} = \frac{v_1}{v_2} = \frac{1}{\gamma + 1}\frac{\mu_2 T_1}{\mu_1 T_2}. \tag{22.8}$$

We recall that here $v_2 = c_s = (\gamma R T_2/\mu_2)^{1/2}$. Substitution of numerical values for ε_0 from Sec. 6 and also of $\gamma = 5/3$, $\mu_1 = 1.5$, and $\mu_2 = 0.7$ leads to the temperature limits $9000 < T_2 < 15{,}000^\circ$ with Eqs. (22.7) and $3000 < T_2 < 6000^\circ$ with Eqs. (22.8). This means that in the case of a D-type wave there follows behind the front a region of monotonic temperature increase up to 10,000°. Such a wave does not differ in principle from the isothermal ionization front discussed above. On the contrary, in the case of an R-type front created by a hot star, the maximum temperature inside is less than 15,000°, but greater than 10,000°. Here the Jouguet-point condition is somewhat modified [387]. However, since this maximum does not differ very much from 10,000°, the wave in question will hardly differ from that described by Eqs. (22.5).

If we abandon the Jouguet condition, further possible solutions of the ionization-front equations will exist, although all these solutions may be classified, as before, as R- or D-type fronts. Whether R and D fronts are strong or weak depends on whether the velocity of the outward flow of gas from the front is subsonic or supersonic. Using this terminology, we call fronts incorporating the Jouguet condition (intermediate between strong and weak) critical R and D fronts. A detailed analysis of ionization-front classification is given in [387, 389]. Their properties are qualitatively the same as those with the Jouguet condition.

The Motion of Ionization Fronts

If a hot star suddenly flares up in an H I region, the stream of L_c quanta produced rapidly ionizes the surrounding stellar gas and creates an H II zone with the radius of a Strömgren zone. The initial formation of a Strömgren zone is not connected with gas-dynamic effects, since its formation time, equal in order of magnitude to its recombination time, is considerably less than the time scale of the motion [34].

After the formation of the initial Strömgren zone, its relatively slow expansion begins—the movement of the ionization front into the H I region described at the beginning of this section. As we have already noted, masses of singly ionized compressed gas now enter the H II region. Once in the H II zone, these masses of hydrogen apparently cause the characteristic peripheral structure of many emission nebulae. A semiquantitative analysis of this expansion was given by R. E. Gershberg [34]. He was able to show that H II zones in turn might expand and form condensations at their peripheries.

A more detailed analysis of the formation of H II zones was made by Vandervoort [389a, b, c], in which he posed the problem in the following way. A hot star flares up initially in an infinite medium of neutral hydrogen, and then continues to radiate with a constant intensity. The increase with time of the H II region thus formed must then be calculated.

As follows from Eq. (22.1), the initial velocity v_H of the ionization front for small r is large. The rapid expansion of the ionization region at this step prevents the gas behind the wave front from reaching a state of noticeable motion and therefore the Jouguet-point condition at the existing R-type discontinuities will not be satisfied. Consequently, Eq. (22.4) is not used in the solution of the system (22.1)–(22.3), but rather N_H is considered given. The solution of this system then allows us to obtain the dependence of N_p and v_i on the distance from the star. The equation of motion of the ionization front, following immediately from (22.1), is to a first approximation

$$v_H = \frac{dr_i}{dt} = \frac{R_*^2 N_L}{4r_i^2 N_H}. \tag{22.9}$$

If we assume that N_L is constant, that is, if we neglect the absorption of ionizing quanta in the path of the front, then the solution is elementary, or

$$r_i^3 = \frac{3R_*^2 N_L}{4N_H} t. \tag{22.10}$$

With these conditions, the values of N_p and v_i behind the wave front can be obtained from the system (22.1)–(22.3), and, in addition, the spatial and temporal dependence of the density and velocity of the gas can be calculated from the ordinary equations of gas dynamics. In the time $\frac{1}{12}(R_*^2 N_L/c_s^3 N_H)^{1/2}$ the velocity of the ionization front falls to a value such that from then on the Jouguet-point condition will be satisfied.

The absorption behind the front of ionizing quanta was taken into account in the work of Axford [389d]. On the other hand, it was assumed in this work that the electron concentration N_e was constant in the H II region. The change in time of the distance r_i of the ionization front from the star is determined by the equation

$$r_i = \zeta_0 \left[1 - \exp\left(-\frac{\alpha N_e}{4\pi} t \right) \right], \tag{22.11}$$

where ζ_0 is the radius of the Strömgren sphere and α is the recombination coefficient. Appreciable motion of the gas begins when the radius of the ionized region reaches 90 percent of the dimension of the Strömgren sphere. This takes place 10^5–10^6 yr after the sudden explosion of an O-type star.

The expansion of a nebula taking into account evolution of the exciting star was discussed by Mathews [389f, g]. This is necessary because the time of expansion is comparable with the time of evolution of hot stars. The intense stellar wind may produce the hollow in the middle of the nebula. The last stage of expansion of an H II region was investigated by Lasker [389h].

If the gas density in H I is close to the value in (22.5), an ionization front in the form of an isolated front of type R can advance into an H I region, leaving behind an H II zone. As follows from (22.5) its velocity is about 22 km/sec and the density of the gas during the transition into the H II region is not changed by more than twice. However, this case is unusual. Even if the condition (22.5) is met during the initial motion, it will finally be broken as the H II zone expands.

Apparently, motions of a D-type ionization front that satisfy Eqs. (22.6) are often encountered. In this case, a shock wave precedes the ionization front, compressing the gas in the H I zone, but leaving the hydrogen neutral, until its density no longer satisfies the conditions (22.6). Such motions arise, in particular, if the ionization front approaches a denser gas region in the H I zone; as we have already noted, it is precisely in this way that hoops occur.

The quantitative study of this motion involves a number of difficulties, partly mathematical and partly due to the uncertainty of the physical data on the changes of L_c emission flux with time and to the absence of proper expressions for calculating the rate of cooling of the gas.

This problem has been discussed in a number of papers in an idealized setting. Similar results have been obtained in every case. Schatzman and Kahn [17:115] and also Savedoff and Green [390] gave the rate of expansion of the ionized region ($\sim$14 km/sec), and determined the parameters of the shock wave compressing the gas. In particular, in [17] the isothermal character of this shock wave was taken into account, which permitted a more accurate value to be obtained for the density increase in the compressed region up to 200 times. Goldsworthy [389] and S. A. Kaplan [8] discussed the self-similar problem of the consistent motion of ionization-shock-wave fronts.

The problem of gaseous motion and the motion of disturbances is self-similar if the motion depends only on two fixed parameters with independent dimensions. In this case, the partial differential equations of gas dynamics can lead to a system of ordinary differential equations, which can be solved, sometimes analytically, and often numerically. The concept of self-similar motion will be considered in more detail in the following section.

Strictly speaking, the motion of a ionization front is determined by many parameters with independent dimensions and is therefore not self-similar. However, in the hypothetical problem we may restrict ourselves to two essential parameters, the flux of ionizing L_c quanta ($N_L R_*^2$), and the velocity of the emergence of the gas from the ionization front. We cannot assess accurately in this case changes in the flux of L_c quanta with time, the rate of gas cooling in the compressed region, and so on. Moreover, if we consider the spherical problem, then, as follows from (22.5) or (22.6), we must assume that the gas density in an undisturbed H I region changes with distance from the ionizing star according to r^{-2} (within the limits of the self-similar solution, $N_L R_*^2/c_s$ = const.). If the ionization front is a symmetric cylinder, we must take $N_H \sim 1/r$ [390]. Only for a flat ionization front can we consider that $N_H \approx$ const.

A detailed solution of the self-similar problem for the motion of a flat ionization front is presented in [8] and for spherical and cylindrical fronts in [390]. The results of these solutions may be summarized as follows. First a shock wave travels through an H I region with a velocity of 13–20 km/sec, compressing it several tens of times, depending on the temperature in the condensed region. Then, with almost the same velocity relative to the noncondensed gas, a D-type ionization front travels in the condensed regions of the H I zone, by which ionization and expansion of the gas take place. The relative thickness of the condensed H I region is inversely proportional to the increase in density, and consequently amounts to several percent of the radius of the H II zone. The velocity of the ionization front relative to the compressed gas is low (see (22.6) for v_H). Thus, we may regard the condensed H I region as a kind of "shell" through which gas is slowly passing, at first being compressed and then ionized and expanded. As time passes, the size of the "shell" is increased ($\sim t$). The case of a flat ionization front is analyzed in more detail with regard to equations and relations in [8].

The influence of the magnetic field on the structure of the ionization front and on the conditions of its motion was studied by Abe, Sakashita, and Ono [389e]. To the system (22.1)–(22.4) were added terms to account for the magnetic pressure. For simplicity, they considered the case of a magnetic field parallel to the front of the ionization wave under the assumption of complete freezing. Considering the influence of the field decreases the compression of the gas and somewhat increases the velocity. These studies were also generalized to deter-

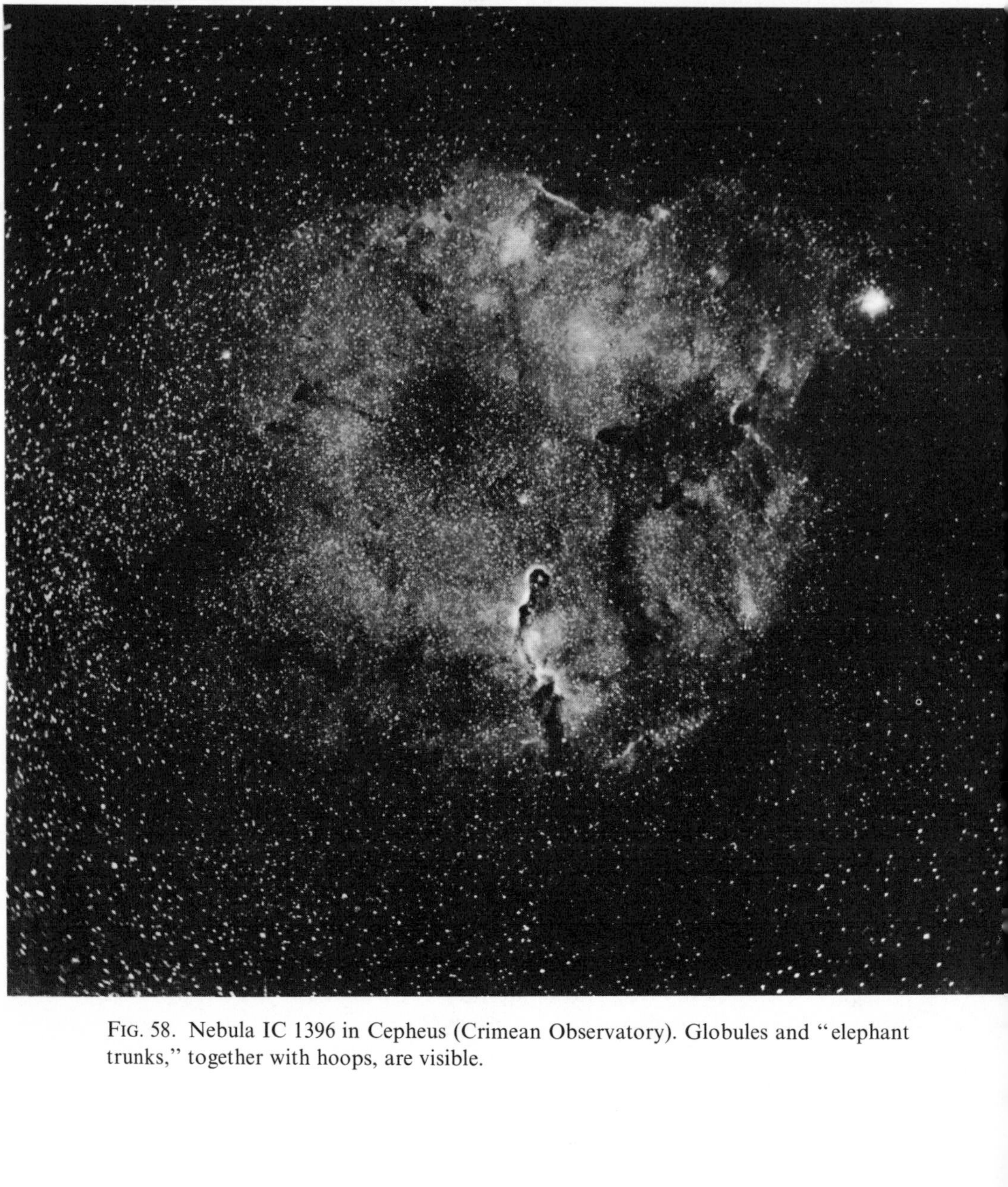

FIG. 58. Nebula IC 1396 in Cepheus (Crimean Observatory). Globules and "elephant trunks," together with hoops, are visible.

mining the effect of the magnetic field on the solution obtained in [8] of the self-similar problem on the motion of the ionization discontinuity. As would be expected, the increase in density between the shock wave and the ionization discontinuity is less here and the relative velocity of these waves is greater. In addition, the form of the front becomes ellipsoidal.

An ionization front can maintain its own form only when motion takes place in a homogeneous or a spherically symmetric medium.

If a gas-dust cloud of high density exceeding N_H from (22.5) or (22.6) is found in the path of the front, it remains unionized. The ionization front appears to bend around this high fluctuation of density. This leads to the formation of inclusions of H I regions inside H II zones. Such inclusions will be compressed, owing to the pressure of hot gas in the H II zones, and in this way they may be the origin of globules, the dense dark formations usually found in emission regions. In agreement with contemporary ideas, these globules are embryonic stars. Compression of the globules by the surrounding hot gas of the H II zone and the formation in them of cumulative convergent shock waves facilitates their gravitational condensation [440, 441] (see also Sec. 24).

If the ionization front does not bend completely around the density fluctuations, the so-called "elephant trunks" occur, usually bordered by hoops (Fig. 58). It was assumed originally that these trunks occurred as a result of instabilities of the Rayleigh–Taylor type at the boundary of the denser cloud, which was accelerated by the pressure of the ionized gas. However, this hypothesis could not interpret the shape and size of the "elephant trunks." Moreover, the observed flow of gas from the hoops restricting these formations indicates clearly that the "elephant trunks," like the globules, consist of dark clouds of neutral hydrogen, compressed by the ionization front and by its attendant shock waves. A calculation by Pottasch [391] shows that this hypothesis explains satisfactorily the brightness distribution in hoops. A flow of gas from a hoop toward the star was recently detected spectroscopically [392].

23. Motion of the Interstellar Gas

These is no basic difficulty in the formulation of gas-dynamic equations that take into account conditions in interstellar space (luminescence, magnetic fields, cosmic-ray pressure, and so forth). However, as is well known, gas-dynamic equations, because of their nonlinearity, are solved only in the rare cases of highly symmetric fluxes or in "favorable" problems with elementary boundary conditions. We can speak only provisionally of symmetric motion in interstellar space, while the boundary conditions are, as a rule, unknown. Therefore, the pursuit of accurate or even approximate solutions of the complete system of the equations of interstellar gas dynamics, involving as a rule much mathematical complexity, is on the whole not justified. The main problem of interstellar gas dynamics is to find an explanation of the basic regularities in the motion of the interstellar gas and some simple quantitative or even qualitative relations that would allow us to study approximately the features of this motion.

The Equations of Motion of the Gas

As in ordinary gas dynamics, we can distinguish here two kinds of flow: ordered laminar flow and turbulent random motion. The latter type of flow is characterized by statistical laws.

Interstellar gas can practically always be considered monatomic and therefore, when it travels adiabatically, we can usually take the value 5/3 for its adiabatic index γ. However, adiabatic conditions are often not realized in interstellar space. For example, during not too rapid gas motion in H II zones, the interaction of the gas with radiation leads to rapid equalization of temperature. In this case, the gas motion is isothermal in nature and it is better to use the value $\gamma = 1$ in its description. On the other hand, during gas motion across a magnetic field, the effect of magnetic pressure becomes important. In particular, if the latter is much greater than the gas pressure because of low temperature, from the adherence principle, $p = H^2/8\pi \sim \rho^2$, and the gas motion can be described by the index $\gamma = 2$ [393]. Therefore, in calculations on the motion of interstellar gas, it is desirable to retain the index γ in an explicit form, choosing its value with regard to the conditions of motion.

The simplest solution is for the case of the one-dimensional dispersion of gas into a vacuum, the so-called Riemann solution. Here the gas velocity v and the velocity of sound c_s are connected by the condition

$$v \pm \frac{2c_s}{\gamma - 1} = v \pm \frac{2}{\gamma - 1}\left(\gamma \frac{p}{\rho}\right)^{1/2} = \text{const.} \tag{23.1}$$

We must emphasize that the solution of Eq. (23.1) is practicable only when the resistance of the intercloud medium can be neglected. It follows from Eq. (23.1) that, if, for example, the initially motionless cloud ($v = 0$) disperses into empty space, the velocity of the gas at the leading edge of the front is

$$v = \frac{2c_s}{\gamma - 1} = \left[\frac{4\gamma}{(\gamma - 1)^2} \frac{k}{\mu m_{\rm H}} T\right]^{1/2}, \tag{23.2}$$

where T is the temperature of the motionless gas. In particular, for the adiabatic dispersion of an H II region, this velocity is three times the velocity of sound and is equal to $\sim$40 km/sec. When cold gas travels across lines of magnetic force, we must set $\gamma = 2$ in Eq. (23.1) and replace p by the magnetic pressure. Then we obtain, for the rate of dispersion, for $H = 3 \times 10^{-6}$ Oe and $N \approx 10$ cm^{-3}, twice the Alfvén velocity $v = 2H\,(4\pi\rho)^{1/2} \approx 4$ km/sec.

We cannot use Eq. (23.1) during isothermal dispersion ($\gamma \to 1$). Here, the

Riemann solution is described by the relation

$$v \pm c_s \ln \rho = v \pm \left(\frac{kT}{\mu m_{\mathrm{H}}}\right)^{1/2} \ln \rho = \text{const.} \tag{23.3}$$

It follows, therefore, that the rate of isothermal dispersion of the leading edge of the front may be high (when $\rho \to 0$). However, we must remember that the interaction of the gas with the radiation is weakened with a decrease in gas density. Therefore, at low densities motion becomes adiabatic again, which also restricts the rate of dispersion in this case [8].

The relations (23.2) and (23.3) are valid for gas dispersion, when we need not allow for external resistance of the medium, that is, in the absence of shock waves. In the conditions of interstellar space this occurs, for example, during the relatively slow ($v \approx 10$–60 km/sec) movement of dense H II emission zones or filaments in the tenuous space between clouds in H I regions. If the motion is faster ($v > 100$ km/sec), we have to consider even the resistance of the tenuous intercloud medium. Considerable interest arises, in particular, over the previously mentioned problem of the compression of interstellar gas through the dynamic pressure of the envelope thrown off by a nova or supernova. Similar problems arise during the motion of a cosmic cloud in a relatively dense complex, or in the collision of two large clouds, as, for example, in the collision of the interstellar media of two galaxies.

The Method of Self-similar Solutions

If the dynamic pressure is high, which is equivalent to the envelope velocity's being much greater than the velocity of sound in the resisting medium, then all these problems may be considered in the self-similar approximation. We recall that motion is called self-similar if all its characteristic thermodynamic parameters (pressure, temperature, velocity, and so forth) and the initial conditions are fixed by only two parameters of independent dimensions. Methods for the solution of self-similar problems are given in many textbooks on gas dynamics [375, 394, 395]. Here we shall give only the simple results that may find application during an analysis of the motion of the interstellar medium.

In all self-similar problems of interstellar-gas dynamics, one of the fixed parameters is the unperturbed gas density ρ_0, or the quantity $\rho_0 r^{+\alpha}$ if we assume that the original density had the dependence $r^{-\alpha}$. The second of the fixed parameters depends on the basic conditions of the problem. In the preceding section, the self-similar problem was described by a study of the motion of an ionization disturbance. There the parameters were the velocity of sound or its equivalent, the gas pressure, and the flux of ionizing radiation, since the gas movement was determined by its heating during ionization. The density of the undisturbed gas is a combination of these parameters. In self-similar problems with the motion of envelopes or the collision of gas clouds, motion is determined either by the energy of the envelope $\mathscr{E}$ or by the momentum $\mathscr{P}$ (similarly by the energy of the colliding clouds or their momentum). If gas luminescence may be neglected, the motion is adiabatic and we must choose the energy of the envelope as the fixed parameter. However, if luminescence is considerable and the total energy is not conserved, we can use as the fixed parameter $\mathscr{P}$ the momentum of the envelope, which hardly changes during luminescence, since the role of radiation pressure is small.

We must emphasize the following point. The interacting movements of interstellar gas are never self-similar in the complete sense of the word, that is, the nature of the movement depends on a large number of parameters, such as gravitational forces, the rate of luminescence, the geometry of the magnetic field, and so on. Therefore, in choosing one of the fixed parameters in addition to the density, we restrict the problem, trying to determine one fundamental regularity of the motion without paying attention to less important characteristics of the problem. Thus, the correct choice of the fixed parameter is most important.

Since in the self-similar problems discussed below both fixed parameters are proportional to the mass, they will occur in the future only as $\mathscr{E}/\rho_0$ or $\mathscr{P}/\rho_0$. We shall call this combination A and let its dimensionality be $[A] = \text{cm}^m/\text{sec}^n$, where m and n are integers. Then we can construct from the parameter A and the coordinates r and time t only the one dimensionless combination

$$\eta = [At^n]^{1/m}/r. \tag{23.4}$$

The total picture of self-similar motion at a given instant depends only on the single independent variable η, which with the passage of time remains similar to itself.

During motion of an envelope or during collisions between clouds, a shock wave is formed which travels through the motionless gas. In self-similar motion, the parameter $\eta = \eta_s$ is constant at the front of the shock wave. This means that the position of the shock front and its velocity are determined by the relations

$$r_s = \frac{A^{1/m}}{\eta_s} t^{n/m}, \qquad v_s = \frac{dr_s}{at} = \frac{n}{m}\frac{r_s}{t}. \tag{23.5}$$

The velocity of the gas, its density, and its pressure behind the shock front are described in the form of the dimensionless functions of the dimensionless quantity η:

$$V = \frac{r}{t} V(\eta), \qquad \rho = \rho_0 R(\eta), \qquad p = \frac{r^2}{t^2} \rho_0 P(\eta). \tag{23.6}$$

Substituting Eqs. (23.6) in the gas-dynamic equation (21.3), we obtain the system of ordinary differential equations for the functions $V(\eta)$, $R(\eta)$, and $P(\eta)$. The boundary conditions for this system are found from the shock-wave parameters, and the value of η_s is determined by calculating the total energy or the total momentum and equating this value to the given value $\mathscr{E}$ or $\mathscr{P}$. Usually η_s is close to unity, but it depends on the adiabatic index γ and, although weakly, on the index of symmetry ν. For example, in the problem discussed below of an adiabatic spherical explosion, the parameter η_s changes from 1.3 to 2 as γ changes from 5/3 to 3.

This kind of problem is called a point-explosion problem, even if energy is liberated over a surface or along a line. The point nature must be taken here in the sense that the release of energy or the transfer of momentum proceeds instantaneously in comparison with the scale of the entire motion. The self-similar method for solving point-explosion problems was proposed by L. I. Sedov. It was used further for the solution of many similar problems. By the method of successive approximations, non-self-similarity was taken into account, in particular, the effect of the pressure of the undisturbed gas, its heterogeneity, the presence of magnetic fields, the role of viscosity and thermal conductivity, and so forth. A suitable illustration of this method for solving problems on point explosions and a review of the results obtained may be found in a series of monographs [394, 395]. The application of this method to the problem of interstellar gas dynamics, in particular to the solution of the problem of the motion of an envelope accompanied by luminescence, is given by S. A. Kaplan [8].

We shall restrict ourselves here to a description of the main results of the solution of these problems that have a direct bearing on interstellar-gas dynamics and lead to a direct comparison with observational data. In particular, from this viewpoint we must first determine the dependence of the shock-wave velocity on time, estimate the relative size and mass of the layer of interstellar gas carried along by the moving envelope, and determine the gas temperature behind the front of the shock wave (or discontinuity in the density).

We consider some concrete examples.

(1) *An adiabatic spherical burst.* If the envelope from a supernova travels at very high velocity (of the order of 1000 km/sec), luminescence does not play an essential part in the initial stages of the expansion of the envelope [380]. At later stages, when luminescence becomes significant, the conditions of motion are approximately isothermal (see below). Restricting ourselves to the initial stages, we shall assume that the total energy is conserved. Under these conditions, the process of compression of the interstellar gas by the envelope may be described in the following manner. At the moment of explosion, the envelope is thrown off with total kinetic energy $\mathscr{E}$. Traveling in the interstellar medium, the envelope seems to "rake together" interstellar gas with the result that the energy of the burst is distributed over a larger mass and the envelope is slowed down. A very strong shock wave, in which the temperature is considerably raised but in which, because of the adiabatic conditions, the density is increased by only four times, travels through the undisturbed gas in front of the envelope.

The entire motion is determined by the energy of the explosion $\mathscr{E}$, which has the dimensions g cm^2/sec^2. The dimensions of the parameter $[A] = [\mathscr{E}/\rho_0]$ are cm^5/sec^2, that is, $m = 5$ and $n = 2$. Setting also $\eta_s = 1$, we have for the law of motion for the forward front of a shock wave created by a supernova outburst

$$r_s = \left(\frac{\mathscr{E}}{\rho_0}\right)^{1/5} t^{2/5}, \qquad v_s = \frac{2}{5}\frac{r_s}{t} = \frac{2}{5}\left(\frac{\mathscr{E}}{\rho_0}\right)^{1/2} \frac{1}{r_s^{3/2}}. \tag{23.7}$$

Knowing the shock-wave velocity, we find from Eq. (21.1) the temperature behind its front:

$$T_2 = \frac{3m_H v_s^2}{16k} \approx \frac{3}{100}\frac{m_H \mathscr{E}}{k\rho_0 r^3} = \frac{3}{100}\frac{\mathscr{E}}{kN_H r^3}. \tag{23.8}$$

Thus, as the shock wave progresses, its intensity rapidly decreases. The physical meaning of this equality lies in the fact that the energy of the outburst is distributed as thermal and kinetic energy over the entire mass $\sim \rho_0 r^3$ involved in the motion. To estimate the numerical values, let us set $\mathscr{E} = 10^{49}$ erg and $\rho_0 = 10^{-24}$ g/cm^3. Then

$$r_s\,(\text{pc}) \approx 0.13(t\ \text{yr})^{2/5};$$
$$v_s \approx \frac{2300}{[r\,(\text{pc})]^{3/2}}\ \text{km/sec}; \qquad T_2 \approx \frac{1.2 \times 10^8}{[r\,(\text{pc})]^3}\ \text{deg}. \tag{23.9}$$

In these relations, we must bear in mind that they are valid only when the mass of the interstellar gas carried along by the motion is considerably greater than the initial mass of the envelope. Assuming that the latter is about 0.1 $M_\odot$, we obtain for the condition under which relation (23.7) and (23.8) apply

$$\frac{4\pi r^3}{3}\rho_0 > 0.1\ M_\odot; \qquad r > 1.2\ \text{pc}. \tag{23.10}$$

Over shorter distances the envelope travels with almost constant velocity, since damping is still negligible. If the mass of the envelope is small, as for example in novae, the value of r is correspondingly reduced.

The distribution of gas density, temperature, and pressure inside the region carried by the motion is found from the solution of the corresponding equations. Because of self-similarity, these distributions depend only on the parameter η and remain the same at different times. Almost the entire mass of the interstellar gas carried by the motion of the envelope is concentrated near the front of the shock wave. The gas density behind the front is reduced by $(r/r_s)^{10/3}$. Pressure changes relatively slowly, since the gas temperature rises deep inside the envelope. Formulas and graphs illustrating the solution are given in the monographs already cited. This solution has been applied to the study of the motion of supernova envelopes. From it, we can determine the ages of envelopes from their velocities. For example, we estimate an age of 70,000 yr for the Loop in Cygnus, instead of the value of 150,000 yr used earlier on the assumption of uniform expansion [380].

(2) *The motion of a spherical envelope accompanied by luminescence.* When the density of the quiescent gas is relatively high and the velocity of the envelope is in the range of hundreds of kilometers per second, compression of the interstellar gas produces intense luminescence. In this case, the total energy of the motion of the envelope and of the shock wave formed by the collision of gaseous masses is no longer conserved, and consequently the energy $\mathscr{E}$ cannot be defined by a parameter. Such a problem is not in principle self-similar. However, if the luminescence is so intense that the gas temperature is fixed only by its interaction with radiation, we can still use self-similar conditions if we choose for the fixed parameter the momentum [8]. In fact, the internal pressure here no longer accelerates the envelope, whose motion is determined only by the original momentum obtained during the outburst. Such conditions occur, for example, during the late stages of expansion of nova and supernova envelopes (velocities less than 100 km/sec), or during collisions of clouds in H I regions at low velocities, $v \approx 20$–30 km/sec. In each of these cases, shock waves, accompanied by luminescence, are formed which also dissipate energy by emission but conserve momentum.

We consider the problem of the motion of a spherical envelope accompanied by luminescence. Let its initial mass be M and its initial velocity v_0. Then the total momentum $\mathscr{P} = MV_0$ has the dimensions g cm/sec. Thus, the dimensions of the parameter $[A]$ are cm^4/sec, that is, $m = 4$, $n = 1$. Substituting $\eta_s = 1$ as before, we obtain from (23.5) for the law of motion of the front of a shock wave

$$r_s = \left(\frac{\mathscr{P}}{\rho_0}\right)^{1/4} t^{1/4}, \qquad v_s = \frac{1}{4}\frac{r_s}{t} = \frac{\mathscr{P}}{4\rho_0}\frac{1}{r_s^3}. \tag{23.11}$$

In shock waves accompanied by luminescence their velocity determines not the temperature, but the discontinuity in pressure. Substituting (23.11) in (21.27), we find (when $\mu = 0.7$)

$$\frac{\rho_1}{\rho_0} = \frac{0.7 m_{\mathrm{H}} v_s^2}{kT_2} = 0.04\frac{m_{\mathrm{H}}\mathscr{P}^2}{k\rho_0^2 T_2}\frac{1}{r_s^6}. \tag{23.12}$$

We should note the very rapid ($\sim 1/r^6$) damping of shock waves with luminescence; this is explained essentially by the fact that a considerable amount of energy is emitted as radiation, reducing the very pressure that sustains the motion of the leading front. The temperature T_2 is determined, as before, by the balance of energy and radiation.

Let us take some numerical examples. The momentum of an envelope thrown off by a type I supernova outburst is $\mathscr{P} \approx 10^{40}$ g cm/sec, and in the case of a type II supernova, 10^{42} g cm/sec. Substituting $\rho_0 = 10^{-24}$ and $T_2 \approx 10{,}000°$

and taking as the mean $\mathscr{P} \approx 10^{41}$ g cm/sec, we have

$$r\,(\mathrm{pc}) = 0.4(t\ \mathrm{yr})^{1/4}, \qquad v_s = \frac{8500}{[r\,(\mathrm{pc})]^3}\ \mathrm{km/sec}, \qquad \frac{\rho_1}{\rho_0} = \frac{5.5 \times 10^5}{[r\,(\mathrm{pc})]^6}. \tag{23.13}$$

Thus, at $r > 10$ pc, the shock wave created by the supernova envelope disappears. The time of expansion of the envelope to this distance is about 4×10^5 yr and its final velocity is ~ 8–9 km/sec. These results depend very little on the choice of $\mathscr{P}(\sim \mathscr{P}^{1/4})$, but they depend critically on the condition of luminescence. These equations can be used only over comparatively large distances ($r \geqslant 5$ pc). A detailed study of the structure and density distribution inside the compressed gas of the envelope is given in [8]. The relative thickness of the compressed gas layer is low, about 1 percent of the radius, and in the course of time grows proportionally to the increase in radius of the envelope. This is shown directly by Eq. (23.12) since compression of the gas is high at the front of a shock wave for $r < 10$ pc. The numerical value of the parameter η_s is ≈ 1.5.

We note that the dependence (23.11) can be obtained to within a numerical factor from the simple relation [396]

$$\left(\frac{4\pi r_s^3}{3}\rho_{\mathrm{H}} + M\right)v_s = MV_0 = \mathscr{P}, \tag{23.14}$$

which expresses the law of conservation of momentum of the envelope and the mass of the "raked up" interstellar gas. Formula (23.11) follows from (23.14) when $\frac{4}{3}\pi r_s^3\rho_0 \gg M$. However, we cannot study the structure of the compressed-gas region in this way, nor, in particular, determine its thickness. The motion of a flat self-similar wave, allowing for luminescence, was investigated by V. G. Gorbatzkii [397].

(3) *Cumulative shock waves.* E. A. Dibai and S. A. Kaplan have studied the problem of the cumulative compression of globules, that is, shock waves converging toward the center of the globule [397a]. In fact, a strong pressure is present at the surface of a globule of ionized hydrogen, which causes the formation of a shock wave converging toward the center of the globule. In order that such a wave shall actually be cumulative, the ionization of hydrogen around the globule must take place more rapidly than the propagation of the wave inside this globule. If the H II zone is formed from the explosion of a hot star (see page 323), then this condition is in fact satisfied, since, although the time of propagation of the ionization discontinuity is small, because of the low temperature, the velocity of sound in globules is low.

Since the equations describing globules are almost spherically symmetric, the cumulative shock waves formed are also spherical. When such waves converge toward the center, the density of the gas inside the globule is 10–1000 times the initial density (depending on the initial and boundary conditions of the problem). After reaching the center, the cumulative wave is reflected and leaves in its wake a region of quiescent gas with a density 2×10^2–10^5 times the initial density. The velocity of the converging cumulative wave is approximately 6–10 km/sec, and the velocity of the wave reflected from the center is three times greater.

Cumulative shock waves are possibly observed in comet-shaped nebulae [442, 397b].

The relations (23.4)–(23.14) enable us to study shock-wave motion in the interstellar gas when we know the initial conditions: an outburst and the throwing off of an envelope, a collision between clouds, or the like. In addition to such motions, such shock waves are propagated in interstellar space owing to the turbulence of the interstellar medium, the frequent but relatively weak collisions between clouds, and so on. Such motion does not have one most important fixed parameter in the initial conditions through which we can express its subsequent properties. In this case, shock-wave propagation depends, as it were, on internal properties such as the intensity of the wave (amplitude) and the amount of energy dissipated at a given point in space as the shock wave passes through that volume.

Magnetohydrodynamics is concerned with the study of motion in magnetic fields. In principle, its methods are similar to the methods of ordinary gas dynamics, including those used here, but the solutions are mathematically complicated. Therefore, we shall restrict ourselves to qualitative comments.

As long as the velocity of envelopes or clouds considerably exceeds the Alfvén velocity $H/(4\pi\rho)^{1/2}$, the effect of the magnetic field is unimportant. We note, incidentally, that an envelope appears to repel the magnetic field. Because of the adherence principle, it cannot penetrate the envelope. Compression of the "raked up" interstellar gas leads, of course, to intensification of the magnetic field inside it.

When shock-wave velocities approximate magnetohydrodynamic velocities, the effect of a magnetic field on the motion, especially on its geometry, becomes important. A spherical envelope acquires an ellipsoidal shape, and thermal dissipation of energy is reduced. During collisions between masses of gas, the magnetic field becomes elastic. We can obtain a quantitative estimate in all these cases, while allowing for the peculiarities of shock waves in a magnetic field described in Sec. 21, and using the formulas given above for various types of motion with the appropriate choice of the parameter γ. Geometric considerations are very helpful here when we use a graphic representation of the entrainment of lines of magnetic force during gas motion and take into account the nonisotropic nature of magnetic pressure.

We have already noted in Sec. 21 that nonlinear magnetic waves can be propagated at velocities lying in the range $H/(4\pi\rho)^{1/2}$ to $H/(\pi\rho)^{1/2}$ in a magnetic field in which no dissipation of energy occurs. These waves can therefore transfer energy and momentum into interstellar space over a great distance without appreciable attenuation.

Finally, we should note that the magnetohydrodynamic wave velocity in a very rarefied medium is close to the velocity of light. Therefore, relativistic magnetohydrodynamic methods may be necessary for the study of motion in the extragalactic medium.

The galactic magnetic field may influence the expansion of a supernova shell [397c, d]. Perhaps the most interesting effect of this interaction is the development of Rayleigh–Taylor instability. It may explain the filamentary structure of this shell.

Interstellar Turbulence

We know that the flow of the gases is described by nonlinear equations. This means that motions on different scales experience mutual interactions, exchanging energy and momentum. For example, a large-scale motion can be broken down into several smaller-scale motions. The reverse process, the combination of small-scale motions into a single large-scale motion, is possible in principle, but it is considerably less probable than the subdivision of motion. In other words, gas as a continuous medium has an infinitely large number of degrees of freedom, and therefore its energy must be broken down into a larger and larger number of degrees of freedom, that is, into smaller- and smaller-scale motions. Thus, if there are no restrictions on the transfer of energy to smaller scales, the motion of the gas must have an intricate turbulent character.

Motion on a small scale is damped by viscosity and, therefore, is not transferred to an even smaller scale. The role of viscosity may be estimated in the following manner. Energy transfer between motions of different scales is described in the equations of gas dynamics by an inertia term $(v\nabla)v$, equal in value to v_l^2/l, where l is the scale of the motion and v_l is the gas velocity at this scale. On the other hand, suppression of motion is described by a kinematic viscosity term, $\nu\Delta v \approx \nu v_l/l^2$. Comparing the two expressions, we find that motion is not damped by viscosity if their ratio, the so-called Reynolds number,

$$R_l = \frac{v_l l}{\nu}, \tag{23.15}$$

is much greater than unity. More precisely, for motion to be turbulent, we must maintain the condition $R_l > R_0$, where R_0 is of the order of several times ten (laminar flow in a pipe can exist when $R_l \approx 10^3$ because the disturbances are carried downstream). The scale of motion at which $R_l \approx R_0$ is called the intrinsic scale. In Eq. (23.15), ν is the maximum value of viscosity (in the presence of a magnetic field in a nonionized plasma it is the quantity ν_m in Sec. 16). In the most unfavorable case of an H I region, where $\nu_m = 10^{23}$ cm^2/sec, $v_l = 10$ km/sec, and $l \approx 10$ pc, we have $R_{10} \approx 300$. In the case of H II zones, the value of R_l is much larger. Evidently, the motion of interstellar gas must nearly always be turbulent.

A magnetic field can be another factor restricting energy transfer to smaller scales. Its effect may be estimated from the ratio of the Alfvén velocity to the gas velocity,

$$A_l = \frac{H}{v_c(4\pi\rho)^{1/2}}. \tag{23.16}$$

If $A_l \ll 1$, a magnetic field cannot suppress motion and thereby restrict the transfer of motion to smaller scales. Since gas velocity also falls with decrease in scale and H in Eq. (23.16) represents the general average field at all scales, on transition to smaller scales the condition $A_l > 1$ must ultimately be satisfied, after which further subdivision of energy may be suppressed.

Unfortunately, the question of the nature of turbulence in a magnetic field still remains far from solved. The difficulty here is that suppression of motion at relatively large scales contradicts the statistical picture of turbulence as motion with a large number of degrees of freedom. Therefore Batchelor [401] proposed that magnetic turbulent motion is possible only when the magnetic field is so weak that it suppresses only motions smaller than the intrinsic motion. In other words, the maximum density of the magnetic energy is approximately equal to the density of the kinetic energy of motion on those scales where the viscosity effect is appreciable, and therefore much less than the mean density of kinetic energy of turbulent motion.

However, it is difficult to imagine why large-scale motions, because of the adherence principle, do not collect lines of magnetic force and thereby intensify the large-scale magnetic energy. Therefore, from another point of view, magnetic energy is comparable with kinetic energy at all scales of motion. It is true that the gas motion itself can be different in this case from that in the absence of the magnetic field. For example, these motions may have the character of statistically distributed vibrations of lines of magnetic force of different amplitude and length [340]. We may also assume a spatial distribution of motion and fields of various scales [18:63].

If turbulent motion proceeds at supersonic velocities, part of the kinetic energy is immediately dissipated into thermal energy in the shock waves. Therefore, the smaller the scale, the smaller the energy transformed here into motion. In other words, shock waves also apparently suppress smaller-scale motions.

The theoretical study of turbulence is possible only for very simple systems. At the present time, only the case of so-called uniform and isotropic turbulence, in which the characteristics of the motion do not depend on position and direction in space, has been more or less fully considered. The theory of such turbulence in an incompressible liquid was developed in papers by A. N. Kolmogorov [398], A. M. Obukhov [399], Heisenberg [400], and others, and was generalized in the case of the magnetic turbulence of an incompressible liquid by Chandrasekhar [403] and in the case of supersonic magnetic turbulence by S. A. Kaplan [407]. A detailed exposition of this theory, bearing in mind its astrophysical uses, is given in [8] and [395].

However, we must stress that interstellar gas turbulence is known not to have an isotropic or homogeneous nature. Therefore, we can draw no further conclusions by comparing theoretical assumptions with observational data. We shall thus restrict ourselves here to describing the theory of this turbulence by only the most general considerations.

For the present we shall neglect the effect of the magnetic field and the compressibility of the medium. Then the energy of large-scale motion will be entirely transformed into smaller-scale motion, whereupon this energy is transformed into still finer-scale motion, and so on, until we reach motions at the intrinsic scale where kinetic energy is converted into heat. Thus, we can define such turbulence by the following physical parameters: a fundamental scale L and a velocity v_L at this scale, an energy per unit mass ε that has traversed the whole hierarchy of motions, and the intrinsic scale of motion l_0.

The quantities L and v_L are determined by the sources of the turbulence, that is, by the mechanisms that set the interstellar medium in motion. The amount of kinetic energy transferred by the sources to 1 g of interstellar medium per second is $v_L^2/t_L \approx v_L^3/L$. Here $t_L \approx L/v_L$ is the characteristic time of motion at the fundamental scale. According to our previous assumption, this energy is transferred through the whole range of scales, almost without loss. Therefore,

$$\varepsilon \approx \frac{v_L^3}{L} \approx \frac{v_l^3}{l} \tag{23.17}$$

for all the values $L > l > l_0$. Hence, we obtain Kolmogorov's [398] well-known law, which determines the dependence of the velocity $v_l \approx (\varepsilon l)^{1/3}$ on the scale. This rule may be obtained from dimensional considerations if we remember that at intermediate scales motion is determined by only a single parameter ε with dimensions $\mathrm{cm}^2/\mathrm{sec}^3$. Only the single dimensionless combination (23.17) is constructed from the three quantities ε, v_l, and l.

The intrinsic scale l_0 is determined both by the energy ε dissipated and by the viscosity with dimensions $\mathrm{cm}^2/\mathrm{sec}$. We may also determine l_0 to within some numerical factor α from dimensional considerations by using a dimensionless combination of l_0, ε, and ν. Such a combination has the form

$$l_0 \approx \alpha(\nu^3/\varepsilon)^{1/4}, \tag{23.18}$$

where the parameter α is determined by the choice of the Reynolds number for the intrinsic scale (R_0).

As we have already noted, the effect of shock waves and suppression of turbulence by a magnetic field must be a reduction in ε during transition to smaller-scale motions and consequently a steep spectral slope at low values of l. A rough theoretical analysis shows that suppression of small-scale motion by a magnetic field leads to a spectrum with $v_l \sim l^{1/2}$, and if energy is dissipated in shock waves the spectrum may be still steeper, approaching $v_l \sim l$ [402]. At very small scales of motion, less than the intrinsic, the kinetic and magnetic energy spectra fall very abruptly, or $v_l \sim l^3$ [400].

Thus, if the motion of interstellar gas is turbulent, a correlation that satisfies Kolmogorov's law at comparatively large scales and falls more abruptly at smaller scales must exist between the scales of motion and their associated velocities. The correlation between the intensity of the magnetic field and its scale is probably more complex. We must also bear in mind that the magnetic field in the arms of a nebula, and possibly between them, has a more or less regular character.

In addition, the energy density of the magnetic field in H I clouds is considerably higher than the kinetic-energy density in the intrinsic motion. Under these conditions the turbulence may have quite another character, a type of oscillating cloud, "strung" along the lines of magnetic force. The correlation between the velocity of the cloud and the scale of motion will be otherwise here, possibly different from that given by Kolmogorov's law. Recently there has been developed in plasma physics the theory of weakly turbulent plasmas in which small oscillations in the motion in a strong magnetic field are considered while pair interactions between these oscillations are taken into account. The spectrum of the oscillations in this case is flatter than in the turbulence with a relatively weak magnetic field [340].

However, the existence under all conditions of definite correlations between velocities and scales of motion, and also between other parameters (for example, the intensity of the magnetic field), indicates that statistical regularities, by which we may also designate turbulence, occur here. If we disregard more detailed questions connected with definite spectra, the parameters of such turbulence, like the type of energy dissipated, the fundamental and intrinsic scales of motion, and so forth have a sufficiently general character and do not depend in the first approximation on its form (that is, its spectrum).

Let us now estimate the values of the turbulence parameters of the interstellar gas. The most probable assumption is that the energy of motion of the entire interstellar-gas system is drawn from expanding H II regions [407]. The distance between them is, on the average, several hundred parsecs. Therefore, we must take for the fundamental scale $L \approx 100$ pc. The number of H II regions in the Galaxy is estimated from the number of class O stars, of which there are approximately 10^4. The evolutionary period of such a star and its corresponding H II zone is about 3×10^6 yr. Thus, on the average, a single H II zone appears once in 300 yr. A class O star usually ionizes a mass of interstellar gas approximately equal to 1000 solar masses. The expansion of an H II zone collects a mass of neutral hydrogen approximately 20–50 times larger, communicating to it a velocity of ~ 10 km/sec. Consequently, each class O star transfers to the interstellar gas about 3×10^{49} erg, or, on the average, $\sim 3 \times 10^{39}$ erg/sec. Taking the mass of gas of the flat subsystem as $1.5 \times 10^9\ M_\odot$, that is, 3×10^{42} g, we find for the amount of energy transferred to 1 g of interstellar gas per second

$$\varepsilon \approx 10^{-3}\ \text{cm}^2/\text{sec}^3. \tag{23.19}$$

Another source of energy for the motion of the interstellar gas may be the outbursts of type II supernovae, but, because of their high expansion rate, a considerable portion of the energy of their envelopes is immediately dissipated as heat or is converted into radiation, and only about 1 percent is transformed into energy of the interstellar gas. For example, in the Loop Nebula in Cygnus approximately 300 solar masses have been collected by the motion at velocities of 100 km/sec. The kinetic energy of this gas is of the order of 3×10^{49} erg, while the initial energy of the envelope of a type II supernova is of the order of 10^{50}–10^{51} erg. Apparently, supernova outbursts do not provide more than 30 percent of the kinetic energy in the energy balance of the interstellar medium.

A direct estimate of the quantity ε may be obtained from (23.17) by substituting $L \approx 100$ pc and $v_L \approx 7$ km/sec. We find $\varepsilon \approx 10^{-3}$ cm^2/sec^3, which agrees qualitatively with (23.19).

The intrinsic scale of the turbulence of the interstellar gas is determined from (23.18) and (23.19). If we use for ν the highest value of magnetic viscosity of the nonionized gas in the H I zone, we obtain for $\alpha \approx 1$ the value $l_0 \approx 2$ pc, which, generally speaking, is comparable with the dimensions of the cloud.

Even more intense turbulence can occur inside H II zones or diffuse nebulae. The fundamental scale of this turbulence is about that of the size of the nebulae. The amount of energy dissipated could be estimated under certain assumptions only for the Orion nebula, where $\varepsilon \approx 0.1$–1 cm^2/sec^3 [405, 406]. We may expect this value to vary considerably in different nebulae. The intrinsic scale of turbulence in H II zones is very small.

Observations of Interstellar Turbulence

The observational study of turbulence leads primarily to the construction of correlation and the structural functions. The quantity

$$B(r) = \overline{[v(r') - v(r'')]^2} \tag{23.20}$$

is called the structural function of the velocity field $v(r)$, where the average is taken over all pairs of points $\mathbf{r}'$ and $\mathbf{r}''$ so chosen that $|\mathbf{r}' - \mathbf{r}''| = r$. The correlation function $D(r)$ is defined by the equations

$$D(r) = \overline{v(\mathbf{r}')v(\mathbf{r}'')}; \qquad B(r) = 2[D(0) - D(r)]. \tag{23.21}$$

We can obtain structural and correlation functions similarly for magnetic fields, pressures, accelerations, and so on. To determine these functions from the usual hydrodynamic or magnetohydrodynamic equations, so-called correlation equations, found first by Carman and Govard, are derived. A detailed description of this method of the statistical theory of turbulence, and also the determination of correlation functions for various parameters, is given in [395]. We shall restrict ourselves here to simple semiqualitative expressions that allow their form to be immediately estimated.

Actually, in the calculation of the average in (23.20), velocities of the motion at scales greater than r do not contribute, since for these scales $v(\mathbf{r}') \approx v(\mathbf{r}'')$. On the other hand, the rate of motion at scales less than r is small and does not contribute essentially to this quantity. Therefore, motion at scales close to r, which have velocities $v_r \approx (\varepsilon r)^{1/3}$, contributes the most to the quantity $B(r)$. Consequently, we can write

$$B(r) \approx \overline{v_r^2} \approx (\varepsilon r)^{2/3} \qquad (r < \mathrm{L}). \tag{23.22}$$

On the other hand, if $r > L$ (fundamental scale), there is no correlation between the velocities $v(\mathbf{r}')$ and $v(\mathbf{r}'')$, and

$$B(r) \approx v_{\mathrm{L}}^2 \approx (\varepsilon L)^{2/3} = \text{const.} \qquad (r > L). \tag{23.23}$$

Equations (23.22) and (23.23) are approximated with sufficient accuracy by the expression

$$B(r) = \frac{(\varepsilon L)^{2/3}}{1 + (L/r)^{2/3}}. \tag{23.24}$$

Structural functions are similarly determined for other parameters. For example, if equidistribution of magnetic and kinetic energy occurs, that is, if $H_l \approx (4\pi\rho)^{1/2} v_l$, the magnetic-field correlation function agrees with Eq. (23.24) to within the factor $4\pi\rho$. The pressure field changes as the dynamic pressure, that is, as the square of the velocity ($p \approx \rho v^2$), and therefore the structural function for pressure fluctuations has the form

$$B_p(r) = \overline{[p(\mathbf{r}') - p(\mathbf{r}'')]^2} \approx \frac{\rho^2 (\varepsilon L)^{4/3}}{1 + (L/r)^{4/3}}. \tag{23.25}$$

The same structural fluctuation also occurs for concentration fluctuations [404a], or,

$$\overline{[N(\mathbf{r}') - N(\mathbf{r}'')]^2} = B_p(r)/(kT)^2.$$

Under terrestrial conditions structural functions are found directly, for example, by velocity measurements at different points of a medium. The determination of structural functions in interstellar conditions is complex, first because we cannot always determine the distance between discrete masses in the medium together with their velocity measurements, and second because measurements usually do not refer to a single element of mass, but consist only of an average along the line of sight of the values of the observed parameter. Methods for determining structural functions of the interstellar medium under these conditions have been developed by Hoerner [405] and by the authors [8, 408].

Let us consider first the following problem. Suppose there is a transparent nebula of linear dimension R along the line of sight. We shall consider that the forward boundary lies in the pictorial plane, or perpendicular to the line of sight, and that the lines of sight are parallel inside the nebula. Let us assume first that we can measure the velocity of a fixed element of mass whose position along the corresponding line of sight is equally probable throughout the extent of the nebula. Then the observed structural function $A(x)$, where x is the distance between the lines of sight in the pictorial plane, is connected with the actual structural function by the obvious relation

$$A(x) = \frac{1}{R^2}\int_0^R\int_0^R B[x^2 + (y_1 - y_2)^2]^{1/2} dy_1\, dy_2, \tag{23.26}$$

where y_1 and y_2 are the coordinates along the lines of sight. We notice that, if $x \to 0$, $A(x)$ converges to a limiting value, since the structural function $B(|y_1 - y_2|)$ from two different elements of mass along the same line of sight then enters the integral (23.26). If, instead of velocities of discrete elements of mass of the gas, we measure uniformly averaged values along the line of sight, we obtain instead of (23.26) [405, 408]

$$\begin{aligned} A(x) &= \frac{1}{R^2}\int_0^R\int_0^R \{B[x^2 + (y_1 - y_2)^2]^{1/2} - B(|y_1 - y_2|)\}\, dy_1\, dy_2 \\ &= 2\int_0^1 \{B(x^2 + R^2u^2)^{1/2} - B(Ru)\}(1 - u)\, du. \end{aligned} \tag{23.27}$$

In particular, when $x \gg R$ we find that $A(x) \approx B(x)$, and when $x \ll R$ we obtain $A(x) \approx 2xB(x)/R$. In other words, the observed structural function agrees with the real one only when the thickness of the nebula is much less than its extent in the pictorial plane. This condition is satisfied, for example, when a nebula is not transparent and only its thin frontal layer is observed. Here R must be taken as a quantity inverse to the extinction coefficient. We can also take absorption into account by inserting the factor $e^{-(y_1+y_2)R}$ into the integral expression (23.26) and changing the limits to infinity.

During the study of turbulence in the interstellar medium it becomes impossible to regard the lines of sight as parallel. It is more convenient here to find correlations of the differences of measured quantities with angular distances on the celestial sphere [404, 8]. Calling R the extent of the region of the interstellar medium under investigation, we have, instead of Eq. (23.26),

$$A(2\alpha) = \frac{1}{R^2}\int_0^R\int_0^R B(y_1^2 + y_2^2 - 2y_1y_2\cos 2\alpha)^{1/2} dy_1\, dy_2, \tag{23.28}$$

where 2α is the angle between two lines of sight and y_1 and y_2 are the coordinates along them. Analogously, if we obtain from observations average values of quantities, we then get for the observed structural function

$$A(x) = \frac{1}{R^2} \int_0^R \int_0^R \{B[(y_1 - y_2)^2 + 4x^2 y_1 y_2]^{1/2} - B(|y_1 - y_2|)\} \, dy_1 \, dy_2, \quad (23.29)$$

where $x = \sin \alpha$. Values of the function $A(x)/(\varepsilon L)^{2/3}$ or of $A(x)/\rho^2(\varepsilon L)^{4/3}$, calculated from Eq. (23.39) with functions $B(r)$ given in Eqs. (23.24) and (23.25), respectively, are shown in Table 23.1 (for the cases $L = 0.1R$ and $L = R$).

Table 23.1.

x	$L = 0.1\,R$		$L = R$	
	(23.24)	(23.25)	(23.24)	(23.25)
0.0	0	0	0	0
.1	0.063	0.075	0.030	0.016
.2	.094	.124	.056	.046
.3	.118	.156	.080	.080
.4	.137	.178	.102	.116
.5	.153	.195	.122	.150
.7	.178	.218	.156	.213
1.0	.206	.239	.199	.292

The data from this table show that, when the fundamental scale is much less than the dimensions of the region, the observed structural function depends only weakly on the real structural function. For example, the mean index of the degree of correlation,

$$\tilde{n} \approx \frac{d \ln A(x)}{d \ln x}, \quad (23.30)$$

calculated for the second and third columns of Table 23.1, is 0.52 and 0.48, respectively, and therefore is not associated with the indices $n = \frac{2}{3}$ in Eq. (23.24) and $n = \frac{4}{3}$ in Eq. (23.25). In other words, we cannot in this case determine the nature of the structural function from averaged observational data. However, if the region under consideration is of the same order as the fundamental scale, the function $A(x)$ reflects more the function $B(r)$. In particular, the mean index of (23.30) calculated for the fourth and fifth columns of Table 23.1 is 0.695 and 1.240, respectively, which is much closer to the indices of Eqs. (23.24) and (23.25). More complete tables of $A(x)$ and the quantity $\tilde{n}$ are published in [405a].

We turn now to the analysis of observational data. Turbulence in an individual nebula has been studied only in the Orion Nebula. Structural functions for the velocity field inside this nebula were obtained by Hoerner [405] and Münch [18:208]. The results are given in Fig. 59. The curve and its asymptotes were calculated by Hoerner from a formula of the type of (23.27) for $R = 0.08$ pc. The observed correlation agrees well with the rule $A(x) \sim x^{2/3}$. However, a difficulty arises here. In fact, if the real velocity-correlation function satisfies

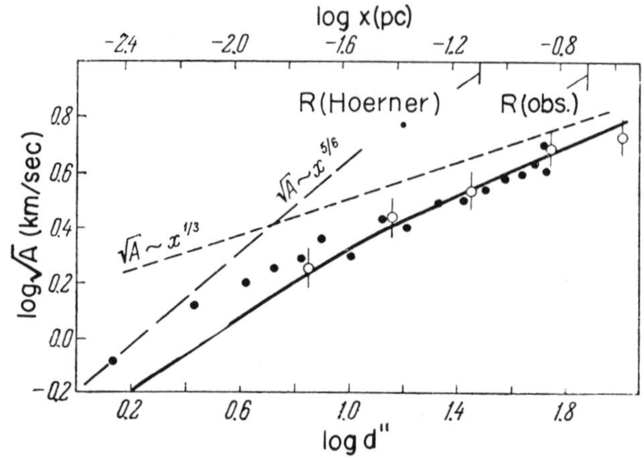

FIG. 59. Structural functions for the velocity field in the Orion Nebula from data by Hoerner (*open circles*) and Münch (*solid circles*) [18:208].

the condition $B(r) \approx (\varepsilon r)^{2/3}$, the observed function will satisfy the same condition if for all points measured $x \gg R$, or, in other words, if only a very thin surface layer, less than one-tenth the dimensions of the nebula, is observed. However, according to (23.22) the velocity range will be small in such a thin layer. Actually $A^{1/2}$ in Fig. 59 corresponds directly to the mean velocity difference in the layer. The value of R, according to Hoerner, corresponds to $\log A^{1/2} \approx 0.5$, and from Münch's data this dependence continues further. This means that the velocity dispersion in the layer must be less than 3 km/sec, while the observed line width is about 10 km/sec. Moreover, the absorption in the nebula measured directly from reddening of the central star θ' Orionis also gives for the thickness of the emitting layer a value of about 0.2 pc, shown in Fig. 59. If the nebula is observed to a sufficient depth, the observed correlation must be closer to the relation $A(x) \sim x^{5/3}$, which is known to contradict observations.

S. B. Pikelner and G. A. Shain [406] studied turbulence in the same nebula from brightness fluctuations, interpreting these as due to turbulent fluctuations in density. The correlation function has here the form of (23.25) and the observed index must be close to $\frac{4}{3}$ if the layer is thin and to $\frac{7}{3}$ if absorption is low. The investigation showed good agreement with Kolmogorov's law also for density fluctuations. However, the difficulty associated with determination of the thickness of the observed layer remains, since here the dependence is also realized for $x = 0.04$ pc. In this connection, the thickness of the layer was determined in [406] from the observed line width giving the velocity dispersion. It is possible that the observed dispersion is somewhat greater than the true one because of the scattering of light. Dust in the nebula has a high albedo, and in a given layer radiation coming from other parts of the nebula is scattered. The turbulence parameter ε for the Orion Nebula is in both cases approximately 1 cm^2/sec^3.

Turbulence in the Orion Nebula has also been investigated by Herschberg [406a]. He showed that, if the scale of the photographic plate is 4′/mm, the index of the structural function is $n \approx \frac{4}{3}$, which corresponds in fact to Kolmogorov's law. However, if a larger plate scale is used (20″/mm), the structural function is better approximated by the linear dependence $A(x) \sim x$. Herschberg explains this peculiarity as a consequence of the random distribution of brightness fluctuations in the pictorial plane, and comes to the conclusion that the observational data do not prove the turbulent character of the motion in the nebula. Herschberg also studied brightness fluctuations in the Crab Nebula. The structural function was determined separately along and perpendicular to the field. In both cases the slope of the correlation curve was found to be close to $\frac{4}{3}$.

Turbulence of the interstellar gas was studied by S. A. Kaplan and V. I. Pronik [409, 404, 8]. If structural functions are constructed from the radial-velocity components of clouds, measured at high dispersions, we may assume that we are investigating velocities of the separate elements of mass. If these clouds were uniformly distributed along the line of sight, the observed structural function would be described by Eq. (23.28). However, as we noted in Secs. 5, 10, and 12, this condition certainly is not satisfied in interstellar space. Kaplan and Pronik [409, 404] assumed that clouds are distributed at approximately the same distance from the observer, which, considering the tendency of clouds to form large complicated complexes, must be a better approximation to reality. From this approximation, the observed structural function must simply coincide with the function (23.24). The catalogue of Adams [142] allows us to build a structural function that is in good agreement with Eq. (23.24) when $L \approx 80$ pc and $\varepsilon \approx 4 \times 10^{-4}$ cm^2/sec^3. When $r < L$, the spectral index is 0.71, which is close to the theoretical value of $\frac{2}{3}$.

For a study of the turbulent properties of an interstellar magnetic field, it is useful to construct structural functions of the fluctuation in interstellar polarization or nonthermal radio emission. True, it is difficult in this case to estimate the effect of other factors, such as density fluctuations of the dust or the concentration of cosmic-ray particles. Therefore, the detection of definite correlations in this case can be considered only qualitative evidence in favor of the hypothesis of the presence of turbulent areas in an interstellar magnetic field.

The structural functions for the degree of interstellar polarization and fluctuations of nonthermal radio emission compiled by I. A. Klimishin and V. M.

Bazilevich [419] are shown in Figs. 60 and 61. Curve I in Fig. 60, drawn from 515 stars (Hall's data), with coordinates $100° < l < 165°$, $-3° \leqslant b \leqslant +2°$, may be simulated in its early stages by the relation $\overline{(p_1 - p_2)^2} \approx 5.1(2\alpha)^{0.54}$ percent, where 2α is the angular distance in degrees between star pairs. Curve II, drawn from 209 stars in the square $355° \leqslant l \leqslant 25°$ and $-4° \leqslant b \leqslant 0°$ (toward the galactic center), is approximated by the relation $\overline{(p_1 - p_2)^2} \approx 3.55(2\alpha)^{0.80}$ percent.

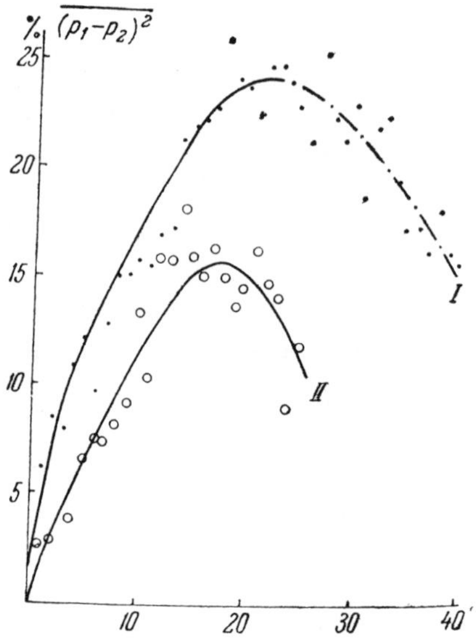

FIG. 60. The structural function for the degree of interstellar polarization toward the galactic anticenter (I) and the galactic center (II) [419].

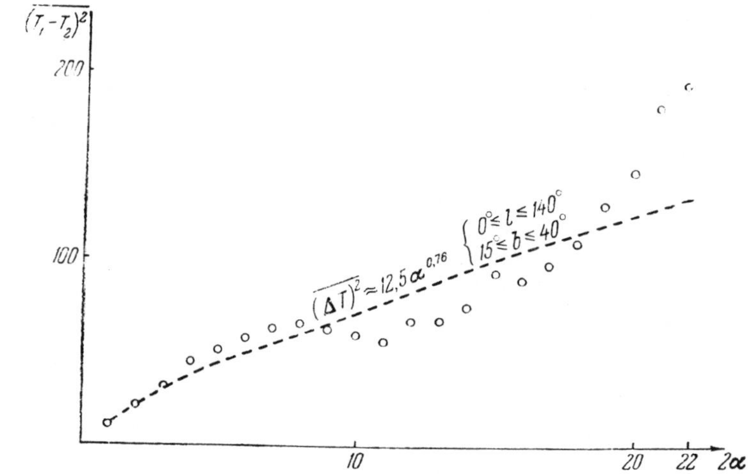

FIG. 61. The structural function of the fluctuations of nonthermal radio emission at $\lambda = 75$ cm [419].

Similar results are obtained for the structural function of brightness-temperature fluctuations of nonthermal radio emission at $\lambda = 75$ cm (using data from [325]). It is better here to construct structural functions at higher galactic latitudes, where the system is less elongated. The structural functions obtained are approximated by the expression $\overline{(\Delta T_b)^2} \approx 12.5\alpha^{0.76}$ deg in a region of the northern sky, $35^\circ \leqslant l \leqslant 175^\circ$ and $15^\circ \leqslant b \leqslant 40^\circ$, and by the expression $\overline{(\Delta T_b)^2} \approx 4.14\alpha^{0.74}$ deg in a region of the southern sky, $35^\circ \leqslant l \leqslant 135^\circ$ and $-35^\circ \leqslant b \leqslant 20^\circ$.

These results must be interpreted with the help of Eq. (23.29) and Table 23.1, since we have in this case correlated values averaged along the line of sight. However, as we have noted, the development of such an interpretation at the present time is not practicable because of difficulties in assessing the role of fluctuations in other parameters. In particular, the reason for the decrease in the structural function of the degree of polarization with large angular distances (Fig. 60) is not clear.

We must emphasize that structural functions of both the velocity fields and magnetic-field characteristics have a spectral index close to that in Kolmogorov's law. This is hardly a chance coincidence.

We may assume, therefore, that the motions of interstellar gas and of nebulae obey certain statistical rules that are characteristic of turbulence.

24. The Evolution of the Interstellar Medium

Galactis Structure

It is impossible to separate the evolution of the interstellar medium from the evolution of other galactic constituents, particularly the stars. This becomes evident from observational and theoretical considerations. The interstellar medium and the stars represent two evolutionary phases of galactic material, and this evolution must be regarded as a single process in the development of the Galaxy. The growth of the Galaxy has left its mark on its structure. Different parts of the Galaxy were formed at various times and have preserved their characteristics. Therefore, in this section we must first consider galactic construction from the point of view of the composition of the various parts, that is, the types of object from which it was formed. We stated previously that gas and hot stars are concentrated toward the galactic plane. This is not true of all stars. Baade [409] was the first to separate the Galaxy into two "population" types. The first type, concentrated toward the galactic plane, includes hot stars of high luminosity from O to A, the red supergiants, and certain other objects. Galactic clusters belong to this type. The second population type contains stars only slightly concentrated toward the plane, but strongly concentrated toward the galactic center. In the second type, there are no white or blue stars from the main sequence; the brightest are the red giants. The spherical clusters belong to this population.

B. V. Kukarkin has amplified these ideas [410]. He showed that, in addition to these extreme types singled out by Baade, subsystems of intermediate types may exist. His separation into subsystems has a bearing on the important case of variable stars. It was shown that different types correspond to different distributions in space and different kinematic peculiarities. The designations flat and spherical subsystems were proposed in place of population types I and II. Further studies, based on more detailed observations and partly connected with contemporary theoretical ideas on stellar evolution, led to the presently accepted division into five subsystems. These subsystems are based on stellar types and above all on their spatial distribution together with their velocity dispersions in the z-coordinate, perpendicular to the galactic plane. The different velocity dispersions are valuable in reconstructing the evolutionary picture of the Galaxy. In fact, there is little interaction between stars, and dispersions cannot have changed appreciably over the cosmogonic period. Therefore, the individual groups of stars conserve the dispersion present in the medium from which they evolved. It is true that interactions between stars and gaseous [411] and stellar clouds [412] are far more effective than pair interactions between stars and can somewhat alter the motion. This may have a bearing on the value of the dispersion in intermediate subsystems, but it cannot transform a flat or intermediate subsystem into a spherical system.

The nomenclature of subsystems and some of their parameters, according to Oort [413], are given in Table 24.1 (with small changes). The average z-coordinate ($\bar{z}$) characterizes the concentration toward the galactic plane. The next column in the table shows the mass relative to hydrogen of elements heavier than helium. A characteristic feature is the decrease in abundance of heavy elements as we move toward the spherical subsystem. The numbers are so far very rough, but there is no doubt about the difference in composition of the subsystems. The differences in chemical composition are responsible for

Table 24.1.

Subsystem	$\bar{z}$ (pc)	Mass of heavy elements (percent)	Approximate age (10^9 yr)	Total mass ($10^9\ M_\odot$)
Flat	120	4	<0.1	2
Old flat	160	3	0.1–1.5	5
Disk	300–450	2	2–7 }	47
Intermediate II	700	1	7–12 }	
Spherical	2000	0.1–0.5	12–14	16
				70

the specific properties of the variable stars and certain other special stars, about which we shall speak later on.

In the flat subsystem are grouped hot class O and B stars, usually forming associations, supergiants of different classes, and type T Taurus stars, usually showing bright lines and characterized by outbursts during which radiation appears in the violet region of the spectrum. Close to these subsystems is the old flat subsystem, including A stars and certain others. Galactic clusters, with

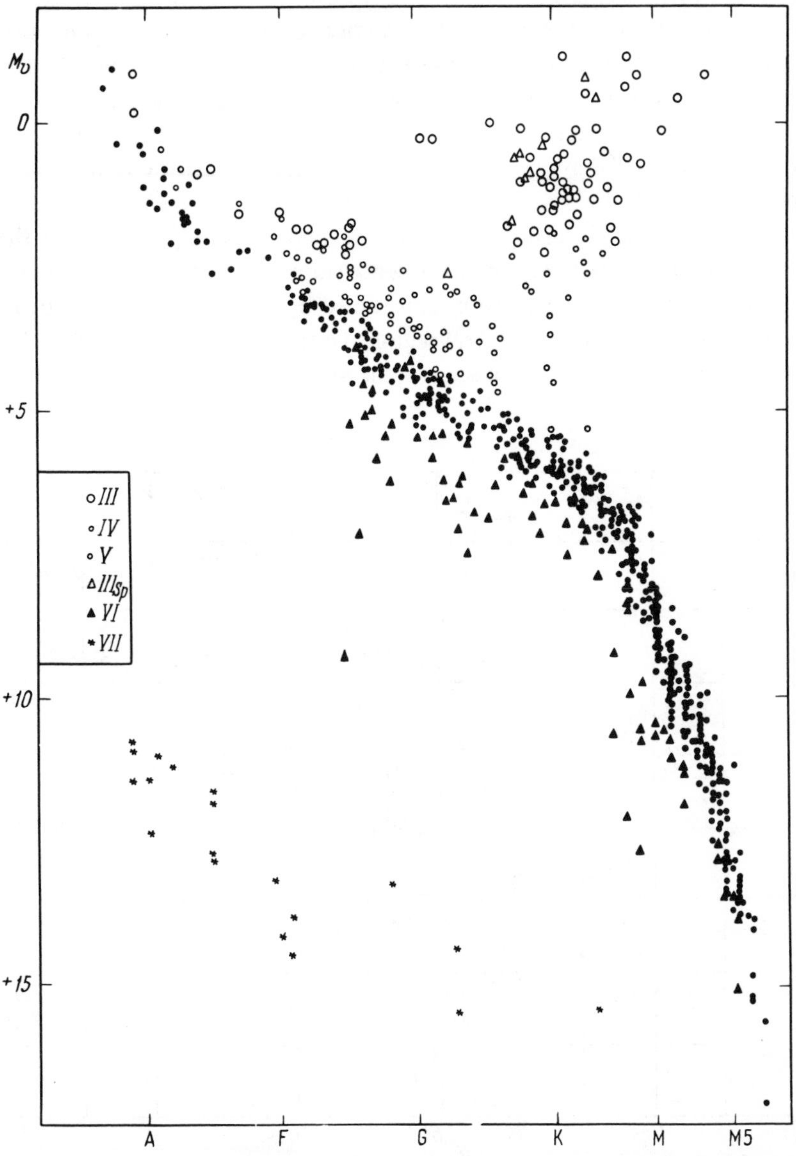

FIG. 62. Hertzsprung–Russell diagram for stars near the galactic plane. Class VI stars (*black triangles*), the subdwarfs, do not belong to the flat subsystem.

a few exceptions, belong to one of the flat subsystems. The flat subsystems, especially the first, are centered in the spiral arms. Interstellar gas, also centered in the arms, and in a number of cases in developing complexes genetically connected with associations, belongs to the flat subsystem.

The disk includes the majority of main-sequence stars, including the Sun. To this subsystem belong also the red giants, new stars, planetary nebulae, and others. To the intermediate II subsystem belong F–M stars of high velocity, the long-period variables, and certain others. In the spherical subsystem we place globular clusters, the subdwarfs, short-period cepheids, and certain others. There is little gas in the spherical subsystem, which includes the tenuous galactic corona and a few fast, low-mass clouds rising from the galactic plane.

The difference between the subsystems shows up very clearly from their Hertzsprung–Russell diagrams, which relate stellar luminosity to radius, surface temperature, spectral class, or color. Such a diagram is shown in Fig. 62 for stars in the solar neighborhood, stars that generally belong to the first three subsystems of Table 24.1. The main sequence (luminosity class V) includes stars of different masses whose emission is sustained by the conversion of hydrogen to helium taking place in the central region. To the right lies a branch of giants (luminosity class III), above (not seen on the diagram), supergiants of all classes

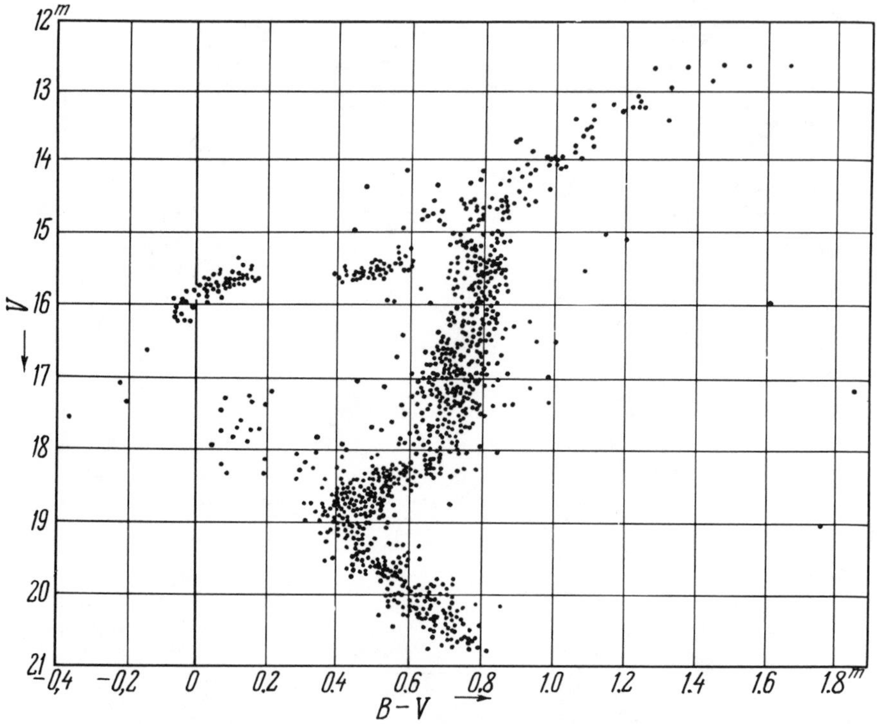

FIG. 63. Diagram of the globular cluster M 3 by Sandage.

(I and II), and below, the white dwarfs (VII). Other less important sequences are not shown in Fig. 62.

A diagram of a typical globular cluster, M 3, of the spherical subsystem is shown in Fig. 63. The main sequence, consisting of subdwarfs, contains hardly any stars earlier than class F. The giant branch rises more steeply than in Fig. 62, that is, these red giants are brighter than in the disk. Moreover, the spherical subsystem contains yellow and white stars of average luminosity, forming the so-called horizontal branch. We should note that the difference in the H–R diagrams, the chemical composition, and other features of the spherical subsystem relative to the flat subsystem is less for globular clusters near the plane, especially at the transition to the intermediate subsystem II.

Stellar Evolution

The principle features of H–R diagrams are well explained by modern ideas on stellar evolution (see, for example, [414]). A newly formed star consists of a chemically homogeneous sphere with smooth density and temperature distributions. The star finds a definite place on the main sequence that depends on its mass. In massive hot stars, the central temperature is higher and the hydrogen-to-helium transformation is accomplished through the carbon cycle, while in stars of medium and low mass, it is by means of the proton-proton reaction. The molecular weight of the nucleus is increased by the hydrogen depletion. To conform with the condition of hydrostatic equilibrium, this leads to increased gas density and temperature. Therefore, the reactions are enhanced and increase the luminosity of the star in spite of the reduction of some of the nuclear hydrogen. As a result, the star is pushed to the right-hand side of the diagram, up from the main sequence toward the red-giant branch. However, hydrogen burning in the nucleus proceeds comparatively slowly, and for a large part of its life the star remains on the main sequence. Stellar luminosity in the main sequence is approximately proportional to M^3. Therefore, the hydrogen burning rate, and consequently the evolutionary pace, become much faster at higher masses. Thus, for solar-type stars, for example, the lifetime on the main sequence is about 10^9 yr, and for massive class O stars it is $(3\text{–}5) \times 10^6$ yr.

The difference in the evolutionary period for various stars is readily seen from a comparison of the H–R diagrams for stellar clusters, whose members were evidently formed at the same time. Young clusters include early-class stars. In older clusters, the youngest stars that appear are not O or B stars, but, for example, A stars. In very old clusters of the main sequence, there are no stars earlier than F.

When the hydrogen in a stellar nucleus is almost completely consumed, a further temperature rise cannot provide the necessary amount of energy, and the nucleus, deprived of sources of energy, begins to contract rapidly. This leads to degeneracy of the electron gas and the pressure now depends not on temperature, but on density. It is proportional to $\rho^{5/3}$ in the case of nonrelativistic degeneracy, and to $\rho^{4/3}$ for relativistic degeneracy, when the velocity of a large number of electrons is comparable with the velocity of light. The gravitational pressure is also proportional to $\rho^{4/3}$. Therefore, during nonrelativistic degeneracy, there exists a value of the density at which equilibrium can be established and the pressures are equalized. During relativistic degeneracy, demanding a higher density, a change in ρ has no effect on the ratio of the gas to the gravitational pressure, so that contraction cannot occur until other factors intervene. Such a contraction will occur if the mass of the nucleus, deprived of sources of energy, exceeds 1.43 $M_{\odot}$, that is, in massive stars. In these stars, after exhaustion of the hydrogen, the nucleus contracts and the temperature rises until it reaches $(120\text{–}140) \times 10^6$ deg. At this stage, the helium reaction begins, the conversion of He into C and O and at higher temperatures into Ne and Mg. The combination of these atoms with hydrogen occurring at the boundary of the nucleus forms N, F, and other elements at the beginning of the periodic table. Stellar evolution during the helium reactions has not yet been studied. Evidently, a star at first becomes a red supergiant and then, if its mass is greater than $\sim 10\ M_{\odot}$, it is transformed into a supernova.

With stars of average mass, $(1\text{–}2)\ M_{\odot}$, a portion of the nucleus is in nonrelativistic degeneracy. It therefore acquires an equilibrium configuration and is hardly compressed further. Its temperature is high, but insufficient for the helium reactions. Radiation from the star is supported by the hydrogen reaction proceeding at the periphery of the nucleus. When the hydrogen is consumed in this layer, the reaction proceeds to the following layer and the first layer collapses into the nucleus. The stellar envelope beyond the energy-liberating layer is extended because of the high temperature of the internal layers, and the star becomes a red giant of high luminosity and low surface temperature. The nuclear mass gradually increases, its temperature rises, and finally the helium reactions can also begin. In stars of still smaller mass, these reactions evidently do not set in.

The evolution of a star depends principally on its mass and chemical composition. An essential factor is also the character of the mixing of the material inside the star between the zone where the reactions proceed and the envelope of original composition. Mixing is caused by convection zones, by rotation, and possibly by a magnetic field. Comparison of calculations with cluster diagrams shows that mixing is of little significance in the majority of stars, but that it may be responsible for certain anomalies. In close binary systems, the internal structure must deviate from spherical symmetry. This can also affect their evolution.

As the result of hydrogen depletion, stars are converted into the extended red giants with dense nuclei consisting of helium and the heavier elements. The dimensions of the envelope are continually increasing and this results finally in escape of the gas. This is assisted by the presence of convection currents, which in the red giants enclose almost the entire envelope. A particularly marked emanation is observed in certain close binaries, for example in the star β Lyrae. In these stars escape is facilitated by the tidal action of the other component. The strong flux of matter from close pairs evidently explains [415] the sharp anomalies of chemical composition in many of these stars. The original envelope is thrown off and the nucleus appears, consisting of He, N, Ne, and other elements, synthesized in stellar nuclei by the hydrogen and helium reactions. The presence of a strong escape mechanism explains, in principle, why certain old stars contain hardly any hydrogen, while other stars of approximately the same age have strong hydrogen lines.

The influence of a close component, disturbing the normal evolutionary process, can evidently lead to instabilities, such as outbursts. In this connection, it is interesting that new stars appear, as a rule, to be the components of close binaries [416]. New stars also appear as binaries, from which outbursts often take place, but with smaller amplitudes. It has been suggested that explosions of stars of this type may be connected with a flow of gas from large to smaller stars and may develop because of the instability of the latter.

Escape of gas from the red supergiant α Herculis has been observed spectroscopically. This star is a binary and in the spectrum of the second star, found inside the extended envelope of the supergiant, are seen displaced absorption lines formed by the envelope. The rate of loss is $\sim 1\ M_\odot$ per 3×10^7 yr [417]. Escape of matter occurs in many unstable stars of types Be, P Cygni, possibly Wolf–Rayet, and others [4], not to mention envelopes separated during outbreaks of novae and supernovae. With most stars of average mass, loss of matter leads to the formation of planetary nebulae [418], while the nucleus of degenerate gas becomes a hot dense star, exciting luminescence in the nebulae. The nebula is dispersed in the course of time and the nucleus cools off, while at least some nuclei become ordinary white dwarfs, which have been cooling for 10^9 yr. In this connection, it is of interest that the white dwarfs are found principally in older clusters. Massive stars throw off some matter during the early stages of evolution, when they appear as stars of different spectral classes or as red supergiants. During the final evolutionary stage, when the nucleus is apparently compressed to a superdense state, a vast amount of gravitational energy is released and, possibly because of this energy, an explosion takes place, a type II supernova outburst, accompanied by the ejection of an envelope, whose mass is several solar masses, with a velocity of 5000–7000 km/sec. In spite of many hypotheses, the problem of the explanation of explosions of supernovae is still far from being solved. However, observations show that this process is accompanied by the release of a vast number of relativistic particles, which by their magnetobremsstrahlung, as shown in [420], account for part of the optical outburst. A large number of relativistic particles occur in expanding nebulae, the remnants of supernova envelopes. In addition to relativistic particles, there are released during the outburst a large number of neutrons, which are captured by atomic nuclei, converting the latter to heavy elements up to uranium or beyond. Thus, supernova envelopes bring primarily heavy elements into the Galaxy, while gas escaping from the giants and the planetary nebulae provides hydrogen located in stellar exteriors and light elements synthesized in the nuclei of the red giants and in the layers bounding these nuclei (see, for example, [421 and 422]).

Stellar evolution explains the absence of hot stars of the subdwarf sequence in the spherical subsystem and in intermediate system II, since the stars of these subsystems and the globular clusters, in particular, are old. Calculations show that the globular clusters exceed 10^{10} yr in age; since then, stellar formation has practically ceased in the spherical subsystem. The similarly estimated age of other subsystems is given in Table 24.1. The study of the evolution and age of the disk stars is difficult, since they contain almost no clusters older than 10^9 yr. This is explained by the dissociation of clusters with time. Mutual approach and exchange of energy mean that one after another stars acquire greater velocities and leave the cluster, while the remainder lose energy and the cluster becomes more compact. Since interactions proceed more frequently in the compact system, cluster disintegration is accelerated with time. Finally, one or several multiple systems of high gravitational energy remain, while all the rest of the stars are dispersed. The dispersion process also takes place in globular clusters. In this connection we point out that globular clusters are observed to have dense nuclei and relatively extended "atmospheres." This "atmosphere" apparently does not consist of "runaway" stars, but was formed directly from the external parts of the large cloud, which was converted into stars of the globular cluster. However because of the high mass of globular clusters, they lose stars relatively slowly and survive, while the younger galactic clusters, having less mass, have a shorter life. Moreover, the distintegration of non-compact galactic clusters is accelerated by tidal forces. A few old galactic clusters, such as M 67, NGC 188, and NGC 7148, have survived and the study of them has provided the basic data for old disk stars on the H–R diagram. The age of these clusters exceeds 5×10^9 yr. It is interesting that they all fall within 300–500 pc of the galactic plane, that is, the older the cluster, the greater the distance. Selection effects are obviously important here. Those clusters that have survived for such a long time have orbits inclined to the plane at a considerable angle and have spent a long while away from the clouds that would assist in their disintegratien by tidal action [415]. There are some individual stars of the disk that are older than these clusters, but it is difficult to choose stars of one particular age and draw diagrams for them.

The Connection between Gas and Stars

The presence of stars in the flat subsystem, whose age is measured in several millions or tens of millions of years, shows that stellar formation is still going on. The youth of hot stars is confirmed not only by their rapid consumption of energy but by dynamic considerations. Hot stars are found in clusters and in more elongated systems such as associations. V. A. Ambartsumyan [2] has pointed out that the density of stars in associations is less than that of stars in the field and therefore that they must be unstable relative to tidal disturbances. The disintegration period of an association is about 10^6 yr, which fixes the upper limit of its age. Ambartsumyan stated further that the total energy of associations is positive, that is, they expand not because of disturbances, but because of the initial conditions [2]. However, proof of this assertion is not yet available.

The majority of associations contain emission nebulae whose luminescence is excited by hot stars. In Sec. 22 we noted that the high pressure of the hot gas caused nebulae to expand with a velocity of 10–15 km/sec. Groups and complexes of such nebulae also disintegrate [184, 425]. The period of expansion and disintegration of nebulae and their groups amounts to several million years, that is, it is of the same order of magnitude as the period over which associations disintegrate. All this led G. A. Shain to propose that young stars and emission nebulae are formed together in a single process [426]. Associated phenomena and the development of stars apparently explains the fact that clusters containing O8 and earlier stars are usually associated with nebulae, while clusters without nebulae usually do not contain stars earlier than O9–B0 [427]. Here, we must recall that B0 stars form a very small ionization zone and at low gas densities their luminosity is difficult to detect.

The expansion of nebulae is accompanied by changes in their appearance. V. F. Gaze and G. A. Shain [97] separated nebulae into classes according to their structural characteristics: structures with peripheral or uniform features, and so on. It is impossible to carry out such a classification with sufficient clarity. R. N. Ikhsanov has shown that if we divide nebulae according to their illuminating stars, then with increase in diameter their classification changes from structured and filamentary through peripheral to large fields. At the same time, nebulae irradiated by O5 stars are usually structured [428]. This is explained by the fact that, during the short evolutionary period of O5 stars, the nebula has not yet been able to expand and to run through its configuration sequence. Consideration of the connection between the diameter and the structural classification of a nebula, on the one hand, and the type of exciting star on the other shows that the nebula and the star evolve simultaneously. As the nebula expands, the hottest stars disappear and the cluster gradually changes to a later spectral class. However, we must not assume that all nebulae start with O5 stars. Young dense nebulae occur that are associated with stars of the later spectral subclasses [33].

The expansion and disintegration of nebulae and their groupings, the movement of masses of neutral hydrogen out of complexes, and other phenomena at boundaries between H I and H II zones are explained by the formation of hot stars in a complex of cold dense gas. When part of the gas recedes because of the radiation energy of the stars in the association, the potential energy of the system is lowered. This lowers the stability of the association to external disturbances and may even result in changing the energy of the stars that have just been formed from negative to positive.

One of the fundamental problems of cosmogony is, from what are the stars formed? The connection between young stars and associations and the interstellar gas is most conspicuous. They are first of all distributed in the spiral arms, which consist mainly of gas. This is clearly visible in the Andromeda Nebula and in other nearby spirals and may be demonstrated in our own Galaxy. As an example of the connection between associations and the arms, we can take the large number of hot stars in the constellation Cygnus where the Orion arm is intersected. There is a still more conspicuous grouping of young stars in the Sagittarius arm, where the density of the diffuse medium is high [429]. The connection between gas and young stars is not well defined. Gas complexes and large parts of the arms, especially in the internal parts of galaxies, are found to be almost devoid of hot stars. However, there are almost no gas-free associations to be found in a spiral arm. We must note that the existence of an arm itself as a comparatively stable formation results from the action of a magnetic field on the gas. The fact that stars are formed in the arms is a powerful argument that gas is the starting material for their formation.

If stars are formed from a diffuse medium, then during their earlier evolutionary stages they must represent objects compressed by internal gravity, without having achieved an equilibrium configuration in which energy losses by emission are compensated by nuclear reactions.

In the initial stages of the contraction, the transparency of these objects to infrared radiation is sufficient to ensure that the contraction is almost unimpeded. When the temperature inside the globule becomes greater than 1300°, the evaporation of dust particles increases the transparency even more, so that further contraction to stellar densities is not hindered by this effect (we neglect the influence of rotation and other similar effects [195a]).

During further contraction the stellar opacity becomes significant. It was initially assumed that the rate of contraction is determined by radiative transfer, from which it follows that the star moves almost horizontally on the Hertzsprung–Russell diagram with a slow ($\sim\rho^{1/4}$) increase in luminosity. At the end of the contraction the luminosity was thought to decrease slightly, and the star settled down toward the lower boundary of the main sequence.

However, more detailed investigations made by Hayashi [429a] of the boundary conditions for the loss of radiation from the protostar have shown that during their contraction until they reach the main sequence the energy is transferred to the surface not by radiation, but by convection. In this case the energy of contraction is liberated quickly, the protostar crosses the H–R diagram almost vertically, and the surface temperature increases by from 2000 to

3000 or 3500 deg. The time of evolution of the star in the process of gravitational contraction is also decreased considerably, although it depends essentially on the mass of the star. For stars like the Sun the contraction time is of the order of 2×10^6 yr, whereas under the assumption of radiative transfer the time for this evolution was 25 times longer. The contraction time of hot stars is of the order of only 10^3–10^4 yr, but for dwarf stars it is comparable with the lifetime of the Galaxy.

In the final stages of contraction the stars are located along the lower boundary of the main sequence, but, since the contraction time of the star depends on the mass, some of the stars in the cluster must fall more to the right of the main sequence, while the younger the cluster, the more massive the contracted stars. Observations have actually shown such stars to be present in clusters. During a certain phase of contraction, bright lines occur in their spectra and they are described as T Sagittarii variables. It is interesting to note that in the youngest clusters, such as the cluster associated with the Orion nebula, A and F stars are in fact T Sagittarii stars, while in clusters of medium age this is true only for K or M stars, whose contraction time is considerably longer. All associations containing O stars contain T Sagittarii stars as well [430], but there are T associations that do not contain hot stars. It is possible that here hot stars either have finished evolving or, owing to peculiar conditions, were never formed.

T Sagitarii stars are unstable objects. Sometimes their brightness suddenly increases, principally in the violet region of the spectrum. With certain stars this emission is observed almost continuously. The presence of the emission lines of hydrogen and helium in such comparatively cool stars also indicates instability. On the one hand, there must be convectional instability in such comparatively cool stars, producing as in the Sun, although more vigorously, a chromosphere and corona. On the other hand, the contraction process itself may terminate irregularly and produce random motion. Such instability occurs in UV Ceti red dwarfs, which during their brief periods of outburst can hardly be distinguished spectroscopically from T Sagittarii stars. The unstable motion of these stars, in conjunction with rotation and magnetic effects also connected with the transfer of angular momentum from the star to the surrounding medium, produces cosmic rays similar to those observed in solar flares. Synchrotron emission from relativistic electrons possibly plays some part in the observed spectral features of these stars [431]. The arrival of shock waves at the surface, leading to envelope formation, may also be responsible for sudden outbursts of violet emission.

It is important to note that T Sagittarii stars, closely associated with diffuse material, are observed, as a rule, in dust clouds. The Herbig–Haro objects seem to represent still earlier stages of stellar contraction. They appear as chains of hazy pinwheellike objects with an emission spectrum and are often found in dark clouds. No stars have been detected inside them, so that they may be only gas clouds, shining because of the energy released during contraction.

Gravitational Condensation

Stellar contraction during the first stages of evolution arises through natural gravitation. However, for a gaseous mass to start contracting spontaneously, certain conditions must be met. The question of gravitational instability was solved first by D. Jeans, who showed that an infinite homogeneous medium is unstable to longitudinal waves. At a sufficiently long wavelength, contraction once initiated will continue because of gravitation. It is true that the problem posed in this way is incorrect. An infinite homogeneous medium may be initially in static equilibrium only when its density is zero [432]. In general, it is necessary to reject the initial condition of equilibrium. However, the introduction of the corresponding changes hardly affects the final result of Jeans, who showed that the critical wavelength l_c is given by the expression

$$l_c = c_s\left(\frac{\pi}{G\rho}\right)^{1/2} = \left(\frac{\gamma\pi kT}{G\mu m_{\mathrm{H}}\rho}\right)^{1/2} = 2 \times 10^{-11}\left(\frac{\gamma T}{\mu\rho}\right) \mathrm{pc}, \tag{24.1}$$

where $c_s = (\gamma p/\rho)^{1/2}$ is the velocity of sound under the given conditions. The mass of gas in a sphere of diameter l_c is

$$M = \frac{\pi}{6}\rho l_c^3 \approx \frac{1}{2}\frac{\gamma\pi kT}{G\mu m_{\mathrm{H}}} l_c \approx 3\frac{\gamma T}{\mu} l_c\,(\mathrm{pc})\, M_\odot, \tag{24.2}$$

where l_c is in parsecs. The criterion of instability of the *finite* mass is determined by the virial theorem, which gives numerically approximately the same condition for the critical mass as the Jeans criterion. As we shall see later, in several cases we must assume that the gas is condensed first by external pressure. This pressure continues to operate in the gravitational condensation process, and therefore we must generalize the virial theorem by introducing external pressure. This calculation was made by Ebert [433] and in a more graphic form by McCrea [434].

The equation of motion for particles under the action of a force $\mathbf{f}$ is $m_a\ddot{\mathbf{r}} = \mathbf{f}$. Transforming it somewhat and then summing similar equations for all the particles, we obtain

$$\frac{1}{2}\frac{d^2}{dt^2}\sum_a m_a\mathbf{r}^2 = \sum_a m_a\dot{\mathbf{r}}^2 + \sum_a \mathbf{r}\cdot\mathbf{f}. \tag{24.3}$$

The quantity $I = \sum_a m\mathbf{r}^2$ is the moment of inertia relative to the origin of coordinates and $\sum m\dot{\mathbf{r}}^2 = 2W = 3MkT/m_{\mathrm{H}}\mu$ is the total kinetic energy of the particles, which we shall assume to be thermal energy. The sum of the moments of force on the right-hand side is called the virial. The forces acting on the particles come from collisions between them (their contribution to the virial is zero), from collisions with external gas particles, that is, from the pressure p acting at the regional boundary, denoted by S, and from the gravitational effect described by the gravitational acceleration $\mathbf{g}$. Thus, we can rewrite Eq. (24.3) in the form

$$\frac{1}{2}\ddot{I} = 2W + \int_S p\mathbf{r}\cdot d\mathbf{S} + \int_V \rho\mathbf{r}\cdot\mathbf{g}\,dV, \tag{24.4}$$

where the vector element at the surface is directed along the inward normal. If a quasi-equilibrium (not necessarily stable) state exists, $\ddot{I} = 0$. Expansion or contraction from the initial state correspond to the condition $\ddot{I} > 0$ or $\ddot{I} < 0$, respectively. Assuming that the external gravitational field may be neglected and that the system is spherical, we can write the last term of Eq. (24.4), the total gravitational energy of the system, in the form $\Omega = -AGM^2/R$, where A is a numerical factor, depending on the density distribution, of approximately unity. If we assume that the external pressure is constant over the total surface of the protostar, and take p outside the integral, we obtain $p\int \mathbf{r}\cdot d\mathbf{S} = -p\int \operatorname{div}\mathbf{r}dV = -3pV$. The equilibrium state, that is, the transformation of $\ddot{I}$ to zero, now takes the form

$$p = \frac{2}{3}\frac{W}{V} + \frac{\Omega}{3V} = \frac{3MkT}{4\pi R^3 m_H\mu} - \frac{AGM^2}{4\pi R^4}. \tag{24.5}$$

When $p = 0$, we obtain the usual condition for the kinetic and gravitational energy of equilibrium systems, $2W = |\Omega|$. We shall now consider a sequence of gaseous spheres of constant mass at constant temperatures, but with different radii, in quasi-equilibrium, while A remains the same. For large values of R, the second term disappears and Eq. (24.5) is converted into a simple relation between pressure and isothermal gas density. When R is reduced, the external pressure must be increased in order to retain the gas. The role of gravity is at the same time intensified, and at a certain value R_c the increase in p is replaced by a decrease. Beginning with $R = R_c$, the sphere will contract owing to gravitation. Differentiating Eq. (24.5), we obtain for the critical values of the parameters (for a homogeneous sphere $A = 3/5$):

$$R_c = \frac{4A}{9}\frac{GMm_H\mu}{kT} = 6\times 10^6\,\frac{\mu}{T}\frac{M}{M_\odot}R_\odot;$$

$$p_c = \left(\frac{9}{4A}\right)^3 \frac{3}{16\pi G^3M^2}\left(\frac{kT}{m_H\mu}\right)^4 = 1.4\times 10^{-13}\left(\frac{T}{\mu}\right)^4\left(\frac{M_\odot}{M}\right)^2 \text{ bar};$$

$$M_c = \left(\frac{9}{4A}\right)^{3/2}\frac{\sqrt{3}}{4\sqrt{\pi}G^{3/2}p^{1/2}}\left(\frac{kT}{m_H\mu}\right)^2; \tag{24.6}$$

$$\rho c = \left(\frac{9}{4A}\right)^3 \frac{3}{4\pi G^3M^2}\left(\frac{kT}{m_H\mu}\right)^3 = 4p_c\frac{m_H\mu}{kT} = 6.4\times 10^{-21}\left(\frac{T}{\mu}\right)^3\left(\frac{M_\odot}{M}\right)^2 \text{ g/cm}^3.$$

We note that p_c is four times less than the isothermal gas pressure at the same density. This means that, at the critical point, three-quarters of all the restraining force is gravitational. If we consider a sequence of spheres of given radius and mass, but of different temperatures (a cooling sphere), we obtain the same conditions as in Eqs. (24.6) for the critical state. The condition of stability is best studied with the aid of a graph showing the dependence of p_c on T at a given value of μ and various values of M. If a point representing some particular condition lies above the corresponding straight line, the condition is a stable one. The graph in Fig. 64 was drawn by McCrea for $\mu = 1.5$. In addition, the mass is decreased here by 1.5 times, because a more accurate theory, which

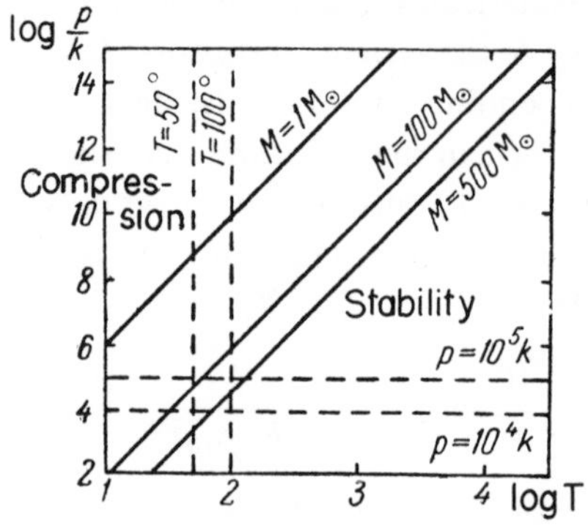

FIG. 64. Stable regions for various masses when $\mu = 1.5$ [434].

takes into account the density distribution inside the sphere, gives almost the same result as the approximate theory for homogeneous spheres, but for a mass 1.5 times greater. In the absence of external pressure, its role can be replaced by the gas pressure of the external layers; Fig. 64 can be used in this case also.

Before applying this graph to concrete examples, we should note a certain peculiarity in the instability of a medium extending in one direction (a cylinder, for example) and in two directions (a plane stratified medium) [434]. When the mass per unit length of a cylinder is M and its radius is R, the virial theorem gives the condition of equilibrium, analogous to Eq. (24.5), in the form

$$p = \left(\frac{kT}{m_H \mu} - \frac{1}{3} BGM \right) \frac{M}{\pi R_2}, \tag{24.7}$$

where B is a numerical factor depending on the density distribution along the radius of the cylinder and p, as before, is the external pressure. Here the sign of p does not depend, contrary to the spherical case, on the radius of the cylinder, but is determined only by the mass per unit length and by the temperature. When $M > 3kT/BGm_H\mu$, equilibrium becomes impossible. If the mass per unit length is less than this value, then the radius, but not the condition for gravitational instability, depends on the external pressure. The physical meaning of this feature is readily understood. The acceleration due to gravity g is practically determined by a segment of the cylinder of length approximately equal to its diameter. Most of the length of the cylinder contributes little to the gravitational acceleration. During contraction of the cylinder, the length of the "effective" section is reduced and the "effective" mass is correspondingly reduced, which does not happen in the contraction of a sphere. The reduction in 'effective" mass compensates for the smaller radius and the instability does not develop, if, of course, it was absent initially. When a layer is compressed, this effect is even stronger. Even a finite layer that is initially unstable reaches gravitational equilibrium under contraction if its temperature is not reduced. This is because the "effective" mass in this case decreases as the square of the thickness. The formation of contracting condensations from stable cylinders or layers is possible only when they disintegrate into fragments, each of which then contracts in three dimensions. This always occurs in practice as long as the smallest dimension exceeds the value l_c.

The Contraction of a Cloud

We shall now consider the further evolution of the contracting cloud. Its character depends essentially on the thermal regime from which the energy released during contraction is liberated. We should remember that at equilibrium the kinetic energy of particle motion is equal to half the gravitational energy. If the cloud contracts while remaining at the limit of stability, this equality must always be approximately satisfied. For this to happen, half the energy (when $\gamma = 5/3$) released during contraction must be converted into energy of internal motion, while the other half leaves the system as, for example, radiation. A star, for example, contracts in this way in the first stages where the thermonuclear sources are still not operating. Here the period of contraction is fixed by the rate of energy loss due to the luminosity of the protostar. If the amount of outgoing energy is less than half or, in particular, is zero, that is, if the contraction is adiabatic, the thermal (kinetic) energy increases more rapidly in order to establish equilibrium at an energy equal to half the gravitational energy. In the other limiting case, the energy of contraction leaves the system entirely, and the thermal energy does not increase but remains relatively low. The period of such a contraction, to which there is no resistance, is almost the same as the time for free fall from the initial to the final dimensions. This value sets a lower limit in every case for the contraction time if only the external compression forces are low. The time of fall from the original radius to the smaller dimension is approximately

$$t_n = \frac{1}{2}\frac{R^{3/2}}{(GM)^{1/2}} \approx \frac{1}{(17G\bar{\rho})^{1/2}} \approx \frac{10^3}{\sqrt{\rho}^{1/2}}\,\text{sec.} \tag{24.8}$$

We assume here that initially the rate of contraction was equal to the parabolic rate at the distance R. If contraction began from rest, for example, owing to rapid cooling, the contraction time is increased by $3\pi/4$. A feature of Eq. (24.8) is the nondependence of t_n on R during isotropic contraction. Under certain conditions, a cloud may disintegrate into parts and the contraction time of the separate parts is equal to the contraction time of a cloud of the same mean density, if, of course, the parts also do not encounter any resistance.

The process of contraction and division of a cloud under isothermal conditions was considered by Hoyle [435]. The gravitational energy liberated during contraction is approximately equal to

$$AGM^2\left(\frac{1}{R_2} - \frac{1}{R_1}\right) = \frac{AGM^2}{R_1}\left(\frac{R_1}{R_2} - 1\right). \tag{24.9}$$

Thermodynamics requires that during isothermal compression of an ideal gas the energy liberated shall be

$$M\int dQ = -M\int p\,dV = M\frac{kT}{m_{\rm H}\mu}\ln\frac{V_1}{V_2} = 3M\frac{kT}{m_{\rm H}\mu}\ln\frac{R_1}{R_2}. \tag{24.10}$$

This energy must be emitted by the system in order that the process be isothermal. It may happen that the energy (24.10) will be less than the energy liberated during contraction. Since the initial state was one of quasi equilibrium for $p \approx 0$, when the equality $AGM^2/R_1 \approx 3MkT/m_{\rm H}\mu$ is approximately satisfied the equality of liberated and emitted energy requires that the relation (when $A = \text{const.}$) $\ln R_1/R_2 = (R_1/R_2) - 1$ be satisfied. The roots of this equation are $R_1 \approx 2R_2$. With further contraction, the rate of release of gravitational energy will be greater than necessary for overcoming the relatively low pressure of the isothermal gas and most of the energy will begin to change into the energy of internal macroscopic motion. If these motions are not damped, the system as a whole will become adiabatic and, with its size one-third what it was originally, it will no longer contract. However, a density increase of 3^3, with the temperature remaining the same, renders the system unstable and it disintegrates into fragments several times smaller than the main system. The presence of motions does not hinder this disintegration if the motions are large scale. Inside each fragment the dispersion of velocities need not be so large. The fragments themselves will move with a velocity dispersion for which the system is stable. The energy of motion must constitute about one-third of the liberated gravitational energy of the system. The system is now in equilibrium and satisfies the virial theorem with $\ddot{I} = 0$. Contraction of the separate fragments follows the same scheme. The process of subdivision is intensified, since the period of gravitational condensation is $\sim(G\rho)^{-1/2}$, that is, several times less for each stage than for the previous stage. The short condensation time of the fragments allows us to assume that the kinetic energy of motion of the separate masses is not dissipated, that is, that there are practically no collisions.

Isothermal contraction requires that the following conditions be satisfied. Emission must depend strongly on temperature, the dependence on density must be weak or such heating and cooling must be compensated for, and the cloud must be sufficiently transparent that energy is able to escape during the period of contraction. If these conditions are not met, the temperature will change. When the temperature rises, the system will contract more rapidly as a whole, and if it falls, the system will break up into parts. The nature of the subdivision depends on the relation between the cooling time and the time of collapse t_n given in Eq. (24.8) [436]. The cooling time is the temperature-relaxation time of the cloud. t_T. If $t_n \ll t_T$, contraction proceeds almost adiabatically, its time is determined by cooling, and there is no fragmentation . If $t_n \approx t_T$, the process is almost isothermal, and in every case where fragmentation occurs the number of fragments is relatively small. For subdivision of the system into a great number of small fragments, the fragmentation must take place when the system is far beyond the limit of stability. For this to happen, we must have $t_n \gg t_T$.

In recent investigations the change of temperature during the contraction of the cloud is taken into account. Gould [436b] has shown that the dissociation of molecular and ionization of atomic hydrogen may keep the temperature low for a considerable time. If $N_H < 10^2$ cm^{-3} and $T \approx 100$°K, the contraction goes with the velocity of free fall. At higher densities the contraction may evolve to oscillation [436c]. It is possible that during the contraction the dense nucleus may be formed in the center of the cloud [436d].

The Role of Rotation and the Magnetic Field

In addition to temperature, density, and the other factors mentioned, rotation and the magnetic field are also significant in the contraction process. The energy of rotation is part of W and must be considered along with the thermal energy, while the magnetic energy, $(1/8\pi) \int H^2\, dV$, is an additional component in the virial equation (25.4) [339]. We shall consider how rotational and magnetic energy change with the radius of the system. The condition of conservation of angular momentum yields $\omega \sim R^{-2}$. Thus, the linear velocity is $V \sim R^{-1}$ and the rotational energy is $W \sim \frac{1}{2}MV^2 \sim M/R^2$. Since the gravitational energy is $\Omega \sim M^2/R$, the rotational effect is increased with contraction and sets a limit on it. When the system reaches an equilibrium state, the value $W = \frac{1}{2}|\Omega|$ is determined either by random motion, which occurs, for example, in globular clusters, or by circular motion, as in the flat galactic subsystems or in binary systems. After separation into several smaller parts, the angular momentum is transformed into relative motion of the fragments, which then contract individually. In each particular case the effect of rotation may be different. Under certain conditions, the angular momentum is not conserved, but may be transferred to the surrounding masses owing to turbulent friction, and particularly the magnetic field. The lines of force of the field formerly in the cloud from which the condensation was formed extend into the external gas. During rotation of the condensation, accelerated by contraction, the lines of force are twisted and, because of their tension, momentum is transferred from the condensation to the surrounding medium. This mechanism was proposed, for example, to explain the distribution of momentum in the solar system [437]. Hoyle calculated [438] that if a cloud of interstellar gas traversed by the field of the spiral arms did not rotate, but only took part in the differential galactic rotation (its angular velocity near the Sun is $\omega_0 \approx 10^{-15}$ sec^{-1}), a star formed in this case with conservation of momentum and with a density 10^{21} times that of the cloud must rotate with an angular velocity $\omega \approx 10^{14}\,\omega_0 \approx 10^{-1}$ sec^{-1}. Furthermore, if rotation of the cloud was damped by the field in the spiral arms until it was one-twentieth of the original value, then the period of rotation of the star must be about 10 hr. This is quite close to the period of rotation of hot stars, but is a hundred times less than the period of rotation of stars of class F and later classes. At the same time, the momentum of the solar system, if we remember that its mass was formerly greater because of hydrogen and helium that have since escaped, corresponds approximately to the momentum of the cloud being considered. In this connection, the idea was long ago expressed that the angular momentum of the Sun was transmitted to the planets, and that the slow rotation of stars of class F and later classes means that they have planetary systems. We might say, with more reservation, that slow rotation means the presence around these stars at the time of their formation of a diffuse medium, the external parts of the same cloud connected with the stellar magnetic field. We cannot say with certainty whether this medium was condensed into planets.

The rapid rotation of stars of early spectral classes or of close pairs is expected if we assume that stars are formed by condensation from the interstellar medium. However, they become completely unexplainable if we assume, as V. Ambartsumyan does, that stars originate from an expanding superdense medium. In the expansion process the star can conserve any loss of momentum, but it can hardly devise a mechanism by which it could be released, thus transmitting a high angular velocity. The rapid rotation of close pairs is even more difficult to explain. We may thus consider that rapid rotation is one proof of stellar formation from the condensation of a more tenuous medium.

We shall now consider the part played by the magnetic energy. During contraction of the medium with long damping times of the field, the flux across material contours hardly changes. For a constant configuration of the lines of force it then follows that the field intensity is $H \sim 1/R^2$, and the total magnetic energy of the system is $H^2R^3 \sim 1/R$, that is, the magnetic energy as well as the gravitational energy changes under isotropic contraction. In order that the field shall not impede contraction, it is sufficient that the magnetic energy of the system in the initial state be noticeably less than the gravitational energy. For a homogeneous sphere, this requirement may be written $H < 8G^{1/2}\rho R \approx 2 \times 10^{-3}\rho R$, or

$$H \lesssim 10^{-3}M^{1/3}\rho^{2/3} = 10^{-3}Ml^{-2}. \tag{24.11}$$

In the contraction of a cylindrically symmetric configuration in the presence of a magnetic field, the intensity of the field parallel to the axis of the cylinder increases proportionally to the density, and consequently the magnetic energy increases faster than the gravitational energy. Therefore, a cylinder with a magnetic field contracts only to some definite limit. In fact, the stability of the spiral arms is connected with this phenomenon.

The Gravitational Condensation of the Interstellar Medium

Formation of stars from a diffuse medium is very obscure at the present time, partly because the observational data are not very certain. Globules—the small dark formations lying inside luminous nebulae and sometimes visible against the background of the Milky Way—are usually thought to be protostars. These globules, whose cosmogonic significance was first pointed out by Bok and Reilly [439], consist of dense masses of dust and gas. Their dimensions vary between 0.01 and 0.1 pc, and their absorption amounts to several stellar magnitudes, while the smaller the globule, the more it absorbs on the average. The mass of the dust in compact globules does not usually exceed $(10^{-3}$–$10^{-2})M_\odot$, but if the globule contains gas in the usual proportion, its mass may be approximately that of the Sun. The less compact, larger globules may contain from 2 to 10 or more solar masses. Globules may be formed in nebulae through disintegration of the "elephant trunks." We know that a sequence of shapes exists between the "elephant trunks" and the globules, and we can see a correlation between the rim formed bordering the "trunk," the distance from the star, its spectral class, and other characteristics [391].

Compression of the gas into a globule arises from the pressure of the surrounding hot gas of the nebula. Therefore, the calculations described above may be applied to an analysis of the gravitational stability of a globule. We shall assume for the temperature of the globule $T = 50^\circ$ and for an H II zone the values $N = 10^2$ cm^{-3} and $T = 10{,}000^\circ$. Then the external pressure is $p/k = NT = 10^6$, and from Fig. 62 we find that cold gas in the H I zone will be compressed if its mass is greater than 22 $M_\odot$. In a less dense H II zone the required mass must be greater still. However, an H II zone formed by an earlier-class star cannot envelop and compress a cloud with a mass greater than several solar masses if this cloud had a normal density before contraction [440]. The observed masses of globules embedded in H II zones do not exceed the mass of the Sun. Thus, the genesis of stars through the gradual compression of neutral hydrogen by a surrounding diffuse nebula demands highly unusual conditions and seems, therefore, to be improbable, though not impossible. We can show that certain phenomena facilitate contraction. First, when dust exists in a dense cloud, hydrogen is converted to the molecular form (see Sec. 9), which increases μ and helps to lower the temperature. Apparently, the temperature can fall to 15–20° if it is assisted by cooling: the transition of part of the carbon to the neutral state, an increase in the amount of dust, which cools the gas directly and absorbs radiation that otherwise would heat the medium, the transition of almost all the elements except helium and some hydrogen into the molecular state, and the formation of particles, small in comparison with the dust grains and therefore not significant for absorption. In combination, all this can lower the critical mass to (1–2) $M_{\odot}$. We observe that large globules have masses up to 10 $M_{\odot}$, but that they are usually not situated in dense nebulae, and are associated with lower external pressure, while the critical mass is correspondingly greater.

One possibility that must be taken into account during calculations on the compression of globules by the surrounding gas is the following. So far, we have considered this compression to be in quasi equilibrium, that is, so slow that the gas inside the sphere is continually in equilibrium. In reality, the pressure of the hot gas induces a shock wave in the globule, which converges at the center, and therefore is a cumulative wave [441, 440] (see also page 417). In principle, such a wave can cause a strong compression, but only for a rather low mass. This process has still not been studied quantitatively. Another possibility is connected with the presence in nebulae of ordinary stars that can become centers of gas condensation. In other words, here we meet not the birth of new stars, but the rejuvenation of old ones [412]. Summing up, we can say that the transformation of an ordinary small globule into a star is improbable, since it requires special conditions, but it is in general not ruled out.

The hypothesis of star condensation from globules cannot be simply discarded, because there exist observations that seem to speak in its favor. These are connected with the "comet-shaped" nebulae about which we spoke in Sec. 13. There is located at the apex of these nebulae a T Sagittarii star, a young star occurring in a state of contraction. Some of the comet-shaped nebulae are found at the periphery of emission regions. In these cases they are usually bordered on the side toward the exciting star by bright emission rims similar to the rims that border the globules and "elephant trunks." The comet-shaped nebulae themselves are also oriented in a certain manner, with their tail directed away from the star. E. A. Dibai [442] proposed that comet-shaped nebulae are the "elephant trunks" in which the young star illuminating the nebula was formed. As time passes, the emission nebula expands and becomes invisible, the hot stars exciting its luminescence also continue their evolution, and the T Sagittarii star and the dust nebula associated with it continue to exist. This may explain the presence of certain comet-shaped nebulae not associated with emission regions or hot stars. It is still not possible to regard comet-shaped nebulae as explained, but the hypothesis stated seems very probable. The location of T Sagittarii stars in the center of the head of the "comet" suggests that condensation has occurred and not the capture of gas by an old star.

Comet-shaped nebulae show apparently that certain stars of medium mass can evolve from globules or "elephant trunks." However, this mechanism is hardly a basic one because of the difficulties mentioned and because it does not explain the formation of stars with very low masses, the formation of the first hot star in a given complex, ionization of gas, and so forth. We must therefore consider the gravitational stability of other gaseous formations. Using the graph in Fig. 64 and the relations (24.2) and (24.6), we can establish that a cloud of gas with $T \geqslant 50^\circ$, a concentration of 10–20 cm^{-3}, and dimensions of 5–10 pc, that is, with a mass of 50–500 $M_\odot$, corresponds to $M_c \approx 3 \times 10^3 M_\odot$ and that therefore such a cloud cannot contract. On the other hand, the larger complexes, whose masses are measured in tens of thousands of solar masses and whose dimensions are given in tens of parsecs, can contract under their own gravity, as we may readily estimate. In addition to thermal motion, we must take into account macroscopic velocities, which impede condensation. A temperature of 50° in neutral hydrogen corresponds to a velocity of about 1 km/sec. Velocities inside gaseous complexes are of the same order of magnitude or slightly larger. If there are no hot stars in the complex whose energy sustains motion, then the motion is rapidly damped and converted into heat. Kinetic energy is transformed into heat rapidly by shock waves. For this to happen, the velocity of the mass of gas must be higher than the velocity of sound, that is, motion is more quickly damped in cold or rapidly cooling gas like that from which the complex is made. Moreover, calculations show that the gravitational energy of the complex exceeds the limit of stability by a certain margin apparently sufficient for overcoming internal motion, particularly if the main portion of the hydrogen is in the molecular state. Thus, we may conclude that large complexes of cold dense gas can provide a start for the birth of stellar systems. In this connection it is useful to remember (Sec. 10) that the mass of the system of young stars, emission nebulae, and expanding gaseous masses in the Orion Nebula exceeds 10^5 solar masses and its radius, after expanding for several years with a velocity of 10 km/sec, is approximately 70 pc. We may readily calculate that before the expansion the radius of the system was less than 40 pc and that, even if the internal velocity reached 1 km/sec, the system was gravitationally unstable.

The agreement of the mass and the dimensions of the system in Orion with those estimated above can hardly be fortuitous. In agreement with the formation of stars in complexes is the fact that most stars are formed in groups, associations, and clusters, while the characteristic mass of such groups, taking into account stars of all classes, amounts to a thousand solar masses, and, if we include nebulae and expanding cold gas, reaches a value of $\sim 10^4 M_\odot$. The region in Orion, the bright regions in Sagittarius, and others are formations of still larger scale, consisting usually of several associations.

As stated before, we can reconstruct the thermal conditions of contraction from the structure of stellar systems. The fact that Orion-type systems disintegrate into several associations apparently indicates that the temperature was approximately constant in the original condensation process. However, when an association disintegrates into hundreds and thousands of stars, instead of forming several tens of more finely divided systems, it may be an indication that the temperature fell rapidly during the condensation process. On the other hand, if the temperature of the cloud does not exceed, say, 50°, the possible cooling is insufficient, since the gas temperature can hardly fall below 20° and the temperature drop is too small. In this case, the following arrangement would be more acceptable. First, cooling proceeds relatively slowly, the cloud temperature is somewhat increased, and contraction can continue without partition. Then, because of a change in the cooling conditions, due to the ionization of certain elements, for example, or for other reasons, the temperature falls rapidly and the cloud separates into smaller fragments. The arguments advanced, of course, are arbitrary, but they show that, in principle, a cloud may separate into hundreds of small fragments, which then form opaque stars or small multiple systems. However, other possibilities cannot be excluded. A cloud may separate into tens of fragments, each of which forms a small system of massive stars, as, for example, the Trapezium type of system or groups of ordinary stars of higher multiplicity. Preheating is not required here. A large fraction of the systems formed may have parameters favorable to rapid disintegration, even favorable to the formation of stars of medium mass. Massive stars are quickly formed and we observe them as Trapezium-type systems. Stars of later classes are able to disperse to the periphery of the association, leaving behind only a few multiple systems and relatively stable clusters. A separate analysis of this problem is required.

The question may be raised why we do not observe complexes in the intermediate stage of development between the gaseous cloud and the stellar system. We must answer that until young stars are formed in the complex it is seen only in absorption. In some cases very dense opaque clouds are seen. The amount of condensation of the individual fragments becomes even more difficult to observe, since, first, their dimensions are reduced and, second, the contraction time is correspondingly limited. However, special observations may allow such dense globs to be observed.

The greater the gas density, the more probable, in general, must be the formation of stars. Schmidt, proceeding from the fact that the thickness of the gas layer is approximately twice the thickness of the subsystem of class O and B stars, came to the conclusion that the probability of star formation is proportional to the square of the density [443]. However, in the Andromeda nebula the region of maximum density of hot stars falls between two maxima of gas density, so that there is no direct well-behaved relation between gas density and the number of stars formed [444]. Apparently the reason is that, once started, star formation continues beyond the time when a large amount of gas has been consumed.

We consider now the possible role of the magnetic field. Using (24.11), we can show that, if a globule of mass $M_\odot$ has a density $\rho \approx 10^{-16}$ g/cm^3, a field of intensity $H < 3 \times 10^{-3}$ Oe will not impede contraction. On the other hand, in a large complex, where $M \approx 10^5 M_\odot$ and $\rho \approx 10^{-22}$ g/cm^3, the magnetic field must not exceed $H \approx 10^{-5}$ Oe. This condition cannot be satisfied, and therefore the field in such a complex will interfere with contraction. The field acts even more effectively when the complex disintegrates, since the limiting value $H \sim M^{1/3}$ decreases with fragmentation. Therefore, a star cannot be formed from a cloud if the magnetic flux is conserved.

Nonconservation of magnetic flux may be due to strong absorption of radiation and a decrease in the degree of ionization in the dense compressed cloud. First, Joule damping at low ionization is increased [445], that is, because of losses connected with the motion of the plasma relative to the neutral medium (Sec. 16), the magnetic flux can diminish. Second, flux conservation, or freezing in of the field, binds the field to the plasma and not to the neutral gas. The plasma and the atoms are connected only by friction. Therefore, in principle, the field may check the plasma, preventing it from contracting further, while the neutral gas will slowly impregnate it [446]. The action of gravitational forces is here counteracted by frictional forces proportional to velocity and to the product of the concentrations, $N_i N_a$. When ionization is low, $N_a \gg N_1$, and the velocity may be considerable. Detailed calculations show that the field of such a contracting protostar has a complex structure [18:183, 447], while only a comparatively small portion of the magnetic flux is enclosed in the star. The remaining lines of force lag behind the gas and seem to form wedges turned toward the equator (Fig. 65). At the "cusp" of these wedges rot **H**, and consequently the current **j**, are large. This produces relatively strong dissociation, especially if we take into account the low conductivity. Therefore, at these points the plasma is moving slowly relative to the field. As the star condenses, the motion is accelerated. The field itself remains stationary, since the magnetic force $(1/c)\mathbf{j} \times \mathbf{H}$ is sufficiently large to overcome the weight of the plasma, together with the frictional force acting on it from the neutral gas. At the second stage of contraction, when a field of the type shown in Fig. 65 has formed, the gas flow into the star proceeds along the magnetic lines of force passing through it. The field does not interfere with the stream, which enters as both ionized and neutral gas. This circumstance, together with the motion of the plasma relative to the field in the equatorial plane, is significant with respect to the chemical

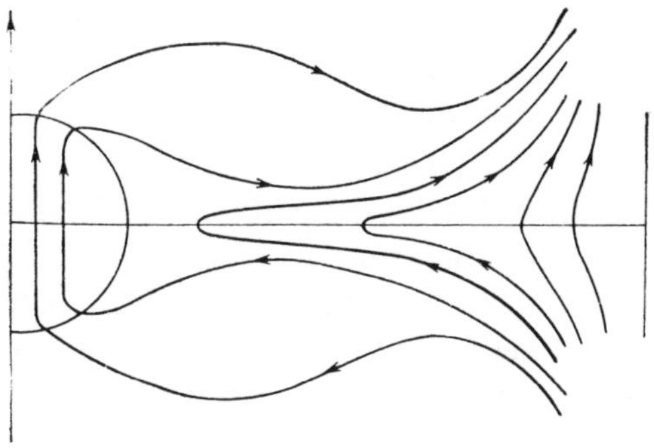

FIG. 65. Lines of force of a contracting star [18:183].

abundances in stars. If the plasma is completely arrested, very few metals and other elements of low ionization potential would enter into the composition of the star.

The picture we have presented is relevant to a nonrotating star. During rotation, the field is twisted and transfers angular momentum from the central part of the cloud, the protostar, to the envelope. The transfer of momentum causes contraction of the internal parts of the cloud and expansion of the external parts, changing the nature of the gravitational condensation. The efficiency of this process depends on how far the lines of force extend from the star. The twisting of the lines of force, which have opposite directions at different sides of the equatorial plane, forms neutral points of type *X*, where the principle of freezing is not fulfilled and a discontinuity in the field may arise, even at high conductivity. The region in which magnetic forces act is then substantially reduced, which leads to reduced transfer of momentum. Some calculations on this subject have been made [448], but the problem of the contraction of a rotating star with a magnetic field has not yet been studied completely.

25. Formation and Evolution of Galaxies

Formation of Galactic Subsystems

Clusters of galaxies and galaxies themselves are formed, according to present ideas, by the condensation of masses of gas. The conditions for the condensation of galaxy clusters and the formation of galaxies of various types will be discussed below, but first we shall discuss ideas about the formation of our own Galaxy.

In recent times the results of M. Schwarzschild, Hoyle, Oort, and other investigators have furnished a qualitative picture of the formation of the Galaxy that explains in general terms the presence of the subsystems with their various kinematic properties, chemical constitution, and other observed peculiarities. With certain additions, we may describe this picture as follows. About (12–15) $\times$ 10^9 yr ago the Galaxy was in the form of a spheroidal cloud. It was rotating, but slowly, so that the rotation did not affect the character of the contraction. This cloud was relatively poor in heavy elements, which is why the heavy-element content relative to hydrogen in very old stars is from 100 to 200 times less than in the Sun. It is also possible that this insignificant quantity of heavy elements was formed in the very early stages of contraction during the explosions of rapidly evolving supermassive stars. In this case the cloud would initially have been pure hydrogen. The initial helium content is difficult to determine at the present time from observations, since it is not observed in cold stars, and in hot stars of the horizontal branch it would have been formed in the process of their more rapid evolution.

Radial contraction of the Galaxy at very early stages is shown by the study of the trajectories of very old stars with low abundances of heavy elements [436a]. These stars move in very extended orbits, from which it follows that they were formed from gas which initially occupied a region approximately ten times larger than the present dimensions, and fell to the center of the Galaxy with low angular momentum. Stars of the spherical subsystem were formed in the process of contraction while the initial velocity dispersion and spatial distribution that reflect the conditions of their formation were conserved. The gas lost its kinetic energy relatively quickly and conserved its angular momentum. Therefore, the dimensions of the system of gas were reduced, but its rotation was accelerated. The system was transformed into a rotating spheroid. After that the system was only flattened. At this time the stars of the spherical subsystem close to the galactic plane were formed, together with stars of the intermediate subsystem that comprise the major portion of the mass of the Galaxy. The distribution of angular momentum in the intermediate subsystem is the same in a uniformly rotating spheroid [450a]. This proves that there was no strong turbulence in the initial cloud, and that it actually began contraction as an almost uniformly rotating spheroid. During further condensation the separate masses of gas did not exchange momentum, since the turbulent friction was low.

The whole process of contraction and flattening proceeded relatively quickly. It depends primarily on the collapse time of the gas t_n given by Eq. (24.8). The flattening time for a medium-dense spheroid with $\bar{\rho} \approx 10^{-24}$ g/cm^3 is 3×10^7 yr. The initial density of the collapsing cloud is $\approx 10^{-27}$ g/cm^3, but in this case the contraction time hardly exceeds 10^9 yr. Stars formed from the initial gas had different masses. The least massive are preserved up to the present time, but the more massive ones quickly proceeded through their evolution, ejected gas, and liberated heavy elements. A particularly large quantity of heavy elements is formed during supernovae explosions of massive stars. The ejected gas as well as that left from the initial gas cannot remain very long in the spherical subsystem. The gas liberated as heavy elements gradually contracts toward the plane and forms new clusters. This explains the increase in the abundance of heavy elements with decrease in the z-coordinate of stars and clusters.

Although we have always been speaking of the formation of stars, it is easy to show that individual stars could not condense at the density of the protogalactic gas. Condensation had to take place as large clouds. The globular clusters are preserved from the early stages. Their masses are of the order of $(10^5–10^6)\ M_\odot$, and their number in the Galaxy is about 220, so that their total mass constitutes less than 1 percent of the mass of the spherical subsystem. Evidently, only a few slowly rotating clusters have been preserved (thus of spherical form), clusters sufficiently compact that they are stable against external disturbances, although not so dense as to cause considerable dissipation during this time. In addition, in order that the cluster be preserved, it must not pass close to the nucleus of the Galaxy, where the disturbing tidal forces are strong [423, 449]. If the initial dimensions of the cluster were of the order of 10 pc, then the temperature of the gas must be less than 1000° [436]. Cooling to such a temperature may take place either because of an abundance of heavy elements or because of the formation of negative hydrogen ions, which act as an element with a low ionization potential and which are extremely effective in partial ionization of the gas. The fact that masses of globular and galactic clusters differ by two to three orders of magnitude indicate that the conditions of their formation were actually extremely different. In particular, the density in the spherical subsystem must have been lower and the temperature higher than during the formation of the galactic clusters. The division of globular clusters into stars has not yet been studied. One might think that, since the system consists of a large number of stars, rapid cooling would continue further (perhaps after some heating). However, it can be shown that, even at densities comparable to that of the nucleus of the cluster, in order to form a star with the mass of the Sun the temperature of the gas must not be higher than a few degrees. Apparently, the cluster first divided into more massive parts, contracted further, and broke up into stars at higher densities. The groups of stars formed would then disintegrate owing to disturbances which are considerable in such dense systems. The considerations introduced here are only of a rough preliminary nature.

We turn now to the evolution of the Galaxy. Random motions of masses of gas generate shock waves, which are quickly damped and transformed into heat. It is easy to show that the cooling time at the average density of the Galaxy over the whole spherical subsystem ($\sim 10^{-24}$ g/cm^3) is many times less than the collapse time t_n, although the latter is less than 10^8 yr. Therefore, we might suppose that during a time less than 10^9 yr the gas remaining from the spherical subsystem not transformed into stars would lose its kinetic energy and fall to the galactic plane, the principal plane of the system which determines its total angular momentum.

Consider next the formation of the Galaxy. Since the contraction time of the gas is relatively short, the ages of almost all the stars of the spherical subsystem must be approximately the same, and do not exceed by much the ages of the oldest stars of the intermediate subsystem II formed from the remaining stars falling to the plane. Stars of the spherical subsystem continue to evolve and eject gas. Calculations show that after 10^{10} yr the mass of the ejected gas must constitute approximately one-half of the initial mass of the subsystem [450]. However, this gas does not form stars there, since its density is low. Apparently it falls toward the plane and the galactic center, or is partially scattered in the galactic corona. The only exception is the gas thrown off by stars in the globular clusters. This gas, mixed with dust, is observed directly as dark lanes and bands, quite conspicuous against the background of the bright nucleus of the cluster [451, 452, 453]. The amount of gas is small. An estimate from absorption of the mass of the dust in the cluster M 13 gives a value of about 2.6 $M_\odot$, and, if the dust amounts to about 1 percent of the mass, the mass of the entire cloud is $\sim 260\ M_{\odot}$. The absence of radio emission at $\lambda = 21$ cm in M 3 and M 13 means that the masses there are less than 700 and 200 $M_{\odot}$, respectively [119]. The small quantity of gas is explained by the fact that the escape velocity from the globular cluster is low and gas is swept from the cluster when it intersects the galactic plane and collides with interstellar clouds. Consequently, gas is accumulated for one revolution, that is, for $\sim 10^8$ yr. During this period, the stars will throw off less than $10^3\ M_{\odot}$ [452]. Some of the gas during this period is able to be converted into stars. This explains the presence of a small number of stars above the point where the subdwarf sequence breaks off (Fig. 63). Thus, in M 3, for example, there are about 20 stars whose age is less than 10^9 yr. Conditions for the formation of stars in the relatively dense cold clouds of the globular clusters must be similar to the conditions for their formation in the flat subsystem. It is interesting to note in this connection that the young stars in M 3 form a group displaced to one side of the center of the cluster [453].

The sinking of the gas from the spherical to the flat subsystem proceeds as long as equilibrium is not established in a certain layer. It is difficult to say what the conditions were in the period of formation of the intermediate subsystem II. In the present flat subsystem gas is maintained against the gravitational force by the magnetic field of the arms and partly by random motions. The formation of stars now proceeds where gas occurs, that is, in the spiral arms. These stars evolve and again eject gas, but in smaller amounts, since during each cycle a portion of gas is converted into white dwarfs and into low-mass stars that evolve slowly and thus is withdrawn from the process. In the process of evolution, the gas is enriched with heavy elements, but now this happens slowly because the number of massive stars formed is relatively small. The main part of the heavy elements was evidently formed in the process of contraction toward the disk and particularly in the first period of the disk's existence, when there was much gas, and stellar formation proceeded at a high rate. We should note that it was principally the massive stars that finished their evolution during the initial period, since stars of average mass exist for a long time. Therefore, in the gas ejected by stars during the first period of galactic evolution, the ratio of heavy elements to light elements, helium in particular, must have been greater than now, when gas is ejected chiefly by stars of average mass in which light elements are synthesized. Observations also seem to show that the difference in stellar composition cannot be described in the form of a sequence of a single parameter, the ratio of the number of hydrogen atoms to the number of all other atoms forming a standard mixture [454].

The stars in the disk were formed from gas that retained its angular momentum. Discrete masses, not having momentum, or having lost it through collisions, sank to the galactic center and there turned into stars. This led to the formation of a dense nucleus. Dissipation of stars from it means that the remaining stars are packed still more tightly. In contrast to stars, there is no concentration of gas toward the center, if we disregard the small nebula in the nucleus itself. This is apparently explained by the fact that when gas is very dense it is rapidly turned into stars or is accreted by them. In certain cases, gas may transfer its angular momentum to the surrounding medium and upon compression does not form a group of stars, but a massive superstar.

Classification of Galaxies

As Hubble first observed, galaxies are divided into spirals and ellipticals. In addition, there are a small number of irregular galaxies. Each class is divided into subclasses, depending on the connection between the spherical subsystem and the arms or spirals or, in the case of elliptical galaxies, on the degree of contraction (Fig. 66). Moreover, the spirals are divided into two parallel sequences, the ordinary spirals and barred spirals (Fig. 67). The more complex present-day classifications depend fundamentally on the degree of luminosity concentration toward the center [455] and other features. These classifications preserve Hubble's scheme in its general features. The giant elliptical galaxies have the greatest luminosity concentration with the luminosity per unit volume of their nucleus a thousand times greater than the average. At the other end of the sequence, we have the irregular galaxies, almost without any concentration

FIG. 66. The elliptical galaxy NGC 205. (Palomar Observatory.)

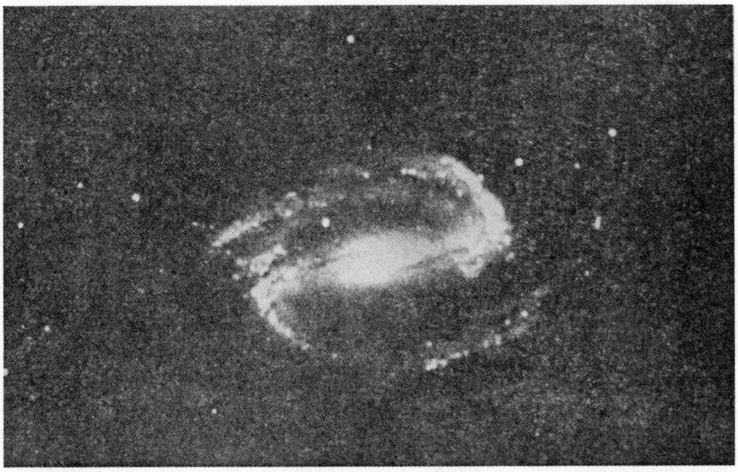

FIG. 67. *Sb* spiral galaxy with bar, NGC 1300. (Palomar Observatory.)

toward the center. Between extreme cases, we have the spirals and barred spirals, containing spherical and flat subsystems in different proportions. *Sa* galaxies have a relatively high concentration toward the center and scarcely noticeable arms, while *Sc* galaxies have a strong flat subsystem (Fig. 68). The stronger the concentration, the greater the relative role of the spherical subsystem. A simplified variation of the new classification is shown in Fig. 69.

Between the ellipticals and the spirals are found the lens-shaped systems *S*0 with and without bars. They resemble the spiral galaxies in their concentration of luminosity toward the plane, but have no spiral structure or dark bands across the equator. Most of them appear as a lens, a relatively uniform disk, usually of extended form, with a weak concentration toward the center and relatively sharp boundaries. It has been suggested that these galaxies were formed from spiral galaxies that have been swept clean of gas by collisions [456b]. In support of this hypothesis is the fact that *S*0 galaxies are found primarily in dense clusters where collisions must be relatively frequent. However, the presence of a bar in many *S*0 galaxies, and certain other pecularities, such as the sharp boundary of the lens, lead one to think that the nature of these galaxies is more complicated.

In addition to these basic subdivisions, the precise classification developed by de Vaucouleur [456a] takes into account the number of spiral arms, the presence of external ring-shaped structures, and certain other peculiarities of spiral and lens-shaped galaxies.

The class of galaxy determined by the luminosity concentration toward the center is strongly connected with the form of the spectrum of the galaxy as a whole and with the nature of its brightest stars. The spectra of several irregular galaxies with weak concentrations resemble the spectrum from the internal part of the Orion Nebula near Trapezium formed by hot class O and B stars

FIG. 68. *Sc* galaxy NGC 628. (Palomar Observatory.)

together with H and He emission from the gaseous nebula. In other cases (the Large Magellanic Cloud), the continuous spectrum is due to B and A stars. There is a great deal of gas in irregular galaxies in addition to many diffuse nebulae excited by numerous hot stars. In composition these galaxies generally resemble spiral arms, that is, the flat subsystem. Their brightest stars, the blue giants, lie at the upper end of the main sequence.

Elliptical galaxies, on the contrary, remind us of the spherical subsystem of the Galaxy. Their Hertzsprung–Russell diagrams resemble the diagrams of the globular clusters. The upper part of the main sequence is almost absent, and hot stars, if they exist, belong mainly to the horizontal arm associated with the end of evolution, while the brightest are the red giants. We should recall that the globular clusters have red giants considerably brighter than those of the flat subsystem and that they are the brightest stars of the cluster.

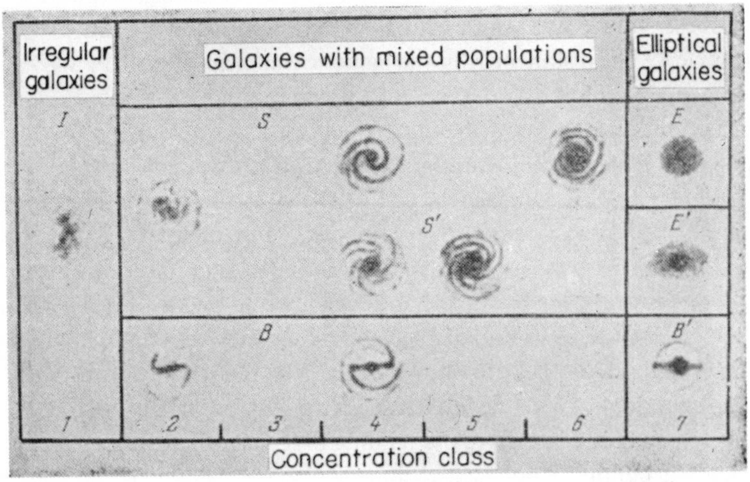

FIG. 69. A simplified scheme of galactic classification according to degree of concentration [455].

There are two reasons for the difference in the diagrams for the flat and spherical subsystems. First, star formation in the spherical subsystem stopped almost completely a long time ago, the gas is practically all consumed, and the majority of the stars are old and evolve slowly. Therefore, there are almost no bright stars on the main sequence. The second factor distinguishing the diagram of the spherical subsystem is the difference in chemical composition, the low abundance of heavy elements. This explains the brightness of the red giants, the presence of RR Lyrae variable stars, the presence of the horizontal arm in the diagram, and other features. Generally speaking, we do not necessarily find these two properties combined in all galaxies. In principle, we may find old systems with a large quantity of heavy elements, and young systems poor in these elements. It is interesting to note that the large elliptical galaxies are like the nucleus of our Galaxy and of the Andromeda Nebula and also like the globular clusters with strong metal lines. Thus, their heavy-element content is by no means low, and they are similar to the old population of the disk. Consequently, the matter of these galaxies has already passed through the stage of star formation and constitutes, as it were, a second or later generation. For this to happen, the gas ejected by stars either was unable to leave the galactic corona but was converted there into stars, or fell, for example, to the center and there formed second-generation stars, some of which, as the result of mutual disturbances, again returned to the corona.

Certain elliptical galaxies, primarily these of low density, such as the "blue" satellite of the Andromeda Nebula, NGC 205, contain hot stars in the horizontal arm; hence the designation "blue." The presence of these stars is an indication of low heavy-element content, as in the globular clusters of the spherical subsystem. These galaxies have, on the whole, preserved first-generation stars. Of course, these stars evolve and eject gas enriched with heavy elements, but this gas is converted into stars only in the galactic center itself.

A comparison of the two types of elliptical galaxy discussed shows that chemical composition is not the most essential characteristic in the evolution of a galaxy of high concentration or the spherical subsystems of spiral galaxies. Their primary feature is the great age of the overwhelming majority of stars and the old age of a system. All systems evolve in such a way that the stars eject gas, which is subsequently again turned into stars. During this process, some of the gas remains in stars that become white dwarfs, or in stars of low mass. Thus, all systems will finally become old and the difference between them consists principally in the rate of evolution. This rate is connected statistically (but not functionally) with the gas density. Consequently, in order that a system run through its life cycle, the gas remaining in it, regardless of how much, must continually accumulate into dense clouds. Apparently, this is the reason why galactic age is connected with concentration toward the center. The region of concentration occurs where gas ejected by stars collects, and the more compact this region, the more rapidly does the process of star formation take place and the gas become exhausted. On the other hand, in nearly uniform systems, such as irregular galaxies, gas remains scattered and is consumed slowly.

Low gas concentration can arise, in principle, for two reasons. The chief one is rotation. In a rotating mass, the gas losing its energy does not travel to the center, but to the principal plane, forming an extended disk. The fragmentation of the disk finally leads to condensation, but this process takes place slowly, especially in the presence of a magnetic field, which inhibits condensation. Therefore, gas is retained for a long time in the spiral arms. The arms are developed differently in different galaxies. In *Sa* galaxies, the condensation process is considerably further advanced than in *Sc* galaxies. However, we must not expect that after a certain period an *Sc* galaxy will be converted into an *Sa* galaxy. It is actually losing gas together with hot stars from the spiral arms, but the spherical subsystem cannot become as strong as in *Sa* galaxies. These galaxies differed in mass from the very start, the *Sa* galaxies being the more massive, which causes the stronger spherical subsystem and the conversion of a large part of the gaseous mass directly into first-generation stars. In addition, only the flat subsystem of an *Sc* galaxy is relatively stronger than that of an *Sb* galaxy, the mass of gas in the two types being comparable [455b], but generally greater than in the irregulars. The connection between the density of the spherical subsystem and the strength of the spiral arms is explained, as was described in Sec. 19, by the influence of stars on the equilibrium of the gas in the arms. Thus the mean density is the second essential factor determining the rate of gas consumption. It is hardly by chance that irregular galaxies with low concentrations toward the center have densities two orders of magnitude less than elliptical galaxies.

If a protogalaxy does not rotate, or rotates slowly, the primary gas escaping in the evolutionary process travels to the center and an elliptical galaxy is formed. Gas is observed in elliptical galaxies as diffuse nebulae and dark clouds; its origin, liberation by stars, is the same as in the globular clusters, but it has not yet been condensed into protostars.

In irregular galaxies, formation of stars proceeds slowly. They contain much gas and many hot stars and resemble the spiral arms. Globular clusters exist in these galaxies. Diagrams of these clusters [456] are of great interest. About half the clusters in the Large Magellanic Cloud have a reddish color and resemble the globular clusters distributed near the center of our Galaxy. The brightest stars in them are red giants, there are no stars on the main sequence bluer than F5, they contain RR Lyrae variables, and so on. The heavy-element content in these clusters, judging by the behavior of the red-giant branch, is somewhat greater than in the spherical subsystem, but less than in the intermediate subsystem II. The other half of the globular clusters are blue, resembling young galactic clusters. They contain hot stars on the main sequence, but their red-giant branch is above the h and χ clusters in Perseus. This difference is explained by lower content of heavy elements. Thus, we see in the formation of globular clusters in the Magellanic Clouds a process that was generally finished in our Galaxy more than 10^6 yr ago.

Radio Galaxies and Typical Galaxies of Various Classes

The elliptical galaxy NGC 4486 or M 87. This gigantic, almost spherical galaxy, of magnitude $9^{m}_{.}9$, is at the same time the powerful radio source known as Virgo A. It is the brightest member of the large Virgo cluster, located at a distance of approximately 11 Mpc. Its radius is about 15 kpc, while its brightness, as in other elliptical galaxies, is highly concentrated in the center. A rough estimate of the mass of this galaxy is $(5 \times 10^{11}\text{–}10^{12})\ M_{\odot}$ [361]; the nucleus contains about 1 percent of the entire mass and its diameter is 50 pc. Consequently, the stellar density amounts to 10^5 solar masses per cubic parsec. This galaxy contains about 1000 globular clusters, characteristic of the spherical subsystem. Certain small areas of condensation, each about 2″ (110 pc) across, elongated into a chain about 20″ long are visible in the central portion of the galaxy (Fig. 70). Light from these areas is bluer than in the galaxy, and the spectrum is continuous, with no emission and absorption lines. Baade [362] found that emission from these condensations was strongly polarized (up to 30 percent), while the planes of polarization in neighboring condensations were almost perpendicular. We note that only relativistic electrons can produce a continuous spectrum with similar polarization. The direction of ejection almost coincides with the direction of the minor axis of the galaxy.

FIG. 70. Central portion of the radio galaxy Virgo A with ejection [362].

In the galactic nucleus itself, of diameter 1″ (50 pc), the strong emission line 3727 Å (O II) was detected [363]. Since this line is appreciably weakened at high density by collisions of the second kind, and since it belongs to lines of "low excitation," it is evidently formed, not in the individual planetary nebulae in which lines [O III] and others usually predominate, but rather in extended H II regions. This conclusion is confirmed by the fact that in certain other elliptical galaxies where [O II] lines are also observed, they are usually not widespread. In this case, from the location of the center of gravity of the line profile, we can estimate roughly the ratio of the components and find the electron density [364]. It proves to be less than 200 cm^{-3}.

In the galaxy NGC 4486, the profile of this line is asymmetric. It consists of a fundamental line, whose frequency corresponds to the radial velocity of the galaxy, and a somewhat weaker line with a displacement corresponding to a velocity of about 900 km/sec [363]. The spatial velocity may be still greater. Thus, gas producing this component leaves the nucleus after 10,000–20,000 yr, which corresponds to the ejection of several solar masses per year [334]. The ejected mass is possibly part of a newly forming system of condensation. The width of a "stationary" line is determined by the velocity dispersion, which is ~500 km/sec. This dispersion is approximately correct if we use in the virial theorem the earlier estimated nuclear mass $\sim 6 \times 10^9\ M_\odot$, but it is generally very high and difficult to explain if we consider that at such velocities energy must rapidly be transformed into the thermal form. Unilateral ejection of gas is an uncommon phenomenon. If such ejection is observed in other radio galaxies, it is symmetric about the center.

In the radio range, the galaxy presents a comparatively extended source ($\sim 10'$, that is, 35 kpc) with the usual radio spectrum. Ejection also occurs with the radio source [365, 366], which in turn consists of two condensations. Its radio luminosity in the (20–30)-cm band constitutes more than half of the total radio emission from the galaxy. At the same time, emission from the ejection was not observed at 3 m. Thus we must conclude that a break occurs in the ejection spectrum in the wavelength range 3 m to 20 cm, where the intensity decreases slowly ($\gamma \approx 2.5$) and then more abruptly, but more slowly than in the galaxy itself. Thus the radio spectrum intersects the optical region. In the visible, the spectrum is much more abrupt ($\gamma \approx 6.2$) than in the intermediate range (from the optical to the radio waves). This means that there is again a single break in the spectrum in the near infrared. This break is obviously determined by the properties of the source of relativistic particles itself. The first break may be induced by de-excitation of electrons and their simultaneous reproduction [367]. From the condition $W_e \approx H^2/8\pi$ applied to ejection, it follows that $H \approx 7 \times 10^{-5}$ Oe and for $W_p \approx 10W_e$, $H \approx 1.3 \times 10^{-4}$ Oe. The lifetime of electrons giving optical bremsstrahlung in such a field is less than 1000 yr. At the same time, the length of the ejection is about 20″, that is, 1100 pc or 3500 light years. Consequently, the electrons could not be ejected from the nucleus, and must be formed in the ejection at the present time, evidently by collisions of fast protons with atoms of the gas. To provide an escape energy of $\sim 5 \times 10^{42}$ erg/sec (emission from the ejection in the optical spectrum is $\sim 10^{42}$ erg/sec, and in the range from the infrared to radio emission it is $\sim 4 \times 10^{42}$ erg/sec), we need $0.1\ W_p V/\tau \approx 5 \times 10^{42}$ erg/sec, where 0.1 is the fraction of the energy transferred from the protons to the electrons, $W_p V$ is the total proton energy in the ejection, and $\tau \approx (Nc\sigma)^{-1}$ is the time of flight. When $\sigma = 2.5 \times 10^{-26}\ \text{cm}^2$ and $N \approx 200\ \text{cm}^{-3}$, we have $W_p V \approx 3 \times 10^{56}$ erg and the energy density in the ejection is $W_p \sim 10^{-6}\ \text{erg/cm}^3$. In reality, W_p must be still larger, since the spectrum of secondary electrons is steeper than the proton spectrum. Equilibrium conditions between particle energy and magnetic energy necessitate a very high field intensity, or $H \sim 5 \times 10^{-3}$ Oe. It is impossible to increase N by very much, since this would lead to the emergence of lines, and, moreover, the suggested estimate agrees with the mass ejected from the nucleus after several hundred thousand years. It is possible that the ejection is not in equilibrium, but is expanding; if so, the field may be weaker. However, this contradicts the fact that the dimensions of the ejections are approximately identical. Why the ejections do not expand is not understood. If there were a field there, it would only assist in the expansion. Only an external field would contain expansion, but it is difficult to assume the necessary intensity. As in the case of our own Galaxy, the same problem arises of renewing gas reserves in the nucleus as gas is ejected. We must note that the phenomenon of ejection is far more powerful than supernova outbursts. In 10^6 yr, more than 3×10^{56} erg are emitted, which is eight orders of magnitude greater than the outburst of energy created by the Crab Nebula. The total energy of relativistic particles, including protons, must be still larger. The original energy of the relativistic particles must still be high, but only 5–10 percent of the proton energy is transferred to the electrons. Certain of the difficulties enumerated are removed if we assume that the ejection consists primarily of relativistic particles ejected from the nucleus, while electrons are accelerated to "optical" energies by the inductive electric fields produced from the energy from heavy particles. However, it is difficult to propose a concrete model. We must note that although NGC 4486 is considered here as an example of an elliptical galaxy, some of its properties, such as radio emission, an optical ejection, and large gas velocities in the nucleus, are certainly not typical. Gas is generally not seen in the majority of elliptical galaxies, and in 20 percent of them it is observed close to the center, but with small velocities.

The Andromeda Nebula (NGC 224, M 31). This neighboring class *Sb* spiral galaxy is located at a distance of somewhat more than 500 kpc. The visual dimensions of its bright parts are $30' \times 15'$, but it may be detected out to an area of $6.^\circ5 \times 2.^\circ5$, which corresponds to a major axis of more than 60 kpc. The plane of the galaxy is inclined to the line of sight by only 12°, making its structure difficult to study. It is approaching us with a velocity of 300 km/sec. Thus it follows that it forms a connected system with our own Galaxy. The Andromeda Nebula is surrounded by globular clusters, of which at the present time about 400 have been observed. They are found out to 27 kpc from the center, symmetrically distributed about and concentrated toward the nucleus.

In conformity with this class, this galaxy has a bright nucleus and relatively well-developed spiral arms. According to Baade, dark bands appear in the arm, close to the nucleus, showing no bright objects but having only gas and dust. At distances of from 1.5 to 2 kpc the first hot stars appear. Later, young associations and H II regions are encountered more frequently, but they are poorly seen in blue light because of strong absorption. These H II regions and associations extend in chains along the axes of the arms, but the uneven distribution of dust causes an apparent irregularity in the distribution of emission objects. The arms here are no longer dark, but are luminous against the background of the galaxy. At even larger distances from the center, more than 10 kpc, there are many hot stars and the absorption decreases. Finally, only separate associations are seen at distances greater than 17 kpc, and the dust is no longer noticeable. The total density of matter in the arms is substantially decreased here. Approximately 700 emission nebulae are observed altogether in the arms. Between the arms absorption is small, less than $0.^m2$. The light from globular clusters passing through the disk at these places is not significantly reddened.

A supernova, called S Andromeda, exploded in M 31 in 1885, and yearly about 30 new stars flare up. These outbursts do not take place in the spiral arms, since they are old stars. A stream of ionized gas with a density $N \approx 50\ \text{cm}^{-3}$ is observed close to the nucleus ($R \approx 250$ pc) with a velocity of 60 km/sec [341]. Radio emission from the Andromeda Nebula, forming an extended corona of irregular shape, is not strong.

Measurements at 21 cm to determine the mass of the gas yield about $3 \times 10^9\ M_\odot$. According to the law of galactic rotation, the total mass is estimated at $3 \times 10^{11}\ M_\odot$ [465]; that is, M 31 is more massive than the Galaxy. The gas constitutes about 1 percent of the total mass, while in our Galaxy this figure is 2 percent. Obviously, our Galaxy is not the typical *Sb* galaxy the Andromeda Nebula is. This is connected, apparently, with the lower masses of systems of old stars.

The Andromeda Nebula has several satellites, which are small elliptical galaxies. The most obvious of them are NGC 221, or M 32, a class *E*2 almost spherical galaxy with a diameter of about 4′ and with strong central concentration, and NGC 205, a class *E*5 galaxy with a ratio of axes of 1:2, a major axis of 8′, and weak concentration toward the center. This latter galaxy is called the blue satellite because of the presence of hot stars in the horizontal branch of the Hertzsprung–Russell diagram. In addition to these galaxies, there is the physical pair of weaker elliptical galaxies NGC 147 and 185 at a distance of 7° from the center of M 31. Their dimensions are 7′ and 14′, respectively. Their low luminosity is not typical for elliptical galaxies, and their low gradient in brightness makes them intermediate between NGC 205 and the dwarf galaxies in Sculptor and in Fornax.

The Magellanic Clouds. The Large (LMC) and small (SMC) Magellanic Clouds, close neighbors to our Galaxy, are its satellites. Their galactic latitude is about −70°, so that they may be studied only from observatories in the Southern Hemisphere. The distance to LMC is approximately 55 kpc and to SMC slightly more. The dimension of the brightest part of LMC is approximately 6 kpc, while the total galaxy is twice as large. The distance between LMC and SMC is approximately 25 kpc, but these galaxies cannot be considered distinctly separated, since their outer parts merge together. Observations at 21 cm also show the presence of a tenuous gas cloud with a diameter of approximately 30 kpc surrounding both galaxies. This cloud possibly encompasses our own Galaxy. It is interesting to notice that at the external sides of this pair of galaxies the cloud is enhanced as if the gas were compressed in front of these mutually separating galaxies.

The Magellanic Clouds belong to the class of irregular galaxies, although G. de Vaucouleurs emphasizes the presence of structure, such as a bar and spiral arm. However, these formations consist of old stars, and their axes of rotation are far from the bar, so that they cannot be compared with the bars and spiral arms of ordinary galaxies. LMC is a plane disk inclined some 30°–45° to the line of sight. The mass of this galaxy is approximately 10^{10} $M_{\odot}$, with gas constituting approximately 7 percent, or considerably more than in our own Galaxy. The gas forms separate condensations dispersed at random against the general background. There are no gas condensations nor any other remarkable objects in the vicinity of the dynamical center of the galaxy.

The most spectacular formation in LMC is the gigantic diffuse nebula 30 Doradus. It has a diameter of approximately 250 pc and a mass of the order of 10^6 $M_{\odot}$. There are no nebulae in the Galaxy even approaching 30 Doradus. In the center of the nebula is a large cluster of blue supergiants furnishing its luminosity. Surrounding the galaxy is an extended strong complex of neutral hydrogen. In addition to 30 Doradus, many weaker nebulae are found in LMC. A large group of stellar associations surrounded by nebulae has been resolved; it was given the name Constellation I, and resembles somewhat the Orion complex. The diameter of the complex is approximately 130 pc, and, according to the estimates of Australian astronomers, the mass of stars is approximately 24,000 $M_{\odot}$, the mass of ionized hydrogen in a region of 500 pc is approximately 5×10^4 $M_{\odot}$, and the mass of the H I region having a diameter of 1000 pc surrounding the complex is approximately 5×10^6 $M_{\odot}$. Radio emission from remnants of supernovae is also present in the complex. Only in LMC are sources found with a nonthermal radio spectrum. Constellation I, 30 Doradus, and other diffuse nebulae are observed not only optically, but in thermal radio emission as well.

Associations of blue stars and red supergiants appear in very young formations. In addition, clusters of medium age are observed. However, they are difficult to study because of the lack of bright stars. As in our own Galaxy, clusters in the Magellanic Clouds can be divided into two types, galactic and globular. However, the properties of the latter class differ significantly from the globular clusters in the Galaxy. Their ages do not, as a rule, exceed 2×10^9 yr, and are often considerably less. Some globular clusters contain massive hot stars, and, as has already been mentioned, are called blue clusters. In contrast to the Galaxy, the globular clusters of LMC are distributed close to the plane, and do not form a spherical subsystem. The chemical composition of young clusters, globular clusters, and associations, and also the composition of bright nebulae, are normal, differing little from that of, for example, the Orion Nebula, although the abundances of He and O are approximately 3 percent less [464a]. However, in old clusters, with ages of approximately 10^9 yr, there are few heavy elements. This indicates that the period of rapid formation of heavy elements in LMC was much later than in our Galaxy.

In the Large Magellanic Cloud there is a magnetic field approximately parallel to the bar, but distributed over all regions. The presence of the field is seen from the weak nonthermal radio emission and from the polarization of starlight, particularly in 30 Doradus. There is little dust in LMC, where the general absorption is less than $0^m.4$, although it increases in regions of associations.

Wesselink [464b] has shown that the law of reddening of stars in the Magellanic Clouds is the same as in the Galaxy. This result is very significant for the theory of dust formation. A hypothesis was mentioned earlier in which the size of dust grains is determined by the equilibrium between their formation and destruction. However, conditions in the Magellanic Clouds are not the same; collisions occur with different frequencies and with different velocities. The same average size of particles in galaxies of different types argues against this hypothesis. Apparently, the size of the particles is determined by several general factors.

The distribution of H I regions in SMC is considerably more regular than in LMC, with few complexes. Profiles of the 21-cm line over the whole galaxy have two peaks separated by the interval $\Delta V \approx 30$–40 km/sec, as if the gas consisted of two masses. The reason for this division is not clear. The mass of SMC is one-third that of LMC.

The radio galaxy Cygnus A. We shall now consider the most powerful radio galaxy, the Cygnus A source. Its stellar magnitude is approximately 17^m. Its distance is about 220 Mpc, so that its optical luminosity is ten times the luminosity of our Galaxy. It consists of two condensations, like a binary galaxy, surrounded by a weaker elliptical cloud. Most of its brightness is concentrated in lines from atoms and ions of [O I], [O II], [O III], [Ne III], [N II], and the like. These lines are broad, with a width corresponding to a dispersion in velocities of $\sim$400 km/sec. The continuous spectrum is relatively weak.

Although the optical luminosity of this galaxy is high, its radio luminosity is several times higher. We recall that the radio luminosity of NGC 4486 is 10^3 times less than its optical luminosity, while the corresponding factor for a normal galaxy is 10^5. The shape of the Cygnus A source in the radio region is completely different from its optical shape. It consists of two sources situated almost symmetrically with respect to the galaxy itself and is noticeably larger. These sources evidently have been ejected from the galaxy. It is characteristic that ejection took place approximately along the minor axis, as in Virgo A. The brightness distribution in each source is not spherical, while the outer periphery is considerably brighter [368]. The galaxy itself is also a weaker source of radio emission.

The binary structure is characteristic of many radio galaxies. In Fig. 71 isophots are given of the radio source Centaurus A, NGC 5128. Its distance is 3.8 Mpc, its magnitude $7^m.6$, that is, it is brighter than our Galaxy by two orders of magnitude. The dimensions of the sources in this case are considerably greater than those of the Cygnus A sources (their diameter is about 0.5 Mpc and they are more spherical), while their radio intensity is 4000 times less. The galaxy itself resembles an ellipse, but with the dark band characteristic of spirals. The individual components in the spiral arms are moving with velocities of

100 km/sec or more [369]. At the limits of the galaxy there are also two small bright sources lying along the minor axis [370]. Calculations by I. S. Shklovskii showed that the energy of cosmic rays in these sources is several hundred times less than the energy of these particles in the large clouds of the Centaurus A source itself. We may assume, therefore, that the latter were formed gradually by the accumulation of comparatively small clouds ejected from the nucleus.

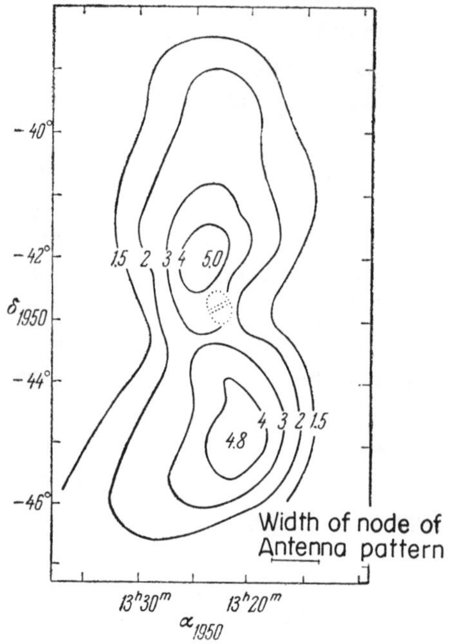

FIG. 71. Isophot contours of the Centaurus radio source [368].

It has been shown [281] that emission from the upper half of the radio source at $\lambda = 10$ cm is strongly polarized—up to 15 percent. This points to the homogeneity of the magnetic field. Polarization in the lower half of the source is weak. The position angle of the polarization vector is different at different wavelengths (from 10 to 31 cm the vector is rotated by 360°). This indicates the presence of Faraday rotation in our Galaxy. The true position angle indicates that the lines of force in the source are directed away from the galaxy. The formation of a homogeneous field on such a scale is one of the problems so far unsolved. Polarization has been observed in other radio galaxies, making it a rather widespread phenomenon.

The radio galaxy Perseus A. Among the interesting radio galaxies is NGC 1275 (stellar magnitude $13^{\mathrm{m}}.3$), the radio source Perseus A. This galaxy belongs to a special type of spiral galaxy discovered by Seyfert [371]. There are approximately ten Seyfert galaxies known, although Perseus A is the only one among them with strong radio emission. In the galaxy NGC 1068, which also belongs to this type, the nucleus is the main source. In Seyfert galaxies, there are characteristic broad emission lines in a small nucleus (~ 100 pc) arising from the ions [O II], [O III], [S II], [S III], [N II] and also from the neutral atoms [O I], [N I], [Ne I]. We have also high-excitation lines from He I, He II, He III, and even [Fe VII]. The width of these lines corresponds to velocities of several thousand kilometers per second. Sometimes bifurcation of lines is seen, corresponding to these same high velocities. Outside the nucleus the lines are narrow and inclined owing to galactic rotation [372]. A line of [O II] in the nucleus of NGC 1275 is displaced relative to the stellar lines, indicating a motion of the gas from the nucleus toward the observer with an average velocity of 300 km/sec. The velocities of random motions in the nucleus, judged from the line widths, are more parabolic, so that gas must be ejected from the nucleus with a velocity of approximately 1000 km/sec. Radiation from gas moving at this side of the center is partially absorbed, so that the average velocity is approximately 300 km/sec. In certain galaxies, the line profile in the nucleus is very complex, which indicates large-scale random motion and rapid rotation. The total energy emitted in the lines by the nucleus reaches 10^{42} erg/sec or more. The gas temperature measured from the ratio $\lambda 4363/(N_1 + N_2)$ is 20,000°–40,000° or more, reaching 100,000° in special cases [373]. However, Woltjer has shown that high temperatures are common for strongly ionized gas (O III and He II). Where hydrogen emission alone is observed, we have regions of gas with high velocity, since the wings of hydrogen lines are often broader than those of other lines. Therefore, the temperature must be lower. Evidently, radiation ionizing O II and He II in the nucleus does not reach these zones at the periphery of the nucleus. From the ratio of the components of a line from [S II], we can estimate by the method of Seaton and Osterbrock that the density N_e lies between 200–500 and 16,000 cm^{-3}. To ionize gas of such density in a nucleus of radius 50–60 pc would require several hundred thousand O stars or more than 10^8 blue stars of the spherical subsystem. Even considering the large number of stars in the nuclei of these galaxies, this many stars seems improbable. The average density of the gas ejected from the nucleus of the galaxy NGC 1068 is 10^3 cm^{-3} and its temperature is approximately 17,000°K [372a]. These data were obtained by a comparison of lines of [O III] and [S II] under the assumption that they originate in the same volume. The total mass of the ejected gas is $(2–8) \times 10^{14}$ $M_\odot$ and its velocity is more than 1000 km/sec. The causes of the phenomena observed in Seyfert galaxies must evidently be associated with processes observed, for example, in the nuclei of Virgo A and other galaxies and to some extent in the nuclei of ordinary galaxies. All these processes are connected with explosions. Although the Seyfert galaxies are few, they constitute approximately 1 percent of large galaxies studied. However, considering the brevity of the explosions, we can assume that each large galaxy experiences from 10 to 100 explosions in its lifetime [343]. Such an explosion took place relatively recently in the nearby galaxy M 82.

Outbursts in the galaxy M 82 (NGC 3034). This galaxy belongs to the irregulars, but is not a typical system. It has an elongated form, and its brightness is not concentrated toward the center, but fluctuates strongly because of the presence of large amounts of absorbing material. A characteristic peculiarity of this galaxy is its thin filaments, well observed in the H_α line. They stretch out to 2′, or 3 kpc, on both sides of the galactic plane approximately parallel to the minor axis, but closed in loops. The spectrum of the filaments resembles that of the Crab Nebula, with strong lines of H_α, [N II], [S II], and weak lines of [O III], H_β, and possibly He 5876 and [N I] 5198 [372b]. Forbidden lines are relatively strong, which points to the collisional character of the excitation. The system of filaments expands along the minor axis of the galaxy. This is seen from radial velocities, though they are strongly diminished owing to projection, but it is appreciably more apparent from proper motions. If the expansion has been constant in time, it has been continuous for 1.5×10^6 yr with velocities of approximately 1000 km/sec. The density of the filaments ($N_e \approx 10\ \mathrm{cm}^{-3}$) was estimated assuming recombination as the origin of the radiation. The total mass is approximately $6 \times 10^6\ M_\odot$, and the kinetic energy is $W \approx 10^{56}$ erg, or eight orders of magnitude larger than in the Crab Nebula. In addition to the line emission, a continuous polarized spectrum is observed, in some filaments, so that radiation from the filaments is in general 10–15 percent polarized [373a]. Consequently, the radiation is synchrotron radiation with the magnetic field directed approximately along the filaments. Weak luminescence is observed from the whole volume of the halo [372b]. Radiation from the galaxy itself is 2–3 percent polarized, but this is apparently connected with the anisotropy of the absorbing dust, as in the case of polarization of starlight in our own Galaxy. M 82 is a radio galaxy with a spectrum whose slope is the same as that of the Crab Nebula ($\alpha = 0.2$). This emphasizes further the similarity between these two outstanding objects in spite of their difference in scale. On the other hand, outbursts in M 82, as in Seyfert and in radio galaxies, are undoubtedly associated with phenomena in galactic nuclei.

The quasistellar radio source 3C 273. The most powerful explosive phenomena observed are in superstars or discrete radio sources. These are very rare (about 35 are known) and therefore are very distant sources having a characteristic radio and optical spectrum. Their spectrum is continuous together with emission lines. The closest source of this type is number 273 in the Third Cambridge catalog. The velocity of recession of this object is 0.16 of the velocity of light [465], and the object is therefore located at a distance of 500 Mpc. 3C 273 consists of two parts, designated A and B. Source B, approximately circular, has a point nucleus ($<0''.5$) and is surrounded by a diffuse halo of approximately 7″. Source A is a weak elongated nebula (length 20″) directed away from source B. At the end of this nebula opposite source B is the center of radio emission, furnishing most of this radiation from 3C273. Source B furnishes the remainder. At the same time, the optical radiation is formed primarily in the nucleus of source B.

We must distinguish three types of radiation: (*a*) optical radiation with a continuous spectrum; (*b*) broad emission lines from hydrogen and weaker lines from [O III], [Ne III], [Mg II], and others; (*c*) radio emission. The three forms of radiation originate in different regions. The optical radiation varies with a period of approximately 13 yr and has an amplitude of approximately $0^{m}.5$. In addition, a systematic decrease in brightness by $0^{m}.2$ has been observed over the past 100 yr, plus irregular changes in brightness, sometimes over 10-day periods. The reason for this is that the dimensions of the optical source are less than 10 light days, or 3×10^{16} cm. This small region emits more than 10^{45} erg/sec, so that its luminosity is 10^{12} times that of the Sun, and ten times that of the Galaxy. The nature of this radiation has not been definitely established. It is most likely synchrotron radiation, although admixture of thermal radiation from a giant star cannot be ruled out. The continuous optical radiation from the extended object A is apparently synchrotron radiation. The gaseous envelope, expanding with velocities, according to the width of the lines, of approximately 1000 km/sec, must have extremely large dimensions, exceeding several parsecs, and may have a radius of approximately 100 pc if the explosion, together with the formation of the electron halo, took place more than 10^5 yr ago. From the weakness of forbidden lines relative to hydrogen lines, it follows that the density of the envelope is approximately 10^7 cm^{-3}. The mass of the envelope is then approximately $10^5 \, M_{\odot}$ and it must be relatively thin. At later stages the envelope may become similar to the expanding filaments in M 82. In radio emission the halo has a radius of approximately 10 kpc, or considerably more than the gaseous envelope. The dimensions of the point radio nucleus must be larger than 200 pc, or otherwise there would be considerable absorption of radio emission by relativistic electrons. Source A is probably connected with source B, which appeared as the result of an explosion earlier than that which produced the expanding envelope.

Other quasistellar sources resemble 3C 273, but because of the variety in physical conditions, such as density, ionization, and so on, their spectra are quite different. For example, forbidden lines are usually stronger, and lines of [O II] are observed. That is, the density is usually lower than 10^7 cm^{-3}. We do not present here the detailed observations and their interpretation, because these questions are only beginning to be explored and the data quickly become antiquated.

The General Properties of Radio Galaxies

From the foregoing short review of radio galaxies, it is apparent that they can be separated into two principal groups: truly radio galaxies, elliptical or spiral, with a strongly developed spherical subsystem, and the clouds ejected from galaxies. I. S. Shklovskii [334, 361] plotted for radio galaxies the dependence of radio luminosity on effective size (Fig. 72). The diagram indicates clearly the

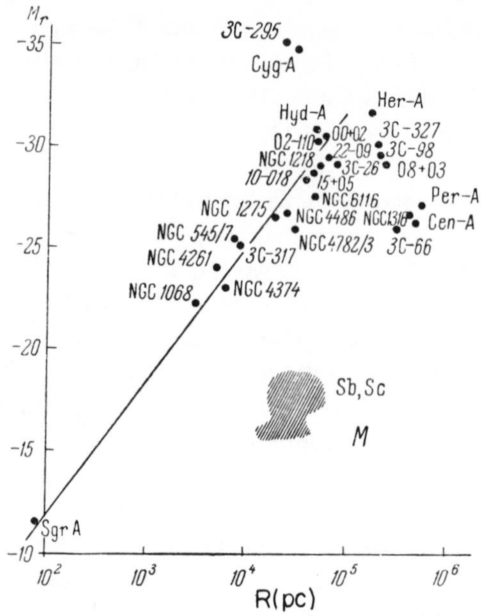

FIG. 72. Diagram of absolute radio luminosity versus size for radio galaxies [334]; hatching shows the region of ordinary spirals.

main sequence in which luminosity increases with size. This sequence is composed of elliptical galaxies and the peculiar spirals. At the top are some of the brightest radio sources. Together with the remote "point" galaxies of high radio luminosity, they form a second sequence.

The galaxy Cygnus A is a characteristic representative of the sequence of "giants." The brightest sources of the Cygnus A type are the youngest. After some time they expand, become weaker, and pass over into a source of the Centaurus A type. However, the latter are overabundant, so that not all weak sources were once similar to Cygnus A. In a short time an enormous number of relativistic particles were ejected. Many "giants" had, even during their formation, moderate luminosities, and the process of their evolution was prolonged, so that they resembled the Centaurus A source for a longer time.

The value of the spectral index entering into the calculation of the attenuation in flux depends statistically, generally speaking, on size. This may be associated with the evolutionary change with time in the spectrum of the relativistic electrons due to the increased number of fast particles. A similar effect was detected in the remnants of supernovae of the Cassiopeia A type [360a].

Thus, radio galaxies are one of a wide class of phenomena beginning with processes in the nuclei of ordinary galaxies. These phenomena include sources of cosmic rays and radio emission in galaxies, explosions creating the radio emission in the corona, Seyfert galaxies, radio galaxies of various strengths and ages, and, finally, ultrapowerful phenomena in quasistellar sources. All these processes are accompanied by the liberation of a large quantity of relativistic particles, expansion of gas at high velocities, and the formation of a magnetic field. The energy freed in various objects reaches 10^{58}–10^{62} erg. Hoyle and Fowler [343] associate these phenomena with the formation and collapse of supermassive stars. The lower limit of the mass of these stars can be estimated from the condition that Mc^2 be the total energy in the explosion. Seyfert galaxies require about $10^5\ M_\odot$, ordinary radio galaxies $(10^6$–$10^7)\ M_\odot$, and, finally, superstars about $10^8\ M_\odot$.

The enhancement of the magnetic field by the coiling of lines of force in gaseous masses compressed because of the impossibility of achieving equilibrium (compression almost to the gravitational radius, where for an external observer time has virtually stopped) and the release of gravitational energy comparable with Mc^2 may in principle lead to an explosion. However, we must emphasize that so far no mechanism has been suggested that will transform the gravitational energy into explosive energy. In addition to the collapse of masses of gas, there is the possibility of the compression of stellar clusters. Upon ejection of individual members of the clusters, the remaining members are compressed all the more strongly, until collisions between stars set in [455a, b]. However, the problem of conversion into explosive energy is also present in this case.

Clusters and Groups of Galaxies

Galaxies form, as a rule, clusters or groups of various dimensions. Our Galaxy, together with the two Magellanic Clouds, forms a triple system. Nearby are found two small groups of dwarf galaxies ($M \approx 10^9\ M_\odot$) in the constellations Fornax and Sculptor. The massive galaxy M 31 with four satellites, and the *Sc* galaxy M 33 in the constellation Triangulum, are located at somewhat larger distances. These galaxies, numbering about 24, form a gravitationally connected system, the Local Group. The average density in the Local Group, estimated from the condition of its equilibrium [331], is approximately 2×10^{-28} g/cm^3. A large part of its mass must be in the form of intergalactic gas, since the mass of M 31 and the Galaxy is one-third of the mass needed to explain their binding (Sec. 19).

The fundamental unit of structure in the universe is the cluster of galaxies. These clusters may contain from several tens to several thousands of members or more. The closest cluster to us ($r = 11$ Mpc), in the constellation Virgo, contains approximately 2000 galaxies in an area of $15^\circ \times 40^\circ$, the brightest and most massive of which is the radio galaxy Virgo A. Among large groups close to us are the clusters in Perseus (about 500 galaxies), in Hercules, and in Ursa Major (300 galaxies), and the large cluster Coma Bernices (about 10,000 galaxies). According to Zwicky and Rudnicki [457a], clusters are the largest stellar systems, with dimensions of up to 1000 Mpc, and superclusters do not exist. The average dimension of the unit of space occupied per galaxy is about

41 Mpc. The diameters of the largest clusters of the various types, compact, intermediate, or open, are the same, and equal on the average 7.8 Mpc. More average clusters have diameters of from 2 to 4 Mpc. These dimensions do not include the weaker extended "atmospheres" of the cluster.

The brightest galaxies in clusters may be the ellipticals as well as the spirals, while there is a systematic difference in different clusters.

There is one radio galaxy in almost every cluster. The radio galaxies are frequently double. Whether they emit radio waves from the space between the galaxies, that is, whether the field there is sufficiently intense and there are relativistic electrons present, is not known at the present time.

Observations at the 21-cm wavelength shifted in agreement with the recessional velocity of the Virgo cluster show that a hydrogen absorption line is present in the spectrum of Virgo A. Its width corresponds to a velocity dispersion of about 500 km/sec. The equivalent width of the line gives for the quantity $(\mathfrak{N}/T_s)$ 3×10^{18} atoms/cm^2 K deg [459a]. To determine $\mathfrak{N}$ and T_s individually, we must also measure the emission from the gas. However, such a measurement is below the sensitivity of the apparatus. The gas can hardly be concentrated in the region close to M 87, since a large radiation flux from this galaxy would give a high spin temperature, and correspondingly, a high value for $\mathfrak{N}$. If we assume that the hydrogen is evenly distributed over a distance of 750 kpc, equivalent to the radius of the galaxy, and that $T_s \approx 20°$ (this value of T_s corresponds to the sensitivity of the apparatus), the authors [459a] obtain $N_H < 2 \times 10^{-5}$ cm^{-3} and a total mass $M < 10^{12}\ M_{\odot}$, which constitutes 10 percent of the total mass of the cluster. The same estimate is obtained under the assumption that hydrogen is concentrated in separate regions projected on the radio galaxy. On the other hand, the mass of neutral hydrogen cannot be much less than the quantity quoted. An estimate of the amount of ionized hydrogen in the cluster has not been made.

Inside clusters there are small groups, sometimes comprising two or three galaxies. Among such groups, interacting galaxies, studied in detail principally by F. Zwicky [458a] and B. A. Vorontsov-Velyaminov [459, 459a], are particularly interesting. These galaxies have long luminous appendages, which sometimes form a bridge connecting the two galaxies. Tails, extending from the side of a galaxy opposite the other galaxy, often appear. These tails and bridges frequently are extensions of the galactic arms, although sometimes galaxies with no noticeable spiral structure are observed interacting. We must keep in mind, of course, that sometimes the spiral structure is very weak and would be difficult to observe at great distances. Sometimes the two interacting galaxies merge into a luminous haze.

The light from the tails and connecting regions is bluish and resembles the light from the spiral arms. These structures are apparently of the same nature as the spiral arms, that is, they consist of gas, connected by a longitudinal magnetic field, and associations of young stars formed from this gas. Old stars formed in the gaseous filaments must have gradually escaped. They constitute, apparently, the luminous background of the cluster. The thickness of the bridges, greater than that of the arms, reaches 1 to 2 kpc, and their length is considerably greater, up to several tens of kiloparsecs. We shall discuss the formation of the bridges below, but here we estimate their parameters from the condition of equilibrium. Since these bridges are often the continuation of the spiral arms, we shall assume the magnetic flux to be the same in both cases, or equal to $10^{-5} \times 5 \times 10^{42} = 5 \times 10^{37}$ G cm^2. This gives, for a bridge with a diameter of 1 kpc, $H \approx 5 \times 10^{-6}$ Oe. The cross section of the bridge is circular, since there is no stellar disk whose attraction combined with rotation flattens the arms. Knowing H and D, the diameter of the bridge, one can determine ρ from the condition of equilibrium of the bridge, Eq. (19.2) [339a]. Setting $D \approx 1$ kpc, we find $\rho \approx 10^{-24}$ g/cm^3. During contraction of a cylindrical structure with a longitudinal field the ratio H/ρ is conserved. Therefore, we may estimate that if the original density in a group of galaxies was, for example, 2×10^{-28} g/cm^3, then the initial intensity of the magnetic field was $H \approx 10^{-9}$ Oe. Compression of the gas during the formation of the Local Group itself can be considered approximately isotropic, for which $H \sim \rho^{2/3}$.

Assuming the initial density of the gas then collected in the cluster to be 3×10^{-31} g/cm^3 (after conversion to the present scale of the universe), we can estimate the average field in the universe to be $H \approx 10^{-11}$ Oe. An intensity of this order must exist today in the space between clusters of galaxies.

The age of the bridges poses a significant question. The most probable assumption is that they were formed simultaneously with the galaxies. However, Hoyle and Harvit [459c] assume that the age of the bridges is not more than 5×10^8 yr because of instabilities of the following nature. If the bridge expands a little at one spot, the gas will move under the action of gravitational forces along the lines of force toward a region where the cross section is less. The decrease in mass then enhances the expansion at that spot, and so on. The gas finally separates into discrete bunches, forming chains of clusters and dwarf galaxies, and the bridge between them disperses. This process must actually take place, but its rate depends significantly on the parameters chosen, which in [459c] were assumed the same as those in the arms. In reality, the cross section of a bridge is larger and the density of gas less than in the arm near the Sun, so that the process must proceed more slowly and it is possible that the age of the bridge may be comparable with the age of the galaxy.

Gravitational Condensation of Clusters and Galaxies

We consider now the formation of clusters from the extragalactic medium. In general we may proceed from the conception of Jeans, which treats the instability of an infinite homogeneous medium. However, the universe is expanding with a velocity of about 100 km/sec Mpc; and consequently each element of its volume is also expanding. At the present time the clusters and groups of galaxies are not expanding, but if they are bound by gravitation this is explained by their relatively high density. At the average density of the universe the deceleration over 10^{10} yr would not be sufficient. Consequently, to explain the existence of clusters, we must assume that there were already initial fluctuations present in the early stages of expansion that were large enough to stop the expansion in local regions. Therefore, we must assume that instead of small disturbances, developing like those considered by Jeans, disturbances with finite amplitudes from the very beginning were significant. The nature of these disturbances remains unknown.

The condensation time is determined by the initial density of the gas. For densities $\rho \approx 10^{-29}$ g/cm^3 the condensation time is, according to Eq. (24.8), $t_n \approx 10^{10}$ yr. Consequently, the initial density may not be less than 10^{-29} g/cm^3, since otherwise the condensation time would be greater than the expansion time of the universe. On the other hand, it may not be considerably larger than 10^{-29} g/cm^3, since the Local Group, with a density $\rho \approx 2 \times 10^{-28}$ g/cm^3, in order to enter into an equilibrium state with its internal motions, must have contracted from a larger volume. The energy of motion was obtained from the liberated gravitational energy. Therefore, we may estimate that the radius in the initial process of condensation was approximately twice that at present, and consequently the density was approximately one order of magnitude less than it is now. It is interesting that a small dispersion in the initial density may lead to a considerable difference in the moment of formation of clusters of galaxies. Therefore, we cannot assume that all clusters visible have the same age. Some of them may still be in the process of condensation. On the other hand, clusters could not have been formed at the earliest stages of the expansion of the universe, since the density then was considerably greater than the present density of clusters.

As we said before, the condensation process and the relative masses of the parts into which the system disintegrates depend in general on the thermal regime of the condensation process. This problem was considered by L. M. Ozernoi [436]. To study processes of condensation, we need a cooling rate that depends on physical conditions. If the medium consists of ionized hydrogen, emission is due to free-free transitions and recombinations, with subsequent ionizations. Using Eqs. (2.25) and (2.29), we can write for the emission per unit mass

$$\frac{dW}{dt} = \frac{4\pi}{\rho}\left(\varepsilon_{ff} + \sum_{n=1}^{\infty} \varepsilon_n\right) = 4.3 \times 10^{20} \rho T^{1/2}\left(1 + \frac{4.7 \times 10^5}{T}\right). \quad (25.1)$$

Hence the temperature relaxation time is equal to

$$t_T = \frac{\dfrac{3}{2}\dfrac{kT}{\mu m_{\rm H}}}{\dfrac{dW}{dt}} \approx \frac{10^4 T^{1/2}}{N\left(1 + \dfrac{4.7 \times 10^5}{T}\right)} \text{ yr.} \quad (25.2)$$

We can estimate from Eq. (25.2) that the initial temperature of the gas from which the Local Group was formed was considerably less than 10^6 °K [331]. Otherwise it would not have cooled enough to permit condensation. Since the Local Group separated into several groups of galaxies, the conditions for contraction must have been almost isothermal, that is, if the temperature decreased in the process of condensation, it was not by much. The condition of quasiequilibrium of the parts of the cloud from which were formed the Galaxy and the Magellanic Clouds ($M \approx 2 \times 10^{11}\ M_{\odot}$, $\rho \approx 10^{-28}$ g/cm^3) shows that at the moment the Local Group split into parts its temperature did not exceed 30,000 °K [436]. The true temperature of the gas may be even lower, since, in the first place, part of the energy is in macroscopic motions, and, in the second place, the state may not be one of quasiequilibrium.

We must keep in mind that Eq. (25.1) refers to ionized hydrogen and therefore Eq. (25.2) is applicable only when $T \gtrsim 20{,}000°$. At lower temperatures, hydrogen starts to change into the neutral state and cooling decreases. A study of the process of the gravitational condensation of galaxies taking into account the thermal conditions during contraction shows [436] that at an early stage neither the magnetic field nor the rotation inhibits condensation. Temperature conditions at this stage could lead to the formation of massive galaxies, but if the original density was higher and cooling proceeded somewhat further the system would disintegrate into dwarf galaxies with masses of (10^8–10^9) $M_{\odot}$. The effect of cooling increases sharply with further contraction. Since $t_n \sim \rho^{1/2}$ and $t_T \sim \rho^{-1}$, the cooling process in the presence of high density outweighs compression and when $T <$ 1000–2000° the galaxy (here a slight contamination with heavy elements is taken into account) must split into a great number of small systems such as the globular clusters. This apparently explains the absence of systems with masses intermediate between dwarf galaxies and globular clusters.

As has already been noted, a small change in the density of the gas strongly influences the condensation time of clusters of galaxies. Therefore, there may be clusters at the present time with massive young galaxies in which stellar formation has recently started, but is progressing rapidly. Such galaxies must have plenty of gas and many hot stars exciting their luminescence. In this connection, the blue galaxies with a strong ultraviolet excess, often embedded in a luminous haze, are of interest. The blue color of these formations is apparently caused by [O II] lines and partly by the radiation from hot stars. As we mentioned previously, bridges and tails of interacting galaxies are of this same nature. However, in this case their youth is determined, as is that of the spiral arms, by the fact that the gas is transformed into stars slowly because of the magnetic field.

It is interesting that systematic differences exist between clusters of galaxies. For example, in all rich clusters, the brightest galaxies are elliptical, with strong concentrations toward the center, while the irregulars are very weak. On the other hand, in the less rich cluster of Ursa Major, the brightest are the spirals with concentration class 5 and many irregulars [455]. These differences are associated in some way with the original conditions of mass, density, the state of rotation of the cloud, and possibly a different chemical composition.

The interstellar medium is concentrated not only inside galaxies, but apparently between them as well, particularly in the clusters. From the fact that the deceleration of the expansion of the universe, if it exists, is not very great, it follows that the average density of the universe is less than or equal to 2×10^{-29} g/cm^3, which is considerably greater than the density, 5×10^{-31} g/cm^3, of all galaxies smeared over all space. Refinements in the determination of distant radio sources will soon establish how close the true density of extragalactic gas is to the upper limit just quoted. At present this gas is apparently not condensing, since the conditions for condensation become unfavorable as expansion proceeds. The rate of cooling is also decreased. Ginzburg and Ozerny [466] have looked at the problem of the heat balance of the intergalactic and extragalactic medium. They took into account heating by cosmic and subcosmic rays, as well as by outbursts from radio galaxies, and cooling by radiation and expansion of the universe. The results showed that the lower limit of the possible temperature of the extragalactic gas at the present epoch is 10^5 °K.

The existence of gas inside clusters follows from considerations of stability. It is true that the total energy was calculated positive, for example, for the system of five galaxies called Stephen's quintet [460]. However, one pair of galaxies was recently shown to be a binary system [461]. In this case, the energy of rotation of the pair drops out of the stability condition, making the total energy of the system negative. This effect must be taken into account in estimating the stability of other systems as well, but in certain cases, such as our Local system, stability certainly requires the presence of an additional mass of intergalactic diffuse medium. We should note that in the case of associations expansion does not contradict the formation of stars from gas, since additional energy is imparted through stellar radiation, while in galactic systems this radiation is not very effective and systems formed by gravitational condensation (that is, with negative total energy) must be dynamically stable. From the fact that 75 percent of all elliptical galaxies and 50 percent of spiral and irregular galaxies in the Shapley–Ames catalog occur in clusters, we can conclude that clusters do not disintegrate in cosmologic periods of time, that is, the majority of clusters are stable [463]. However, certain clusters, such as the very loose cluster in Ursa Major, may be in a state of disruption.

The Magnetic Fields of Galaxies and the Formation of Spiral Arms

Efforts have been made for a long time to explain the magnetic fields of galaxies by internal processes. It was supposed that weak fields appeared because of thermal currents, that is, diffusion of electrons over closed contours. This field would then become entangled and strengthened by random motions of the gas. However, analyses have shown that the strong and relatively regular fields of the spiral arms could hardly have been formed in this manner. Therefore, a more probable hypothesis at present seems to be that the field of the galaxy was compressed together with the gas of the extragalactic medium. This concept, first expressed by Hoyle and then developed by Kardashov, explains the distortions of the density in the arms at the periphery of the galaxy resulting from the extragalactic field [112]. Further development of these ideas was presented in the work of Piddington [339e]. He drew attention to the significance of the angle between the axis of rotation of the galaxy and the direction of the extragalactic magnetic field. If this angle is close to zero, the contraction of the system to the plane of the galaxy proceeds along the lines of force, and, if we neglect the small contraction connected with the decrease of the radius of the disk itself, the gas does not absorb the field [Fig. 73(*a*)]. Differential rotation in the disk twists the radial component of the field, but, since this component formed from the radial contraction is small, the field of the galaxy will be relatively weak, as, for example, in ordinary spiral galaxies. If the field and the axis of rotation are mutually perpendicular, the contraction of the gas to the disk

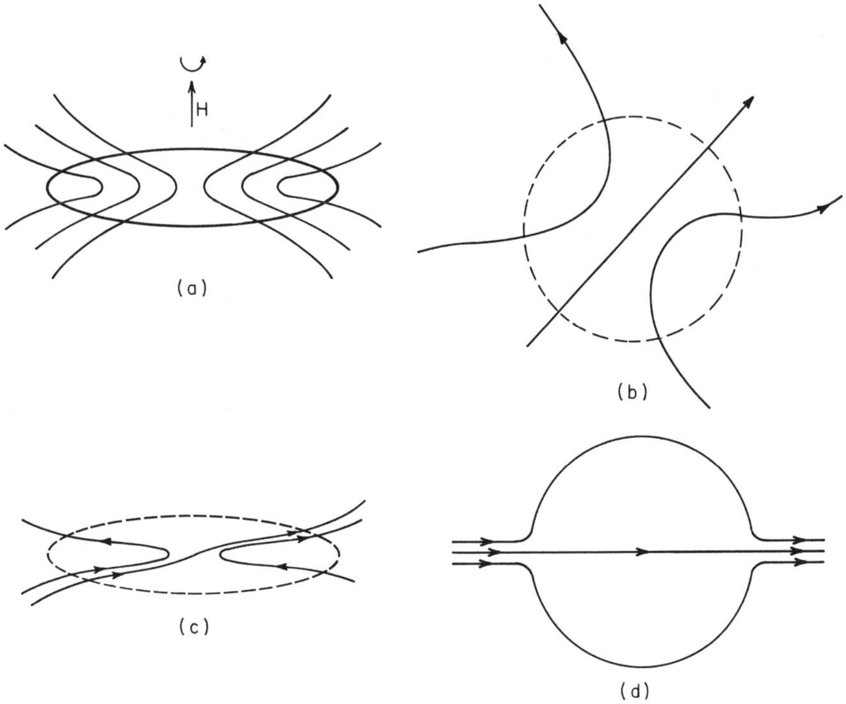

FIG. 73. Configurations of lines of magnetic force in galaxies: (*a*) after contraction, with the extragalactic magnetic field parallel to the axis of rotation; (*b*) after contraction, with the extragalactic magnetic field making an actue angle with the plane of the galaxy; (*c*) lines from (*b*) after flattening of the disk; (*d*) field in the disk of a barred spiral.

also contracts the lines of force. A strong initial field is formed which is further strengthened by rotation. In this case a galaxy may be formed with a strong field, apparently a radio galaxy. Intermediate orientations were not considered by Piddington.

The formation of arms is connected with compression of the gas. Lin and Shu [463c, d] considered a flat disk of gravitating stars, on which there are superimposed linear perturbations of the spiral type with a small pitch angle, and they showed that such perturbations can be self-sustaining, that is, once a perturbation with definite parameters has originated, it can continue to exist. The density wave is quasi-stationary, that is, it rotates as a solid with an angular velocity $\Omega_0 < \Omega$, the angular velocity of circular rotation in this part of the Galaxy. A condensation of stars attracts the gas. The density of the gas in the arm increases considerably as its elasticity is low. The gas crosses the arms with a velocity that is lower than the magneto-sound velocity. Therefore the thickness of the gas layer is determined, as in stationary conditions, by hydrostatic equilibrium in a magnetic field. The higher the density of stars, the greater the gravitation and the less the thickness of the arms (Pikelner [464]). The density of old stars strongly increases toward the galactic center; hence the thickness of the gas layer increases toward the periphery of the Galaxy. The same effect explains why in *Sa* galaxies with a dense spherical subsystem the thickness of the arms is less than in *Sc* galaxies.

Lines of magnetic force in the Galaxy are tightly wound up owing to differential rotation of the gas. When the gas is compressed into a spiral arm the lines incline almost along the arm, but with a small angle between them. The strength of the field increases here by several times.

In the general case the field forms an acute angle with the plane of the Galaxy. After the first stage, when the contraction of the Galaxy is almost radial, it assumes a configuration as in Fig. 73(*b*). After flattening of the disk, some of the lines of force will proceed along the diameter, and some will approach the center and then reverse in the opposite direction [(Fig. 73(*c*)]. We recall that a field of this character is observed in our own Galaxy [339f]. It follows from Fig. 51 that the RM in the southern hemisphere is from 1.5 to 2 times larger than in the northern hemisphere. This is explained naturally from Fig. 73(*c*). The line of division need not be symmetric with respect to the mass of the arm, although because of the flattening there of the gas it will pass very close to the galactic plane.

Barred Spirals and the Annular Structure of Galaxies

A particular type of spiral galaxy, as abundantly distributed as ordinary spirals, is the barred spiral *Sb* (Fig. 67). They are also divided into subclasses *a*, *b*, *c*, and do not differ in stellar population from the corresponding ordinary spirals. The basic difference is that in the central region in place of the spiral arms there is a straight bar which transforms at its end into arms or an annular structure. The transition into spiral arms is often very sharp. Sometimes the Galaxy itself is in the form of an *S*. The nature of these galaxies has been studied by Pikelner [464]. In order to preserve the straight bar, differential rotation must be practically absent. This condition holds for a homogeneous sphere or spheroid with very weak concentration toward the center. The distribution of mass is determined not by the bright bar, but by the subsystem of old stars having low luminosities. These subsystems in ordinary galaxies are strongly concentrated toward the center, which explains the coiling of the arms. To explain the bar we must assume that the distribution of old stars in the *Sb* spiral is approximately homogeneous out to some distance, after which it drops off sharply. The distribution of old stars reflects the distribution in the density of the gas from which they were formed. For some reason gas was driven away from the central regions of the Galaxy before the formation of most of the stars, while the boundary of the removed gas was sharp. Only a shock wave could have produced such an effect.

A hypothesis that explains *Sb* galaxies is the following. At the very early stage of contraction of the Galaxy, when the contraction was almost radial but when the central condensation of gas had already formed, explosions similar to those in M 82 or in strong radio sources took place in the center of the Galaxy. If the system was still spherical, a shock wave would form a spherical envelope almost hollow inside. When the energy of the wave is completely consumed by the heating of the gas and the increase in its potential energy in the gravitational field, the envelope is checked. With an energy $W > 10^{58}$ erg, an envelope with a mass $M > 10^{10}\ M_{\odot}$ and radius $R \approx 10$ kpc may be formed. The thickness of the envelope is approximately $0.1R$, and the density of the gas inside it is of the order of 10^{-24} g/cm^3 or higher. The radius of the envelope remains almost constant in the course of 10^8 yr. During this time the gas is cooled to a few thousand degrees so that the envelope is fragmented, breaking into parts that upon contraction form clusters. Since the density of the envelope is high, the mass of the clusters must be low. These clusters resemble galactic clusters, and not the massive globular clusters formed in the tenuous gas of the spherical protogalaxy.

Stars formed in the envelope have relatively low momenta, but high energies. Therefore, they must move in extended trajectories that depend on the initial momentum. The inside parts of the envelope formed from the central condensation have low momenta, and the outside parts formed from the gas, displaced only a little by the wave, have momenta corresponding to almost circular motion. Turbulent viscosity in the envelope is apparently low, so that angular momentum is conserved. If we assume for the distribution of momentum that of the intermediate subsystem of the Galaxy, which represents the major part of the mass, we can then calculate the distribution of star density inside the sphere. The density will be highest at the center and at the boundaries of the sphere, but will decrease by a factor of 2 at medium distances. This differs sharply from the strong concentration of density toward the center in ordinary galaxies. The angular velocity ω also decreases somewhat at median distances from the center, but does not differ much from the angular velocity ω_0 of points outside the sphere.

That ω be constant is a necessary but not a sufficient condition for the existence of the bar. To determine the other conditions we examine the spirals that emerge from the ends of the bar. The ends of the bar must inhibit the intensity of the lines of force and give it the *S* shape. In those cases where this is not so, large masses of the order of $10^8\ M_\odot$ must be concentrated at the ends of the bar. This mass cannot be in the form of gas, since there would be many hot stars and the brightness would be many times the brightness of the bar, which is not observed. It is supposed that clusters of old stars are found at the ends of the bar, and that inside the clusters are gas clouds with masses of approximately $10^6\ M_\odot$. The intensity of lines of force for such a mass cannot extract gas from the cluster.

The intensity of the lines of force between clusters at the ends of the bar and in the center stabilizes the bar, causing it to rotate as a solid body. The velocity at median radii is greater than the local angular velocity. This causes the gas to flow along the bar toward the ends with velocities of 20–30 km/sec. Velocities of this order of magnitude are actually observed spectroscopically. In the central parts of the bar the gas must flow toward the center, forming hot stars there. This explains the increase in brightness of the bar toward the center. In ordinary galaxies the brightness of the central regions in blue light is less than the brightness of the arms at median galactic radii.

The presence of three clusters of old stars along the same magnetic bundle is not a coincidence. Apparently the formation of the central cluster induced the formation of the clusters at the ends of the bar. The bar formed in this case may be described as follows. After the formation of stars in the envelope, the remaining gas gradually loses its energy, and condenses into a disk about the plane of the Galaxy. Momentum is conserved and transferred to circular orbits with radii corresponding to this momentum. Without the magnetic field, a symmetric disk with central condensation would be formed. The action of the magnetic field causes the disk to contract to a diameter coinciding with the projection of the initial direction of the field.

In certain cases internal parts of the envelope may collect into the bar, but the outer parts, which have almost exclusively circular motion, may remain in the form of a ring condensed toward the plane. The magnetic field, compressed by the shock wave in the envelope, also remains in this ring, giving it stability. Now the field in the disk may have the form shown in Fig. 73(*d*). Upon leaving the ring, the lines of force are compressed by the gas and collect in the clusters, so that they leave in the form of thin arms almost tangent to the ring. Such a structure is often observed in barred spirals.

The large percentage of barred spirals observed suggests that during the short period of initial contraction of galaxies when there is much gas present the probability of explosions of great force is very high. This supports the hypothesis that explosions occur as a result of the gravitational condensation of large masses of gas.

The Extragalactic Magnetic Field

Since galactic morphology points to a connection between galactic fields and the extragalactic field, we should discuss the structure of the extragalactic magnetic field. We shall assume to begin with that the initial extragalactic field was quasihomogeneous over scales considerably larger than the distances between clusters of galaxies. During the condensation of clusters and galaxies the lines of force were compressed and the field became enhanced in these objects. However, the lines of force that connect the galaxies and clusters of galaxies with one another were originally in the same bundle of lines of force. As is well known, the field over relatively small scales can become tangled because of the motion of matter. However, such is not the case over scales as large as the universe or clusters of galaxies. In fact, the velocity of any distant cluster of galaxies due to universal expansion is greater than its peculiar velocity, and the lines of force connected with the cluster can only be stretched. Inside the cluster the average time required for a galaxy to traverse the cluster is of the order of t_n from Eq. (24.8), which in this case is only a few times less than the age of the cluster. Therefore, the field inside the cluster does not become appreciably tangled. The greatest amount of stirring of the field takes place directly in the galaxies and their surroundings.

The structure of the magnetic field is very useful in the theory of the origin of cosmic rays. If cosmic rays can emerge from a galaxy while conserving their adiabatic invariant, they will move in a weak field along the lines of force as long as they do not enter the strong field of another galaxy, where the motion again approaches isotropy. However, this representation meets with the difficulty that the anisotropy of the motion of the particles causes bunching instability, braking of the cosmic rays, and heating of the extragalactic gas. In addition, it is difficult to explain in this scheme the extreme isotropy of galactic cosmic rays about the Sun. Apparently, the escape of cosmic rays from the Galaxy is difficult. It should be emphasized that, although it may be possible to inhibit the escape of cosmic rays from the Galaxy, it is much more difficult to contain them inside the cluster, since the field there is weakly tangled. In this connection the hypothesis of Sciama [333], in which it is supposed that cosmic rays are bound within the Local Group, does not seem probable. This gives favor to the galactic theory of the origin of cosmic rays. Any hypothesis on the extragalactic origin of cosmic rays must provide for an isotropic density of cosmic rays, a density approximately equal to the galactic density. With anisotropies present the density required would be less, but the bunching instability makes this case improbable. After the radio emission from clouds of radio galaxies is evenly distributed, their cosmic-ray density is less than in the Galaxy, and therefore can hardly be a significant source of the observed cosmic rays.

The problem of the origin of the extragalactic magnetic field is connected with the over-all cosmological problem, and therefore this problem is far from solution.

The discovery of the relic black-body radiation ($T = 3^\circ$) gives some indication of the early hot stage of evolution of the metagalaxy. The high radiation pressure prevents condensation of ionized opaque gas and the formation of galaxies and stars in the early stage [463a]. When ionization decreased and the condensations became possible, the first formations might have been clouds with masses of $10^5\ M_\odot$ [463b]. These objects were not yet galaxies, but they might have been similar to quasistellar objects. Gravitational energy released in their contraction in the form of shock waves or cosmic rays heated the uncondensed gas and prevented the formation of small condensations. Therefore massive condensations similar to clusters of galaxies were formed. Contraction and fragmentation of such condensations produced galaxies and stellar systems. It may be added to this analysis that galaxies not in clusters could have been formed only in selected regions where the gas was relatively cool. The small number of such separate galaxies shows that the gas was heated uniformly.

Addendum

The Russian edition of this book appeared in 1963; additions for the English translation were mainly completed in 1965. A great many of the results described retain their importance, but there has also been considerable progress, arising chiefly from space researches in the ultraviolet, x-ray, and infrared regions of the spectrum. Much information has also resulted from the increase in sensitivity and especially in the resolving power of radiotelescopes. We cannot describe these results in detail, but we shall mention briefly directions of development in this field.

The temperatures of nebulae calculated on the basis of investigations of radio lines of highly excited hydrogen sometimes disagree with those determined by other methods. The discrepancy was explained in general by means of nonuniformities of temperature, which was assumed to decrease from about 8000° in the central region to about 4000° at the outer edge. Lines in the far infrared, excited by electron impact, become important in connection with the thermal balance of nebulae. Measurements of lines of highly excited hydrogen permit one to determine the radial velocities and hence the distances of heavily obscured nebulae. Especially interesting in a cosmogonical aspect are compact high-density H II regions; their density is about 10^4 cm^{-3}, and they have diameters of less than 0.5 pc and total masses of a few solar masses. These nebulae apparently are surrounded by neutral hydrogen and dust and are the inner ionized parts of globules with very young high-temperature stars. Their outer parts form shells or "cocoons" which reradiate the emission of the star and nebula in the infrared region. Such "cocoon stars" are situated usually in large H II regions ionized by class O stars that are a little older.

There has been some modification in ideas on the nature of H I regions. At the end of Sec. 8 it is stated that it follows from Faraday rotation that the average electron concentration N_e is about 0.01–0.03 cm^{-3}. This is too high for H I and too low for H II. It follows from this that hydrogen should be partly ionized, that is, there must be an agent with a small cross section for ionization. Low-energy cosmic rays might be such an agent, but more probably (according to J. Silk and M. Weiner and to R. A. Sunjaev) it is the recently discovered soft x-ray emission (40–70 Å) originating in galactic and extragalactic sources. The physical processes here are similar to those with cosmic rays: high-velocity electrons detached from atoms by quanta move through the interstellar medium and produce ionization and heating. The thermal instability divides the gas into dense and rarefied phases if the average density is higher than some threshold. If a large, dense complex of clouds contains more than 10^{21} atoms/cm^2, its optical thickness for soft x-rays is $\tau > 1$. Quanta of higher energy penetrate into the inner parts of the complex, but their efficiency is much lower. Ultraviolet radiation from stars is also absorbed by dust. Therefore the temperature in this inner part decreases to 10–15°. This effect facilitates the formation of stellar clusters and associations (S. B. Pikelner).

Large clouds—complexes—originate from the instability of the gas supported by the pressure of the magnetic field and of cosmic rays. When a random fluctuation occurs, the gas slides "downhill" under gravity along the lines of magnetic force, the mass on the "hill" decreases, and the pressure of the field and the cosmic rays raises the hill still higher. Some kind of complex grows at the foot of such a hill. After the appearance of young stars gas may be dispersed. If the valley persists, the gas continues to collect there and stars will form again after several million years.

Systematic observations in the 21-cm line have changed the general picture of the structure of the interstellar gas. Instead of statistically uniform standard clouds, large-scale structures with fluctuations of various magnitudes and densities have been found (C. Heiles). But there is a general division into dense and rarefied phases.

The distribution of gas has been investigated also by means of absorption lines, especially L_α. In a few cases this method shows a smaller number of atoms in a column of gas between the observer and the star than the observations of the 21-cm line. It might be related to a fine structure of the interstellar gas. Strong lines of Mg II and other elements are also observed. Absorption of soft x-rays from the extragalactic background also gives some information about the number of atoms in different directions. This method makes it possible in principle to determine the chemical composition of the interstellar gas from jumps of absorption at the limits of the *K*-series. A number of different methods give an average concentration in the galactic plane of about $N_H \approx 0.3\ cm^{-3}$, that is, a little lower than was accepted earlier.

Radiofrequency lines of OH molecules have been discovered in pointlike sources, and their anomalous spectrum was explained qualitatively by a maser mechanism of emission. Later, emission lines of OH were found with thermodynamic relative intensities of their components. Such thermal emission is observed in a massive condensation with no 21-cm emission. Condensations with 21-cm emission do not show thermal OH lines. Apparently molecules of OH and H_2 are both formed and destroyed in similar conditions; they are decomposed by ultraviolet radiation. Radiofrequency lines of other molecules, such as H_2O, NH_3, SO_3, and H_2COH, have also been observed recently.

Observations of interstellar extinction, polarization, and scattering with high albedo support the model of grains consisting of graphite cores with dielectric mantles of dirty ice. The observations are best accounted for by a size distribution of grains with two maxima, at about 0.1 and 0.6 μ.

Interstellar dust is apparently responsible for part of the infrared emission of the galactic nucleus and some other sources. The half-width of the nucleus is about 20 pc at $\lambda = 1.6$–$3.4\ \mu$, and there is a point source inside it. Especially interesting is a powerful source of long-wavelength emission ($\lambda = 100\ \mu$) in the central region which emits about 3 per cent of the total integrated energy of the Galaxy. Extended sources are observed in some nebulae, or they are not identified with any optical source. Point sources of infrared emission sometimes coincide with OH sources and may be like "cocoon stars" with a dust shell. The emission from some clouds reaches 3000 times the solar luminosity; the size of such sources is about 0.1 pc and the temperature of the dust is about 70°. The infrared emission of T Tauri stars corresponds to a size of about 10 a.u. and a temperature of 600–800°. Apparently it is the emission of a protoplanetary system. Infrared emission may also be connected with emission lines and in some cases with strong interstellar absorption or with synchrotron emission. The nature of the powerful long-wavelength source is unknown. From investigations of the infrared and scattered emission from dust we can find a value of the albedo of the grains and improve our understanding of their nature.

Measurements of the Zeeman effect in the 21-cm line show the presence of a radial component of the magnetic field in several clouds of from 3 to 20 μG. Pulsars give a new possibility for investigations. The pulses of their radioemission undergo dispersion in the interstellar medium; at the higher frequencies they come earlier than at the lower ones. The time lag enables us to calculate a value of $\int N_e dl$. This measurement confirmed that the electron density N_e is about 0.02–0.03 cm^{-3}. Absorption of the same pulses in the 21-cm line gives a value of $\int N_H dl$ in the same column of gas. Pulses from some pulsars are linearly polarized; the Faraday rotation gives a value of $N_e H_{\|} dl$. From comparison of these values we can find an average radial component of the field. In some cases it is about $3\mu G$, sometimes less, apparently owing to the effects of projection and of the opposite directions of the fields on the way.

Pulsars may be partly responsible for the formation of relativistic electrons and cosmic rays. The youngest pulsar, with a period of 0.033 sec, is in the center of the Crab Nebula. The nebula has a flat spectrum and the presence of electrons leads to the emission of optical and ultraviolet radiation.

There is no adequate theory of pulsars, but the following picture seems rather probable. A pulsar is a neutron star with a radius of about 10 km, in very rapid rotation and with a magnetic field of up to 10^{12} G. This field may be a result of compression of the stellar field; it proves, incidentally, that inside stars there is a regular field of about 10^4 G. The rotating magnetic dipole creates a solenoidal electric field in a region where magnetohydrodynamics is not valid. This field accelerates particles up to ultrarelativistic energies. A relativistic plasma surrounds a pulsar and in it there is a low-frequency wave of enormous amplitude. The properties of such plasmas have not yet been investigated.

Some x-ray sources show a flat spectrum like that of a plasma heated to $T \approx 10^8 {}^\circ K$. It is possible that such a plasma results from accretion of interstellar gas or gas flowing from the second companion of a binary system to the neutron star (Ya. B. Zeldovich). The emission from such high-temperature gas has been calculated.

Measurements of the ultraviolet and x-ray background determine an upper limit for the temperature and density of the intergalactic gas. The presence of rarefied neutral hydrogen at the periphery of our Galaxy and of some others shows that the density of the hot intergalactic gas should be lower than the critical one, which would stop the expansion of the universe. Otherwise the ultraviolet emission of this gas would ionize the hydrogen (R. A. Sunjaev).

Appendix I. The Probability of Elementary Processes for Hydrogenlike Atoms and Ions

A summary is given below of the main equations and numerical values of various transition probabilities for hydrogen. These expressions may also be used to obtain an approximate estimate of transition probabilities for other atoms. The symbol Z is used here for the charge of the nucleus or the atomic core (for a hydrogen atom and for single-stage ionization $Z = 1$, for two-stage ionization $Z = 2$, and so on).

Probability of Discrete Transitions

In a pure Coulomb field, the energy χ_n of a term depends only on the principal quantum number n, or

$$\chi_n = \frac{\chi_1}{n^2} = \frac{2\pi^2 me^4}{h^2 n^2} Z^2 = \frac{13.6}{n^2} Z^2 \text{ eV}. \tag{I.1}$$

We note that when $Z = 1$ the energy of ionization from the first term, $\chi_1 = 13.6$ eV, is equivalent to a temperature $T = \chi_1/k = 158{,}000°$. From Eq. (I.1), the quantum frequency corresponding to a transition between terms n and n' is

$$\nu_{nn'} = \frac{\chi_n - \chi_{n'}}{h} = \frac{\chi_1}{h}\left(\frac{1}{n^2} - \frac{1}{n'^2}\right) = 3.289 \times 10^{15} Z^2 \frac{n'^2 - n^2}{(nn')^2} \text{ sec}^{-1}. \tag{I.2}$$

Transition probabilities depend also on the azimuthal quantum number l. As is customary, we denote by $A_{nln'l'}$ the probability of a spontaneous transition from sublevel nl to sublevel $n'l'$. In most problems dealing with the physics of interstellar gas we may limit ourselves to the average values for the transition probability from level n to level n', denoted by $A_{nn'}$. When $A_{nln'l'}$ is averaged, it is assumed that the distribution of atoms according to sublevels is proportional to their statistical weight $\tilde{\omega}_l = 2(2l + 1)$. That is, if N_{nl} is the number of atoms in level nl and N_n is the number of atoms in the whole term n ($\tilde{\omega}_{n'} = 2n^2$), then

$$N_{nl} = \frac{2l + 1}{n^2} N_n. \tag{I.3}$$

This condition is satisfied, in particular, during thermodynamic equilibrium. Gas does not occur in thermodynamic equilibrium in interstellar space, but nevertheless condition (I.3) is often assumed to be satisfied. According to this equation, the average spontaneous transition probability is

$$A_{nn'} = \sum_{l=0}^{n-1} \sum_{l'=0}^{n'-1} \frac{2l+1}{n^2} A_{nln'l'}. \tag{I.4}$$

We introduce first the data for the probabilities $A_{nn'}$ as the most common and then write down later expressions for $A_{nln'l'}$. We recall the well-known relation (see, for example, [1, 9])

$$A_{nn'} = \frac{8\pi^2 e^2 \nu_{nn'}^2}{mc^3} \frac{\tilde{\omega}_{n'}}{\tilde{\omega}_n} f_{n'n} = \frac{8\pi^2 e^2 \nu_{nn'}^2}{mc^3} \frac{n'^2}{n^2} \cdot \frac{2^5}{3\sqrt{3}\pi} \frac{n'^3 n}{(n^2 - n'^2)^3} g_{nn'}. \tag{I.5}$$

Here $f_{nn'}$ is the oscillator strength for absorption and $g_{nn'}$ is the so-called Gaunt factor, which is usually written in the form of the series

$$g_{nn'} = 1 - \frac{0.1728(1 + (n'/n)^2)}{n'^{2/3}(1 - (n'/n)^2)^{2/3}} - 0.0496 \frac{1 - \frac{4}{3}(n'/n)^2 + (n'/n)^4}{n'^{4/3}(1 - (n'/n)^2)^{4/3}} + \cdots. \tag{I.6}$$

For transitions $2 \to 1$, the Gaunt factor is equal to 0.717, and for large values of n and n' it is nearly unity ($g_{nn'} \approx 0.95$). For transitions in which $n' = n - 1$, the Gaunt factor is $g_{n(n-1)} = 0.78$. A table of Gaunt factors is given in [1]. Substituting numerical values in Eq. (I.5) and introducing the required Gaunt factor, we obtain for the most common values (see also [1]):

$$\begin{gathered} A_{21} = 4.68 \times 10^8 \text{ sec}^{-1}, \qquad A_{31} = 5.54 \times 10^7 \text{ sec}^{-1}, \\ A_{32} = 4.39 \times 10^7 \text{ sec}^{-1}, \qquad A_{42} = 8.37 \times 10^6 \text{ sec}^{-1}, \\ A_{nn'} = \frac{1.47 \times 10^{10}}{n^3 n'(n^2 - n'^2)} \text{ sec}^{-1} \quad (n \gg n' \gg 1), \\ A_{n(n-1)} = \frac{6.05 \times 10^9}{n^5} \text{ sec}^{-1} \quad (n \gg 1), \\ A_{n1} = \frac{1.2 \times 10^{10}}{n^5} \text{ sec}^{-1}. \end{gathered} \tag{I.7}$$

When $Z \neq 1$, we must multiply all the $A_{nn'}$ values in (I.7) by Z^4.

We note that the probability $A_{n(n-1)}$ for transitions with $n' = n - 1 \gg 1$ may be calculated classically by use of the correspondence principle. The oscillator strength for this transition is $f_{n,n-1} = n/6$ [96]. The direct use of $A_{nn'}$ is convenient for calculating transitions between close and, in particular, low-lying levels. However, for transitions between levels with widely different principal quantum numbers, cascade processes become essential, where an atom passes from the state n to the state n' through intermediate levels by different means.

Because downward cascade transitions in interstellar space are not interrupted by absorption of quanta or by collisions (when n is not too large), so that they are determined only by atomic properties, we can use the so-called cascade matrix introduced by Seaton [42]. In contrast to $A_{nn'}$, which determines the probability of direct transition from level n to level n', the element $C_{nn'}$ of the cascade matrix represents the probability of a cascade transition from level n to the level n' by all possible routes through intermediate states p $(n > p \geqslant n')$. From the definition we obtain for the matrix equation that the cascade matrix must satisfy

$$C_{nn'} = \sum_{p=n'}^{n-1} \left(\frac{A_{np}}{\sum_{i=2}^{n-1} A_{ni}} \right) C_{pn'}. \tag{I.8}$$

Actually, the probability of a transition $n \to n'$ by any of the possible means is equal to the sum of all possible products of the individual probabilities of transitions $n \to p$ and $p \to n'$ (summation over all possible chains). The quantity $A_{np}/\sum_{i=2}^{n-1} A_{ni}$ is the probability that atoms in level n are converted to level p relative to the total probability of a transition to all lower levels. Summation over i is started from the second level for the reasons given in Sec. 2. It follows from the definition of the matrix $\|C_{nn'}\|$ that all its elements $C_{nn'}$ are less than 1 when $n > n'$. To normalize the matrix in Eq. (I.8), we set $C_{n'n'} = 1$. The matrix $\|C_{nn'}\|$ has been calculated by Seaton. Numerical values for elements of the $\|C_{nn'}\|$ matrices for eight values of the lower level n' and ten values of $n - n'$, as well as the limiting value $C_{\infty n'}$, are given in Table I.1. Obviously, the values of the cascade-matrix elements do not depend on Z.

Table I.1.

	n'							
$n - n'$	3	4	5	6	7	8	9	10
1	0.5164	0.3633	0.2890	0.2454	0.2163	0.1961	0.1804	0.1683
2	.4838	.3224	.2458	.2015	.1726	.1531	.1382	.1269
3	.4714	.3060	.2282	.1836	.1550	.1351	.1206	.1094
4	.4651	.2975	.2187	.1737	.1451	.1253	.1107	.0997
5	.4614	.2923	.2128	.1676	.1389	.1190	.1045	.0936
6	.4591	.2888	.2090	.1635	.1346	.1148	.1003	.0894
7	.4575	.2866	.2062	.1605	.1315	.1117	.0971	.0863
8	.4564	.2848	.2041	.1584	.1292	.1094	.0948	.0840
9	.4555	.2833	.2025	.1567	.1274	.1076	.0930	.0822
10	.4547	.2822	.2012	.1553	.1260	.1062	.0916	.0808
∞	.4481	.2721	.1895	.1431	.1130	.0934	.0787	.0680

So far, unresolved terms have been considered. We now look at the dependence of transition probabilities on azimuthal quantum number. Unfortunately, these data cannot be represented here in such a complete and simple form. The equation analogous to (I.5) that takes into account the azimuthal quantum numbers l is

$$A_{nln'l'} = \frac{8\pi^2 e^2 \nu_{nn'}^2}{mc^3} \frac{2l' + 1}{2l + 1} f_{n'l'nl}$$

$$\approx \frac{32\pi^6 m e^{10} Z^4}{3h^6 c^3} \frac{l}{2l + 1} \frac{p(n, l)}{p(n', l - 1)} \frac{1.701 g_{nln'(l-1)}}{(nn')^{7/3}(n^2 - n'^2)^{1/3}}. \tag{I.9}$$

In the second equation it is assumed that $n > n' \gg 1$ and $l' = l - 1$. If $l' = l + 1$, we must substitute instead of $[l/(2l + 1)][p(n, l)/p(n', l - 1)]$ the expression $[(l + 1)/(2l + 1)][p(n', l + 1)/p(n, l)]$. We use here the designation

$$p(n, l) = \prod_{r=0}^{r=l} \left(1 - \frac{r^2}{n^2}\right), \tag{I.10}$$

and the Gaunt factor

$$g_{nln'(l-1)} = 1 - 0.1728 \frac{5l + 1 - (n'/n)^2(5l - 1)}{(1 - (n/n')^2)^{2/3} n^{2/3}} + \cdots. \tag{I.11}$$

The representation of $A_{nln'l'}$ by Eq. (I.9) was given by Burgess [49]. Strictly speaking, it is correct when $n \gg n' \gg l$. Then $p(n, l)$ may be written in the form $p(n, l) \approx 1 - l(l + 1)(2l + 1)/n^2$. However, Eqs. (I.9) and (I.10) may be used for estimates when n, n', and l are comparable. The numerical value of $A_{nln'(l-1)}$ for the Gaunt factor equal to 1 is

$$A_{nln'(l-1)} = \frac{5.45 \times 10^9 Z^4}{(nn')^{7/3}(n^2 - n'^2)} \frac{l}{2l + 1} \frac{p(n, l)}{p(n', l - 1)} \sec^{-1}. \tag{I.12}$$

The most complete tables of coefficients $A_{nln'l'}$ are those published by Green, Rush, and Chandler [49a]. The cascade matrix $C_{nln'l'}$ analogous to (I.8), but taking into account azimuthal degeneracy, has been calculated by Pengelly [49b]. Because of its large size, it has not been published.

One particular kind of transitions with change in azimuthal quantum numbers is transitions in which the principal quantum number does not change, that is, transitions from levels nl to levels $n(l \pm 1)$. The probability of these transitions is

$$A_{nln(l\pm 1)} = \frac{3Z^4}{(me)^2} \left(\frac{h\nu}{c}\right)^3 \frac{\max(l, l \pm 1)}{2l + 1} n^2(n^2 - l^2)$$

$$= 1.95 \times 10^8 \left(\frac{912\text{Å}}{\lambda}\right)^3 Z^4 \frac{\max(l, l \pm 1)}{2l + 1} n^2(n^2 - l^2) \sec^{-1}. \tag{I.13}$$

We note that Eq. (I.13) is an exact expression. In a Coulomb field, transitions for atomic hydrogen with no change in principal quantum number produce radiation in the radio band. Because of the third-power dependence of probability on frequency, values of $A_{nln(l\pm 1)}$ are small in this case, although these transitions are permitted according to the selection rules. The wavelengths and probabilities for three such transitions are given in Table I.2.

Table I.2.

Transition	λ (cm)	A (sec^{-1})
$2s - 2p$	3.03	6.5×10^{-7}
$3s - 3p$	10.2	1.02×10^{-7}
$3p - 4d$	27.7	3.8×10^{-9}

Since the number of excited atoms of interstellar hydrogen is small (the probability is high for transitions $n = 2 \rightarrow n = 1$, $n = 3 \rightarrow n = 2$, and so forth), it has not been possible so far to observe these lines.

When the atomic field is no longer a Coulomb field, because, for example, of incomplete shielding of the nuclear charge by the electrons, the energy splitting of the terms with identical values of n and different values of l is considerably increased and the quanta of the corresponding transitions fall into the visible spectral region. Transition probabilities are correspondingly increased. A large part of the observed emission lines from the interstellar gas belong to transitions with no change in principal quantum number. In cases where direct calculation of an allowed transition probability is difficult, an estimate of it may be obtained directly from (I.13).

Of particular importance in the radio band is the transition in the hyperfine structure of the ground state of atomic hydrogen, giving the line $\lambda = 21.11$ cm (1420.4 MHz). The sublevels here are characterized by the quantum number $F = S + I$, where S is the electron spin and I is the spin of the nucleus. The nuclear and electron spins at the lower sublevel are antiparallel and $F = 0$, while at the excited sublevel, the spins are parallel and $F = 1$. The smallness of the nuclear moment in comparison with the electron moment puts this transition into the radio band. The transition probability was calculated by I. S. Shklovskii [15] and is

$$A_{10} = 2.85 \times 10^{-15}\ \text{sec}^{-1}. \tag{I.14}$$

Dipole radiation is forbidden here by the selection rules, so that the transition is a magnetic-dipole transition, which leads to a further (in addition to the dependence on ν^3) considerable reduction in probability.

The transition $2s \rightarrow 1s$ is forbidden for dipole radiation by the selection rules, making the $2s$ level metastable. Here, a magnetic dipole transition to the ground state is also possible, but the so-called two-photon ($2q$) dipole transition $2s \rightarrow 1s$ is more probable. The total probability of this transition is $A_{2s1s} =$

8.226 $\sec^{-1}$. According to calculations by A. Ya. Kipper [55] and by Spitzer and Greenstein [50], the probability of the simultaneous emission of one quantum of frequency ν in the range $d\nu$ and another of frequency $\nu_\alpha - \nu$ in the range $d\nu$, where ν_α is the frequency of the transition $2s \to 1s$ ($\nu_\alpha = 2.467 \times 10^{15}$ $\sec^{-1}$) is determined by the equation

$$dA_{2s1s}(\nu) = \frac{3}{4}\left(\frac{\pi e^2}{ch}\right)^6 \psi\left(\frac{\nu}{\nu_\alpha}\right) d\nu = 1.9 \times 10^{-15}\psi\left(\frac{\nu}{\nu_\alpha}\right) d\nu. \tag{I.15}$$

The function $\psi(\nu/\nu_\alpha)$ is given in Table I.3. This function is symmetric relative to the frequency $\nu = \nu_\alpha/2$.

Table I.3.

λ (Å)	24 313	12 157	8105	6078	4362	4052	3473	3039	2702	2403
$\psi(\nu/\nu_\alpha)$	1.725	2.783	3.481	3.961	4.306	4.546	4.711	4.824	4.889	4.907

Ionization and Recombination

We shall now consider transitions between discrete and continuous states. We shall pay no attention, at first, to the azimuthal quantum number. The probabilities of transitions between discrete and continuous states are described by the atomic absorption coefficients $a_n(\nu)$ and the effective recombination cross section $\sigma_n^i(v)$, which depends on the velocity of the recombining electron. These quantities are connected by the Milne relation

$$\sigma_n^i(v) = \left(\frac{h\nu}{mcv}\right)^2 \frac{\tilde{\omega}_n}{u_l} a_n(\nu), \qquad h\nu = \chi_n + \frac{mv^2}{2}, \tag{I.16}$$

which is derived from the condition of detailed balance at thermodynamic equilibrium. As is well known, this relation is also correct in the absence of thermodynamic equilibrium, since it contains only atomic parameters. The quantity u_i in (I.16) describes the statistical weight of an ion, taking into account the populations of the different levels. For a hydrogen atom, $\tilde{\omega}_n/u_i = n^2$.

The absorption coefficient depends not only on the principal quantum number n, but also on l, that is, it must be designated as $a_{nl}(\nu)$. However, in this case we are determining the mean atomic absorption coefficient, or

$$a_n(\nu) = \sum_{l=0}^{n-1} \frac{2l+1}{n^2} a_{nl}(\nu). \tag{I.17}$$

The mean atomic absorption coefficient for hydrogenlike atoms is described by the simple expression

$$a_n(\nu) = \frac{64\pi^4 m e^{10} Z^4}{3\sqrt{3}\, ch^6 n^5 \nu^3} g_n = \frac{2.82 \times 10^{29} Z^4}{n^5 \nu^3} g_n(\nu). \tag{I.18}$$

Here the Gaunt factor depends on frequency and may be written in the form of the expansion

$$g_n(\nu) = 1 - 0.1728\left(\frac{\nu}{3.3 \times 10^{15}Z^2}\right)^{1/3}\left[\frac{2}{n^2}\left(\frac{3.3 \times 10^{15}Z^2}{\nu}\right) - 1\right] + \cdots. \tag{I.19}$$

Near the ionization limit ($\nu_n = \chi_n/h$) the absorption coefficient is equal to

$$a_n(\nu_n) = a_n^{(0)} = \frac{8h^3 n g_n^{(0)}}{3\sqrt{3}\pi^2 c(mZe)^2} = 7.91 \times 10^{-18}\frac{n g_n^{(0)}}{Z^2}\ \text{cm}^2. \tag{I.20}$$

Table (I.4) provides the values of $g_n^{(0)}$ at the ionization limit and values of $\bar{g}_n$ averaged over frequency [1]. We see from the table that we may set the Gaunt

Table I.4.

n	1	2	3	4	5	6	7
$g_n^{(0)}$	0.80	0.89	0.91	0.92	0.93	0.94	0.95
$\bar{g}_n$	0.89	1.00	1.04	1.06	1.07		

factor for free-free transitions equal to unity without serious inaccuracies (except for absorption from the ground level).

Substituting Eq. (I.18) into Eq. (.16), we obtain for the effective recombination cross section for hydrogenlike ions the expression

$$\sigma_n^i(v) = \frac{64\pi^4 e^{10} Z^4}{3\sqrt{3}m(chn)^3 v^2}\,\frac{g_n}{\chi_n + mv^2/2}. \tag{I.21}$$

The recombination cross section is rarely used directly in the physics of the interstellar gas. In most cases we must know the recombination coefficient $\alpha_n(T)$ obtained by averaging (I.21) with respect to the Maxwell velocity distribution function, or

$$\begin{aligned}\alpha_n(T) &= 4\pi\left(\frac{m}{2\pi kT}\right)^{3/2}\int_0^\infty \sigma_n^i(v)v^3 e^{-mv^2/2kT}\,dv \\ &= \frac{2^9\pi^5 e^{10}Z^4}{m^2(ch)^3}\left(\frac{m}{6\pi kT}\right)^{3/2}\frac{\bar{g}_n}{n^3}e^{\chi_n/kT}\,\mathrm{Ei}\left(\frac{\chi_n}{kT}\right) \\ &= \frac{3.262 \times 10^{-6}Z^4}{n^3 T^{3/2}}\bar{g}_n e^{\chi_n/kT}\,\mathrm{Ei}\left(\frac{\chi_n}{kT}\right)\ \text{cm}^3/\text{sec}.\end{aligned} \tag{I.22}$$

Here $\mathrm{Ei}(x) = \int_x^\infty e^{-u}\,du/u$. When $\chi_n > 5kT$, which for hydrogen is equivalent to the condition that $n^2T < 32{,}000°$, Eq. (I.22) may be simplified to

$$\alpha_n(T) = 3.26 \times 10^{-6} \frac{kZ^4\bar{g}_n}{nT^{1/2}\chi_1} = 2.1 \times 10^{-11} \frac{Z^2\bar{g}_n}{nT^{1/2}} \text{ cm}^3/\text{sec}. \quad \text{(I.23)}$$

Thus, for recombinations to the ground level $\alpha_n(T) \sim T^{-1/2}$. For higher-level recombinations $\chi_n \ll kT$, while in calculating the coefficient $\alpha_n(T)$ we can assume $e^{\chi_n/kT}$ Ei (χ_n/kT) in (I.22) is equal to ln $(1.781\ \chi_n/kT)$. In this case, $\alpha_n(T) \sim T^{-3/2}$. As the temperature falls, the part played by recombinations to excited levels is increased. Tables for finding more accurate values of $\alpha_n(T)$, taking into account two terms in the Gaunt-factor series, are published by Seaton [58]. An approximate equation given there for calculating the total recombination coefficient to all levels is

$$\alpha_t(T) = \sum_{n=1}^{\infty} \alpha_n(T) = 5.20 \times 10^{-14} Z \left(\frac{\chi_1 Z^2}{kT}\right)^{1/2}$$
$$\times \left[0.43 + \frac{1}{2}\ln\left(\frac{\chi_1 Z^2}{kT}\right) + 0.47\left(\frac{kT}{\chi_1 Z^2}\right)^{1/3}\right]. \quad \text{(I.24)}$$

When $T \lesssim 10^5 Z^2$, the error of this equation does not exceed 1 percent. The formulas

$$\alpha_t = 4 \times 10^{-13}(10^4/T)^{0.73} \text{ cm}^3/\text{sec}$$

and

$$\alpha_t - \alpha_1 = 2.6 \times 10^{-13}(10^4/T)^{0.85} \text{ cm}^3/\text{sec}$$

are also sufficiently accurate.

It is well known that the electron temperature in ionized-hydrogen regions is approximately 10,000°. Therefore, values often used in the literature for the hydrogen recombination coefficients when $T = 10{,}000°$ are (cm³/sec):

$$\alpha_1 = 1.55 \times 10^{-13}, \qquad \alpha_2 = 0.74 \times 10^{-13}, \qquad \alpha_3 = 0.42 \times 10^{-13},$$
$$\alpha_t = 4.0 \times 10^{-13}. \quad \text{(I.25)}$$

For different temperatures, we can easily divide by $(10^{-4}T)^{1/2}$.

We shall now take the azimuthal quantum number into account in the transition coefficients between discrete and continuous spectra. We can do this approximately by assuming that the Gaunt factor in Eqs. (I.18) to (I.23) is also dependent on l. We note that now this factor changes more sharply with frequency. Table I.5 gives values for the Gaunt factor g_{nl} and also the exponent γ relating the atomic absorption coefficient to frequency (that is, $a_{nl}(\nu) = a_{nl}^{(0)}(\nu_n/\nu)^\gamma$, where ν_n is the frequency of the ionization limit).

Table I.5.

Factor	Level							
	$1s$	$2s$	$2p$	$3s$	$3p$	$3d$	$4s$	$4p$
$g_{nl}^{(0)}$	0.80	0.95	0.89	1.08	1.10	0.76	1.19	1.25
$\bar{g}_{nl}$	0.89	1.47	0.80	1.79	1.32	0.61	2.02	1.75
γ	3	2.5	3	2	3	3	2	2

Burgess [49] obtained as an approximate equation for the absorption coefficient at the nl level, where $n \gg l \gg 1$, the expression

$$a_{nl}(\nu) = \frac{2.7h^3n^{5/3}}{6\pi^2c(me)^2}\left(\frac{\nu_n}{\nu}\right)^{7/3}$$

$$\times\left[\frac{l+1}{2l+1}\frac{p(i/k, l+1)}{p(n, l)}g_{n(l+1)} + \frac{l}{2l+1}\frac{p(n, l)}{p(i/k, l-1)}g_{n(l-1)}\right]. \quad \text{(I.26)}$$

Here $k = (mv^2/2\chi_1)^{1/2}$, $h\nu = \chi_n + \chi_1 k^2$, the function $p(n, l)$ is determined by Eq. (I.10), and the expression $p(i/k, l)$ is equal to $\prod_{r=0}^{l}[1 + (kr)^2]$. The Gaunt factors are also represented by series. We shall not reproduce them since they are always close to unity. Taking into account numerical values of atomic constants, we have at the ionization limit

$$a_{nl}^{(0)} = 1.4 \times 10^{-17}\frac{n^{5/3}}{2l+1}\left[\frac{l+1}{p(n, l)} + lp(n, l)\right]\text{cm}^2. \quad \text{(I.27)}$$

In particular, if $p(n, l) \approx 1$, then $a_{nl}^{(0)} \approx 1.4 \times 10^{-17} \times n^{5/3}$ cm^2. Moreover, at low values of l, the dependence on frequency of the factor in square brackets in (I.26) need not be taken into account, so that the absorption coefficient decreases with frequency approximately as $\nu^{-7/3}$. However, when the values of l are comparable with those of n, the absorption coefficient falls more rapidly with decrease in frequency (when $l \approx n - 1$, we have approximately $a_{n(n-1)} \sim \nu^{-(7+n)/3}$). We must stress that Eqs. (I.26) and (I.27) are approximate and cannot be used for accurate quantitative calculations.

A similar equation may be obtained for recombination coefficients. The Milne relation, allowing for the azimuthal quantum number, has the form

$$\sigma_{nl}^{i} = \left(\frac{h\nu}{mcv}\right)^2\frac{\tilde{\omega}_l}{u_i}a_{nl}(\nu). \quad \text{(I.28)}$$

For hydrogen, $\tilde{\omega}_l/u_i = 2l + 1$. Substituting here Eq. (I.26) and then putting the expression obtained into an integral of the type of (I.26), we find, using the numerical values of the atomic constants,

$$\alpha_{nl}(T) \approx \frac{4.0 n^{5/3} \chi_n^{7/3}}{(kT)^{5/6}} \left[\frac{l+1}{p(n, l)} + l p(n, l) \right] e^{-\chi_n/kT} \int_{\chi_n/kT}^{\infty} \frac{e^{-u} du}{u^{1/3}}. \tag{I.29}$$

Since the condition for the validity of Eq. (I.26) is that $-n \gg 1$, we have when $\chi_n \gg kT$, that is, at low temperatures,

$$\alpha_{nl}(T) = \frac{4.0 n^{5/3} \chi_n^2}{(kT)^{1/2}} \left[\frac{l+1}{p(n, l)} + l p(n, l) \right] \text{cm}^3/\text{sec}. \tag{I.30}$$

On the contrary, when $\chi_n \ll kT$, the integral in Eq. (I.29) is approximately equal to $\frac{3}{2}(\chi_n/kT)^{2/3} e^{-\chi_n/kT}$. Then we obtain

$$\alpha_{nl}(T) \approx \frac{6.0 n^{5/3} \chi_n^3}{(kT)^{3/2}} \left[\frac{l+1}{p(n, l)} + l p(n, l) \right] \approx \frac{6.2 \times 10^{-9}}{n^{13/3} T^{3/2}} \left[\frac{l+1}{p(n, l)} + l p(n, l) \right] \text{cm}^3/\text{sec}. \tag{I.31}$$

This approximate equation may be used for estimating the contribution of the various excited sublevels to the recombination coefficients at large values of n. As we should expect, the recombination coefficient to an excited sublevel is considerably reduced with an increase in temperature.

More precise values of the coefficients of absorption and recombination, taking into account azimuthal degeneracy, are given by Burgess in tabular form [49e].

Free-free Transitions

We turn now to the calculation of transitions in continuous spectra. Electrons under the conditions of interstellar space have a Maxwellian velocity distribution. Therefore, we need only determine the equation for the atomic absorption coefficient, and the emission probability can then be calculated from Kirchhoff's law.

The absorption coefficient in the continuous spectrum, calculated on the basis of an encounter between a single ion and a single electron moving with velocity v, is equal to

$$a(\nu, v) = \frac{4\pi}{3\sqrt{3}} \frac{Z^2 e^6}{chm^2 \nu v^3} g(\nu). \tag{I.32}$$

In the optical band, the Gaunt factor is about unity as before ($g = 1.1$–1.4 [1] for $h\nu/kT = 1$–7), but for radio transitions this factor is much larger, which is connected with the divergence of the transition integral for distant collisions, during which, naturally, radio waves are absorbed. Ginzburg's calculation [81] in the radio band, which takes into account the tenuity of the interstellar gas (the wavelength is less than the Debye length), yields

$$g = \frac{\sqrt{3}}{\pi} \ln \frac{(2kT)^{3/2}}{13.5\nu e^2 m^{1/2}} = 0.55[17.7 + \ln (T^{3/2}/\nu)]. \qquad \text{(I.33)}$$

In the decimeter band, $g \approx 6$–10. Averaging Eq. (I.32) with respect to the velocity distribution, we obtain for the mean absorption coefficient calculated on the basis of one ion and one electron the expression

$$k(\nu) = 4\pi\left(\frac{m}{2\pi kT}\right)^{3/2} \int_0^\infty a(\nu, v)v^2 e^{-mv^2/2kT}\, dv$$

$$= \frac{8\pi}{3} \frac{Z^2 e^6 \bar{g}}{mhc(6\pi mkT)^{1/2}} \frac{1}{\nu^3} = 3.2 \times 10^8 \frac{Z^2 g}{T^{1/2}\nu^3} \text{ cm}^2. \qquad \text{(I.34)}$$

Before finishing our description of radiative transition probabilities, let us examine induced emission. According to the well-known relations between the Einstein coefficients, induced emission (negative absorption) can be taken into account by multiplying the true absorption coefficient by a factor

$$1 - \frac{\tilde{\omega}_{n'} N_n}{\tilde{\omega}_n N_{n'}}, \qquad \text{(I.35)}$$

where N_n and $N_{n'}$ are the numbers of atoms in the upper and lower states, respectively. We consider now whether induced emission must be taken into account in any concrete conditions of interstellar space. During absorption from the ground level in the optical band, $N_n \ll N_{n'}$ and the factor (I.35) is almost equal to unity. If the excited level is so close to the ground state that electron collisions are more frequent than radiative transitions, a Boltzman distribution is established,

$$\frac{\tilde{\omega}_{n'} N_n}{\tilde{\omega}_n N_{n'}} = \exp\left(-\frac{h\nu_{nn'}}{kT}\right). \qquad \text{(I.36)}$$

The relation (I.36) is realized in the case of transitions between neighboring levels with large principal quantum numbers because of the low probability of radiative transitions and the high probability of electron collisions. In the second case, the transition frequency falls in the radio band, and instead of the factor (I.35) we can write

$$1 - e^{-h\nu/kT} \approx \frac{h\nu}{kT}, \qquad \text{(I.37)}$$

since in this case $h\nu \ll kT$. The condition (I.36) also applies to the continuous spectrum if the velocity distribution function is Maxwellian. In the optical region of the continuous spectrum, $h\nu \gg kT$ and the factor (I.35), as before, is almost unity. The relation (I.37) also holds in the continuous radio spectrum.

Finally, during transitions between low excited levels, the quantity $\tilde{\omega}_{n'}N_n/\tilde{\omega}_n N_{n'}$ may be either less than or greater than unity (that is, a laser effect is possible here). However, the absorption coefficient in this case is usually very small, so that we need not take into account here the true or the induced absorption. Thus, induced absorption is effective only in the radio band. The corresponding total absorption coefficient will be designated by a prime. Consequently,

$$k'(\nu) = k(\nu)\frac{h\nu}{kT} = \frac{0.01782Z^2g}{\nu^2T^{3/2}} = \frac{0.98 \times 10^{-2}Z^2}{\nu^2T^{3/2}}\left[17.7 + \ln\left(\frac{T^{3/2}}{\nu}\right)\right]. \quad \text{(I.38)}$$

Excitation and Ionization of Hydrogen Atoms by Electron Collision

Calculations of the probability of excitation by electron collision, using the methods of quantum mechanics, can still not be made as accurately and completely as for radiative transitions. Thus we are often compelled in this case to use semiempirical methods.

The classical equation for the effective cross section for the excitation of a level with energy χ during collision with an electron of energy $\mathscr{E}$ has the form

$$\sigma(\mathscr{E}) = \frac{3\pi e^4 f}{\mathscr{E}\chi}\left(1 - \frac{\chi}{\mathscr{E}}\right), \quad \text{(I.39)}$$

where f is the oscillator strength during absorption. It is convenient to introduce the probability for excitation by electron collision as

$$q = 4\pi\left(\frac{m}{2\pi kT}\right)^{3/2}\int_{\sqrt{2\chi/m}}^{\infty} \sigma(v)v^3e^{-mv^2/2kT}\,dv. \quad \text{(I.40)}$$

We have for the classical effective cross section

$$q = \frac{24\pi^2e^4mf}{(2\pi mkT)^{3/2}}\left[\frac{kT}{\chi}e^{-\chi/kT} - \mathrm{Ei}\left(\frac{\chi}{kT}\right)\right]. \quad \text{(I.41)}$$

We should stress that Eq. (I.39), and consequently Eq. (I.41), are accurate only to within several tens percent and sometimes differ by several orders of magnitude from real values. For fast electrons whose energies are much greater than the excitation or ionization limit, a more accurate equation may be obtained, using the Born approximation, from quantum mechanics. In particular, the effective cross section for ionization of a hydrogen atom by a fast electron is determined by the Bates equation

$$\sigma_i(\mathscr{E}) = 0.285\frac{\pi e^4}{\mathscr{E}\chi_1}\ln\left(\frac{100\mathscr{E}}{\chi_1}\right). \quad \text{(I.42)}$$

The corresponding ionization probability is [36]

$$q_{1i} = \frac{1.14\pi e^4}{(2\pi mkT)^{1/2}\chi_1}\left[e^{-\chi_1/kT}\ln 100 + \mathrm{Ei}\left(\frac{\chi_1}{kT}\right)\right] \approx \frac{5.25\pi e^4}{(2\pi mkT)^{1/2}}\frac{e^{-\chi_1/kT}}{\chi_1}. \quad (I.43)$$

This equation is correct only for electron temperatures of several hundreds of thousands of degrees.

Similarly, we can calculate the probability of excitation of hydrogenic levels by collisions with fast electrons. A very simple estimate, which is usually sufficiently accurate, may be obtained from the fact that the excitation probability is proportional to the oscillator strength [see (I.39)]; this condition is preserved in the Born approximation. At high temperatures, the energy intervals between levels of different n values are small in comparison with kT, and therefore are unimportant for calculating q_{1n}. Therefore,

$$q_{1i}:q_{12}:q_{13}:q_{14}\cdots = 0.436:0.416:0.079:0.029\cdots. \quad (I.44)$$

As a rule, the excitation of the higher discrete levels of a hydrogen atom by electron collision is unimportant.

For the determination of probabilities of excitation by slow electrons we must know in detail the behavior of the effective cross section near the excitation or ionization threshold, the exact location of the maximum of this quantity, and so on. Many calculations have been made, particularly for the transitions $1s \rightarrow 2s$, $2p$ [46a, b, c]. Theoretical calculations and experimental data frequantly differ considerably [46d, e], while the accuracy of both the theory and the experiments remains in doubt. Nevertheless, excitation coefficients for a number of low levels in atomic hydrogen and its ionization coefficient, obtained with the help of theoretical and experimental data, are useful in astrophysical calculations where a particularly high accuracy is not necessary. According to the most recent data, these coefficients may be approximated by the expressions

$$\begin{aligned} q_{1n}(T) &= q_{1n}^{(0)}\left(\frac{kT}{\chi_1 - \chi_n}\right)^{1/8} e^{-(\chi_1-\chi_n)/kT}, \\ q_{1i}(T) &= q_{1i}^{(0)}\left(\frac{kT}{\chi_1}\right)^{1/2} e^{-\chi_1/kT}, \end{aligned} \quad (I.45)$$

where the values of the quantities $q_{1n}^{(0)}$ and $q_{1i}^{(0)}$ obtained experimentally by Pronik and theoretically by Hammer [49f] are presented in Table I.6.

The theoretical estimates are approximately twice the experimental values, which is also evident in Fig. 74. Preliminary estimates of excitation coefficients [45, 46] differ from those presented here by a factor of 4 or more. We may assume that the relative increase in the excitation coefficients with temperature described for Eqs. (I.45) are in error by scarcely more than a few tens percent, and that the error in their absolute values does not exceed a factor of 2.

Table I.6.

Date	Coefficients					
	$q_{12}^{(0)}$	$q_{13}^{(0)}$	$q_{14}^{(0)}$	$q_{15}^{(0)}$	$q_{16}^{(0)}$	$q_{1i}^{(0)}$
Experimental	2×10^{-8}	4×10^{-9}	1.5×10^{-9}	—	—	2×10^{-8}
Theoretical	5×10^{-8}	8×10^{-9}	3×10^{-9}	1.5×10^{-9}	7.5×10^{-10}	2×10^{-8}

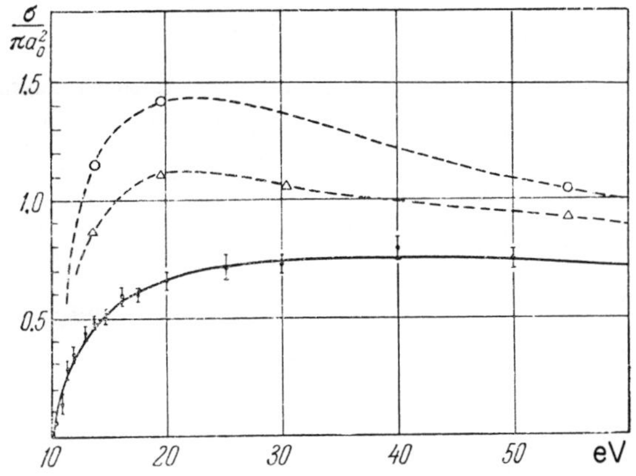

FIG. 74. The cross section of the excitation $1s - 2p$ of hydrogen by electron collision; circles show the Born approximation, triangles give the distorted wave approximation, and the points are experimental values.

Together with the excitation coefficient $q_{n'n}$, we introduce the coefficients of de-excitation $q_{nn'}$ due to collisions of the second kind. The two coefficients are connected by the relation

$$\frac{q_{nn'}}{q_{n'n}} = \frac{\tilde{\omega}_{n'}}{\tilde{\omega}_n} e^{-(\chi_n - \chi_{n'})/kT}, \tag{I.46}$$

which follows from the condition of detailed balance. The calculation of de-excitation coefficients is often simpler than that of excitation coefficients.

Excitation and de-excitation coefficients are especially large for transitions between the higher levels. In particular, the excitation and de-excitation coefficients for transitions $n' = n \pm 1$ when $n \gg 1$ are equal to one another and are determined by the equation

$$q_{n(n\pm1)} = \frac{9 \times 10^{-6}}{T^{1/2}} n^4 \ln 0.0471n \text{ cm}^3/\text{sec}, \tag{I.47}$$

which follows from (I.41) when $\chi = \chi_{n-1} \ll kT$ and $f = n/6$, but with the corrections of Seaton [49d]. The proportionality to the fourth power of the principal quantum number arises because the radius of the corresponding electron orbit is $\sim n^2$. The excitation coefficient for the transition $2s$–$2p$, according to Seaton's calculations [18], is equal to 5.3×10^{-4} cm^3/sec for $T = 10{,}000°$ and 4.7×10^{-4} cm^3/sec for $T = 20{,}000°$. Proton collisions, which provide almost 90 percent of the total excitation coefficient, are also taken into account here. More general calculations for other transitions with no change in principal quantum number have been made by Seaton and Pengelly [49c].

In calculating excitation and de-excitation of the sublevel $F = 1$ in the hyperfine structure of atomic hydrogen, we must consider mutual collisions between hydrogen atoms. From recent calculations by Dalgarno [18], the effective cross section for this process depends only slightly on atomic velocities. For a kinetic temperature change from 2° to 10,000°, the average effective cross section σ_F goes from 6.6×10^{-15} to 2.9×10^{-15} cm^2. Thus, the excitation coefficient is

$$q_{01} = \bar{\sigma}_F\left(\frac{8}{\pi}\frac{kT}{m_{\mathrm{H}}}\right)^{1/2} \approx 7 \times 10^{-11} T^{1/2} \text{ cm}^3/\text{sec}. \tag{I.47}$$

The de-excitation coefficient q_{10} is three times smaller than q_{01}.

Appendix II. Coefficients of Excitation and De-excitation of Metastable Levels by Electron Collisions

In studies on inelastic collisions, that is, collisions of the second kind, between electrons and atoms, it is often convenient to start with the calculation of de-excitation coefficients q_{ki}, since the uncertainty in determining the dependence of the effective cross section on the energy of the colliding electron is less in this method. The corresponding excitation coefficient q_{ik} is then obtained from Eq. (7.17) or (I.45).

Let us denote by $\sigma_{ki}(v)$ the effective de-excitation cross section of level k. Then the probability of quenching is calculated from the expression

$$q_{ki} = \int_0^{\infty} vf(v)\sigma_{ki}(v)dv = 4\pi\left(\frac{m}{2\pi kT}\right)^{3/2} \int_0^{\infty} \sigma_{ki}(v)v^3 e^{-mv^2/2kT}\, dv, \qquad \text{(II.1)}$$

where T is the temperature of the electron gas.

Values of the effective cross section for allowed transitions in atomic hydrogen are given in Appendix I [Eqs. (I.42) and (I.39)]. These expressions do not apply here, however. The classical explanation for this difference is as follows. When a level is excited by an allowed transition, the colliding electron yields some of its energy to an atomic electron, which is transferred from the ground state to an excited state. This process is most probable when the energy of the free electron escaping after collision is comparable with the energy yielded to the atomic electron. In other words, the effective excitation cross section of allowed transitions is maximum when the energy of the incident electron is approximately twice the excitation limit. This condition is taken tino account in Eq. (I.39). Equation (I.42) is valid only for high energies.

In the excitation of metastable levels an atomic electron can no longer pass from the ground state to a metastable state. Here the excitation process consists of the incident electron itself "sitting" in the excited state, while the atomic electron escapes (strictly, in quantum-mechanical language we must speak not of electron exchange—electrons are indistinguishable—but of spin exchange). The longer the incident electron spends near the atom, that is, the less its energy, the more probable is the exchange process. Therefore, the effective cross section of metastable levels increases very rapidly from the excitation limit up to a maximum and then falls approximately in the same way as in the excitation of allowed transitions. Both Eqs. (I.39) and (I.42) show that, for high energies of the incident electron, the effective cross sections diminish as $1/\mathscr{E} \sim 1/v^2$. We can assume that excitation and de-excitation cross sections of a metastable level at high electron energies are similarly related to velocity. Hence we can write

$$\sigma_{ki} = \left(\frac{h}{mv}\right)^2 \frac{\Omega_{ki}}{4\pi\tilde{\omega}_k}. \tag{II.2}$$

Here Ω_{ki} is some numerical parameter of order unity, called the *collision efficiency*. Equation (II.2) is readily interpreted in terms of quantum mechanics. The quantity $h/2\pi mv$ is the de Broglie wavelength for the incident electron, so that Eq. (II.2) states that the effective cross section is proportional to the square of the electron wavelength. The proportionality factor in Eq. (II.2) obviously takes into account the possibility of interaction during the passage of the electron at various distances of impact from the atom. From the point of view of quantum mechanics, this means that interaction can occur for various values of the azimuthal quantum number l of a free electron. If l is restricted, limits must also be imposed on Ω_{ki}. The appropriate limit is not difficult to obtain. If ρ is the impact parameter, the angular momentum of the incident electron is equal to $mv\rho$. From quantum mechanics, the azimuthal quantum number is determined by the relation

$$mv\rho = \frac{h}{2\pi}[l(l+1)]^{1/2}. \tag{II.3}$$

When the azimuthal quantum number changes by unity, the effective cross-section value is changed by

$$2\pi\rho\Delta\rho = \frac{h^2}{2\pi(mv)^2}[l(l+1)]^{1/2}\frac{2l+1}{2[l(l+1)]^{1/2}} = \frac{h^2(2l+1)}{4\pi(mv)^2}, \tag{II.4}$$

where $\Delta\rho$ corresponds to a change in l in (II.3) of unity. Comparing Eqs. (II.2) and (II.4), we obtain the following rule: the collision efficiency for electrons of azimuthal quantum number l must not exceed $2l + 1$, since not every electron incident with a specific impact parameter causes excitation or de-excitation of the level. Since spin changes are possible only at a few, usually low, values of l, we can thus obtain an upper limit for the effective cross section, important for estimating the role of collisions.

Equation (II.2) holds for electron energies greater than the energy at which the maximum effective cross section is reached. However, taking into account, first, that the maximum of σ_{ik} occurs at energies close to the excitation limit and, second, that a slow electron in a collision with an ion is accelerated in its neighborhood, so that the interaction takes place in any case at a high velocity, we shall use Eq. (II.2) with some average value of Ω_{ki} valid over the whole range of permissible velocities. For excitation of ions the accuracy of such an assumption is sufficient. In the case of atomic excitation, error may arise through an overestimate (see below). Substituting Eq. (II.2) into Eq. (II.1), we obtain

$$q_{ki} = \frac{h^2}{2\pi m^2}\left(\frac{m}{2\pi kT}\right)^{1/2}\frac{\Omega_{ki}}{\tilde{\omega}_k} = \frac{8.63 \times 10^{-6}}{\tilde{\omega}_k T^{1/2}}\Omega_{ki} \text{ cm}^3/\text{sec}. \tag{II.5}$$

Table II.1.

Configuration	Ion	Ω_{12}	Ω_{13}	Ω_{23}
$2p^2$	N II	2.17	0.31	0.50
	O III	1.59	.22	.64
	F IV	*1.21*	*.17*	*.58*
	Ne V	*0.84*	*.16*	*.53*
$2p^3$	O II	1.28	.58	2.12
	F III	*1.00*	*.22*	*3.11*
	Ne IV	*0.68*	*.23*	*3.51*
	Na V	*.43*	*.25*	*3.49*
$2p^4$	F II	*.95*	*.057*	*0.17*
	Ne III	*.76*	.077	.27
	Na IV	*.61*	*.092*	*.30*
	Mg V	*.54*	*.112*	*.30*
$3p^3$	S II	2.02	.383	12.7

Table (II.1) shows values of $\Omega_{ik} = \Omega_{ki}$, calculated by Seaton [18] for transitions in the p^2, p^3, and p^4 configurations. The italicized values are approximate. For a rough estimate of the collision efficiency in states with configurations $2p^n$ and $3p^n$, Gould [46g] has suggested the formulas $\Omega_{ik}(2p) \approx \frac{1}{8}\tilde{\omega}_i\tilde{\omega}_k$ and $\Omega_{ik}(3p) \approx \tilde{\omega}_i\tilde{\omega}_k$. The numerical coefficients are chosen such that for ionized carbon and silicon, for which Seaton found

$$\text{C II},\quad \Omega(P_{3/2}, P_{1/2}) = 1.28; \qquad \text{Si II},\quad \Omega(P_{3/2}, \mathrm{P}_{1/2}) = 7.7,$$

the correct values are obtained.

For neutral atoms, as we have already noted, the assumption of the applicability of Eq. (II.2) to the whole velocity range can give a more substantial error. Therefore for O I and N I atoms, Seaton calculated directly de-excitation probabilities for a series of temperatures. The probability values are given in Table II.2.

Table II.2.

T (°K)	O I $10^9 q_{21}$	O I $10^9 q_{31}$	O I $10^9 q_{32}$	N I $10^9 q_{21}$	N I $10^9 q_{31}$	N I $10^9 q_{32}$
500	0.9	0.8	0.4	0.5	0.4	0.9
1 000	1.6	1.2	0.6	0.8	0.6	1.5
5 000	5.0	3.5	1.9	2.6	1.8	4.4
10 000	6.7	5.1	3.0	3.5	2.7	6.9
50 000	7.6	7.2	5.6	4.0	3.7	13
100 000	6.8	6.6	6.0	3.5	3.4	14

In the foregoing tables, values of Ω_{ik} and q_{ik} are given for the levels 1, 2, 3, as a whole, without taking into account their multiplet character. Transitions between individual sublevels were calculated for the O II ion. The results are given in Table II.3.

Table II.3.

i	k	Ω_{ik}
$^2P_{1/2}$	$^2P_{3/2}$	0.33
$^2P_{1/2}$	$^2D_{3/2}$	.33
$^2P_{1/2}$	$^2D_{5/2}$	.38
$^2P_{1/2}$	$^4S_{3/2}$	.19
$^2P_{1/2}$	$^2D_{3/2}$	.52
$^2P_{3/2}$	$^2D_{5/2}$	.89
$^2P_{3/2}$	$^4S_{3/2}$	.39
$^2D_{3/2}$	$^2D_{5/2}$	.85
$^2D_{3/2}$	$^4S_{3/2}$	.51
$^2D_{5/2}$	$^4S_{3/2}$	.77

Osterbrock [464] made estimates of the collision strengths for some of the fine-structure transitions, Table II.4; estimates for Si II are by Seaton. For the excitation of Fe II levels, Seaton obtained the following values of this parameter: $\Omega(^6D_{7/2} - {}^6D_{9/2}) = 16$; $\Omega(^6D_{5/2} - {}^6D_{9/2}) = 12$.

To estimate the value of the parameter Ω in other cases we can use the graphs in Fig. 75, in which are plotted the dependence of Ω on the energy of a metastable level for various terms [57].

The atomic parameters A_{ik}, q_{ik}, and Ω_{ik} introduced here, when substituted in Eqs. (7.17), (7.18), (7.22), and (II.5), allow us to calculate the populations of levels, emission coefficients, and, subsequently, intensities of the corresponding forbidden lines.

Table II.4.

Configuration	Ion	Transition			
		$^2P_{3/2}$–$^2P_{1/2}$	$^3P_1$–$^3P_0$	$^3P_2$–$^3P_0$	$^3P_2$–$^3P_1$
$2p^1$	C II	1.44	—	—	—
	N III	0.92	—	—	—
	O IV	.92	—	—	—
$2p^2$	N II	—	0.43	0.19	1.04
	O III	—	.31	.25	0.96
	Ne V	—	.14	.21	.64
$2p^4$	Ne III	—	.13	.10	.42
$2p^5$	Ne II	.32	—	—	—
$3p^1$	Si II	7.7	—	—	—

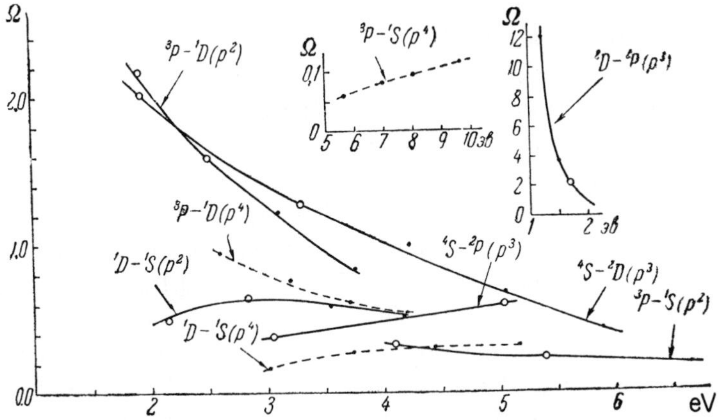

FIG. 75. The dependence of Ω on electron energy for various transitions [O II].

Appendix III. Electronic Terms of Simple Diatomic Molecules

The quantum number Λ, representing the projection on the axis of the molecule of the orbital angular momenta of the electrons, which is analogous to the azimuthal quantum number L in atoms, is the basis for the classification of electronic terms in diatomic molecules. The numerical values of $\Lambda = 0, 1, 2, 3, \ldots$ are denoted by the Greek letters Σ, Π, Δ, Φ, ..., respectively, by analogy with the atomic symbols S, P, D, F, ...

The various indices in the term notation have the following meanings. Let us assume that a molecule consists of two atoms, with their own total azimuthal quantum numbers L_1 and L_2 and total spins S_1 and S_2. The vector sum of the momenta, $L_1 + L_2$, and of the spins, $S_1 + S_2$, may be projected along the molecular axis in various ways. Suppose the projection of the sum $L_1 + L_2$ on the molecular axis has the quantum number Λ, changing from 0 to $L_1 + L_2$, and the projection of the sum of the spins, the quantum number Σ (not to be confused with the symbol of the term for $\Lambda = 0$). By analogy with the system of atomic terms, we can introduce the multiplet nature of the term by $2S + 1$ (the upper left-hand index in the term designation) and the total internal quantum number $\Omega = \Lambda + \Sigma$ (the lower right-hand index); Ω has only two values, since it is composed not of the vectors, but of their projections along the axis. The internal quantum number loses its meaning for the terms Σ ($\Lambda = 0$), since here we cannot speak about the orientation of the spin relative to the orbital angular momentum, so that the corresponding index does not arise.

The Σ term ($\Lambda = 0$) may have a wave function that is either odd or even relative to a change of sign in all the particle coordinates. In this case we must distinguish between two states, which are given a plus or a minus sign. The properties of a term also depend on whether the sum of the azimuthal quantum numbers of all the electrons is odd or even. The first of these states is described by the letter u, the second by g.

If the molecule is rotating, the sign of the direction of the projection of the total orbital quantum number Λ is also significant, since the orbital angular momentum of the electrons interacts with the molecular rotation. Two close components of the term correspond to the two different directions of the Λ projection. This is the so-called Λ splitting which occurs if $\Lambda > 0$ (the states Π, Δ, and so on). We note that all the statements made above are correct if the mutual interactions between the orbital angular momenta of electrons are stronger than the interactions with the spins, that is, if we have the analogue of Russell-Saunders coupling. When the opposite is true, the molecular electronic term system is somewhat complex.

The electronic term diagram of a particular molecule may be obtained from data on the electron terms of its constituent atoms. We can illustrate this by, for example, the CH molecule. If the atoms are separated from one another by a large distance, the electronic terms of this system are simply the sums of the terms of the separate atoms, in this case ^2S (hydrogen) and ^3P, ^1D (carbon), that is, $^2S + {}^3P$ or $^2S + {}^1D$. Upon approach, the electric field of each atom splits the levels of the other atom and the various combinations of the Stark components of the atomic electron terms give the molecular terms shown graphically in Fig. 76. Briefly, the sum of $^2S + {}^3P$ gives the terms $^2\Sigma$, $^4\Sigma$, $^2\Pi$, $^4\Pi$, while the sum $^2S + {}^1D$ gives $^2\Sigma$, $^2\Pi$, $^2\Delta$. In fact, the first combination ($L_1 = 0$, $L_2 = 1$) can only give $\Lambda = 0, 1$, that is, Σ and Π, while the second ($L_1 = 0$, $L_2 = 2$) can give $\Lambda = 0, 1, 2$. Similarly, we have for the spins in the first case ($S_1 = \frac{1}{2}$, $S_2 = 1$) $\Sigma = \frac{1}{2}, \frac{3}{2}$, and in the second case ($S_1 = \frac{1}{2}$, $S_2 = 0$), $\Sigma = \frac{1}{2}$ only. On the other hand, if the C and H atoms unite, we obtain a nitrogen atom with terms ^4S, ^2D, ^2P, which determines the character of the terms when the atoms are close together. If we now separate the C and H atoms, these levels are also split and give molecular terms. From the ^4S state we obtain the $^4\Sigma^-$ term, and from the ^2D level, terms $^2\Sigma^-$, $^2\Pi$, $^2\Delta$ are formed, while the ^2P level gives the terms $^2\Sigma^+$ and $^2\Pi$. This is readily understood from the following considerations. The spin, and hence the multiplicity, are not altered by the splitting, while the value of Λ can be either equal to or less than the value of L in the fused atom. The location of levels at intermediate internuclear distances is obtained by interpolation. Which of these terms is the lowest can often be determined only by experiment. In the CH molecule the lowest is the $^2\Pi$ term.

Another simple example is the formation of the electronic terms of an H_2 molecule ($^1\Sigma$ and $^3\Sigma$) from the sum of the electronic terms of H atoms ($^2S + {}^2S$). Actually, in this case $L_1 = 0$ and $L_2 = 0$, and consequently Λ must equal 0 (the Σ state). The two electron spins can give Σ equal to either zero or unity (see Fig. 18).

A molecular transition from one electronic state to another, accompanied by absorption of a quantum of radiation, may be represented as a transition between potential curves that describe the energies of the given terms for various internuclear distances. The selection rules for allowed transitions are

$$\Delta\Lambda = 0, \pm 1; \qquad \Delta\Sigma = 0; \qquad \Delta\Omega = 0, \pm 1. \qquad \text{(III.1)}$$

Moreover, we must remember that positive terms combine with negative ($^+$

and $^-$) and odd with even (u and g). In addition to electronic terms, molecules also have rotational and vibrational terms. The first are characterized by the wave number K, the second by the wave number v. The electron states are split into vibrational terms, the intervals between which are usually much less than those between electron terms. The vibrational terms are split in turn by the closely placed rotational terms. Intervals between the latter are measured in hundredths of an electron volt. Transitions between combinations of rotational, vibrational, and electron terms give the complex band systems characteristic of molecular spectra. Bands cannot occur in interstellar molecular systems, since these molecules are always found at the lowest rotational as well as vibrational and electron levels and absorption is possible only in the allowed lines arising from the lowest level of the ground state. Transitions between rotational and vibrational terms give lines located in the infrared region of the spectrum.

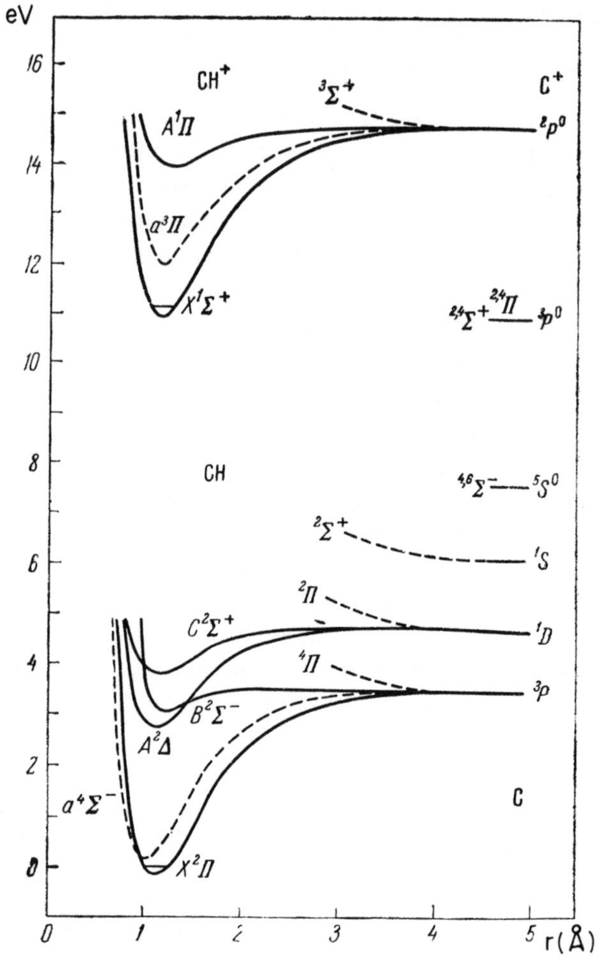

FIG. 76. The electron states of CH and CH$^+$ molecules.

These lines cannot occur in absorption because, first, stellar infrared spectra are weak, and, second, infrared radiation is strongly absorbed in the Earth's atmosphere. In the radio band, we can observe, at least in principle, transitions between the components of Λ splitting for certain molecules. This question is briefly discussed in Sec. 9 (see also [15]).

Transitions between electronic terms give lines in the visible and ultraviolet spectral regions. A list of lines observed has already been given in Sec. 9. McKellar [168] and Herzberg [169] have compiled a long list of lines from a number of molecules, which they have not yet observed, but whose discovery is very probable.

Unfortunately, the majority of the most abundant molecular lines arising during absorption from the ground state occur in the ultraviolet region of the spectrum. The upper limits (Å) of a number of molecular bands [169] are: H_2, 1100; N_2, 1500; CO, 2500; NO, 2300; O_2^+, 2600; O_2, 2400; SiO, 2500. During an electron transition, the rotational and vibrational quantum numbers are usually changed because, during a rapid electron transition, the internuclear distance cannot be appreciably altered, while the potential curve of the upper electron state is different. The transition from a lower to a higher state may be shown as a vertical line in Fig. 76. Therefore, if the molecule was originally at the bottom of a potential well, it moves up the potential curve in the excited state, and this produces a vibration.

There are no selection rules for the vibrational quantum number v and its change is denoted by the parentheses (v, v') in Table 9.1. Generally speaking, the change Δv during an electron transition is arbitrary, but, according to the Franck–Condon rule, a transition is most probable in the vicinity of the turning points of the original state, that is, at the ends of the segment defined by the vibrational term. Physically this means that the vibrational system spends much of its time near the turning points.

If the potential curves for the upper and lower states differ markedly, a molecule after a transition can appear above the value of the potential curve at infinity, which leads to dissociation of the molecule.

A change in the rotational quantum number K satisfies for all transitions the selection rule $\Delta K = 0, \pm 1$, where $K \rightarrow K + 1$ transitions are given the symbol $R(K)$, $K \rightarrow K - 1$ transitions the symbol $P(K)$, and a transition with no change in rotational quantum number is denoted by the symbol $Q(K)$. Since the interval between the rotational terms is small, a change in K leads only to a small splitting of an electron transition.

References

Books and Reviews

1. K. U. Allen, *Astrophysical values* (IL, 1960).
2. V. A. Ambartsumyan, *Scientific works* (Erevan, 1960), 1:2.
3. G. R. Burbidge, F. D. Kahn, R. Ebert, S. v. Hörner, and S. Temesvary, *Die Entstehung von Sternen durch Kondensation diffuser Matter* (Springer Verlag, Berlin, 1960).
4. B. A. Vorontsov-Velyaminov, *Gaseous nebulae and new stars* (Izd. AN SSSR, 1948).
5. V. L. Ginzburg and S. I. Syrovatskii, *The origin of cosmic rays* (Izd. AN SSSR, 1963).
6. G. A. Gurzadyan, *Planetary nebulae* (Fizmatgiz, 1962).
7. V. G. Gorbatskii and I. N. Minin, *Nonfixed stars* (Fizmatgiz, 1963).
8. S. A. Kaplan, *Interstellar gas dynamics* (Fizmatgiz, 1958).
9. D. H. Menzel *et al.*, *Physical processes in gaseous nebulae* (1948).
10. S. B. Pikelner, "Interstellar light polarization," *Usp. Fiz. Nauk 58*, 285 (1956).
11. S. B. Pikelner, *The physics of the interstellar medium* (Izd. AN SSSR, 1959).
12. S. B. Pikelner, *The fundamentals of cosmic electrodynamics* (Fizmatgiz, 1960).
13. J. L. Pawsey and E. R. Hill, "Cosmic radiowaves and their interpretation," *Rept. Progr. Phys. 24*, 69 (1961).
14. M. J. Seaton, "Planetary nebulae," *Rept. Progr. Phys. 23*, 313 (1960).
15. I. S. Shklovskii, *Cosmic radio waves* (Gostekhizdat, 1956; English trans., Harvard University Press, Cambridge, Massachusetts, 1960).
16. *Radioastronomy*. Paris symposium, 1958 (1961).
17. *Gas dynamics of cosmic clouds* (Amsterdam, 1955).
18. *Cosmic gas dynamics*. Third symposium, 1957 (1960).
19. G. A. Shain and V. F. Gaze, *An atlas of diffuse gaseous nebulae* (Izd. AN SSSR, 1952).
20. V. G. Fesenkov and D. A. Rozhkovskii, *An atlas of gas-dust nebulae* (Izd. AN SSSR, 1953).
21. Zh. Sh. Khavtasi, *An atlas of dark galactic nebulae* (Tbilisi, 1960).

Books and Articles on Individual Problems

22. B. Strömgren, "The physical state of interstellar hydrogen," *Ap. J. 89*, 526 (1939).
23. A. B. Underhill, "Four B-type model atmospheres," *Publ. Dominion Astrophys. Obs., Victoria, B.C. 10*, No. 19 (1957).

24. A. B. Underhill, "On model atmospheres for the high-temperature stars," *Publ. Med. Københavns Obs.*, No. 151 (1950).
25. A. B. Underhill, "A model atmosphere for an early O-type star," *Publ. Dominion Astrophys. Obs., Victoria, B.C. 8*, 357 (1951).
26. J. C. Pecker, "Contribution à la théorie du type spectral," *Ann. Astrophys. 13*, 433 (1950).
27. S. Saito, "On the model atmospheres of early-type stars," *Contrib. Inst. Astrophys. Kwasan Obs.*, No. 69, 55 (1956).
28. S. Saito and A. Uesugi, "On the model of the O-type stars," *Publ. Astron. Soc. Japan 11*, 90 (1958).
29. G. Traving, "Die Atmosphäre des B0-Sternes τ Scorpii," *Z. Astrophys. 36*, 1 (1955).
30. G. Traving, "Die Atmosphäre des O9V-Sternes 10 Lac.," *Z. Astrophys. 41*, 215 (1955).
31. E. R. Mustel, "Energy distribution in the continuous spectrum of various classes of stars," *Astron. Zh. 21*, 133 (1944).
32. R. E. Hershberg and V. I. Pronik, "The theory of the Strömgren zone," *Astron. Zh. 36*, 902 (1959).
33. R. N. Ikhsanov, "Certain problems on the interconnection of stars and nebulae and their evolution," *Astron. Zh. 37*, 642 (1960).
33a. G. Münch, "The linear dimensions of H II regions," *Astron. J. 65*, 495 (1960).
34. R. E. Hershberg, "The expansion of H II regions and the formation of peripheral structures in diffuse nebulae," *Izv. Krymsk. Astrofiz. Observ. 25*, 76 (1961).
35. B. Strömgren, "On the density distribution and chemical composition of the interstellar gas," *Ap. J. 108*, 242 (1948).
36. I. S. Shklovskii, *The physics of the solar corona* (Fitzmatgiz, 1962).
37. G. Elwert, "Über die Ionisations- und Rekombinations Prozesse in einem Plasma und die Ionisationsformel der Sonnenkorona," *Z. Naturforsch. 7a*, 432 (1952).
37a. R. E. Williams, "The size of solar H II regions," *Ap. J. 142*, 314 (1965).
37b. A. M. Lenchek, "The solar H II region and the radio spectrum of nonthermal radiation of the Galaxy," *Ann. Astrophys. 27*, 219 (1964).
37c. J. K. Alexander and R. C. Stone, "Rocket measurement of cosmic noise intensities below 5 Mc/s," *Ap. J. 142*, 1327 (1965).
37d. G. R. A. Ellis and P. A. Hamilton, "Cosmic radio noise survey at 4.7 Mc/s," *Ap. J. 143*, 227 (1966).
37e. G. R. A. Ellis and P. A. Hamilton, "Ionized hydrogen in the plane of the Galaxy," *Ap. J. 146*, 28 (1966).
38. H. Zanstra, "Luminosity of planetary nebulae and stellar temperature," *Publ. Dominion Astrophys. Obs., Victoria, B.C. 4*, 209 (1931).
39. H. Zanstra, "Untersuchungen über planetarische Nebel," *Z. Astrophys. 2*, 1 (1931).
40. G. Cillié, "The hydrogen emission in gaseous nebulae," *Monthly Notices Roy. Astron. Soc. 92*, 820 (1932).
41. G. Cillié, "The theoretical capture spectrum of hydrogen," *Monthly Notices Roy. Astron. Soc. 96*, 771 (1936).
42. M. J. Seaton, "The solution of capture-cascade equations for hydrogen," *Monthly Notices Roy. Astron. Soc. 119*, 90 (1959).
43. S. Miyamoto, "On the Balmer emission of the planetary nebulae," *Mem. Coll. Sci. Univ. of Kyoto 21*, 173 (1938).
44. S. Miyamoto, "Collisional emission spectra of cosmic clouds," *Z. Astrophys. 38*, 245 (1956).

45. J. W. Chamberlain, "Collisional exitation of hydrogen in a gaseous nebula," *Ap. J. 117*, 387 (1953).
46. S. A. Kaplan and S. I. Gopasyuk, "The excitation of interstellar hydrogen luminescence by electron collisions," *Tsirk. L'vovsk Astron. Observ.*, No. 25 (1953).
46a. R. Damburg and R. Peterkop, "Excitation of hydrogen by electron collision allowing for exchange and 1*s*-2*s*-2*p* strong coupling," *Proc. Phys. Soc. (London) 80*, 563 (1962).
46b. *Atomic collisions* (Trudy Instituta Fiziki Akademiya Nauk Latviiskoi SSR, Riga, 1963).
46c. P. G. Burke, H. M. Schey, and K. Smith, "Collisions of slow electrons and positrons with atomic H," *Phys. Rev. 129*, 1258 (1963).
46d. G. F. Drukarev, *Theory of collisions of electrons with atoms* (Moscow, 1963).
46e. R. F. Stelling, W. L. Fite, D. G. Hummer, and R. T. Brackman, "Collisions of electrons with hydrogen atoms. Excitation of metastable 2*s* hydrogen atoms," *Phys. Rev. 119*, 1939 (1960).
46f. D. G. Hummer, "The ionization structure of planetary nebulae, II. Collisional cooling of pure hydrogen nebulae," *Monthly Notices Roy. Astron. Soc. 125*, 461 (1963).
46g. R. J. Gould, "Infrared emission lines from H II regions," *Ap. J. 138*, 1308 (1963).
47. E. R. Rynidina, "The Balmer decrement in the spectra of planetary nebulae," *Tr. Astron. Observ. LGU 17*, 18 (1957).
48. L. Searle, "The recombination spectrum of nebular hydrogen," *Ap. J. 128*, 489 (1958).
49. A. Burgess, "The hydrogen recombination spectrum," *Monthly Notices Roy. Astron. Soc. 118*, 477 (1958).
49a. L. C. Green, P. P. Rush, and C. D. Chandler, "Oscillator strengths and matrix elements for the electric dipole moment for hydrogen," *Ap. J. Suppl. Ser. 3*, 37 (1957).
49b. R. M. Pengelly, "Recombination spectra. I. Calculation for hydrogenic ions in the limit of low density," *Monthly Notices Roy. Astron. Soc. 127*, 145 (1964).
49c. R. M. Pengelly and M. J. Seaton, "Recombination spectra. II. Collisional transitions between states of degenerate energy levels," *Monthly Notices Roy. Astron. Soc. 127*, 165 (1964).
49d. M. J. Seaton, "Recombination spectra. III. Populations of highly excited states," *Monthly Notices Roy. Astron. Soc. 127*, 177 (1964).
50. L. Spitzer and J. L. Greenstein, "Continuous emission from planetary nebulae," *Ap. J. 114*, 407 (1951).
51. M. J. Seaton, "Deactivation by collisions of the 2*s* metastable state of hydrogen in planetary nebulae," *Ann. Astrophys. 17*, 296 (1954).
52. D. E. Osterbrock, "On para- and orthohydrogen molecules in interstellar space," *Ap. J. 136*, 359 (1962).
53. S. R. Pottasch, "Balmer decrements: the diffuse nebulae," *Ap. J. 131*, 202 (1960).
54. S. R. Pottasch, "Comments on some physical processes in diffuse nebulae," *Ann. Astrophys. 23*, 749 (1960).
55. A. Ya. Kipper, "The continuous spectrum of planetary nebulae," in *Development of Soviet science in the Estonian SSR, 1940–1950* (Tallin, 1950).
56. M. J. Seaton, "Continuum intensities in planetary nebulae," *Monthly Notices Roy. Astron. Soc. 115*, 279 (1955).
57. A. A. Boyarchuk, R. E. Hershberg, and V. I. Pronik, "Formulae, graphs and nomograms for the quantitative spectral analysis of emitting objects," *Izv. Krymsk. Astrofiz. Observ. 29*, 292 (1963).

58. M. J. Seaton, "H I, He I, and He II intensities in planetary nebulae," *Monthly Notices Roy. Astron. Soc. 120*, 326 (1960).
58a. A. V. Phelps, "Absorption studies of helium metastable atoms and molecules," *Phys. Rev. 99*, 307 (1955).
58b. S. R. Pottasch, "The effect of optical depth in the spectrum of helium (triplets) in nebulae," *Ap. J. 135*, 385 (1962).
58c. C. R. O'Dell, "The interaction of He I and Lyman-radiation," *Ap. J. 142*, 1093 (1965).
59. V. A. Ambartsumyan, "On the radiative equilibrium of a planetary nebula," *Izv. GAO 13*, No. 114, 1 (1933); [2], *1*, 17.
60. H. Zanstra, "Radiation pressure in an expanding nebula," *Monthly Notices Roy. Astron. Soc. 95*, 84 (1934).
61. V. V. Sobolev, *Radiant energy transfer in the atmospheres of stars and planets* (Gostekhizdat, 1956).
62. R. E. Hershberg, "Temperature in the transition region between the H I and H II zones of the interstellar medium," *Izv. Krymsk. Astrofiz. Observ. 26*, 234 (1961).
63. I. S. Shklovskii, "The corpuscular emission of different spectral type stars as a possible source of nebula emission in the 1225–1350 Å region," *Astron. Zh. 36*, 579 (1959).
64. J. E. Milligan and T. P. Stecher, "Stellar spectrophotometry from above the atmosphere," *Ap. J. 136*, 1 (1962).
64a. A. B. Meinel, "On the ultraviolet continuous spectrum of B stars," *Ap. J. 137*, 321 (1963).
64b. J.-C. Pecker, "Effets de l'équilibre des envelopes de poussières circumstellaires," *Compt. Rend. 254*, 821 (1962).
65. L. M. Biberman, "The theory of the diffusion of resonance radiation," *Zh. Eksperim. i Teor. Fiz. 17*, 416 (1957).
66. V. V. Sobolev, "The diffusion of L_α radiation in nebulae and stellar envelopes," *Astron. Zh. 34*, 694 (1957).
67. B. Yada, "Study of an H I-envelope in a galactic nebula—interpretation of the spectrum of the Orion-nebula," *Publ. Astron. Soc. Japan 12*, 449 (1960).
68. H. Zanstra, "On scattering with redistribution and radiation pressure in a stationary nebula. I," *Bull Astron. Inst. Neth. 11*, No. 401 (1949).
69. H. Zanstra, "On radiative equilibrium and radiation pressure in a stationary nebula. II." *Bull. Astron. Inst. Neth. 11*, No. 429 (1951).
70. O. Koelbloed, "An accurate solution of the integral equation for Lyman alpha emission in a stationary nebula," *Bull. Astron. Inst. Neth. 12*, No. 465 (1956).
71. S. Miyamoto, "On the radiation field of the planetary nebulae," *Publ. Astron. Soc. Japan 2*, 23 (1950).
72. W. Unno, "On the radiation pressure in a planetary nebula. I," *Publ. Astron. Soc. Japan 2*, 53 (1950).
73. W. Unno, "On the radiation pressure in a planetary nebula. II," *Publ. Astron. Soc. Japan 3*, 158 (1951).
74. W. Unno, "Note on the Zanstra redistribution in planetary nebulae," *Publ. Astron. Soc. Japan 4*, 100 (1952).
75. V. S. Safronov, "Light pressure on dust and gas in the vicinity of stars of various types," *Vopr. Kosmogonii, Akad. Nauk. SSSR 4*, 87 (1955).
76. G. B. Field, "The time relaxation of a resonance-line profile," *Ap. J. 129*, 557 (1959).
77. V. V. Ivanov, "The diffusion of radiation, redistributed according to frequency, in a one-dimensional case," *Vest. LGU*, No. 19, 117 (1960).

78. V. V. Ivanov, "The diffusion of radiation, redistributed according to frequency, in a semi-infinite medium," *Tr. Astron. Observ. LGU 19*, 52 (1962).
79. V. V. Ivanov, "The diffusion of resonance radiation in the atmospheres of stars and nebulae," *Astron. Zh. 39*, 1020 (1962).
80. L. Spitzer, Jr., *The physics of fully ionized gases* (2nd ed.; Wiley, New York, 1962).
81. V. L. Ginzburg, *Electromagnetic field propagation in plasma* (Fizmatgiz, 1960).
82. A. M. Karachun, A. D. Kuzmin, and A. E. Salomonovich, "Observations of certain discrete sources of radio emission at 3.2 cm," *Astron. Zh. 38*, 83 (1961).
83. C. M. Wade, "On the radio emission of hydrogen nebulae," *Australian J. Phys. 11*, 388 (1958).
84. Yu. N. Pariiskii, "Model of the Orion nebula from radio observations," *Astron. Zh. 38*, 798 (1961).
85. Yu. N. Pariiskii, "The relation between gaseous-nebula radiation in hydrogen lines and in the radio band. A new method of measuring the distance to nebulae," *Izv. GAO 21*, No. 164, 54 (1961).
86. G. Westerhout, "A survey of the continuous radiation from the galactic system at a frequency of 1390 Mc/s," *Bull. Astron. Inst. Neth. 14*, No. 488, 215 (1958).
87. H. C. van de Hulst, "Radiogolven mit hat wereldrium. I. Outvangst der radiogolven, II. Herkomst der radiogolven," *Ned. Tijdschr. Natuurk. 11*, 201, 210 (1945).
88. I. S. Shklovskii, "Monochromatic galactic radio emission and its possible observation," *Astron. Zh. 26*, 10 (1949).
89. H. J. Ewen and E. M. Purcell, "Observations of a line in the galactic radio spectrum. I. Radiation from galactic hydrogen at 1420 Mc/sec," *Nature 168*, 356 (1951).
90. C. A. Muller and J. H. Oort, "Observation of a line in the galactic radio spectrum. II. The interstellar hydrogen line at 1420 Mc/sec and an estimate of galactic rotation," *Nature 168*, 357 (1951).
91. W. N. Christiansen and I. V. Hindman, "A preliminary survey of 1420 Mc/s line emission from galactic hydrogen," *Australian J. Sci. Res. (A) 5*, 437 (1952).
92. E. M. Purcell and G. B. Field, "Influence of collisions upon population of hyperfine states in hydrogen," *Ap. J. 124*, 542 (1956).
93. S. A. Wouthuysen, "On the excitation mechanism of the 21-cm (radio-frequency) interstellar hydrogen emission line," *Astron. J. 57*, 31 (1952).
94. H. C. van de Hulst, C. A. Muller, and J. H. Oort, "The spiral structure of the outer part of the galactic system derived from the hydrogen emission at 21-cm wave length," *Bull. Astron. Inst. Neth. 12*, No. 452 (1954).
94a. W. L. H. Shuter and G. L. Vershur, "A high resolution investigation of 21 cm absorption spectra," *Monthly Notices Roy. Astron. Soc. 127*, 387 (1964).
94b. G. B. Field and R. B. Partridge, "Stimulated emission of the 3.04 cm fine structure line of hydrogen in diffuse nebulae." *Ap. J. 134*, 959 (1961).
95. J. P. Wild, "The radio-frequency line spectrum of atomic hydrogen and its applications in astronomy," *Ap. J. 115*, 206 (1952).
96. N. S. Kardashev, "The possible detection of allowed atomic hydrogen lines in the radio band," *Astron. Zh. 36*, 838 (1959).
96a. Z. V. Dravski and A. F. Dravski, "Attempts to observe radio lines of excited hydrogen," *Astron. Circ.*, No. 282, 2 (1964).
96b. Z. V. Borozdich and R. L. Sorochenko, "Radio lines of excited hydrogen," *Dokl. na 12 Sezde MAS* (1963).
96c. B. Höglund and P. G. Mezger, "Hydrogen emission line detection at 5009 megahertz in galactic H II regions," *Science 150*, 339 (1965).

96d. A. E. Lilley, P. Palmer, M. Penfield, and B. Zuckerman, "Radio astronomical detection of helium," *Nature (London) 211*, 174–175 (1966): Harvard College Observatory Series I, No. 735.
96e. J. P. Hollinger, "Observations of the region of the galactic center at 2.07 centimeters," *Ap. J. 142*, 609 (1965).
96f. D. Downes and A. Maxwell, "Radio observations of the galactic center region," *Ap. J. 146*, 653 (1966).
96g. P. Palmer and B. Zuckerman, "Detection of hydrogen emission line 166 in M 17," *Nature (London) 209*, 1118 (1966).
96h. L. Goldberg, "Stimulated emission of radio-frequency lines of hydrogen," *Ap. J. 144*, 1225 (1966).
97. G. A. Shain and V. F. Gaze, "Some results of a study of luminous galactic nebulae," *Izv. Krymsk. Astrofiz. Observ. 6*, 3 (1951).
98. G. A. Shain and V. F. Gaze, "A special class of diffuse gaseous nebulae," *Izv. Krymsk. Astrofiz. Observ. 7*, 87 (1951).
99. G. A. Shain and V. F. Gaze, "A new system of fine filamentary nebulae in Auriga," *Izv. Krymsk. Astrofiz. Observ. 9*, 123 (1952).
100. G. A. Shain and V. F. Gaze, "The structure and mass of the diffuse gaseous nebulae NGC 6523, 6618, 2237," *Izv. Krymsk. Astrofiz. Observ. 8*, 80 (1952).
101. G. A. Shain and V. F. Gaze, "The mass of gaseous objects in certain extragalactic nebulae," *Izv. Krymsk. Astrofiz. Observ. 9*, 13 (1952).
102. V. F. Gaze and V. G. Shain, "A catalogue of emission nebulae," *Izv. Krymsk. Astrofiz. Observ. 15*, 11 (1955).
103. G. Westerhout, "The distribution of atomic hydrogen in the outer parts of the galactic system," *Bull. Astron. Inst. Neth. 13*, No. 475 (1957).
103a. D. S. Mathewson, J. R. Healey, and J. M. Rome, "A radio survey of the Southern Milky Way at a frequency of 1440 Mc/s. II. The continuum emission from the galactic disk," *Australian J. Phys. 15*, 354 (1962).
103b. R. W. Wilson, *Owens Valley Radio Observatory, Observations*, No. 3 (1963).
104. V. I. Pronik, "A maximum in the galactic thermal radio emission in the direction $l = 353°$, $b = 0°$," *Astron. Zh. 37*, 436 (1960).
105. A. Blaauw, "The velocity distribution of the interstellar calcium clouds," *Bull. Astron. Inst. Neth. 11*, No. 436 (1952).
106. A. Ollongren and H. C. van de Hulst, "Connection of 21 cm line profiles," *Bull. Astron. Inst. Neth. 13*, No. 475, 196 (1957).
107. M. Schmidt, "Spiral structure in the inner parts of the galactic system derived from hydrogen emission at 21 cm wave length," *Bull. Astron. Inst. Neth. 13*, No. 475, 247 (1957).
108. K. K. Kwee, C. A. Muller, and G. Westerhout, "The rotation of the inner parts of the galactic system," *Bull. Astron. Inst. Neth. 12*, No. 458, 211 (1954).
108a. T. A. Agekyan and E. V. Klosovskaya, "The determination of the galactic rotation rule from radio observation data," *Vestn. LGU*, No. 3, 3 (1962).
108b. T. A. Agekyan, I. V. Petrovskaya, and B. I. Fesenko, "Rotation of the Galaxy from radio observations," *Astron. Zh. 51*, 1027 (1964).
109. G. W. Rougoor and J. H. Oort, "Distribution and motion of interstellar hydrogen in the galactic system with particular reference to the region within 3 kpc of the center," *Proc. Nat. Acad. Sci. U.S.A. 46*, 1 (1960).
109a. W. W. Shane and G. P. Bieger-Smith, "The galactic rotation curve derived from observations of neutral hydrogen," *Bull. Astron. Inst. Neth. 18*, 263 (1966).
109b. I. I. Pronik, "On the possible deviation of circular motion of hydrogen clouds in outer spiral arms of the Galaxy," *Astron. J. USSR 42*, 923 (1965).

110. M. Schmidt, "A model of the distribution of mass in the galactic system," *Bull. Astron. Inst. Neth. 13*, 15 (1956).
111. C. S. Gum, F. J. Kerr, and G. Westerhout, "A 21-cm determination of the principal plane of the Galaxy. II," *Monthly Notices Roy. Astron. Soc. 121*, 132 (1960).
112. T. A. Lozinskaya and N. S. Kardashev, "Deformation of the gaseous disk of the Galaxy," *Astron. Zh. 39*, 5 (1962).
113. T. A. Lozinskaya and N. S. Kardashev, "Thickness of the gaseous galactic disk from observations in the 21-cm line," *Astron. Zh. 40*, 209 (1963).
114. F. J. Kerr, "Galactic velocity models and the interpretation of 21-cm surveys," *Monthly Notices Roy. Astron. Soc. 123*, 327 (1962).
115. B. J. Bok, "The spiral structure of our Galaxy," *Observatory 79*, 58 (1959).
115a. J. Isserstedt and Th. Schmidt-Kaler, "Interstellare Verfärbung und galaktische Spiralstruktur," *Z. Astrophys. 59*, 182 (1964).
116. J. H. Oort, F. J. Kerr, and G. Westerhout, "The galactic system as a spiral nebula," *Monthly Notices Roy. Astron. Soc. 118*, 379 (1958).
117. I. L. Genkin, "The K-effect and the galactic spiral structure," *Astron. Zh. 39*, 15 (1962).
117a. L. L. E. Braes, "A test for general expansion or contraction of the hydrogen in the galactic disk," *Bull. Astron. Inst. Neth. 17*, 132 (1963).
118. P. X. McGee and J. D. Murray, "A sky survey of neutral hydrogen at 21 cm. I. The general distribution and motions of the local gas," *Australian J. Phys. 14*, 260 (1961).
118a. K. Takakubo, "On the neutral hydrogen in the solar neighbourhood," *Sci. Rept. Tohoku Univ., First Ser. 47*, 65 (1963).
118b. R. Terauti, "Models of the interstellar cloud derived from 21 cm line emission," *Sci. Rept. Tohoku Univ., First Ser. 47*, 114 (1963).
118c. H. Scheffler, "Interstellar absorption H II," *Z. Astrophys. 63*, 267 (1966).
118d. G. B. Field and W. G. Saslaw, "Statistical model of the formation of stars and interstellar clouds," *Ap. J. 142*, 568 (1965).
119. *Annual Report, National Radio Astronomy Observatory*, Green Bank, U.S. (1959).
119a. G. Münch and H. Zirin, "Interstellar matter at large distances from the galactic plane," *Ap. J. 133*, 11 (1961).
120. F. D. Drake, "A high-resolution radio study of the galactic center," *Astron. J. 64*, 329 (1959).
121. Yu. N. Pariiskii, "Observation of the radio source Sagittarius A, at high resolving power," *Dokl. Akad. Nauk SSSR 129*, 126 (1959).
122. Yu. N. Pariiskii, "Observations on certain galactic radio emission sources on the large Pulkovo radio telescope," *Izv. GAO 21*, 45 (1960).
123. Yu. N. Pariiskii, "The structure of the galactic nucleus," *Astron. Zh. 38*, 242 (1961).
124. H. van de Woerden, W. Rougoor, and J. H. Oort, "Expansion d'une structure spirale dans le noyau du Système Galactique et position de la radiosource Sagittarius A," *Compt. Rend. 224*, 1691 (1957).
125. N. F. Ryzhkov, T. M. Egovova, I. V. Gosachinskii, and N. V. Bystrova, "The absorption by interstellar neutral hydrogen of emission from the Sagittarius A source," *Astron. Zh. 40*, 17 (1963).
126. G. Courtes and P. Cruvellier, "Mesures interférentielles de la vitesse radiale des régions H II," *Compt. Rend. 251*, 2470 (1960).
127. G. Field, "An attempt to observe neutral hydrogen between the galaxies," *Ap. J. 129*, 525 (1959).
128. G. Field, "The spin temperature of intergalactic neutral hydrogen," *Ap. J. 129*, 536 (1959).

129. G. Field, "Absorption by intergalactic hydrogen," *Ap. J. 135*, 684 (1962).
129a. S. J. Goldstein, "An attempt to observe 21-cm line emission," *Ap. J. 138*, 978 (1963).
129b. I. S. Shklovskii, "New method for estimating accurately the intergalactic gas," *Astron. Circ.* No. 303, 3 (1964).
130. H. Lambrecht, "A new calculation of the interstellar radiation field," in *Les particles solides dans les astres* (Conte, Liège, 1955), p. 562.
131. S. A. Kaplan, "The energy of over-all stellar emission," *Astron. Zh. 29*, 649 (1952).
131a. S. A. Kaplan, "Some dynamical problems in the interstellar medium," *Tr. Astrophys. Inst. Akad. Nauk Kaz. SSR* (1964).
132. P. P. Pavenago, *A course in stellar astronomy* (Gostekhizdat, 1954).
132a. T. P. Stecher and J. E. Milligan, "An interim interstellar radiation field," *Ann. Astrophys. 25*, 268 (1962).
133. S. B. Pikelner, "Helium ionization in nebulae and the temperature of O stars," *Izv. Krymsk. Astrofiz. Observ. 10*, 183 (1953).
134. V. I. Pronik, "Diffuse nebulae and various star models," *Astron Zhurn. 37*, 1001 (1960).
135. V. I. Pronik, "The electron temperature, density and mass of nebula NGC 6523," *Izv. Krymsk. Astrofiz. Observ. 23*, 3 (1960).
135a. G. R. Burbidge, R. J. Gould, and S. R. Pottasch, "Excitation conditions in H II regions in spiral and irregular galaxies," *Ap. J. 138*, 945 (1963).
136. L. M. Biberman, G. E. Norman, and K. N. Ulyanov, "The photoionization of excited multielectron atoms and ions," *Astron Zh. 39*, 107 (1962).
137. M. J. Seaton, "The kinetic temperature of the interstellar gas in the regions of neutral hydrogen," *Ann. Astrophys. 18*, 188 (1955).
138. S. A. Kaplan and T. T. Tsap, "Ionization functions of the elements C I, Na I, K I, Ca I, Ca II in interstellar space," *Astron. Circ.*, No. 137, 6 (1953).
139. A. Burgess, G. B. Field, and R. W. Michie, "On the possibility of observing interstellar aluminum," *Ap. J. 131*, 529 (1960).
140. V. I. Pronik, "The determination of the temperature of diffuse gaseous nebulae," *Izv. Krymsk. Astrofiz. Observ. 17*, 14 (1957).
140a. D. C. Morton and L. Spitzer, "Spectra of Delta and Pi Scorpii in the far ultraviolet," *Ap. J. 144*, 1 (1966).
141. C. S. Beals, "Report on the progress of astronomy: Interstellar matter," *Monthly Notices Roy. Astron. Soc. 102*, 96 (1942).
142. W. S. Adams, "Observations of interstellar H and K, molecular lines and radial velocities in the spectra of 300 O and B stars," *Ap. J. 109*, 354 (1949).
143. G. Münch, "Interstellar absorption lines in distant stars. I. Northern Milky Way," *Ap. J. 125*, 42 (1957).
143a. K. Takakubo, "Internal motions within interstellar clouds," *Sci. Rept. Tohoku Univ., First Ser. 47*, 108 (1963).
144. S. A. Kaplan, "Turbulent density fluctuations in interstellar space," *Uch. Zap. L'vovskovo Univ. Astron. Sbornik*, No. 2, 53 (1954).
145. S. A. Kaplan, "The absorption lines of interstellar gas," *Tsirk. L'vovsk. Astron. Observ.*, No. 33, 1 (1957).
146. O. C. Wilson and P. W. Merrill, "Analysis of the intensities of the interstellar D lines," *Ap. J. 86*, 44 (1937).
147. C. Ientsch and A. Unsold, "Zur Deutung der interstellaren Kalzium und Natrium Linien," *Z. Astrophys. 125*, 370 (1948).
147a. W. E. Howard, D. G. Wentzel, and R. X. McGee, "On correlation between the radial velocities of optical and radio interstellar lines," *Ap. J. 138*, 989 (1963).

148. I. S. Bowen, "The origin of the chief nebular lines," *Publ. Astron. Soc. Pacific 39*, 295 (1927).
149. I. S. Bowen, "The origin of the nebular lines and the structure of the planetary nebulae," *Ap. J. 67*, 1 (1928).
150. R. H. Garstang, "Energy levels and transition probabilities in p^2 and p^4 configurations," *Monthly Notices Roy. Astron. Soc. 111*, 115 (1957).
151. R. H. Garstang, "Transition probabilities for forbidden lines of Ne IV," *Monthly Notices Roy. Astron. Soc. 120*, 201 (1960).
152. R. H. Garstang, "Multiplet intensities for the lines ^4S–^2D of S II, O II and N II," *Ap. J. 115*, 506 (1952).
153. S. Pasternak, "Transition probabilities of forbidden lines," *Ap. J. 92*, 129 (1940).
154. A. M. Naqvi and S. P. Talwar, "The effect of departures from thermodynamic equilibrium on the intensity ratio of the ^2D–^4S doublet in gaseous nebulae," *Monthly Notices Roy. Astron Soc. 117*, 463 (1957).
154a. S. J. Czysak and T. K. Krueger, *Monthly Notices Roy. Astron. Soc. 126*, 177 (1963).
155. V. A. Ambartsumyan, *Theoretical astrophysics* (Gostekhizdat, 1939).
156. M. J. Seaton and D. E. Osterbrock, "Relative [O II] intensities in gaseous nebulae," *Ap. J. 125*, 66 (1957).
157. L. H. Aller, *Abundance of the elements* (Wiley, New York, 1961).
158. V. V. Sobolev, "The determination of electron temperatures of planetary nebulae and increased precision in the nebular method for determining the temperatures of their nuclei," *Tr. Astron. Observ. LGU 12*, 3 (1941).
159. L. Spitzer, "The temperature of interstellar matter," *Ap. J. 107*, 6 (1948).
160. L. Spitzer, "The temperature of interstellar matter. II," *Ap. J. 109*, 337 (1948).
161. L. Spitzer and M. P. Savedoff, "The temperature of interstellar matter. III," *Ap. J. 111*, 593 (1950).
162. L. Spitzer, "Behavior of matter in space," *Ap. J. 120*, 1 (1954).
163. M. J. Seaton, "Electron temperatures and electron densities in planetary nebulae," *Monthly Notices Roy. Astron. Soc. 114*, 154 (1954).
164. M. J. Seaton, "Radiative recombination of hydrogenic ions," *Monthly Notices Roy. Astron. Soc. 119*, 81 (1959).
164a. D. E. Osterbrock, "Temperature in H II regions and planetary nebulae," *Ap. J. 142*, 1423 (1965).
164b. R. M.Hjellming, "Physical processes in H II regions," *Ap. J. 143*, 420 (1966).
164c. C. R. O'Dell, "Electron temperature derived from observation of low-density H II regions and planetary nebulae," *Ap. J. 143*, 168 (1966).
165. B. G. Clark, V. Radhakrishnan, and R. W. Wilson, "The hydrogen line absorption," *Ap. J. 135*, 151 (1962).
166. K. Takayanagi and S. Nishimura, "Cooling of the interstellar clouds in region of neutral hydrogen," *Publ. Astron. Soc. Japan 12*, 77 (1960).
166a. W. L. H. Shuter and G. L. Verschuur, "A high resolution investigation of 21 cm absorption," *Monthly Notices Roy. Astron. Soc. 127*, 387 (1964).
166b. S. Marx, "Zur Temperatur des interstellar H I Gases," *Astron. Nachr. 288*, 155 (1965).
166c. S. B. Pikelner, "Interstellar clouds and magnetic fields," *Astron. J. USSR 44*, N 5 (1967).
166d. K. Takayanagi, *Joint Inst. Lab. Astroph. Publ., No. 19* (1964).
166e. S. B. Pikelner, *Astron. J. USSR 44*, 915 (1967); *Astrophys. Letters 1*, 43 (1967).
166f. G. B. Field in *The distribution and motion of interstellar matter in galaxies*, ed. L. Woltjer (New York, 1962).

166g. G. R. Ellis and P. S. Hamilton, *Ap. J. 143*, 227 (1966).

166h. J. K. Alexander and R. G. Stone, *Ap. J. 142*, 1327 (1965).

167. T. Dunham, "The concentration of interstellar molecules," *Publ. Amer. Astron. Soc. 10*, 123 (1941).

168. D. R. Bates and L. Spitzer, "The density of molecules in interstellar space," *Ap. J. 113*, 441 (1951).

168a. G. Münch, "An interstellar cloud with a high concentration of CN molecules," *Ap. J. 140*, 107 (1964).

168b. J. M. Greenberg and P. R. Lichtenstein, "Absorption lines produced in interstellar grains," *Ap. J. 68*, 279 (1963).

168c. G. H. Herbig, "The diffuse interstellar band. I. A possible identification of λ4430," *Ap. J. 137*, 200 (1963).

169. G. Herzberg, "Laboratory investigation of the spectra of interstellar and cometary molecules," in *Les particles solide dans les astres* (Conte, Liège, 1955), p. 291.

169a. G. B. Field and J. L. Hitchcock, "Cosmic black-body radiation at $\lambda = 2.8$ mm," *Phys. Rev. Letters 16*, 817 (1966).

169b. I. S. Shklovskii, "Relict radiation in the universe and the population of rotation levels of interstellar molecules," *Astron. Circ. USSR*, No. 364 (1966).

170. W. H. McCrea and D. McNally, "The formation of population I stars. Part II. The formation of molecular hydrogen in interstellar matter," *Monthly Notices Roy. Astron. Soc. 121*, 238 (1960).

170a. R. J. Gould and E. E. Salpeter, "The interstellar abundance of the hydrogen molecule. I. Basic process," *Ap. J. 138*, 393 (1963).

170b. R. J. Gould, T. Gold, and E. E. Salpeter, "The interstellar abundance of the hydrogen molecule. II. Galactic abundance and distribution," *Ap. J. 138*, 408 (1963).

170c. R. J. Gould and M. Harwit, "Expected near infrared radiation from interstellar molecular hydrogen," *Ap. J. 137*, 694 (1963).

171. B. J. Bok, "Gas and dust in interstellar clouds," *Astron. J. 60*, 146 (1955).

171a. S. Weinreb, A. H. Barrett, M. L. Meeks, and J. C. Henry, "Radio observations of OH in the interstellar medium," *Nature (London) 200*, 829 (1963).

171b. J. G. Bolton, K. J. van Damme, F. F. Gardner, and B. J. Robinson, "Observation of OH absorption lines in the radio spectrum of the galactic centre," *Nature (London) 201*, 279 (1964).

171c. N. H. Dieter and H. J. Ewen, "Radio observations of the interstellar OH line at 1.667 Mc/s," *Nature (London) 201*, 279 (1964).

171d. H. F. Weaver and D. R. W. Williams, "OH absorption profile in the direction of Sagittarius A," *Nature (London) 201*, 380 (1964).

171e. A. H. Barrett, M. L. Meeks, and S. Weinreb, "High resolution microwave spectra of H and OH absorption lines of Cassiopeia A," *Nature (London) 202*, 475 (1964).

171f. J. C. Raich and R. H. Good, Jr., "Ortho-para transition in molecular hydrogen," *Ap. J. 139*, 1004 (1964).

171g. L. Spitzer, K. Dressler, and W. L. Upson, "Theoretical equivalent widths of the interstellar H_2 lines," *Publ. Astron. Soc. Pacific 76*, 387 (1964).

171h. I. V. Gosachinsky, "Relative abundance of gas and dust in the direction of the nebula NGC 6618 (Omega)," *Astron. Zh. 42*, 929 (1965).

171i. J. Dorschner, J. Gurtler, and K. H. Schmidt, "Zur Beschleunigung senkrecht zur galaktischen Ebene und zur Haufigkeit des interstellaren H_2 Molekules," *Astron. Nachr. 288*, 149 (1965).

171j. H. F. P. Knaap, C. J. N. van den Meijdenberg, J. J. M. Ebeenakker, and H. C. van der Hulst, "Formation of molecular hydrogen in interstellar space," *Bull. Astron. Inst. Neth. 18*, 256 (1966).

171k. T. P. Stecher and D. A. Williams, "On the interstellar H_2 abundance," *Publ. Astron. Soc. Pacific 78*, 76 (1966).

171l. H. F. Weaver, D. R. W. Williams, N. H. Dieter, and W. T. Lum, "Observations of a strong unidentified microwave line and of emission from the OH molecule," *Nature (London) 208*, 29 (1965).

171m. S. Weinreb, M. I. Meeks, J. C. Carter, A. H. Barrett, and A. C. E. Rogers, "OH emission in the direction of radio source W49," *Nature (London) 208*, 440 (1965).

171n. F. Perkins, T. Gold, and E. E. Salpeter, "Maser action in interstellar OH," *Ap. J. 145*, 361 (1966).

171o. I. S. R. Shklovskii, "Emission of 'misterium' as a mazer effect," *Astron. Circ. USSR*, No. 372 (1966).

171p. M. M. Goss and H. Spinrad, "A search for the ground-state interstellar optical lines of OH molecules in the direction of W750," *Ap. J. 143*, 989 (1966).

172. G. A. Shain, "Diffuse nebulae and an interstellar magnetic field," *Astron. Zh. 32*, 110 (1955).

173. G. A. Shain and V. F. Gaze, "Diffuse nebulae with a peripheral concentration of matter and their interpretation," *Astron. Zh. 30*, 135 (1953).

174. V. G. Fesenkov, "Some structural features of gaseous nebulae and their association with stars," *Astron. Zh. 28*, 215 (1951).

175. T. K. Menon, "Interstellar structure of the Orion region I," *Ap. J. 127*, 28 (1958).

176. A. Blaauw, "On the origin of the O and B stars with high velocities and some related problems," *Bull. Astron. Inst. Neth. 15*, 265 (1961).

176a. P. O. van der Voort, "The age of the Orion Nebula," *Ap. J. 138*, 312 (1963).

177. H. Rishbeth, "Radio emission from Orion," *Monthly Notices Roy. Astron. Soc. 118*, 591 (1958).

178. O. Osterbrock and E. Flather, "Electron densities in the Orion nebula II,," *Ap. J. 129*, 26 (1959).

179. T. K. Menon, "A model of the Orion nebula derived from radio observations," *Publ. Nat. Radio Astron. Observ. 1*, No. 1 (1961).

180. L. H. Aller and W. Liller, "Photoelectric spectrophotometry of gaseous nebulae. I. The Orion nebula," *Ap. J. 130*, 45 (1959).

181. R. M. Sloanaker and J. H. Nichols, "Positions, intensities and sizes of bright celestial sources at a wavelength of 10.2 cm," *Astron. J. 65*, 109 (1960).

181a. A. G. Little, "Observation at 9.1 cm of Tau A, the Orion Nebula, Virgo A, Cen A, Sag A and the Omega Nebula with a 2.′3 fan beam," *Ap. J. 137*, 169 (1963).

182. R. E. Hershberg, V. F. Esipov, V. I. Pronik, and P. V. Shcheglov, "A study of the emission nebula NGC 6618," *Izv. Krymsk. Astrofiz. Observ. 26*, 313 (1961).

183. R. E. Hershberg and V. I. Pronik, "Absolute photometry of the emission nebula NGC 6618," *Izv. Krymsk. Astrofiz. Observ. 26*, 363 (1961).

183a. R. E. Hershberg and P. V. Shcheglov, "Investigations of rapid internal radial motions in diffuse nebulae by means of the Fabry-Perot étalon," *Astron. Zh. 41*, 425 (1964).

183b. R. E. Hershberg, "Investigation of NGC 6618. Eruption of supernovae in diffuse nebulae," *Izv. Krymsk. Astrofiz. Observ. 30*, 90 (1963).

184. G. A. Shain and V. F. Gaze, "Probable motions in emission nebulae," *Astron. Zh. 31*, 305 (1954).

185. G. A. Shain and V. F. Gaze, "Motions in diffuse emission nebulae," *Dokl. Akad. Nauk SSSR 96*, 945 (1954).

186. A. G. Velghe, "The bright knots in the gaseous nebula M8," *Ap. J. 125*, 822 (1957).

187. D. A. Rozhkovskii, "Luminous hoops in diffuse nebulae and their association with central stars," *Astron. Zh. 31*, 318 (1954).

188. R. E. Hershberg and V. I. Pronik, "The absolute spectrophotometry of nebula NGC 7000 using light filters," *Izv. Krymsk. Astrofiz. Observ. 21*, 215 (1959).
189. R. J. Trumpler, "Absorption of light in the galactic system (spectrophotometric measures of interstellar light absorption)," *Publ. Astron. Soc. Pacific 42*, 267 (1930).
190. J. Stebbins and A. E. Whitford, "Six-color photometry of stars, I. The law of space reddening from the colors of O and B stars," *Ap. J. 98*, 20 (1943).
191. A. E. Whitford, "The law of interstellar reddening," *Astron. J. 63*, 201 (1958).
191a. A. Boggess and J. Borgman, "Interstellar extinction in the middle ultraviolet" (preprint, 1964).
192. V. I. Moroz, "An attempt to observe the infrared emission of the galactic nucleus," *Astron. Zh. 38*, 487 (1961).
193. G. A. Melnikov, "The spectrophotometry of the stars of δ Cephei, η Aquilae, and the K-effect for the Cepheids," *Tr. GAO 64* (1950).
194. V. M. Blanco and C. J. Lenon, "The ratio of total to selective absorption," *Astron. J. 66*, 524 (1961).
194a. E. J. Wampler, "Systematic variations in the slope of the correlation between the intensity of λ4430 and color excess," *Ap. J. 137*, 1071 (1963).
194b. R. Wilson, "The relation between interstellar extinction and polarization," *Monthly Notices Roy. Astron. Soc. 120*, 51 (1960).
194c. J. D. Fernie and J. M. Marlborough, "The variation of the total to selective absorption ratio with galactic longitude," *Ap. J. 137*, 700 (1963).
194d. G. M. Aivazyan, "Complex refractive index of particles in the interstellar medium," *Astron. Zh. 42* (1965).
195. H. C. van de Hulst, "Optics of spherical particles," *Rech. Astron. Observ. Utrecht 11*, Part I (1946), Part II (1949).
195a. J. E. Gaustad, "The opacity of diffuse cosmic matter and the early stage of star formation," *Ap. J. 138*, 1050 (1963).
196. H. C. van de Hulst, *Light scattering by small particles* (Wiley, New York, 1957).
197. C. Schalen, "Uber Probleme der interstellaren Absorption," *Medd. Astron. Observ. Uppsala 64*, (1936).
198. C. Schalen, "Beiträge zur Theorie der interstellaren Absorption," *Ann. Astron. Observ. Uppsala 1*, No. 2 (1939).
199. J. M. Greenberg, "Scattering by nonspherical particles," *J. Appl. Phys. 31*, 82 (1960).
200. J. H. Oort and H. C. van de Hulst, "Gas and smoke in interstellar space," *Bull. Astron. Inst. Neth. 10*, No. 376 (1946).
201. S. A. Kaplan, "An elementary method of determining the physical characteristics of cosmic dust," *Tsirk. L'vovsk. Astron. Observ.*, No. 23 (1952).
202. E. Schoenberg and B. Jung, "Uber die Lichtzerstreuung in interstellaren Raum durch Wolken metallischer Partikel," *Astron. Nachr. 253*, 261 (1934).
203. A. Guttler, "Uber die Materie der interstellaren Körner," *Z. Astrophys. 31*, 1 (1952).
204. V. A. Ambertsumyan, "Fluctuations in the number of extragalactic nebulae and galactic absorption," *Byul. Abastumansk. Observ.*, No. 4, 17 (1940).
205. V. A. Ambartsumyan, "Toward a theory on brightness fluctuations in the Milky Way," *Dokl. Akad. Nauk SSSR 44*, 244 (1944).
206. S. Chandrasekhar and G. Münch, "The theory of the fluctuations in brightness of the Milky Way. I," *Ap. J. 112*, 380 (1950).
207. S. Chandrasekhar and G. Münch, "The theory of the fluctuations in brightness of the Milky Way. II," *Ap. J. 112*, 393 (1950).

208. S. Chandrasekhar and G. Münch, "The theory of the fluctuations in brightness of the Milky Way. III," *Ap. J. 114*, 110 (1951).
209. S. Chandrasekhar and G. Münch, "The theory of the fluctuations in brightness of the Milky Way. IV," *Ap. J. 115*, 94 (1952).
210. S. Chandrasekhar and G. Münch, "The theory of the fluctuations in brightness of the Milky Way. V," *Ap. J. 115*, 103 (1952).
211. G. Münch, "The theory of the fluctuations in brightness of the Milky Way. VI," *Ap. J. 121*, 291 (1955).
212. T. A. Agekyan, "Fluctuations in the apparent distribution of stars," *Astron. Zh. 32*, 416 (1955).
213. T. A. Agekyan, "A theory of the fluctuations of the number of observed galaxies," *Astron. Zh. 34*, 371 (1957).
214. B. E. Markaryan, "Fluctuations in the apparent distribution of stars and cosmic absorption," *Soobshch. Byurakansk. Observ., Akad. Nauk Arm. SSR,* I (1946).
215. G. I. Rusakov, "Fluctuations of brightness in the Milky Way and the physical characteristics of diffuse nebulae," *Tr. Astron. Observ. LGU 13*, 53 (1949).
216. K. Serkowski, "Statistical analysis of the polarization and reddening of the double cluster in Perseus," *Acta Astron. 8*, 135 (1958).
217. *Problems of cosmic aerodynamics* (IL, 1953).
218. B. A. Vorontsov-Velyaminov, "The distribution of supergiants and of dust in M33 and their association," *Astron. Zh. 32*, 401 (1955).
219. B. J. Bok, *The analysis of star counts*, Harvard Coll. Observ. Circ. No. 371 (1932).
220. B. J. Bok, *Irregularities in the galactic absorbing layers*, Harvard Coll. Observ. Bull. No. 920 (1951).
221. I. I. Pronik, "A study of interstellar light absorption in the region centered on $l = 343°$, $b = 0°$," *Izv. Krymsk. Astrofiz. Observ. 22*, 152 (1960).
222. I. I. Pronik, "Photographic values and light indexes of 79 early O–B2 stars in the area centered on $a = 18^h54^m$, $\delta = +5\overset{\circ}{.}0$," *Izv. Krymsk. Astrofiz. Observ. 25*, 37 (1961).
223. I. I. Pronik, "The question of the structure of the Orion arm," *Astron. Zh. 39*, 362 (1962).
224. A. B. Numerova, "A study of interstellar absorption in the constellation Cygnus in the area centered on $a = 20^h04^m$, $\delta = +36°$," *Izv. Krymsk. Astrofiz. Observ. 25*, 46 (1961).
225. R. N. Ikhsanov, "A study of absorption in an area of the Milky Way centered on $a = 20^h16^m$, $\delta = +42°\,30'$," *Izv. Krymsk. Astrofiz. Observ. 21*, 257 (1959).
226. R. N. Ikhsanov, "Nebulae IC 1318a, b, c and interstellar absorption," *Astron. Zh. 37*, 275 (1960).
227. L. P. Metik, "The spatial distribution of O–B2 stars and absorbing material in the constellation Cygnus (center $a = 20^h44^m$, $\delta = 45°$, 1950)," *Izv. Krymsk. Astrofiz. Observ. 26*, 386 (1961).
228. V. A. Dombrovskii, "The distribution of stars, of dust and interstellar polarization in Cygnus," *Leningradsk. Inst.*, No. 7, 142 (1961).
229. E. S. Brodskaya, "A study of interstellar absorption in Cassiopeia," *Izv. Krymsk. Astrofiz. Observ. 16*, 162 (1956).
230. E. S. Brodskaya, "The distribution of absorbing material near the galactic equator in a region of longitude 91–107°," *Izv. Krymsk. Astrofiz. Observ. 26*, 375 (1961).
231. L. Binnendijk, "The space distribution of interstellar material in the Milky Way," *Ap. J. 115*, 428 (1952).
232. A. E. Lilley, "Association of gas and dust from 21-cm hydrogen radio observations," *Ap. J. 121*, 559 (1955).

233. H. L. Helfer and H. E. Tatel, "A 21-cm survey around the Pleiades," *Ap. J. 129*, 565 (1959).
234. V. A. Abartsumyan and Sh. G. Gordeladze, "Problem of diffuse nebulae and cosmic absorption," *Bull. Abastumansk. Astron. Observ.*, No. 2, 37 (1938).
235. V. A. Ambartsumyan, "The question of the nature of the association between diffuse nebulae and illuminating stars," *Dokl. Akad. Nauk Arm. SSR 2*, 67 (1945).
236. D. A. Rozhkovskii, "A photometric study of reflecting nebulae, I, II," *Izv. Astrofiz. Inst. Akad. Nauk Kaz. SSR 10*, 15 (1960); *13*, 17 (1961).
237. J. Dufay, *Galactic nebulae and interstellar matter* (London, 1957).
238. J. L. Greenstein, "The colors of reflection nebulae," *Ap. J. 87*, 581 (1938).
238a. G. Münch and H. Zirin, "Interstellar matter at large distances from the galactic plane," *Ap. J. 133*, 11 (1961).
238b. G. A. Müller, J. H. Oort, and E. Raimond, "Hydrogène neutre dans la couronne galactique?," *Compt. Rend. 257*, 1661 (1963).
238c. A. Blaauw, A. N. M. Hulsbosch, G. A. Müller, J. H. Oort, E. Raimond, and C. M. Tolbert, "21-cm line observations at high galactic latitudes," *Bull. Astron. Inst. Neth.* (1965).
239. L. G. Henyey, "On the polarization of light in reflection nebulae," *Ap. J. 84*, 609 (1936).
240. M.-T. Martel, "Polarisation de la lumière diffusée de NGC 7023," *Compt. Rend. 232*, 2183 (1951).
241. L. G. Henyey, "The illumination of reflection nebulae," *Ap. J. 85*, 107 (1937).
242. C. Schalen, "Studies of reflection nebulae," *Uppsala Astron. Obs. Ann. 1*, No. 9 (1945).
243. V. V. Sobolev, "The luminescence of spherical nebulae," *Astron. Zh. 37*, 3 (1960).
243a. I. N. Minin, "The optical properties of dust nebulae," *Astron. Zh. 38*, 641 (1961).
243b. I. N. Minin, "Light scattering by nebular particles," *Astron. Zh. 41*, 662 (1964).
244. G. A. Shain, V. F. Gaze, and S. B. Pikelner, "The presence of dust and gas in diffuse nebulae," *Izv. Krymsk. Astrofiz. Observ. 12*, 64 (1954).
245. C. T. Elwey and F. E. Roach, "A photoelectric study of the light from the night sky," *Ap. J. 85*, 213 (1937).
246. L. G. Henyey and J. L. Greenstein, "Diffuse radiation in the Galaxy," *Ann. Astrophys. 3*, 117 (1940).
247. Wang Shih-Ky, "Diffusion de la lumière dans la Galaxie," *Compt. Rend. 201*, 1326 (1935).
248. Wang Shih-Ky, "Recherches sur la diffusion de la lumière dans la Voie lactée," *Publ. Obs. Lyon 1*, No. 19 (1936).
249. S. A. Kaplan, "Light reflection from dust nebulae," *Astron. Zh. 19*, 326 (1952).
250. S. A. Kaplan and I. A. Klimishin, "Light scattering in spherical nebulae," *Tsirk. L'vovsk. Astron. Observ.* No. 27, 11 (1953).
251. R. E. Hershberg, "The fine filamentary structure and polarization of reflecting nebulae," *Izv. Krymsk. Astrofiz. Observ. 23*, 21 (1960).
251a. S. B. Pikelner, "Spiral arms and interacting galaxies," *Astron. J. USSR 42*, 515 (1965).
252. V. F. Gaze, "The structure of nebula IC 405 in H_α-radiation and in integral light," *Izv. Krymsk. Astrofiz. Observ. 10*, 213 (1953).
253. G. H. Herbig, "The spectrum of the nebulosity at AE Aurigae," *Publ. Astron. Soc. Pacific 70*, 468 (1958).
254. D. A. Rozhkovskii, "Light polarization in the Trifid Nebula," *Izv. Astrofiz. Inst. Akad. Nauk Kaz. SSR 12*, 21 (1961).

255. W. A. Hiltner, "On the presence of polarization in the continuous radiation of stars," *Ap. J. 109*, 471 (1949).
256. W. A. Hiltner, "Polarization of stellar radiation. III. The polarization of 841 stars," *Ap. J. 114*, 241 (1951).
257. J. S. Hall, "Observations of the polarized light from stars," *Science 109*, 166 (1949).
258. V. A. Dombrovskii, "A study of stellar polarization in the Cepheus association I," *Dokl. Akad. Nauk Arm. SSR 12*, 4, 103 (1950).
259. L. O. Loden, "A polarimetric investigation of 1300 stars," *Stockholm Observ. Ann. 21*, No. 73 (1961).
260. L. O. Loden, "Contribution to the discussion of interstellar polarization," *Stockholm Observ. Ann. 22*, No. 3 (1961).
261. T. Gehrels, "The wavelength dependence of polarization, II. Interstellar polarization," *Astron. J. 65*, 470 (1960).
262. A. Behr, "Beobachtungen zur Wellenlangenabhängigkeit der interstellaren Polarization," *Z. Astrophys. 47*, 54 (1959).
263. G. A. Shain, "Orientation of the general galactic magnetic field in the solar neighborhood, relative to the galactic equator," *Astron. Zh. 34*, 3 (1957).
264. H. C. van de Hulst, "The amount of polarization by interstellar grains," *Ap. J. 112*, 1 (1950).
264a. J. M. Greenberg, A. C. Lind, R. T. Wang, and L. F. Lilelo, "The polarization of starlight by oriented nonspherical particles," *ICES Electr. Scat.* (Pergamon Press, New York, 1963), p. 123.
265. J. M. Greenberg and A. S. Meltzer, "The effect of orientation of non-spherical particles on interstellar extinction," *Ap. J. 132*, 667 (1960).
266. E. J. Wampler, "An investigation of the dependence of interstellar reddening on galactic longitude," *Ap. J. 134*, 861 (1961).
266a. E. J. Wampler, "Observed variations in the law of interstellar reddening," *Ap. J. 136*, 100 (1962).
266b. K. Serkowski, "Slopes of the reddening trajectories and intrinsic colors of early-type stars," *Ap. J. 138*, 1035 (1963).
267. J. Borgman and H. L. Johnson, "The law of interstellar reddening," *Ap. J. 135*, 306 (1962).
268. L. Spitzer and J. W. Tukey, "A theory of interstellar polarization," *Ap. J. 114*, 187 (1951).
269. L. Davies and J. L. Greenstein, "The polarization of starlight by alignment of dust grains," *Ap. J. 114*, 206 (1951).
270. L. Davies, "Polarization of starlight: the torque on a nutating grain," *Ap. J. 128*, 508 (1958).
271. L. Davies, "Present status of the theoretical interpretation of the polarization of starlight," *Lowell Observ. Bull. 4*, No. 17, 266 (1960).
272. A. A. Hoag, "Polarimetry of open clusters field," *Astron. J. 58*, 42 (1953).
273. V. A. Dombrovskii, "An analysis of polarization of the stellar emission in the region of the binary cluster h and χ Perseus," *Vestn. Leningr. Univ.*, No. 19, 135 (1959).
274. A. G. Pacholczyk, "Statistical investigations of the polarization of starlight in the Milky Way," *Acta Astron. 9*, 1 (1959).
275. A. G. Pacholczyk, "Interstellar λ4430 band and polarization of starlight in the Milky Way," *Acta Astron. 9*, 90 (1959).
276. S. A. Kaplan and I. A. Klimishin, "The correlation between the observed difference in the degree of interstellar polarization and the angular distance on the celestial sphere," *Astron. Zh. 36*, 370 (1959).

277. A. Elvius, "Polarization of light in the spiral galaxy NGC 7331 and the interpretation of interstellar polarization," *Stockholm Observ. Ann. 19*, 1 (1956).
278. M. T. Martel, "Polarisation de la nébuleuse du Crabe. Polarisation et couleur des nébuleuses diffusantes," *Ann. Astrophys. Suppl. 7*, 82 (1958).
279. D. A. Rozhkovskii, "Light polarization in nebulae," *Izv. Astrofiz. Inst. Akad. Nauk Kaz. SSR 3*, No. 4, 68 (1956).
280. A. Elvius and J. Hall, "Photoelectric measurements of polarization in the irregular galaxy M 82, NGC 5128, the Pleiades and other objects," *Astron. J. 67*, 271 (1962).
281. F. Kerr, "First results of the 210 foot radiotelescope," *Sky and Telescope 24*, 254 (1962).
282. D. ter Haar, "On the origin of smoke particles in the interstellar gas," *Bull. Astron. Inst. Neth. 10*, 361 (1943).
283. S. A. Kaplan, "The condensation of interstellar gas on cosmic dust particles," *Uch. Zap. L'vovsk. Univ. 22*, No. 5, 111 (1953).
284. P. R. Lichtenstein and J. M. Greenberg, "Temperatures of dielectric interstellar grains," *Ap. J. 67*, 275 (1962).
285. F. D. Kahn, "On the formation of interstellar dust," *Monthly Notices Roy. Astron. Soc. 112*, 518 (1952).
286. J. R. Platt, "On the optical properties of interstellar dust," *Ap. J. 123*, 486 (1956).
286a. Wesselink, draft report IAU (1964), p. 511.
287. J. M. Greenberg, "The sizes of interstellar grains," *Ap. J. 132*, 672 (1960).
287a. R. Cayrel and E. Schatzman, "Sur la polarization interstellaire par des particules de graphite," *Ann. Astrophys. 17*, 555 (1954).
287b. T. P. Stecher, "Interstellar extinction in the ultraviolet," *Ap. J. 142*, 1683 (1965).
287c. A. Bogges, III, and J. Borgman, "Interstellar extinction in the middle ultraviolet," *Ap. J. 140*, 1636 (1964).
287d. T. P. Stecher and B. Donn, "On graphite and interstellar extinction," *Ap. J. 142*, 1681 (1965).
288. R. D. Richtmyer and E. Teller, "On the origin of cosmic rays," *Phys. Rev. 75*, 1729 (1949).
289. E. Fermi, "On the origin of the cosmic radiation," *Phys. Rev. 75*, 1169 (1949).
289a. S. B. Pikelner, "An analysis of the possible mechanisms of field formation in radio sources," *Astron. Zh. 40*, 601 (1963).
289b. V. N. Tsitovich, "Acceleration of radiation and matter generates fast particles under cosmic conditions," *Astron. Zh. 40*, 612 (1963).
290. R. Lüst and A. Schlüter, "Kraftfreie Magnetfelder," *Z. Astrophys. 34*, 263 (1954).
291. S. Chandrasekhar, "On force-free magnetic fields," *Proc. Nat. Acad. Sci. U.S. 42*, 1 (1956).
292. S. Lundquist, "Magneto-hydrostatic fields," *Arkiv. Fysik. 2*, 361 (1950).
293. S. A. Kaplan, "Cosmic Bessel fields," *Astron. Zh. 36*, 800 (1959).
294. L. H. Woltjer, "On hydromagnetic equilibrium," *Proc. Nat. Acad. Sci. U.S. 44*, 489, 833 (1958).
295. J. G. Bolton and J. P. Wild, "On the possibility of measuring interstellar magnetic fields by 21-cm Zeeman splitting," *Ap. J. 125*, 296 (1957).
296. R. D. Davies, C. H. Slater, W. L. Shuter, and P. A. T. Wild, "A new limit to the galactic magnetic field set by measurements of the Zeeman splitting of the hydrogen line," *Nature (London) 187*, 1088 (1960).
296a. R. D. Davies, G. L. Verschuur, and P. A. T. Wild, *Nature (London) 196*, 543 (1962).
296b. R. D. Davies, W. L. H. Shuter, C. H. Slater, and P. A. T. Wild, "An experiment to measure the magnetic field of the Galaxy. II. Results and derivation of magnetic field intensities," *Monthly Notices Roy. Astron. Soc. 126*, 353 (1963).

296c. R. D. Davies and W. L. H. Shuter, "An experiment to measure the magnetic field of the Galaxy. III. Discussion of magnetic field measurements," *Monthly Notices Roy. Astron. Soc. 126*, 369 (1963).

296d. D. Morris, B. G. Clark, and R. W. Wilson, "An attempt to measure the galactic magnetic field," *Ap. J. 138*, 889 (1963).

297. J. A. Earl, "Cloud chamber observations of primary cosmic ray electrons," *Phys. Rev. Letters 6*, 125 (1961).

298. P. Meyer and R. Vogt, "Electrons in the primary cosmic radiation," *Phys. Rev. Letters 6*, 193 (1961).

299. L. I. Dorman and G. I. Freidman, "The possible acceleration of charged particles by shock waves in a magnetized plasma," in *Problems of plasma magnetohydrodynamics and physics* (Riga, 1959).

299a. D. G. Wentzel, "Fermi acceleration of charged particles," *Ap. J. 137*, 135 (1963).

299b. D. G. Wentzel, "Motion across magnetic discontinuities and Fermi acceleration of charged particles," *Ap. J. 140*, 1013 (1964).

300. V. P. Shabanskii, "Particle acceleration by the passage of a hydromagnetic shock wave front," *Zh. Eksperim. i Teor. Fiz. 41*, 1107 (1961).

301. E. Schatzman, *Ann. Astrophys. 26*, No 3, 234 (1963).

302. E. Lohrman and M. W. Teucher, "Heavy nuclei and α-particles between 7 and 100 Bev/nucleon. Interaction mean free paths and fragmentation probabilities," *Phys. Rev. 115*, 636 (1959).

303. G. G. Getmantsev, "The isotropy of cosmic rays," *Astron. Zh. 39*, 607 (1962).

303a. R. Firkowski, J. Cavin, A. Zawadzki, and R. Mase, "Evidence for the existence of high-energy photon-initiated EAS," *Nuovo Cimento 29*, 19 (1963).

303b. R. Giacconi, H. Gursky, F. R. Paolini, and B. Rossi, "Evidence for x-ray sources outside the solar system," *Phys. Rev. Letters 9*, 439 (1962).

303c. W. L. Kraushaar and G. W. Clark, "Search for primary cosmic gamma rays with satellite Explorer XI," *Phys. Rev. Letters 8*, 106 (1962).

303d. R. J. Gould and G. R. Burbidge, "X-rays from the galactic center, external galaxies and the intergalactic medium," *Ap. J. 138*, 963 (1963).

303e. V. L. Ginzburg and S. I. Syrovatskii, "Gamma and Roentgen radiation connected with galactic and metagalactic cosmic radiation," *Zh. Eksperim. i Teor Fiz. 46*, 1865 (1964).

304. L. Davis, "The diffusion of cosmic rays in the Galaxy," in *Proceedings of an international conference on cosmic rays*, (Izd. Akad. Nauk SSSR, 1960), 3:225.

305. *Proceedings of the III conference on cosmogonic problems* (Izd. Akad. Nauk SSSR 1953).

306. A. A. Korchak and S. I. Syrovatskii, "Possible preeminent acceleration of heavy elements in cosmic ray sources," *Dokl. Akad. Nauk SSSR 122*, 792 (1958).

307. S. I. Syrovatskii, "Cosmic ray spectra and their role in cosmic gas dynamics," in *Problems of plasma magnetohydrodynamics and physics* (Riga, 1959).

308. A. A. Logunov and Ya. P. Terletskii, "The acceleration of charged particles by a traveling magnetized medium," *Zh. Eksperim. i Teor. Fiz. 26*, 129 (1954).

309. S. A. Kaplan, "The theory of charged particle acceleration by isotropic gas magnetic turbulent fields," *Zh. Eksperim. i Teor. Fiz. 29*, 406 (1955).

309a. S. A. Kaplan, "The theory of statistical charged particle acceleration by isotropic gas-magnetic fields," *Tsirk. L'vovsk. Astron. Observ.*, No. 33 (1957).

310. N. S. Kardeshev, "The nonsteady nature of the spectra of nonthermal cosmic radio emission sources," *Astron. Zh. 39*, 393 (1962).

310a. V. N. Tsytovich, "Statistical acceleration of particles in a turbulent plasma," *Usp. Fiz. Nauk 89*, 89 (1966).

310b. V. L. Ginzburg, "Cosmic rays and plasma phenomena in the Galaxy and the Metagalaxy," *Astron. Zh. 42*, 1129 (1965).

310c. T. R. Hartz, "Spectrum of the galactic radioemission between 1.5 and 6 MHz from the satellite," *Nature* (*London*) *203*, 173 (1964).

311. L. D. Landau and E. M. Lifshitz, *Field theory* (Fizmatgiz, 1960).

312. D. D. Ivanenko and A. A. Sokolav, *Classical field theory* (Fizmatgiz, 1951).

313. K. C. Westfold, "The polarization of synchrotron radiation," *Ap. J. 130*, 241 (1959).

314. A. A. Korchak and S. I. Syrovatskii, "Polarization of radiation and the magnetic field structure of cosmic sources of magnetic bremsstrahlung," *Astron. Zh. 38*, 885 (1961).

314a. V. I. Slish, "Angular size of radio stars," *Nature* (*London*) *199*, 682 (1963).

314b. A. J. Tulle, "The spectrum of the galactic radio emission. II. Interpretation of low resolving power observations," *Monthly Notices Roy. Astron. Soc. 126*, 31 (1963).

315. N. N. Tsytovich, "The problem of radiation by fast electrons in a magnetic field in the presence of a medium," *Vestn. MGU*, No. 11, 29 (1951).

316. V. A. Razin, "The spectrum of nonthermal cosmic radio emission," *Radiofizika 3*, 921, 584 (1960).

317. P. Mathewson, J. R. Healey, and J. M. Rome, "A radio survey of the southern Milky Way at 1440 Mc/s," *Australian J. Phys. 15*, 369 (1962).

318. V. Ya. Eidman, "Radiation by an electron moving in a magnetoactive plasma," *Zh. Eksperim. i Teor. Fiz. 34*, 131 (1958).

319. G. G. Getmantsev, "Cosmic electrons as a source of galactic radio emission," *Dokl. Akad. Nauk SSSR 83*, 557 (1952).

319a. G. M. Garibyan and I. I. Goldman, "The polarization of relativistic electron radiation in magnetic fields," *Izv. Arm. SSR 7*, 31 (1954).

320. A. A. Korchak, "The electromagnetic radiation of cosmic particles in the Galaxy," *Astron. Zh. 34*, 365 (1957).

320a. J. E. Baldwin, "Discussion in the RAS," *Observatory 83*, 153 (1963).

321. C. L. Seeger, G. Westerhout, and R. Conway, "Andromeda Nebula at a wave length of 75 cm," *Ap. J. 126*, 585 (1957).

321a. J. W. Leibacher, "1417 Mc/sec observation of M 31," *Astron. J. 67*, 580 (1962).

321b. J. D. Kraus, "Does the Andromeda Nebula (M 31) have a halo?" *Nature* (*London*) 202, 1202 (1964).

322. J. E. Baldwin and C. H. Costain, "The radio spectrum of the Andromeda Nebula," *Monthly Notices Roy. Astron. Soc. 121*, 413 (1960).

323. G. B. Field, "Radio continuum emission by spiral arms," *Publ. Astron. Soc. Pacific 72*, 303 (1960).

324. V. A. Razin, "The polarization of cosmic radio emission at 1.45 and 3.3 m," *Astron. Zh. 35*, 241 (1958).

325. G. Westerhout, C. L. Seeger, W. N. Brouw, and J. Tindbergen, "Polarization of the galactic 75 cm radiation," *Bull. Astron. Inst. Neth. 16*, 187 (1962).

325a. W. N. Brouw, C. A. Muller, and J. Tindbergen, "Further polarization measurements at 75 cm," *Bull. Astron. Inst. Neth. 16*, 213 (1962).

325b. C. A. Muller, E. M. Berkhnijsen, W. N. Brouw, and J. Tindbergen," Galactic background polarization at 610 Mc/s," *Nature* (*London*) *200*, 155 (1963).

325c. D. S. Mathewson and D. K. Milne, "A pattern in the large scale distribution of galactic polarized radio emission," *Nature* (*London*) *203*, 1273 (1964).

326. S. B. Pikelner, "Interstellar gas and a magnetic field," *Izv. Krymsk. Astrofiz. Observ. 10*, 74 (1953).

327. S. B. Pikelner and I. S. Shklovskii, "A study of the properties and the energy dissipation of the galactic gas corona," *Astron. Zh. 34*, 145 (1957).
327a. S. B. Pikelner and I. S. Shklovskii, "On the problem of the nature of the gas corona of the Galaxy," *Ann. Astrophys. 22*, 913 (1959).
328. L. Spitzer, "On a possible interstellar galactic corona," *Ap. J. 124*, 20 (1956).
328a. F. F. Gardner and R. D. Davies, "Faraday rotation of the emission from linearly polarized sources," *Australian J. Phys. 19*, 129 (1965).
328b. J. M. Bologna, E. F. McClain, W. K. Rose, and R. M. Sloanaker, "Observation of the effective linear polarization of discrete sources at 21.2 centimeter wavelength," *Astrophys. J. 142*, 106 (1965).
328c. P. Maltby, "On the connection between linear polarization and galactic latitude of discrete radio sources," *Astrophys. J. 142*, 621 (1965).
328d. I. I. Pronik, "Magnetic field in the vicinity of the Sun," *Astron. Zh. 43*, 291 (1966).
328e. J. M. Hornby, "The structure of the magnetic field in the local spiral arm," *Monthly Notices Roy. Astron. Soc. 133*, 213 (1966).
328f. L. Woltjer, in *Stars and stellar systems*, vol. V, *Galactic structure*, ed. A. Blaauw and M. Schmidt (University of Chicago Press, Chicago, 1965).
328g. S. B. Pikelner, in *Radio astronomy and galactic systems*, ed. H. van Woerden (Reidel, Dordrecht, 1967).
329. V. A. Razin, "The galactic corona," *Astron. Zh. 35*, 829 (1958); *Radiofizika 3*, 921 (1960).
330. S. B. Pikelner, "The effect of cosmic rays on the nature of a magnetic field and filament formation in supernovae envelopes," *Astron. Zh. 38*, 21 (1961).
331. F. D. Kahn and L. Woltjer, "Intergalactic matter and the Galaxy," *Ap. J. 130*, 705 (1959).
332. C. Hoffmeister, "Uber eine intergalaktische Absorptionswolke," *Z. Astrophys. 55*, 46 (1962).
333. S. Sciama, "Consequences of an open model of the galactic magnetic field," *Monthly Notices Roy. Astron. Soc. 123*, 317 (1962).
334. I. S. Shklovskii, "The nature of the radio galaxies," *Astron. Zh. 39*, 591 (1962).
335. V. M. Byakov, "The possible role of the extragalactic medium in maintaining movement of gas in the halo," *Astron. Zh. 40*, 384 (1963).
336. G. A. Shain, "Diffuse nebulae and the interstellar magnetic field," *Astron. Zh. 32*, 110 (1955).
337. G. A. Shain, "Magnetic fields in interstellar space and in nebulae," *Astron. Zh. 32*, 381 (1955).
338. S. B. Pikelner and L. P. Metik, "Possible anistropy of the cloud velocities of interstellar gas," *Izv. Krymsk. Astrofiz. Observ. 18*, 198 (1958).
339. S. Chandrasekhar and E. Fermi, "Problems of gravitational stability in the presence of a magnetic field," *Ap. J. 118*, 116 (1953).
339a. S. B. Pikelner, "Spiral arms and interacting galaxies," *Astron. Zh. 42*, No. 3 (1965).
339b. F. Hoyle and J. G. Ireland, "On magnetic structure of the galaxy," *Monthly Notices Roy. Astron. Soc. 122*, 35 (1961).
339c. W. Heisenberg and C. F. Weizsäcker, "Die Gestalt der Spiralnebel," *Z. Phys. 125*, 290 (1948).
339d. N. S. Kardashev, "Magnetic collapse and the nature of the remanent source of cosmic radio emission," *Astron. Zh. 41*, 807 (1964).
339e. J. H. Piddington, "The magnetic field and radio emission of galaxies," *Monthly Notices Roy. Astron. Soc. 128*, 345 (1964).

339f. D. Morris and G. L. Burge, "Direction of the galactic magnetic field in the vicinity of the sun," *Ap. J. 139*, 1388 (1964).
339g. Gardner and F. Whiteoak, "Polarization of radiosources and the effect of Faraday rotation in the galaxy," *Nature* (*London*) *197*, 1162 (1963).
340. R. S. Iroshnikov, "Turbulence of a conducting liquid in a strong magnetic field," *Astron. Zh. 40*, 74 (1963).
340a. A. Blaauw and C. R. Tolbert, "Intermediate-velocity features in the local hydrogen layer," *Bull. Astron. Inst. Neth. 18*, 405 (1966).
340b. A. N. M. Hulsbosch and E. Raimond, "Neutral hydrogen with high velocities at high galactic latitudes," *Bull. Astron. Inst. Neth. 18*, 413 (1966).
340c. N. H. Dieter, "Neutral hydrogen near the galactic poles," *Astron. J. 70*, 552 (1965).
340d. J. M. Oort, "Possible interpretations of the high-velocity clouds," *Bull. Astron. Inst. Neth. 18*, 421 (1966).
340e. I. S. Shklovskii, "The possibility of the maser effect in clouds of interstellar hydrogen," *Astron. Zh. 44*, 304 (1967).
340f. V. V. Zeleznyakov, "On a coherent synchrotron mechanism of radio emission of some cosmic sources," *Astron. Zh. 44*, 42 (1967).
340g. S. A. Kaplan, "On the theory of coherent synchrotron radiation of the cosmic radio sources," *Astrofizika 2*, 409 (1966).
340h. L. M. Kovriznych and V. N. Tsytovich, "On the interaction of longitudinal and transverse," *Soviet Phys. JETP* (*English Transl.*) *46*, 2212 (1964).
340i. S. A. Kaplan, "Possible interpretation of the radio emission of cosmic sources of small dimensions," *Astron. Zh. 44*, No. 3 (1967).
341. G. Münch, "Expanding motions of interstellar gas in the nuclear region of Messier 31," *Ap. J. 131*, 250 (1960).
342. Yu. N. Pariiskii, "Exchange of gas between the nucleus and the galactic corona," *Astron. Zh. 38*, 377 (1961).
343. F. Hoyle and W. A. Fowler, "On the nature of strong radio sources," *Monthly Notices Roy. Astron. Soc. 125*, 169 (1962); *Nature* (*London*) *197*, 533 (1963).
343a. G. R. Burbidge and F. Hoyle, "The galactic halo—a transient phenomenon" (preprint, 1962).
343b. S. Van den Bergh, "Optical investigation of possible remnants of supernovae," *Z. Astrophys. 31*, 15 (1960).
344. L. Woltjer, "The polarization and intensity distribution in the Crab nebula derived from plates taken with the 200 inch telescope by W. Baade," *Bull. Astron. Inst. Neth. 13*, 301 (1957).
345. D. E. Osterbrock, "Electron densities in the filaments of the Crab nebula," *Publ. Astron. Soc. Pacific 69*, 227 (1957).
345a. E. S. Brodskaya, "Interstellar absorption in the direction of the Crab nebula," *Izv. Krimsk. Astrofiz. Observ. 30*, 126 (1963).
346. W. Baade, "The polarisation of the Crab nebula on plates taken with the 200 inch telescope," *Bull. Astron. Inst. Neth. 12*, 312 (1956).
347. J. H. Oort and T. Walraven, "Polarisation and composition of the Crab nebula," *Bull. Astron. Inst. Neth. 12*, 285 (1956).
348. N. S. Soboleva, N. A. Prozorov, and Yu. N. Pariiskii, "The distribution of polarized and nonpolarized radio emission in the Crab nebula," *Astron. Zh. 40*, 1 (1963).
348a. D. Morris and V. Radhakrishnan, "Tests for linear polarization in the 1390 Mc/s radiation from six intense radio sources," *Ap. J. 137*, 147 (1963).
349. A. E. Chudakov, "A search of photons with the energy 10^{13} ev from discrete cosmic sources of radiation," *J. Phys. Soc. Japan, Suppl. A III, 106* (1962).

350. A. A. Korchak, "Estimation of cosmic ray energy and magnetic field intensity in radio sources," *Radioastronomiya, Tr. Fiz. Inst. Akad. Nauk SSSR 17*, 149 (1962).
351. S. B. Pikelner, "The magnetic field of the Crab Nebula and of the central star," *Astron. Zh. 33*, 785 (1956).
352. V. I. Moroz, "Infrared observations of the Crab Nebula," *Astron. Zh. 40*, No. 6 (1963).
353. L. Woltjer, "The Crab Nebula," *Bull. Astron. Inst. Neth. 14*, 39 (1958).
353a. E. T. Byram, T. A. Chubb, and H. Friedman, "X-ray sources in the Galaxy," *Nature (London) 201*, 1307 (1964).
353b. B. H. Andrew, N. J. B. Brauson, and D. Wills, "Radio observation of the Crab Nebula during a lunar occultation," *Nature (London) 203*, 171 (1964).
353c. S. Bowyer, E. T. Byram, T. A. Chubb, and H. Friedman, "The lunar occultation of x-ray emission from the Crab Nebula," *Science 146*, 912 (1964).
353d. L. Woltjer, "X-rays and Type I supernova remnants," *Ap. J. 140*, 1309 (1964).
354. W. Baade, "The Crab Nebula," *Ap. J. 96*, 188 (1942).
354a. I. S. Shklovskii, "Remarks on the spectrum of the synchrotron radiation of the Crab Nebula," *Astron. Zh. 43*, 10 (1966).
355. W. Baade and R. Minkowsky, "Identification of the radio sources in Cassiopeia, Cygnus A, Puppis A," *Ap. J. 119*, 206 (1954).
356. R. Minkowsky and L. H. Aller, "The spectrum of the radio source in Cassiopeia," *Ap. J. 119*, 232 (1954).
357. R. E. Hershberg, "The associated filamentary nebulae of type II supernova remains, together with the emitting stars," *Astron. Zh. 39*, 1033 (1962).
358. I. S. Shklovskii, "The possible secular change in radio emission flux and intensity from certain discrete sources," *Astron. Zh. 37*, 256 (1960).
359. D. E. Osterbrock, "Electron densities in filamentary nebulae," *Publ. Astron. Soc. Pacific 70*, 180 (1958).
360. G. B. Sholomitskii, "The mass of a filamentary nebula," *Astron. Zh. 40*, 223 (1963).
360a. S. Kenderline, "Radio emission from the Cygnus Loop," *Monthly Notices Roy. Astron. Soc. 126*, 55 (1963).
361. I. S. Shklovskii, "Radio galaxies," *Usp. Fiz. Nauk 77*, 3 (1962).
362. W. Baade, "Polarization in the jet of Messier 87," *Ap. J. 123*, 550 (1956).
363. D. E. Osterbrock, "Interstellar matter in elliptical galaxies," *Ap. J. 132*, 325 (1960).
364. R. Minkowsky and D. Osterbrock, "Interstellar matter in elliptical galaxies," *Ap. J. 129*, 583 (1959).
365. R. Biraud, J. Lequeux, and E. de Reux, "Interferometric measurements of Cygnus A, Sagittarius A, Virgo A and the supernovae of Tycho Brahe and Kepler," *Observatory 80*, 116 (1960).
366. Yu, N. Pariiskii, "Features of the radio galaxy NGC 4486," *Dokl. Akad. Nauk SSSR 137*, 49 (1961).
367. R. C. Bless, "The non-thermal radiation from NGC 4486," *Ap. J. 135*, 187 (1962).
368. C. M. Wade, "The extended component of Centaurus," *Australian J. Phys. 12*, 471 (1959).
369. E. M. Burbidge and G. R. Burbidge, "Rotation and internal motions in NGC 5128," *Ap. J. 129*, 271 (1959).
370. R. Q. Twiss, A. W. L. Carter, and A. G. Little, "Brightness distribution over some strong radio sources at 1427 MHz," *Observatory 80*, 153 (1960).
371. G. K. Seyfert, "Nuclear emission in spiral nebulae," *Ap. J. 97*, 28 (1943).
372. G. R. Burbidge, E. M. Burbidge, and K. H. Prendergast, "Mass distribution and physical conditions in the inner region of NGC 1068," *Ap. J. 130*, 26 (1959).

372a. E. A. Dibai and V. I. Pronik, "Spectrophotometric investigation of the nucleus of NGC 1068," *Astrofizika 1*, 78 (1965).
372b. C. R. Lynds and A. R. Sandage, "Evidence for an explosion in the center of the galaxy M 82," *Ap. J. 137*, 1005 (1963).
373. L. Woltjer, "Emission nuclei in galaxies," *Ap. J. 130*, 38 (1959).
373a. A. Elvius and J. S. Hall, "Photoelectric observation of polarization in the irregular galaxies M 82, NGC 5128," *Ap. J. 67*, 271 (1962).
374. A. R. Sandage, "First true radio star?" *Sky and Telescope 21*, 148 (1961).
375. L. D. Landau and E. M. Lifshitz, *The mechanics of continuous media* (Gostekhizdat, 1953).
376. H. M. Mott-Smith, "The solution of the Boltzmann equation for a shock wave," *Phys. Rev. 82*, 885 (1951).
377. B. Tidman, "Structure of a shock wave in fully ionized hydrogen," *Phys. Rev. 111*, 1439 (1958).
378. Y. Jukes, "Structure of a shock wave in fully ionized gas," *J. Fluid Mech. 3*, 275 (1957).
379. S. B. Pikelner, "A spectrophotometric study of the excitation mechanism of filamentary nebulae," *Izv. Krymsk. Astrofiz. Observ. 12*, 93 (1954).
379a. S. B. Pikelner, "Dissipation of energy, the heating and ionization of interstellar gas by shock waves," *Astron. Zh. 34*, 314 (1957).
379b. S. A. Kaplan and T. S. Podstrigach, "Parameters of shock waves in a partially ionized gas," *Astron. Zh. 42*, No. 4 (1965).
379c. T. S. Podstrigach, "On the structure of shock waves in an incompletely ionized gas," *Ukr. Fiz. Zh.* (1965).
380. I. S. Shklovskii, "Supernova outbursts and the interstellar medium," *Astron. Zh. 39*, 209 (1962).
380a. G. Heiles, "Supernova shells and galactic x-rays," *Ap. J. 140*, 470 (1964).
381. S. S. Moiseev and R. Z. Sagdeev, "Shock waves in a rarefied plasma, located in a weak magnetic field," *Dokl. Akad. Nauk SSSR 146*, 329 (1962).
382. A. A. Galeev and V. I. Karpman, "The turbulence theory of slightly out-of-equilibrium rarefied plasma and shock wave structure," *Zh. Eksperim. i Teor. Fiz. 44*, 592 (1963).
383. A. A. Blank and H. Grad, "Steady one-dimensional fluid—magnetic collision-less shock theory," in *Aerodynamic phenomena in stellar atmospheres* (1961), p. 459.
384. H. E. Petschek, "Summary introduction: collision-free plasma," in *Aerodynamic phenomena in stellar atmospheres* (1961), p. 448.
384a. S. A. Kaplan and N. S. Kardashev, "Relativistic shock waves in intergalactic space," *Astron. Circ.*, No. 303 (1964).
384b. E. Y. Gidalevich and S. A. Kaplan, "On the shock waves in a gas-dust medium," *Astrofizika 1*, 475 (1965).
384c. E. Y. Gidalevich, "Propagation of shock waves in a gas-dust medium," *Astron. Zh. 43*, 553 (1966).
384d. E. Y. Gidalevich, "Propagation of ionization waves in a gas-dust medium," *Astron. Zh. 43*, 1018 (1966).
385. J. H. Adlam and J. E. Allen, "The structure of strong collision-free hydromagnetic waves," *Phil. Mag. 3*, 448 (1958).
386. F. D. Kahn, "The acceleration of interstellar clouds," *Bull. Astron. Inst. Neth. 12*, 187 (1954).
387. W. I. Axford, "Ionization fronts in interstellar gas: the structure of ionization fronts," *Phil. Trans. Roy. Soc. London, Ser. A, 253*, 301 (1961).

388. P. A. Oliinik, "The theory of ionization waves in interstellar space," *Tsirk. L'vovsk Astron. Observ.*, No. 35–36 (1960).
389. F. A. Goldsworthy, "Ionization fronts in interstellar gas and the expansion of H II regions," *Phil. Trans. Roy. Soc. London, Ser. A, 253*, 278 (1961).
389a. P. O. Vandervoort, "The formation of H II regions," *Ap. J. 137*, 381 (1963).
389b. P. O. Vandervoort, "The formation of H II regions. II," *Ap. J. 138*, 426 (1963).
389c. P. O. Vandervoort, "The stability of ionization fronts and the evolution of H II regions," *Ap. J. 138*, 599 (1963).
389d. W. I. Axford, "The initial development of H II regions," *Ap. J. 139*, 761 (1964).
389e. Y. Abe, S. Sakashita, and Y. Ouo, "Effects of the magnetic field on ionization front," *Progr. Theoret. Phys. (Kyoto) 30*, 816 (1963).
389f. W. G. Mathews, "The time evolution of an H II region," *Ap. J. 142*, 1120 (1965).
389g. W. G. Mathews, "On the central hole in NGC 2237–2246," *Ap. J. 144*, 206 (1966).
389h. B. M. Lasker, "An investigation of the dynamics of an old H II region," *Ap. J. 143*, 700 (1966).
390. M. P. Savedoff and J. Greene, "Expanding H II regions," *Ap. J. 122*, 477 (1955).
391. S. Pottasch, "Dynamics of bright rims in diffuse nebulae," *Bull. Astron. Inst. Neth. 14*, 29 (1958).
392. G. Courtez, P. Cruvellier, and S. R. Pottasch, "Mouvement des bordures brillantes dans les nébuleuses diffuses," *Ann. Astrophys. 25*, 214 (1962).
393. S. A. Kaplan and K. P. Stanyukovich, "The solution of magneto-gas dynamic equations for one-dimensional motion," *Dokl. Akad. Nauk SSSR 95*, 769 (1954).
394. L. I. Sedov, *Congruent and dimensional methods in mechanics* (Gostekhizdat, 1954).
395. F. A. Baum, S. A. Kaplan, and K. P. Stanyukovich, *An introduction to cosmic gas dynamics* (Gostekhizdat, 1958).
396. J. H. Oort, "Some phenomena connected with interstellar matter," *Monthly Notices Roy. Astron. Soc. 106*, 159 (1946).
397. V. G. Gorbatskii, "The dynamics of new stellar envelopes," *Astron. Zh. 39*, 198 (1962).
397a. E. A. Dibai and S. A. Kaplan, "Cumulative shock waves in interstellar space," *Astron. Zh. 41*, 652 (1964).
397b. E. A. Dibai, "On the origin of comet-shaped nebulae," *Astron. Zh. 40*, 795 (1963).
397c. I. B. Bernstein and R. M. Kulsrud, "On the explosion of a supernova into the interstellar magnetic field, I," *Ap. J. 142*, 479 (1965).
397d. R. M. Kulsrud, I. B. Bernstein, M. Kruskal, J. Fanucci, and N. Ness, "On the explosion of a supernova into the interstellar magnetic field, II," *Ap. J. 142*, 491 (1965).
398. A. N. Kolmogorov, "Localized structure of turbulence in an incompressible liquid at very large Reynolds numbers," *Dokl. Akad. Nauk SSSR 30*, 229 (1941).
399. A. M. Obukhov, "Energy distribution in the spectrum of turbulent flow," *Izv. Akad. Nauk SSSR, Ser. Geog. i Geofiz.*, No. 4–5, 453 (1941).
400. W. Heisenberg, "On the theory of statistical and isotropic turbulence," *Proc. Roy. Soc. (London), Ser. A, 195*, 402 (1958).
401. G. K. Batchelor, "On the spontaneous magnetic field in a conducting liquid in turbulence motion," *Proc. Roy. Soc. (London), Ser. A, 201*, 405 (1950).
402. S. A. Kaplan, "A system of spectral equations of magneto-gas-dynamic isotropic turbulence," *Dokl. Akad. Nauk SSSR 94*, 33 (1954).
403. S. Chandrasekhar, "Hydromagnetic turbulence. II. An elementary theory," *Proc. Roy. Soc. (London), Ser. A, 233*, 330 (1955).
404. S. A. Kaplan, "Structure, correlation, and spectral functions of interstellar gas turbulence," *Astron. Zh. 32*, 255 (1955).

404a. A. M. Obukhov and A. M. Yaglom, *Prikl. Matem. i Mekh. 15*, 3 (1951).
405. S. von Hoerner, "Eine Methode zur Untersuchung der Turbulenz der interstellaren Materie," *Z. Astrophys. 30*, 17 (1951).
405a. S. A. Pikelner and I. A. Klimishin, "Methods of analysis of interstellar turbulence," *Astron. Zh. 41*, No. 2 (1964).
406. S. B. Pikelner and G. A. Shain, "A study of turbulence in the Orion Nebula from brightness fluctuations," *Izv. Krymsk. Astrofiz. Observ. 11*, 22 (1954).
406a. R. E. Hershberg, "On the character of internal motion in the interstellar gas," *Izv. Krymsk. Astrofiz. Observ. 31*, 100 (1964).
407. S. A. Kaplan, "Quantitative characteristics of interstellar gas turbulence," *Dokl. Akad. Nauk SSSR 89*, 80 (1953).
408. S. B. Pikelner, "A method for studying turbulence from brightness fluctuations in nebulae," *Izv. Krymsk. Astrofiz. Observ. 11*, 34 (1954).
409. S. A. Kaplan and V. I. Pronik, "The problem of the turbulent nature of motion in interstellar gas clouds," *Dokl. Akad. Nauk SSSR 89*, 643 (1953).
410. B. V. Kukarkin, *A survey of the development and growth of stellar systems based on the study of variable stars* (Gostekhizdat, 1949).
411. L. Spitzer and M. Schwarzschild, "The possible influence of interstellar clouds on stellar velocities," *Ap. J. 114*, 385 (1951).
412. A. I. Lebedinskii, "A hypothesis of stellar formation," *Cosmogonic Problems 2*, 5 (1954).
413. J. H. Oort, "Dynamics and evolution of the Galaxy insofar as relevant to the problem of population," in *Stellar population* (Vatican Observatory, 1958), p. 415.
414. M. Schwarzschild, *Structure and evolution of the stars* (Princeton University Press, Princeton, N.J., 1958; Dover, New York, 1965).
415. E. M. Burbidge and G. R. Burbidge, "Stellar evolution," *Handbuch der Physik*, vol. 51 (1958), p. 134.
416. R. Kraft, "Exploding stars," *Sci. Am. 206*, 54 (1962).
417. A. J. Deutsch, "The dead stars of Population I," *Astron. J. 61*, 174 (1956).
418. I. S. Shklovskii, "The nature of planetary nebulae and their nuclei," *Astron. Zh. 33*, 315 (1956).
419. I. A. Klimishin and V. M. Bazilevich, *Tsirk. L'vovsk Astron. Observ.*, No. 39–40, 3 (1963).
420. I. M. Gordon, "Supernova outbreaks and the origin of cosmic rays," *Astron. Zh. 37*, 246 (1960).
421. E. M. Burbidge, G. R. Burbidge, W. A. Fowler, and F. Hoyle, "Synthesis of the elements in stars," *Rev. Mod. Phys. 29*, 547 (1957).
422. D. A. Frank-Kamenetskii, *Physical processes inside stars* (Fitzmatgiz, 1959).
423. S. van den Bergh, "The old galactic cluster NGC 7142 and stellar evolution," *J. Roy. Astron. Soc. Can. 56*, 41 (1962).
424. F. Hoyle and N. C. Wickramasinghe, "On graphite particles as interstellar grains," *Monthly Notices Roy. Astron. Soc. 124*, 417 (1962).
424a. N. C. Wickramasinghe, "On graphite particles as interstellar grains. II," *Monthly Notices Roy. Astron. Soc. 126*, 99 (1963).
424b. F. Hoyle and N. C. Wickramasinghe, "On the deficiency in the ultraviolet fluxes from early type stars," *Monthly Notices Roy. Astron. Soc. 126*, 401 (1963).
424c. E. Schatzman and R. Cayrel, *Ann. Astrophys. 17*, 555 (1964).
425. G. A. Shain, "The instability and disintegration of diffuse emission nebulae," *Astron. Zh. 2*, 209 (1955).
426. G. A. Shain, "Groups of emission nebulae," *Astron. Zh. 31*, 217 (1954).
427. G. A. Shain and V. F. Gaze, "The connection between diffuse gaseous nebulae

and hot stars," *Izv. Krymsk. Astrofiz. Observ. 10*, 152 (1953); *Astron. Zh. 32*, 492 (1955).

428. G. A. Shain, "The association of hot stars and emission nebulae," *Astron. Zh. 32*, 492 (1955).
429. I. I. Pronik, "A study of the spatial distribution of stars of different spectral classes in the region centered on $l = 343°$, $b = 0°$," *Izv. Krymsk. Astrofiz. Observ. 23*, 46 (1960).

429a. C. Hayashi, "Stellar evolution in early phases of gravitational contraction," *Publ. Astron. Soc. Japan 13*, 450 (1961).

430. P. N. Kholopov, "*T*-associations," *Astron. Zh. 27*, 233 (1950).

430a. R. D. Davies, "The galactic spur," *Monthly Notices Roy. Astron. Soc. 128*, 173 (1964).

430b. M. J. S. Quigley and C. G. T. Haslam, "Association of the galactic radio spur with the Scorpio x-ray source," *Nature (London) 203*, 1282 (1964).

431. I. M. Gordon, "The nonthermal component of the radiation from nonfixed stars and fundamental features of their spectra," *Astron. Zh. 34*, 739 (1957).
432. Ya. B. Zeldovich, "The gravitational condensation of galaxies in an expanding universe," *Cosmogonic Problems 9* (1963).
433. R. Ebert, "Uber die Verdichtung von H I Gebieten," *Z. Astrophys. 37*, 217 (1955).
434. W. H. McCrea, "The formation of population I stars. I. Gravitational contraction," *Monthly Notices Roy. Astron. Soc. 117*, 562 (1957).
435. F. Hoyle, "On the formation of galaxies and type II stars," in *Stellar populations* (Vatican Observatory, 1958), p. 435.
436. L. M. Ozernoi, "The theory of the gravitational condensation of galaxies and of globular clusters," *Scientific Work of Students of MGU 33* (1962).

436a. O. J. Eggen, D. Linden-Bell, and A. R. Sandage, "Evidence from the motions of old stars that the Galaxy collapsed," *Ap. J. 136*, 748 (1962).

436b. R. J. Gould, "The contraction of molecular hydrogen protostars," *Ap. J. 140*, 638 (1964).

436c. M. Simoda, S. Kikuchi, and W. Unno, "On the condensation of interstellar gas. IV. Gravitational contraction," *Publ. Astron. Soc. Japan 18*, 31 (1966).

436d. D. McNally, "The collapse of interstellar clouds," *Ap. J. 140*, 1088 (1964).

437. H. Alfvén, *On the origin of the solar system* (Oxford, 1954).
438. F. Hoyle, "On the origin of the solar nebula," Russian trans., *Vopr. Kosmogonii, Akad Nauk SSSR 7*, 15 (1960).
439. B. Bok and E. F. Reilly, "Small dark nebulae," *Ap. J. 105*, 255 (1947).
440. R. E. Hershberg, "The expansion of H II regions and the formation of gravitationally stable protostars," *Astron. Zh. 38*, 819 (1961).
441. E. A. Dibai, "The evolution of globules in the vicinity of hot stars," *Astron. Zh. 35*, 469 (1958).
442. E. A. Dibai, "The origin of comet-shaped nebulae," *Astron. Zh. 37*, 16 (1960).
443. M. Schmidt, "The rate of star formation," *Ap. J. 129*, 243 (1959).
444. V. C. Reddish, "The rate of star formation as a function of gas density," *Observatory 82*, 14 (1962).
445. L. E. Gurevich, "Evolution of stellar systems," *Cosmogonic Problems 2*, 151 (1954).
446. L. Mestel and L. Spitzer, "Star formation in magnetic dust clouds," *Monthly Notices Roy. Astron. Soc. 116*, 503 (1956).
447. L. Mestel, "The magnetic and dynamical fields outside a protostar," *Monthly Notices Roy. Astron. Soc. 119*, 223 (1959).
448. L. Mestel, "A note on the magnetic braking of a rotating star," *Monthly Notices Roy. Astron. Soc. 119*, 249 (1959).

449. G. M. Idlis, "Criteria for tidal resistance and the distribution of globular clusters in galaxies and of stars in globular clusters," *Dokl. Akad. Nauk SSSR 91*, 1305 (1953).
450. S. van de Bergh, "Interstellar gas and star creation," *Z. Astrophys. 43*, 236 (1957).
450a. D. J. Champin and F. Hoyle, "On the angular-momentum distribution in the disc of spiral galaxies," *Ap. J. 140*, 99 (1964).
451. G. M. Idlis and G. M. Nikolskii, "Diffuse media in globular star clusters," *Astron. Zh. 36*, 668 (1959).
452. M. S. Roberts, "Dust and gas in globular clusters," *Astron. J. 65*, 457 (1960).
453. O. Struve, "Gas and dust in globular clusters," *Sky and Telescope 19*, 456 (1960).
454. S. van de Bergh, "The halo phase of galactic evolution," *Publ. Astron. Soc. Pacific 73*, 135 (1961).
455. W. W. Morgan, "Some characteristics of galaxies," *Ap. J. 135*, 1 (1962).
455a. E. E. Epstein, "Atomic hydrogen in galaxies," *Astron. J. 69*, 490 (1964).
455b. R. M. Muller and E. N. Parker, "Formation of massive quasistellar objects," *Ap. J. 140*, 50 (1964).
455c. Ya. B. Zeldovich, "Dynamic systems of point mass" (preprint, Moscow, 1964).
456. P. W. Hodge, "Globular clusters in the Magellanic Clouds," *Sky and Telescope 21*, 72 (1961).
456a. G. de Vaucouleurs, "Classification of galaxies," *Handbuch der Physik 53* (1959).
456b. L. Spitzer and W. Baade, "Stellar populations and collisions of galaxies," *Ap. J. 113*, 413 (1951).
457. G. Haro, "Nota preliminar sobre galaxias azules conlineas de emission," *Bol. Obs. Tanantzintla y Takubava*, No. 14, 8 (1956).
457a. F. Zwicky and K. Rudnicki, "Area of the sky covered by clusters of galaxies," *Ap. J. 137*, 707 (1963).
458. F. Zwicky and M. L. Humason, "Spectra and other characteristics of interconnected galaxies and of galaxies in groups and in clusters, II," *Ap. J. 133*, 794 (1961).
458a. F. Zwicky, "Luminous intergalactic matter," *Publ. Astron. Soc. Pacific 64*, 242 (1952).
459. B. A. Vorontsov-Velyaminov, "Galactic morphology, V. Interactions between galaxies and the nature of their arms, bridges and tails," *Astron. Zh. 35*, 858 (1958).
459a. B. A. Vorontsov-Velyaminov, "Evidence for magnetlike phenomena in galactic structure," *Astron. Zh. 41*, 814 (1964).
459b. B. J. Robinson, K. J. Van Damme, and J. A. Kochler, "Neutral hydrogen in the Virgo cluster," *Nature* (*London*) *199*, 1176 (1963).
459c. F. Hoyle and M. Harwit, "On the fate of intergalactic bridges," *Publ. Astron. Soc. Pacific 74*, 202 (1962).
460. G. R. Burbidge and E. M. Burbidge, "Kinetic and potential energy in Stephan's quintet," *Ap. J. 130*, 15 (1959).
461. D. N. Limber and W. G. Mathews, "The dynamical stability of Stephan's quintet," *Ap. J. 132*, 286 (1960).
462. M. P. Savedoff, "Observational consequences of O star formation," *Ap. J. 124*, 533 (1956).
463. S. van de Bergh, "The stability of clusters of galaxies," *Z. Astrophys. 55*, 21 (1962).
463a. P. J. E. Peebles, "The black-body radiation content and the formation of galaxies," *Ap. J. 142*, 1317 (1965).
463b. A. G. Doroshkevich, Ya. B. Zeldovich, and I. D. Novikov, "Formation of galaxies in an expanding universe," *Astron. Zh. 44*, 295 (1967).
463c. C. C. Lin and F. H. Shu, *Ap. J. 140*, 646 (1964).

463d. C. C. Lin and F. H. Shu, *Proc. Nat. Acad. Sci. 55*, 229 (1966).
464. S. B. Pikelner, "Barred spirals (*Sb*) and the formation of spiral arms," *Astron. Zh. 42*, 3 (1965).
464a. L. M. Aller and D. J. Faulkner, "The helium to hydrogen ratio in the small Magellanic Cloud," *Publ. Astron. Soc. Pacific 74*, 219 (1962).
464b. H. C. Van de Hulst, E. Raimond, and H. Van Woerden, "Rotation and density distribution of the Andromeda Nebula derived from observations of the 21 cm line," *Bull. Astron. Inst. Neth. 14*, 1 (1957).
464c. M. Schmidt, "Distribution of mass in M 31," *Bull. Astron. Inst. Neth. 14*, 17 (1957).
464d. Wesselink, *Trans. Intern. Astron. Union* (1964).
465. M. Schmidt, "3C 273—a star-like object with large red-shift," *Nature* (*London*) *197*, 4872, 1040 (1963).
466. V. L. Ginzburg and L. M. Ozernoi, "On the temperature of the intergalactic gas," *Astron. Zh. 42*, No. 5 (1965).

Index